THE
INTERNATIONAL SERIES
OF
MONOGRAPHS ON PHYSICS

INTERNATIONAL SERIES OF MONOGRAPHS ON PHYSICS

104. D. Bardin, G. Passarino: *The Standard Model in the making*
103. G. C. Branco, L. Lavoura, J. P. Silva: *CP violation*
102. T. C. Choy: *Effective medium theory*
101. H. Araki: *Mathematical theory of quantum fields*
100. L. M. Pismen: *Vortices in nonlinear fields*
99. L. Mestel: *Stellar magnetism*
98. K. H. Bennemann: *Nonlinear optics in metals*
97. D. Salzmann: *Atomic physics in hot plasmas*
96. M. Brambilla: *Kinetic theory of plasma waves*
95. M. Wakatani: *Stellarator and heliotron devices*
94. S. Chikazumi: *Physics of ferromagnetism*
93. A. Aharoni: *Introduction to the theory of ferromagnetism*
92. J. Zinn-Justin: *Quantum field theory and critical phenomena*
91. R. A. Bertlmann: *Anomalies in quantum field theory*
90. P. K. Gosh: *Ion traps*
89. E. Simánek: *Inhomogeneous superconductors*
88. S. L. Adler: *Quaternionic quantum mechanics and quantum fields*
87. P. S. Joshi: *Global aspects in gravitation and cosmology*
86. E. R. Pike, S. Sarkar: *The quantum theory of radiation*
84. V. Z. Kresin, H. Morawitz, S. A. Wolf: *Mechanisms of conventional and high T_c superconductivity*
83. P. G. de Gennes, J. Prost: *The physics of liquid crystals*
82. B. H. Bransden, M. R. C. McDowell: *Charge exchange and the theory of ion–atom collision*
81. J. Jensen, A. R. Mackintosh: *Rare earth magnetism*
80. R. Gastmans, T. T. Wu: *The ubiquitous photon*
79. P. Luchini, H. Motz: *Undulators and free-electron lasers*
78. P. Weinberger: *Electron scattering theory*
76. H. Aoki, H. Kamimura: *The physics of interacting electrons in disordered sys*
75. J. D. Lawson: *The physics of charged particle beams*
73. M. Doi, S. F. Edwards: *The theory of polymer dynamics*
71. E. L. Wolf: *Principles of electron tunneling spectroscopy*
70. H. K. Henisch: *Semiconductor contacts*
69. S. Chandrasekhar: *The mathematical theory of black holes*
68. G. R. Satchler: *Direct nuclear reactions*
51. C. Møller: *The theory of relativity*
46. H. E. Stanley: *Introduction to phase transitions and critical phenomena*
32. A. Abragam: *Principles of nuclear magnetism*
27. P. A. M. Dirac: *Principles of quantum mechanics*
23. R. E. Peierls: *Quantum theory of solids*

F. P. Bowden, D. Tabor: *The friction and lubrication of solids*
J. M. Ziman: *Electrons and phonons*
M. Born, K. Huang: *Dynamical theory of crystal lattices*
M. E. Lines, A. M. Glass: *Principles and applications of ferroelectrics and related materials*

The Theory of Polymer Dynamics

M. DOI
Department of Applied Physics, Nagoya University

and

S. F. EDWARDS
Cavendish Laboratory, University of Cambridge

CLARENDON PRESS · OXFORD

OXFORD
UNIVERSITY PRESS

Great Clarendon Street, Oxford OX2 6DP

Oxford University Press is a department of the University of Oxford.
It furthers the University's objective of excellence in research, scholarship,
and education by publishing worldwide in

Oxford New York

Athens Auckland Bangkok Bogotá Buenos Aires Calcutta
Cape Town Chennai Dar es Salaam Delhi Florence Hong Kong Istanbul
Karachi Kuala Lumpur Madrid Melbourne Mexico City Mumbai
Nairobi Paris São Paulo Singapore Taipei Tokyo Toronto Warsaw

with associated companies in Berlin Ibadan

Oxford is a registered trade mark of Oxford University Press
in the UK and in certain other countries

Published in the United States
by Oxford University Press Inc., New York

First published 1986
First published in paperback (with corrections) 1988
Reprinted 1989, 1992, 1994, 1995, 1998, 1999

British Library Cataloguing in Publication Data
Doi, M.
The theory of polymer dynamics
1. Polymers and polymerization
I. Title II. Edwards, S. F.
547.7'045 QD381

Library of Congress Cataloging in Publication Data
Doi, Masao.
The theory of polymer dynamics.
Bibliography: p.
Includes index.
1. Polymers and polymerization.
I. Edwards, S. F. II. Title.
QD281.P6E27 1986 547.7 85-29854

ISBN 0 19 852033 6 (Pbk)

Printed in Great Britain
on acid-free paper by
Biddles Ltd., King's Lynn and Guildford

PREFACE

Many people have helped us write this book. We had very helpful comments during the writing of the book from R. C. Ball, F. Boue, J. des Cloizeaux, W. Griffin and J. Klein and the manuscript was read by G. Berry, R. Hayakawa, M. Kurata, Y. Oono, D. Pearson, and E. Samulski who all made useful improvements to it. We had meticulous help with the proofs from Fiona Miller and Ed Samulski.

M. D. would like to give special thanks to Dr. Osaki who gave continual help and encouragement throughout and to the Japan Society for the Promotion of Science and the Royal Society for making his stay in Cambridge with S. F. E. in 1984/5 possible when the book was written.

It is a pleasure for S. F. E. to record, with gratitude, that it is now twenty years since he was introduced to this field by Professor Geoffrey Gee and Sir Geoffrey Allen; it has proved of abiding interest.

Tokyo and
Cambridge
March 1986

M. D.
S. F. E.

CONTENTS

1

INTRODUCTION

Macromolecules play a central role in chemical technology and indeed in biology. Their role and the richness of their properties mean that a whole series of monographs would be required to constitute a comprehensive treatise. This book concentrates on one aspect, the dynamics of polymers in the liquid state. That is, the dynamics of polymer solutions and melts, where in the last decade it has become possible to offer theories which explain the salient features of these systems.

Among various dynamical properties of polymeric liquids, an important and conspicuous property is their mechanics. As one knows in everyday life, polymeric liquids (chewing gum, dough, or egg white, for example) show quite distinct flow behaviours from the usual liquids like water: a polymeric liquid is usually quite viscous and has visible elasticity. For example if one stretches chewing gum and releases it quickly, then it will shrink like rubber, yet chewing gum is a liquid and can fill a container of any shape. This property, called the viscoelasticity, is just one of the many distinctive properties of polymeric liquids. The purpose of this book is to understand such properties from the molecular point of view.

Given the complexity of the polymer molecules, the theories are astonishingly simple, and before embarking on the main text it is worth explaining why it is reasonable to attack the problem of interacting macromolecules with a confidence which would not be justified for say liquid benzene, let alone water.

Polymer molecules are formed when the condition required to add one chemical unit (monomeric unit) to a system is almost independent of its size. If one has a small hydrocarbon, the synthesis of the next homologue can be a distinct process from the one after that, i.e., to go from (A) to $(A)-(A)$ involves a rather different pathway than going from $(A)-(A)$ to $(A)-(A)-(A)$. But if one starts with $(A)-(A)-\cdots(A)$, (A_n, say), the energy required for $A_n \rightarrow A_{n+1}$ is almost identical to that of $A_{n+1} \rightarrow A_{n+2}$, so that the process continues (polymerization reaction) and in principle can go to indefinitely long chains. Mixtures of species will often polymerize together:

$$(A)-(A)-(B)-(A)-(A)-(B)-(B)-(B)$$

and special agents can branch polymers

$$(A)-(A)-(C)<\begin{matrix}(A)-(A)-(A)\\(A)-(A)-(A)\end{matrix}$$

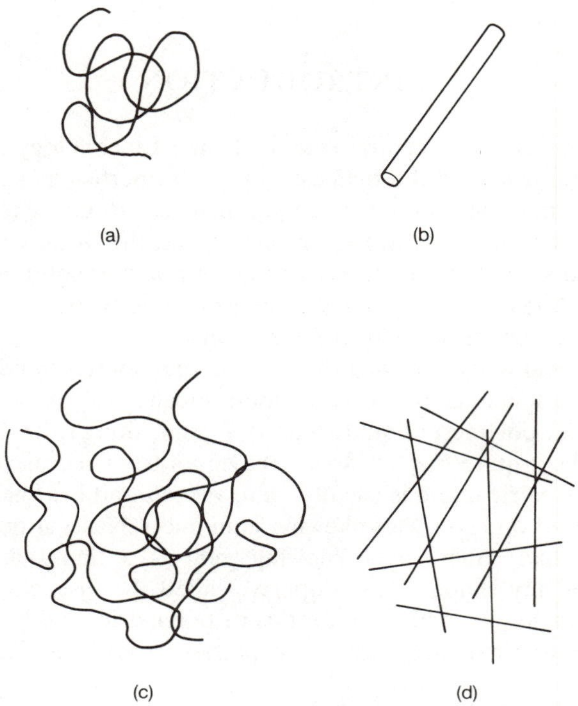

Fig. 1.1. (*a*) Flexible polymer, (*b*) rodlike polymer, (*c*), (*d*) their concentrated solutions.

An enormous number of variants are possible, are formed, and can indeed be designed.

There are two simple cases and these are the two studied in this book: (a) highly flexible polymers, where, on a sufficiently large scale, the polymer appears as a random walk, or a spaghetti-like shape (Fig. 1.1*a*), and (b) a rigid rod (Fig. 1.1*b*). These are, of course, extreme cases and cases intermediate to (a) and (b) can be found.

Now consider a closely packed assembly of these molecules (see Fig. 1.1*c* and *d*). At a sufficiently high temperature, they are in a high state of thermal agitation and form a viscoelastic liquid. (The reader will have a good intuitive feeling for the macroscopic behaviour of a rubbery liquid or a syrup.) Obviously the systems are quite complicated, but there are reasons why we can say that such systems are easier to understand than normal liquids.

One reason is the multiplicity of the interactions: one polymer is simultaneously interacting with many others, perhaps hundreds. Each one of these interactions has only a small effect so that a sound starting point is to add up their effects independently. This is in contrast with a normal liquid where each molecule has only a few, say eight or ten, neighbours and it is quite invalid to believe that these neighbours can move independently of one another. For polymeric liquids, any polymer molecule experiences an average of its surrounding and the problem reduces to deducing the mean properties from the behaviour of individual molecules.

This task is made (comparatively) easy because of the essential feature of polymers, i.e., the molecule itself is very large and the macroscopic behaviour is dominated by this large scale property of the molecule. Let us give a simple example of this.

Suppose a single molecule sits in solution and contributes to the viscosity of a surrounding fluid. Let us compare two polymers of the same size and conformation, but one is composed of spherical segments and the other is composed of triangular segments (see Fig. 1.2*a* and *b*). If the segments are separated from each other as in Fig. 1.2*c* and *d*, the shape of the segments matters; the viscosities of (*c*) and (*d*) are different. However, if the segments are connected to form a large polymer, the

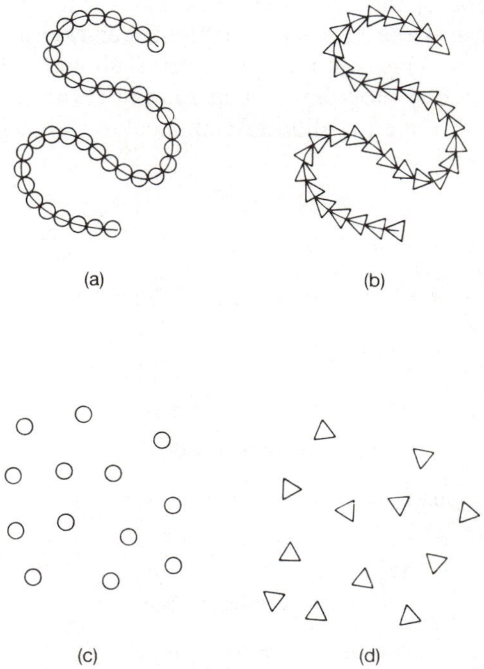

Fig. 1.2. Polymers composed of (*a*) spherical segments and (*b*) triangular segments with (*c*) and (*d*) their segments separated.

difference between (*a*) and (*b*) is not important. Indeed it will be shown later that the macroscopic viscosity is mainly determined by the average size of the polymer coil, and is quite insensitive to the shape of the segmental size (the difference in the viscosity of (*a*) and (*b*) is of the order of $N^{-1/2}$, where N is the number of segments in the polymer).

The above example demonstrates the universality in the properties of polymeric liquids. The macroscopic properties depend only on a few parameters specifying the molecular characteristics, and in so far as these parameters are the same, different systems behave in the same way. An important feature of this is that if experimental data are plotted in a way which eliminates the dependence of the parameters, various systems behave in the same way. Indeed such reduced plots have been found for various physical properties of polymer solutions and melts. An example is shown in Fig. 1.3, where the viscosity of polymer solutions is plotted against the shear rate for various concentrations. First note that the viscosity is not constant, but depends on the shear rate. This is typical in polymeric liquids, where the linear relation between stress and shear rate is valid only under rather limited circumstances. Now, though the nonlinearity is a source of complexity, one can see a simple structure in Fig. 1.3: the shape of the curves are alike, so they can be superimposed into a single curve if the viscosity and the shear rate are normalized for each concentration. This is indeed possible, as shown in Fig. 1.4. Various types of such reduced plots have been found. Those plots can superimpose experimental data for different molecular weights, concentration,

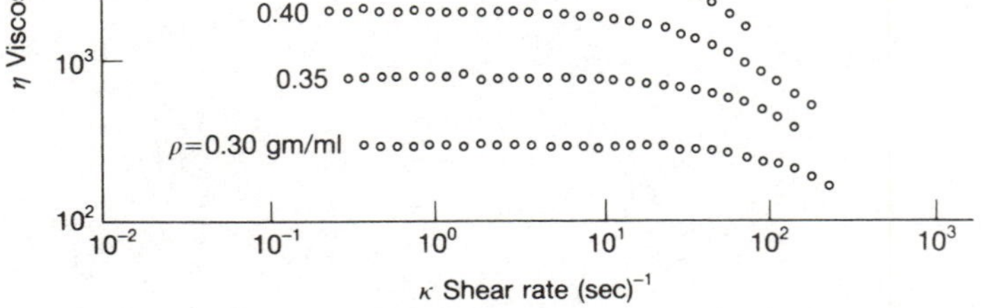

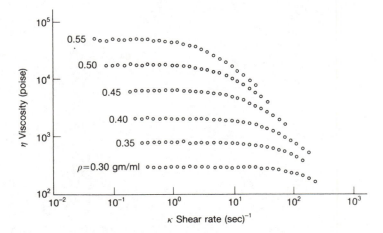

Fig. 1.3. Logarithmic plot of non-Newtonian viscosity against shear rate for solutions of polystyrene, $M_w = 4.11 \times 10^5$ in n-butyl benzene with various concentrations as shown. Reproduced from Graessley, Hazleton and Lindeman, *Trans. Soc. Rheol.* **11,** 267 (1967).

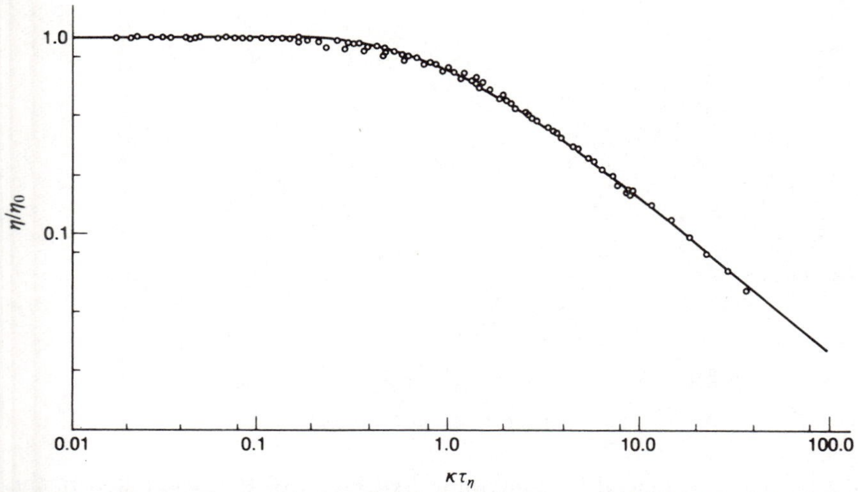

Fig. 1.4. Reduced plot of the non-Newtonian viscosity shown in Fig. 1.3, where η_0 is the viscosity at zero shear rate, and the time constant τ_η is chosen empirically for each solution. Reproduced from W. W. Graessley *Adv. Polym. Sci.* **16,** 1 (1974).

temperature, and chemical species. The ability of such superimposition indicates the existence of an inherent simplicity hidden behind the apparent complexity of the polymer systems. It is this feature that we would like to discuss in this book.

Now, in concentrated solutions such as shown in Fig. 1.1*c* and *d,* the key concept which allows us to express the relationship of the behaviour of a single molecule to the average behaviour of its neighbours is that of a surrounding tube. This is clearly specified for rodlike polymers. In order to draw a picture and without loss of generality, let us consider the freedom of a rod to move in a plane containing itself. Other molecules are represented in this picture by dots. Then in a high density of rods, the picture is as shown in Fig. 1.5.

Suppose other rods are fixed for the moment. The rod is free to move along itself, but rotations and displacement perpendicular to its length permit it to move only in the shaded region; i.e., our rod is confined to a tube as far as rotations and lateral motions are concerned. Of course the tube is made of other rods which are diffusing, but for a high enough density this is not consequential. Thus what we must understand is how a rod moves in a fixed, or a very slowly diffusing tube made of its neighbours.

In the case of rodlike polymers it is easy to calculate the size of the tube, but it is not so easy for flexible polymers where the picture of Fig.

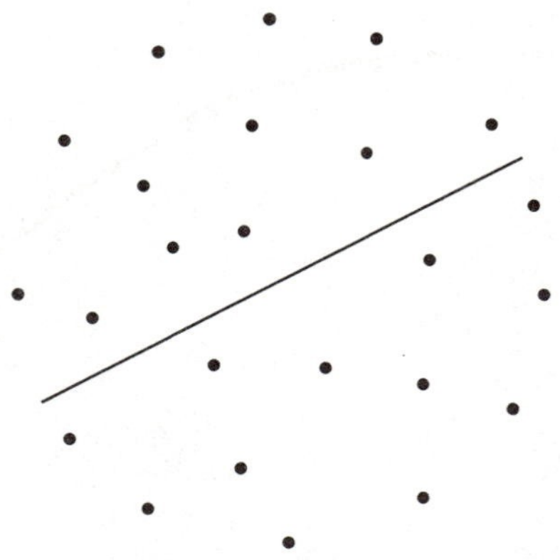

Fig. 1.5. Tube for a rod.

1.2 has to be replaced by something like Fig. 1.6. It is clear that in this case the polymer can go for extensive 'excursions' (Fig. 1.7) so that the tube is a fuzzy concept. It is also difficult to calculate the effective radius and the step length of the tube. However, by just calling it a, and assuming that a is a certain function of concentration, one can predict various dynamical properties in the molten state.

At this stage, however, the reader has to be warned that the 'fine structure' of the behaviour of real polymers is by no means resolved. The contents of this book will produce answers to many questions, and these answers will have qualitatively the right forms. But they may not be correct in detail. For example, the viscosity of a monodisperse melt of flexible polymers is shown to be proportional to M^3 whereas experimentally it is known to be $M^{3.4}$. Possible explanations of the discrepancy are discussed, but the problem is not entirely resolved. Thus the authors believe they are surveying real progress, but by no means the whole

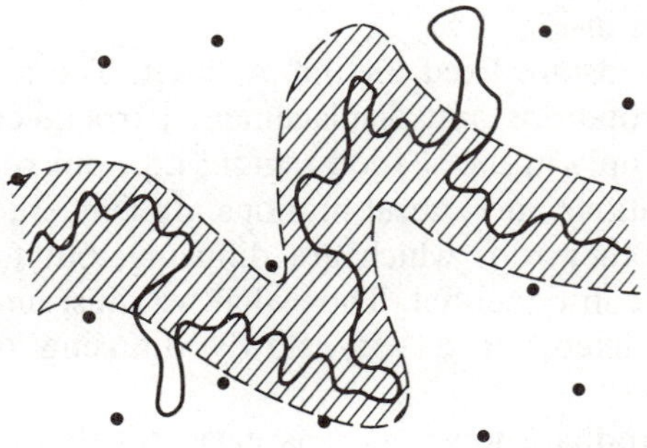

Fig. 1.6. Tube for a flexible polymer.

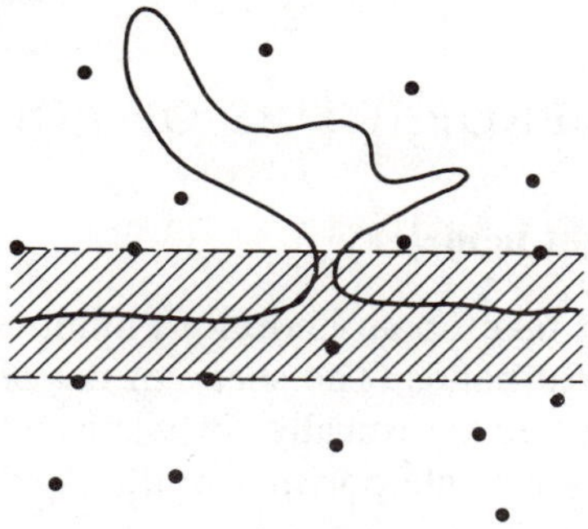

Fig. 1.7. Excursion of a flexible polymer.

story, and have tried in this book to make it clear where approximations are being made and to give a reasoned assessment of their validity.

This book is oriented to one aspect of polymer theory, but in order to reach it the authors have felt it necessary to give brief accounts of the basis of the subject in discussions of single chains and of polymer solutions. These topics too have made major advances in the last twenty years and deserve books of their own, but can only be given a minimal treatment in this book.

2

STATIC PROPERTIES OF POLYMERS

2.1 The random flight model

Flexible polymers can take up an enormous number of configurations by the rotation of chemical bonds. The shape of the polymers can therefore only be usefully described statistically. In this chapter we shall study the statistical properties of a single polymer in the equilibrium state.

2.1.1 The freely jointed model

To study the statistical properties of flexible polymers, let us start from a very simple model: a chain consisting of N links, each of length b_0 and able to point in any direction independently of each other (see Fig. 2.1). Such a model is called the freely jointed chain.

The conformation of the freely jointed chain is represented by the set of $(N+1)$ position vectors $\{\boldsymbol{R}_n\} \equiv (\boldsymbol{R}_0 \ldots \boldsymbol{R}_N)$ of the joints, or alternatively by the set of bond vectors $\{\boldsymbol{r}_n\} \equiv (\boldsymbol{r}_1 \ldots \boldsymbol{r}_N)$, where

$$\boldsymbol{r}_n = \boldsymbol{R}_n - \boldsymbol{R}_{n-1}, \qquad n = 1, 2, \ldots, N. \tag{2.1}$$

Since the bond vectors $\boldsymbol{r}_n$ are independent of each other, the distribution function for the polymer conformation is written as

$$\Psi(\{\boldsymbol{r}_n\}) = \prod_{n=1}^{N} \psi(\boldsymbol{r}_n) \tag{2.2}$$

where $\psi(\boldsymbol{r})$ denotes the random distribution of a vector of constant length b_0:

$$\psi(\boldsymbol{r}) = \frac{1}{4\pi b_0^2}\,\delta(|\boldsymbol{r}| - b_0). \tag{2.3}$$

This distribution is normalized to

$$\int \mathrm{d}\boldsymbol{r}\,\psi(\boldsymbol{r}) = 1. \tag{2.4}$$

To characterize the size of a polymer, we consider the end-to-end vector $\boldsymbol{R}$ of the chain,

$$\boldsymbol{R} = \boldsymbol{R}_N - \boldsymbol{R}_0 = \sum_{n=1}^{N} \boldsymbol{r}_n. \tag{2.5}$$

Since $\langle \boldsymbol{r}_n \rangle = 0$, $\langle \boldsymbol{R} \rangle$ is zero, but $\langle \boldsymbol{R}^2 \rangle$ has a finite value, which can be

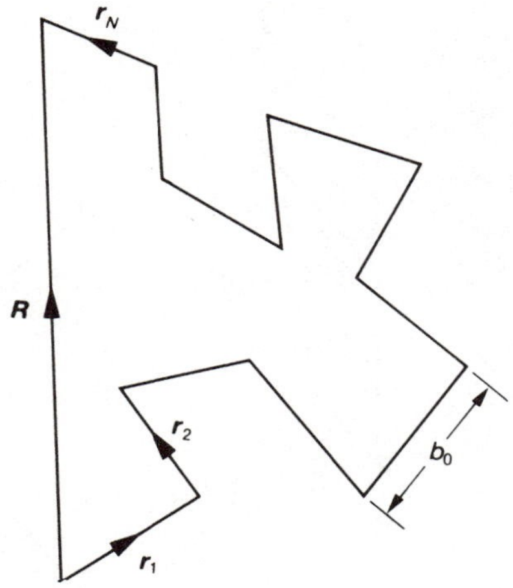

Fig. 2.1. Freely jointed chain.

used as a characteristic length of the chain. Let $\bar{R}$ be defined by

$$\bar{R} \equiv \langle \boldsymbol{R}^2 \rangle^{1/2} = \langle (\boldsymbol{R}_N - \boldsymbol{R}_0)^2 \rangle^{1/2}. \tag{2.6}$$

From eqn (2.5) $\langle \boldsymbol{R}^2 \rangle$ is given by

$$\langle \boldsymbol{R}^2 \rangle = \sum_{n,m=1}^{N} \langle \boldsymbol{r}_n \cdot \boldsymbol{r}_m \rangle = \sum_{n=1}^{N} \langle \boldsymbol{r}_n^2 \rangle + 2 \sum_{n>m} \langle \boldsymbol{r}_n \cdot \boldsymbol{r}_m \rangle = N b_0^2 \tag{2.7}$$

because for $n \neq m$, $\langle \boldsymbol{r}_n \cdot \boldsymbol{r}_m \rangle = \langle \boldsymbol{r}_n \rangle \cdot \langle \boldsymbol{r}_m \rangle = 0$. Thus $\bar{R}$ is given by

$$\bar{R} = \sqrt{N}\, b_0. \tag{2.8}$$

2.1.2 General random flight models

Though the freely jointed chain is a very simple model, the result $\langle \boldsymbol{R}^2 \rangle \propto N$ holds for more general models. Consider for example the model shown in Fig. 2.2, called the freely rotating chain, in which the n-th bond is connected to the $(n-1)$-th bond with a fixed angle θ and can rotate freely around the $(n-1)$-th bond.

For such a model, $\langle \boldsymbol{r}_n \cdot \boldsymbol{r}_m \rangle$ does not vanish for $n \neq m$. However, $\langle \boldsymbol{r}_n \cdot \boldsymbol{r}_m \rangle$ decreases rapidly as $|n-m|$ increases, and the relationship $\langle \boldsymbol{R}^2 \rangle \propto N$ again holds for large N. To see this we calculate $\langle \boldsymbol{r}_n \cdot \boldsymbol{r}_m \rangle$. If the average of $\boldsymbol{r}_n$ is taken with the rest of the chain (i.e., $\boldsymbol{r}_m, \boldsymbol{r}_{m+1}, \ldots, \boldsymbol{r}_{n-1}$) fixed ($n > m$ being assumed), we obtain (see Fig. 2.2b)

$$\langle \boldsymbol{r}_n \rangle_{\boldsymbol{r}_m, \boldsymbol{r}_{m+1}, \ldots, \boldsymbol{r}_{n-1} \text{fixed}} = \cos\theta\, \boldsymbol{r}_{n-1}. \tag{2.9}$$

Multiplying both sides of this equation by $\boldsymbol{r}_m$ and taking the average over

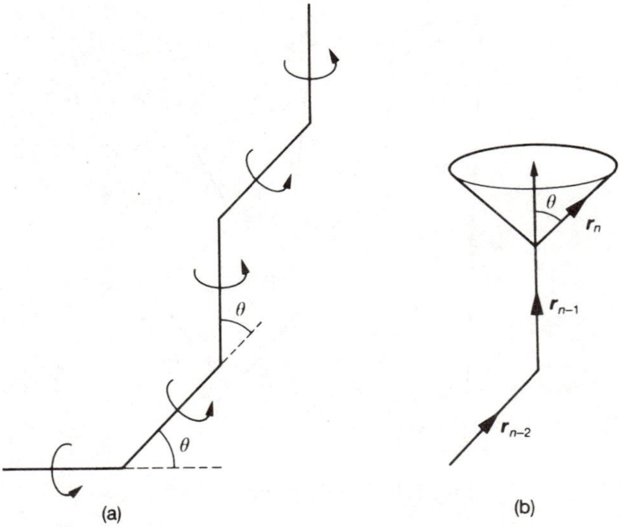

Fig. 2.2. (*a*) Freely rotating chain. (*b*) The average of $\boldsymbol{r}_n$ with $\boldsymbol{r}_{n-1}$ fixed gives $\boldsymbol{r}_{n-1}\cos\theta$.

$\boldsymbol{r}_m, \boldsymbol{r}_{m+1}, \ldots, \boldsymbol{r}_{n-1}$, we have

$$\langle \boldsymbol{r}_n \cdot \boldsymbol{r}_m \rangle = \cos\theta \langle \boldsymbol{r}_{n-1} \cdot \boldsymbol{r}_m \rangle. \tag{2.10}$$

This recursion equation, with the initial condition $\langle \boldsymbol{r}_m^2 \rangle = b_0^2$, is solved by

$$\langle \boldsymbol{r}_n \cdot \boldsymbol{r}_m \rangle = b_0^2 (\cos\theta)^{|n-m|} \tag{2.11}$$

which decreases exponentially with $|n-m|$. Thus for large N, $\langle \boldsymbol{R}^2 \rangle$ is given by

$$\langle \boldsymbol{R}^2 \rangle = \sum_{n=1}^{N} \sum_{m=1}^{N} \langle \boldsymbol{r}_n \cdot \boldsymbol{r}_m \rangle = \sum_{n=1}^{N} \sum_{k=-n+1}^{N-n} \langle \boldsymbol{r}_n \cdot \boldsymbol{r}_{n+k} \rangle$$

$$\simeq \sum_{n=1}^{N} \sum_{k=-\infty}^{\infty} \langle \boldsymbol{r}_n \cdot \boldsymbol{r}_{n+k} \rangle. \tag{2.12}$$

From eqn (2.11)

$$\sum_{k=-\infty}^{\infty} \langle \boldsymbol{r}_n \cdot \boldsymbol{r}_{n+k} \rangle = b_0^2 \left(1 + 2 \sum_{k=1}^{\infty} \cos^k\theta\right) = b_0^2 \frac{1+\cos\theta}{1-\cos\theta}. \tag{2.13}$$

Hence

$$\langle \boldsymbol{R}^2 \rangle = N b_0^2 \frac{1+\cos\theta}{1-\cos\theta}, \tag{2.14}$$

which again shows $\langle \boldsymbol{R}^2 \rangle \propto N$.

In general if the distribution function of $\boldsymbol{r}_n$ is written in the following form

$$\Psi(\{\boldsymbol{r}_n\}) = \prod_n \psi(\boldsymbol{r}_n, \boldsymbol{r}_{n+1}, \boldsymbol{r}_{n+2}, \ldots, \boldsymbol{r}_{n+n_c}), \tag{2.15}$$

$\langle \boldsymbol{R}^2 \rangle$ is written, for large N, as

$$\langle \boldsymbol{R}^2 \rangle = Nb^2. \tag{2.16}$$

The constant b is called the effective bond length. The ratio

$$C_\infty \equiv b^2/b_0^2 \tag{2.17}$$

represents the stiffness† of the polymer and can be calculated from the local structure of the chains. The results are summarized in refs 1–3.

2.1.3 *Distribution of the end-to-end vector*

Next we consider the statistical distribution of the end-to-end vector of the random flight model. Let $\Phi(\boldsymbol{R}, N)$ be the probability distribution function that the end-to-end vector of the chain consisting of N links is $\boldsymbol{R}$. Given the conformational distribution for $\Psi(\{\boldsymbol{r}_n\})$, $\Phi(\boldsymbol{R}, N)$ is calculated by

$$\Phi(\boldsymbol{R}, N) = \int \mathrm{d}\boldsymbol{r}_1 \int \mathrm{d}\boldsymbol{r}_2 \ldots \int \mathrm{d}\boldsymbol{r}_N\, \delta\left(\boldsymbol{R} - \sum_{n=1}^{N} \boldsymbol{r}_n\right) \Psi(\{\boldsymbol{r}_n\}), \tag{2.18}$$

which is rewritten using the identity

$$\delta(\boldsymbol{r}) = \frac{1}{(2\pi)^3} \int \mathrm{d}\boldsymbol{k}\, \mathrm{e}^{\mathrm{i}\boldsymbol{k}\cdot\boldsymbol{r}}, \tag{2.19}$$

as

$$\Phi(\boldsymbol{R}, N) = \frac{1}{(2\pi)^3} \int \mathrm{d}\boldsymbol{k} \int \mathrm{d}\boldsymbol{r}_1 \int \mathrm{d}\boldsymbol{r}_2 \ldots \int \mathrm{d}\boldsymbol{r}_N \times \exp\left(\mathrm{i}\boldsymbol{k} \cdot \left(\boldsymbol{R} - \sum_{n=1}^{N} \boldsymbol{r}_n\right)\right) \Psi(\{\boldsymbol{r}_n\}). \tag{2.20}$$

For the freely jointed chain, eqns (2.2) and (2.20) give

$$\begin{aligned} \Phi(\boldsymbol{R}, N) &= \frac{1}{(2\pi)^3} \int \mathrm{d}\boldsymbol{k}\, \mathrm{e}^{\mathrm{i}\boldsymbol{k}\cdot\boldsymbol{R}} \int \mathrm{d}\boldsymbol{r}_1 \ldots \mathrm{d}\boldsymbol{r}_N \prod_{n=1}^{N} \exp(-\mathrm{i}\boldsymbol{k} \cdot \boldsymbol{r}_n) \psi(\boldsymbol{r}_n) \\ &= \frac{1}{(2\pi)^3} \int \mathrm{d}\boldsymbol{k}\, \mathrm{e}^{\mathrm{i}\boldsymbol{k}\cdot\boldsymbol{R}} \left[\int \mathrm{d}\boldsymbol{r} \exp(-\mathrm{i}\boldsymbol{k} \cdot \boldsymbol{r}) \psi(\boldsymbol{r}) \right]^N. \end{aligned} \tag{2.21}$$

† In the literature, the stiffness is often represented by the Kuhn statistical segment length b_K defined by

$$b_K \equiv \langle \boldsymbol{R}^2 \rangle / R_{\max}$$

where $R_{\max}$ is the maximum length of the end-to-end vector.

The integral over $\boldsymbol{r}$ is evaluated by introducing polar coordinates (r, θ, ϕ; the reference axis of θ being taken along the vector $\boldsymbol{k}$), giving

$$\int d\boldsymbol{r} \exp(-i\boldsymbol{k}\cdot\boldsymbol{r})\psi(\boldsymbol{r})$$

$$=\frac{1}{4\pi b^2}\int_0^\infty dr r^2 \int_0^{2\pi} d\phi \int_0^\pi d\theta \sin\theta \exp(-ikr\cos\theta)\,\delta(r-b)$$

$$=\frac{\sin kb}{kb} \tag{2.22}$$

where $k=|\boldsymbol{k}|$ and b_0 is replaced by b (since $b_0=b$ for a freely jointed chain). From eqns (2.21) and (2.22)

$$\Phi(\boldsymbol{R}, N)=\frac{1}{(2\pi)^3}\int d\boldsymbol{k} e^{i\boldsymbol{k}\cdot\boldsymbol{R}}\left(\frac{\sin kb}{kb}\right)^N. \tag{2.23}$$

If N is large, $((\sin kb)/kb)^N$ becomes very small unless kb is small. For $kb \ll 1$, $((\sin kb)/kb)^N$ can be approximated as

$$\left(\frac{\sin kb}{kb}\right)^N \simeq \left(1-\frac{k^2b^2}{6}\right)^N \simeq \exp\left(-\frac{Nk^2b^2}{6}\right). \tag{2.24}$$

This approximation holds also for $kb \gtrsim 1$ since both sides of eqn (2.24) are nearly zero in such a case. Thus $\Phi(\boldsymbol{R}, N)$ is calculated as

$$\Phi(\boldsymbol{R}, N)=\frac{1}{(2\pi)^3}\int d\boldsymbol{k} e^{i\boldsymbol{k}\cdot\boldsymbol{R}} \exp\left(-\frac{N\boldsymbol{k}^2b^2}{6}\right). \tag{2.25}$$

The integral over $\boldsymbol{k}$ is a standard Gaussian integral. (Some useful Gaussian integrals are summarized in Appendix 2.I.) If k_α and R_α ($\alpha=x, y, z$) denote the components of the vectors $\boldsymbol{k}$ and $\boldsymbol{R}$, then using eqn (2.I.2)

$$\Phi(\boldsymbol{R}, N)=(2\pi)^{-3}\prod_{\alpha=x,y,z}\left[\int_{-\infty}^{\infty} dk_\alpha \exp(ik_\alpha R_\alpha - Nk_\alpha^2 b^2/6)\right]$$

$$=(2\pi)^{-3}\prod_{\alpha=x,y,z}\left(\frac{6\pi}{Nb^2}\right)^{1/2}\exp\left(-\frac{3}{2Nb^2}R_\alpha^2\right)$$

$$=(3/2\pi Nb^2)^{3/2}\exp\left(-\frac{3\boldsymbol{R}^2}{2Nb^2}\right). \tag{2.26}$$

Thus the distribution function of the end-to-end vector is Gaussian.

The distribution (2.26) has the unrealistic feature that $|\boldsymbol{R}|$ can be larger than the maximum extended length Nb of the chain. A more realistic

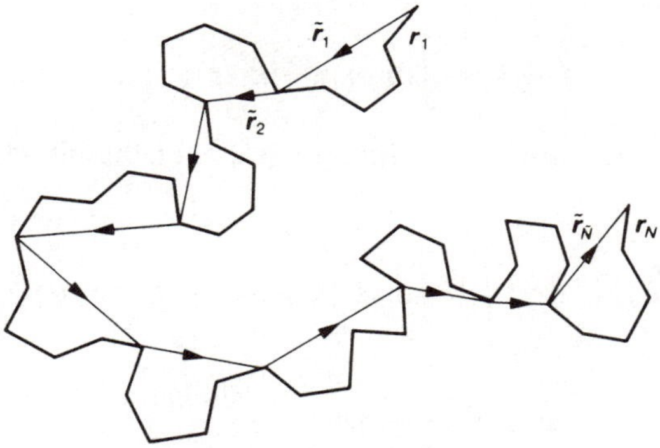

Fig. 2.3. A chain divided into $\tilde{N}$ submolecules.

distribution function which does not have this feature is available in the literature.[1,2,4] In this book, however, such highly extended states of polymers are not considered and eqn (2.26) is sufficient for our purpose.

Although the above derivation is for the freely jointed chain, the result actually holds more generally. In general it can be shown that provided the conformational distribution is described by eqn (2.15), the distribution of the end-to-end vector $\boldsymbol{R}$ of a long chain ($N \gg 1$) is given by eqn (2.26). This is a result of the central limit theorem in statistics.[5]

To prove this, suppose that the chain is divided into $\tilde{N}$ submolecules, each consisting of λ links (see Fig. 2.3). Clearly

$$\tilde{N} = N/\lambda. \tag{2.27}$$

Let $\tilde{\boldsymbol{r}}_n$ be the end-to-end vector of the n-th submolecule. The end-to-end vector of the chain is written as

$$\boldsymbol{R} = \sum_{n=1}^{\tilde{N}} \tilde{\boldsymbol{r}}_n. \tag{2.28}$$

Now if N is very large, both $\tilde{N}$ and λ can be taken large enough so that $\tilde{N} \gg 1$, and $\lambda \gg 1$. If $\lambda \gg 1$, the vectors $\tilde{\boldsymbol{r}}_n$ become independent of each other, and the distribution of $\{\tilde{\boldsymbol{r}}_n\}$ can be written as

$$\Psi(\{\tilde{\boldsymbol{r}}_n\}) = \prod_{n=1}^{\tilde{N}} \tilde{\psi}(\tilde{\boldsymbol{r}}_n). \tag{2.29}$$

From eqns (2.28) and (2.29), the distribution of $\boldsymbol{R}$ is derived without knowing the actual form of $\tilde{\psi}(\tilde{\boldsymbol{r}})$. Again using the identities (2.21) and (2.29), we have

$$\Phi(\boldsymbol{R}, N) = \frac{1}{(2\pi)^3} \int \mathrm{d}\boldsymbol{k}\, \mathrm{e}^{\mathrm{i}\boldsymbol{k}\cdot\boldsymbol{R}} A(\boldsymbol{k})^{\tilde{N}} \tag{2.30}$$

where

$$A(\boldsymbol{k}) = \int \mathrm{d}\boldsymbol{r} \exp(-\mathrm{i}\boldsymbol{k}\cdot\boldsymbol{r})\tilde{\psi}(\boldsymbol{r}). \tag{2.31}$$

Since $\tilde{\psi}(\boldsymbol{r})$ depends only on $|\boldsymbol{r}|$, the integral over the direction of $\boldsymbol{r}$ can be carried out as

$$\begin{aligned} A(\boldsymbol{k}) &= \int_0^\infty \mathrm{d}r r^2 \int_0^\pi \mathrm{d}\theta \sin\theta \int_0^{2\pi} \mathrm{d}\phi \exp(-\mathrm{i}kr\cos\theta)\tilde{\psi}(\boldsymbol{r}) \\ &= \int_0^\infty \mathrm{d}r 4\pi r^2 \frac{\sin kr}{kr}\tilde{\psi}(\boldsymbol{r}) = \left\langle \frac{\sin kr}{kr} \right\rangle_{\tilde{\psi}}, \end{aligned} \tag{2.32}$$

where

$$\langle \ldots \rangle_{\tilde{\psi}} \equiv \int_0^\infty \mathrm{d}r 4\pi r^2 \tilde{\psi}(r) \ldots. \tag{2.33}$$

Since we are interested in the small $\boldsymbol{k}$ region, we may evaluate $A(\boldsymbol{k})$ by expanding $(\sin kr)/kr$ with respect to k

$$A(\boldsymbol{k}) = \langle 1 - \tfrac{1}{6}k^2r^2 + \ldots \rangle_{\tilde{\psi}} = 1 - \tfrac{1}{6}\boldsymbol{k}^2\langle \boldsymbol{r}^2 \rangle_{\tilde{\psi}}. \tag{2.34}$$

For $\lambda \gg 1$, $\langle \boldsymbol{r}^2 \rangle_{\tilde{\psi}}$ is given by λb^2, whence

$$A(\boldsymbol{k}) = 1 - \tfrac{1}{6}\lambda b^2 \boldsymbol{k}^2. \tag{2.35}$$

Thus $\Phi(\boldsymbol{R}, N)$ is evaluated as

$$\begin{aligned} \Phi(\boldsymbol{R}, N) &= \frac{1}{(2\pi)^3}\int \mathrm{d}\boldsymbol{k} \mathrm{e}^{\mathrm{i}\boldsymbol{k}\cdot\boldsymbol{R}}(1 - \tfrac{1}{6}\lambda b^2\boldsymbol{k}^2)^{\tilde{N}} \\ &= \frac{1}{(2\pi)^3}\int \mathrm{d}\boldsymbol{k} \mathrm{e}^{\mathrm{i}\boldsymbol{k}\cdot\boldsymbol{R}} \exp(-\tfrac{1}{6}\lambda\tilde{N}b^2\boldsymbol{k}^2) \\ &= \left(\frac{3}{2\pi\lambda\tilde{N}b^2}\right)^{3/2} \exp\left(-\frac{3\boldsymbol{R}^2}{2\lambda\tilde{N}b^2}\right). \end{aligned} \tag{2.36}$$

Since $\lambda\tilde{N} = N$, eqn (2.36) agrees with eqn (2.26).

2.2 The Gaussian chain

We have seen that, in the statistical distribution of the end-to-end vector, the local structure of the chain appears only through the effective bond length b. This is generally true: the local structure affects only the effective bond length but does not otherwise appear in the problem. Therefore, if we are interested in the global properties of polymers, we can start from the simplest model available.

We consider a chain whose bond length has the Gaussian distribution

$$\psi(\boldsymbol{r}) = \left[\frac{3}{2\pi b^2}\right]^{3/2} \exp\left(-\frac{3\boldsymbol{r}^2}{2b^2}\right) \tag{2.37}$$

so that

$$\langle \boldsymbol{r}^2 \rangle = b^2. \tag{2.38}$$

The conformational distribution function of such a chain is given by

$$\begin{aligned} \Psi(\{\boldsymbol{r}_n\}) &= \prod_{n=1}^{N} \left[\frac{3}{2\pi b^2}\right]^{3/2} \exp\left[-\frac{3\boldsymbol{r}_n^2}{2b^2}\right] \\ &= \left[\frac{3}{2\pi b^2}\right]^{3N/2} \exp\left[-\sum_{n=1}^{N} \frac{3(\boldsymbol{R}_n - \boldsymbol{R}_{n-1})^2}{2b^2}\right]. \end{aligned} \tag{2.39}$$

Such a chain is called the Gaussian chain. The Gaussian chain does not describe correctly the local structure of the polymer, but does correctly describe the property on large length-scale. The advantage of the Gaussian chain as a model is that it is mathematically much easier to handle than any other of the models considered in Section 2.1.

The Gaussian chain is often represented by a mechanical model (see Fig. 2.4): $(N+1)$ 'beads' are considered to be connected by a harmonic spring whose potential energy is given by

$$U_0(\{\boldsymbol{R}_n\}) = \frac{3}{2b^2} k_B T \sum_{n=1}^{N} (\boldsymbol{R}_n - \boldsymbol{R}_{n-1})^2. \tag{2.40}$$

At equilibrium, the Boltzmann distribution for such a model is exactly the same as eqn (2.39).

An important property of the Gaussian chain is that the distribution of the vector $\boldsymbol{R}_n - \boldsymbol{R}_m$ between any two units n and m is Gaussian, being given by

$$\Phi(\boldsymbol{R}_n - \boldsymbol{R}_m, n - m) = \left[\frac{3}{2\pi b^2 |n-m|}\right]^{3/2} \exp\left[-\frac{3(\boldsymbol{R}_n - \boldsymbol{R}_m)^2}{2|n-m| b^2}\right]. \tag{2.41}$$

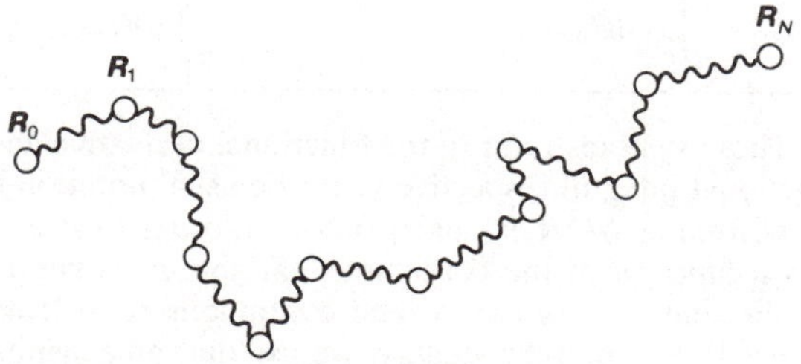

Fig. 2.4. Gaussian chain.

This follows from the properties of the Gaussian integral (see Appendix 2.I). Especially for any n and m

$$\langle (\boldsymbol{R}_n - \boldsymbol{R}_m)^2 \rangle = |n - m|\, b^2. \tag{2.42}$$

The suffix n of the Gaussian chain is often regarded as a continuous variable. In such cases $\boldsymbol{R}_n - \boldsymbol{R}_{n-1}$ is replaced by $\partial \boldsymbol{R}_n/\partial n$ and eqn (2.39) is written as

$$\Psi[\boldsymbol{R}_n] = \text{const}\exp\left[-\frac{3}{2b^2}\int_0^N \mathrm{d}n \left(\frac{\partial \boldsymbol{R}_n}{\partial n}\right)^2\right]. \tag{2.43}$$

This distribution is known as the Wiener distribution.

Mathematically, there is some subtlety in going from discrete n to continuous n, but for the present purpose it is sufficient to understand that eqn (2.43) is a formal rewriting of eqn (2.39). In this book, we use the discrete n and continuous n interchangeably. The transformation rules from the discrete variables to continuous variables are summarized in Table 2.1.

Table 2.1

Discrete		Continuous
$\sum_{n=k}^{k'}$	$\rightarrow$	$\int_k^{k'} \mathrm{d}n$
$\boldsymbol{R}_n - \boldsymbol{R}_{n-1}$	$\rightarrow$	$\partial \boldsymbol{R}_n/\partial n$
$\boldsymbol{R}_{n+1} + \boldsymbol{R}_{n-1} - 2\boldsymbol{R}_n$	$\rightarrow$	$\partial^2 \boldsymbol{R}_n/\partial n^2$
δ_{nm}	$\rightarrow$	$\delta(n-m)$
Kronecker delta		Dirac delta function
$\dfrac{\partial}{\partial \boldsymbol{R}_n}$	$\rightarrow$	$\dfrac{\delta}{\delta \boldsymbol{R}_n}$†
$\int \Pi\, \mathrm{d}\boldsymbol{R}_n$	$\rightarrow$	$\int \delta \boldsymbol{R}_n$‡

†,‡ These symbols refer to the functional derivative and the functional integral respectively. In the usual notation these are written as $\delta/\delta \boldsymbol{R}(n)$ and $\int \delta \boldsymbol{R}(n)$ in order to stress that $\boldsymbol{R}$ is a *function* of the continuous variable n. However, as the discrete representation and continuous representation are used here interchangeably, we use the same symbol $\boldsymbol{R}_n$ in both representations.

2.3 Chain conformation under an external field

2.3.1 *The Green function*

If there is an external field $U_e(\boldsymbol{r})$ acting on each segment, the equilibrium distribution of the Gaussian chain is modified by the Boltzmann factor

$$\exp\left[-\frac{1}{k_BT}\int_0^N \mathrm{d}n U_e(\boldsymbol{R}_n)\right], \tag{2.44}$$

and the conformational distribution function becomes

$$\Psi[\boldsymbol{R}_n] \propto \exp\left[-\frac{3}{2b^2}\int_0^N \mathrm{d}n\left(\frac{\partial \boldsymbol{R}_n}{\partial n}\right)^2 - \frac{1}{k_BT}\int_0^N \mathrm{d}n U_e(\boldsymbol{R}_n)\right]. \tag{2.45}$$

To discuss the statistical properties of such a system it is convenient to consider the 'Green function' defined by,[6,7]

$$G(\boldsymbol{R}, \boldsymbol{R}'; N) \equiv \frac{\displaystyle\int_{\boldsymbol{R}_0=\boldsymbol{R}'}^{\boldsymbol{R}_N=\boldsymbol{R}} \delta \boldsymbol{R}_n \exp\left[-\frac{3}{2b^2}\int_0^N \mathrm{d}n\left(\frac{\partial \boldsymbol{R}_n}{\partial n}\right)^2 - \frac{1}{k_BT}\int_0^N \mathrm{d}n U_e(\boldsymbol{R}_n)\right]}{\displaystyle\int \mathrm{d}\boldsymbol{R}\int_{\boldsymbol{R}_0=\boldsymbol{R}'}^{\boldsymbol{R}_N=\boldsymbol{R}} \delta \boldsymbol{R}_n \exp\left[-\frac{3}{2b^2}\int_0^N \mathrm{d}n\left(\frac{\partial \boldsymbol{R}_n}{\partial n}\right)^2\right]} \tag{2.46}$$

where $\boldsymbol{R}_N = \boldsymbol{R}$ and $\boldsymbol{R}_0 = \boldsymbol{R}'$ in the functional integral means that the integral is taken for all the conformations which start from $\boldsymbol{R}'$ and end at $\boldsymbol{R}$. (Notice that the denominator of eqn (2.46) is independent of $\boldsymbol{R}$ and $\boldsymbol{R}'$.)

For $U_e = 0$, $G(\boldsymbol{R}, \boldsymbol{R}'; N)$ reduces to the Gaussian distribution function

$$G(\boldsymbol{R} - \boldsymbol{R}'; N) = \left(\frac{2\pi N b^2}{3}\right)^{-3/2} \exp\left(-\frac{3(\boldsymbol{R} - \boldsymbol{R}')^2}{2Nb^2}\right). \tag{2.47}$$

In the general case of $U_e \neq 0$, $G(\boldsymbol{R}, \boldsymbol{R}'; N)$ represents the statistical weight (or the partition function) of the chain which starts from $\boldsymbol{R}'$ and ends at $\boldsymbol{R}$ in N steps. The partition function for all possible conformations is given by

$$Z = \int \mathrm{d}\boldsymbol{R}\, \mathrm{d}\boldsymbol{R}' G(\boldsymbol{R}, \boldsymbol{R}'; N). \tag{2.48}$$

From the definition of $G(\boldsymbol{R}, \boldsymbol{R}'; N)$, the following identity holds:

$$G(\boldsymbol{R}, \boldsymbol{R}'; N) = \int \mathrm{d}\boldsymbol{R}'' G(\boldsymbol{R}, \boldsymbol{R}''; N-n) G(\boldsymbol{R}'', \boldsymbol{R}'; n), \quad (\text{for } 0 < n < N). \tag{2.49}$$

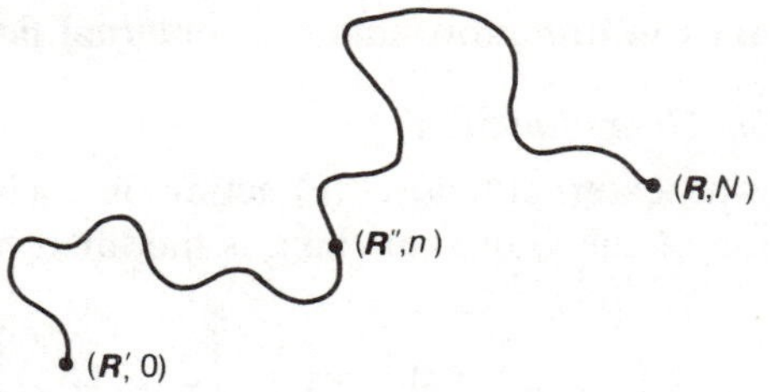

Fig. 2.5. A chain which starts at $\boldsymbol{R}'$, passes through $\boldsymbol{R}''$ in n steps, and ends at $\boldsymbol{R}$ in N steps.

The physical significance of this equation is clear. The factor $G(\boldsymbol{R}, \boldsymbol{R}''; N-n)G(\boldsymbol{R}'', \boldsymbol{R}'; n)$ represents the statistical weight of the chain which starts at $\boldsymbol{R}'$, passes through $\boldsymbol{R}''$ in n steps, and ends at $\boldsymbol{R}$ in N steps (see Fig. 2.5). The integration of this statistical weight over all $\boldsymbol{R}''$ gives the statistical weight of the chain which starts at $\boldsymbol{R}'$ and ends at $\boldsymbol{R}$.

Given $G(\boldsymbol{R}, \boldsymbol{R}'; N)$, the average of an arbitrary physical quantity A is easily calculated. If A depends only on the position of the n-th segment, then

$$\langle A(\boldsymbol{R}_n)\rangle = \frac{\int \mathrm{d}\boldsymbol{R}_N\,\mathrm{d}\boldsymbol{R}_n\,\mathrm{d}\boldsymbol{R}_0 G(\boldsymbol{R}_N, \boldsymbol{R}_n; N-n)G(\boldsymbol{R}_n, \boldsymbol{R}_0; n)A(\boldsymbol{R}_n)}{\int \mathrm{d}\boldsymbol{R}_N\,\mathrm{d}\boldsymbol{R}_0 G(\boldsymbol{R}_N, \boldsymbol{R}_0; N)}. \tag{2.50}$$

Likewise, if A depends on $\boldsymbol{R}_n$ and $\boldsymbol{R}_m$ ($n>m$ being assumed), then

$$\langle A(\boldsymbol{R}_n, \boldsymbol{R}_m)\rangle = \int \mathrm{d}\boldsymbol{R}_N\,\mathrm{d}\boldsymbol{R}_n\,\mathrm{d}\boldsymbol{R}_m\,\mathrm{d}\boldsymbol{R}_0 G(\boldsymbol{R}_N, \boldsymbol{R}_n; N-n)G(\boldsymbol{R}_n, \boldsymbol{R}_m; n-m) \times G(\boldsymbol{R}_m, \boldsymbol{R}_0; m)A(\boldsymbol{R}_n, \boldsymbol{R}_m) \times \left[\int \mathrm{d}\boldsymbol{R}_N\,\mathrm{d}\boldsymbol{R}_0 G(\boldsymbol{R}_N, \boldsymbol{R}_0; N)\right]^{-1} \tag{2.51}$$

Though the Green function $G(\boldsymbol{R}, \boldsymbol{R}'; N)$ has a physical meaning only for $N>0$, it is convenient to *define* $G(\boldsymbol{R}, \boldsymbol{R}'; N)$ for $N<0$ in such a way that

$$G(\boldsymbol{R}, \boldsymbol{R}'; N) = 0 \quad \text{for} \quad N<0. \tag{2.52}$$

With this definition, $G(\boldsymbol{R}, \boldsymbol{R}'; N)$ satisfies a simple differential equation

$$\left(\frac{\partial}{\partial N} - \frac{b^2}{6}\frac{\partial^2}{\partial \boldsymbol{R}^2} + \frac{1}{k_B T}U_e(\boldsymbol{R})\right)G(\boldsymbol{R}, \boldsymbol{R}'; N) = \delta(\boldsymbol{R}-\boldsymbol{R}')\,\delta(N). \tag{2.53}$$

The derivation is given in Appendix 2.II. The product of the delta functions $\delta(\boldsymbol{R}-\boldsymbol{R}')\,\delta(N)$ on the right-hand side takes into account the boundary conditions

$$G(\boldsymbol{R}, \boldsymbol{R}', N)=0 \quad \text{for} \quad N<0 \quad \text{and} \quad G(\boldsymbol{R}, \boldsymbol{R}', N=0)=\delta(\boldsymbol{R}-\boldsymbol{R}'). \tag{2.54}$$

2.3.2 *Example—chain confined in a box*

As an example of the application of the Green function, let us consider a polymer confined in a box of volume $V=L_xL_yL_z$ (Fig. 2.6).[8] The confinement is expressed by an external potential U_e which is infinite outside the box and zero inside the box. Alternatively the effect is expressed by the boundary condition

$$G(\boldsymbol{R}, \boldsymbol{R}'; N)=0 \quad \text{if } \boldsymbol{R} \text{ is on the boundary of the box.} \tag{2.55}$$

The solution of the equation

$$\left(\frac{\partial}{\partial N}-\frac{b^2}{6}\frac{\partial^2}{\partial \boldsymbol{R}^2}\right)G(\boldsymbol{R}, \boldsymbol{R}'; N)=\delta(\boldsymbol{R}-\boldsymbol{R}')\,\delta(N) \tag{2.56}$$

under the boundary condition (2.55) is obtained by the standard method:[8]

$$G(\boldsymbol{R}, \boldsymbol{R}'; N)=g_x(R_x, R_x'; N)g_y(R_y, R_y'; N)g_z(R_z, R_z'; N), \tag{2.57}$$

with

$$g_x(R_x, R_x'; N)=\frac{2}{L_x}\sum_{p=1}^{\infty}\sin\left(\frac{p\pi R_x}{L_x}\right)\sin\left(\frac{p\pi R_x'}{L_x}\right)\exp(-p^2\pi^2Nb^2/6L_x^2). \tag{2.58}$$

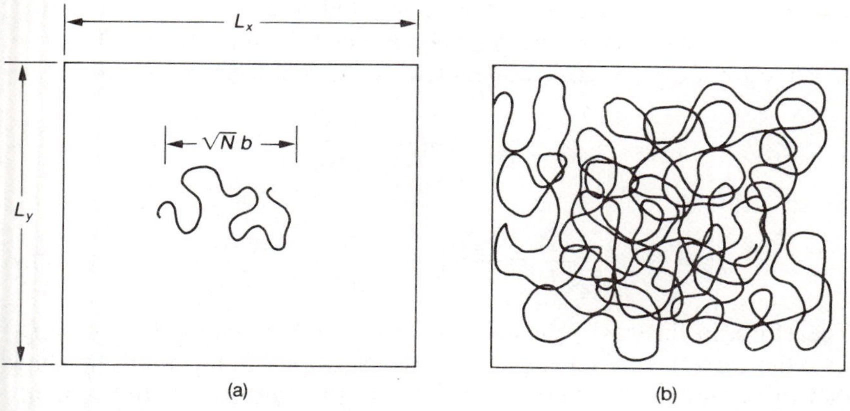

Fig. 2.6. A polymer confined in a box, (*a*) $\sqrt{N}b \ll L_x, L_y, L_z$, and (*b*) $\sqrt{N}b \gg L_x, L_y, L_z$.

The partition function is then given by

$$Z = \int \mathrm{d}\boldsymbol{R}\, \mathrm{d}\boldsymbol{R}' G(\boldsymbol{R}, \boldsymbol{R}'; N) = Z_x Z_y Z_z \tag{2.59}$$

with

$$\begin{aligned} Z_x &= \int_0^{L_x} \mathrm{d}R_x \int_0^{L_x} \mathrm{d}R'_x g_x(R_x, R'_x; N) \\ &= \frac{8}{\pi^2} L_x \sum_{p=1,3,\ldots}^{\infty} \frac{1}{p^2} \exp\left[-\frac{\pi^2 N b^2}{6} \frac{p^2}{L_x^2} \right]. \end{aligned} \tag{2.60}$$

The free energy of the system is calculated by

$$A = -k_B T \ln Z, \tag{2.61}$$

so that the pressure acting on the wall normal to the x axis is

$$P_x = -\frac{1}{L_y L_z} \frac{\partial A}{\partial L_x}. \tag{2.62}$$

Let us consider the two limiting cases:

(i) In the case when the polymer is much smaller than the box (i.e., $\sqrt{N} b \ll L_x, L_y, L_z$), Z_x is approximated by

$$Z_x = \frac{8}{\pi^2} L_x \sum_{p=1,3,\ldots} \frac{1}{p^2} = L_x \tag{2.63}$$

whence

$$P_x = \frac{k_B T}{L_x L_y L_z} = \frac{k_B T}{V} \quad \text{and similarly} \quad P_y = P_z = \frac{k_B T}{V}, \tag{2.64}$$

which is simply the equation of state of an ideal gas.

(ii) In the case when the polymer is much larger than the box (i.e., $\sqrt{N} b \gg L_x, L_y, L_z$), Z_x is dominated by the first term in the sum

$$Z_x = \frac{8}{\pi^2} L_x \exp\left[-\frac{\pi^2 N b^2}{6 L_x^2} \right], \tag{2.65}$$

which yields

$$P_x = -\frac{1}{L_y L_z} \frac{\partial A}{\partial L_x} = \frac{k_B T}{V} \left(1 + \frac{\pi^2 N b^2}{3 L_x^2} \right) = \frac{\pi^2 N b^2}{3 L_x^2} \frac{k_B T}{V}. \tag{2.66}$$

Therefore, in this case P_x is not equal to P_y and P_z (unless $L_x = L_y = L_z$), i.e., the force acting on the wall is not isotropic. Later it will be shown that this anisotropy in the pressure due to the anisotropic confinement is responsible for a variety of mechanical properties peculiar to polymers.

The Green function method has been applied in various problems related to polymers in nonhomogeneous situations such as adsorption of polymers, polymers near the phase boundary, polymers in a porous media, etc. Many such examples can be found in the literature.[9,10]

2.4 Scattering function

The size of a polymer can be measured by various scattering experiments (light scattering, small angle X-ray scattering and neutron scattering). The principles and the practical details of such scattering experiments are given in refs 11–13.

Suppose that the polymer is modelled by a series of units at each $\boldsymbol{R}_n$ which have a scattering amplitude a_n. Then the scattering intensity at the scattering vector $\boldsymbol{k} \equiv \boldsymbol{k}_f - \boldsymbol{k}_i$ ($\boldsymbol{k}_i$ and $\boldsymbol{k}_f$ being the wave vectors of the incident and the scattered beam) is written as

$$\sum_{n,m=1}^{N_0} a_n a_m^* \exp[\mathrm{i}\boldsymbol{k} \cdot (\boldsymbol{R}_n - \boldsymbol{R}_m)] \tag{2.67}$$

where N_0 is the number of scattering units in the system and the sum is taken over all the scattering units in the system. In particular, for a homogeneous polymer one has

$$|a|^2 \sum_{n,m=1}^{N_0} \exp[\mathrm{i}\boldsymbol{k} \cdot (\boldsymbol{R}_n - \boldsymbol{R}_m)]. \tag{2.68}$$

This is proportional to N_0 since the sum $\sum_{m=1}^{N_0} \exp[\mathrm{i}\boldsymbol{k} \cdot (\boldsymbol{R}_n - \boldsymbol{R}_m)]$ remains finite for $N_0 \rightarrow \infty$. We shall use the structure factor $g(\boldsymbol{k})$ defined by

$$g(\boldsymbol{k}) \equiv \frac{1}{N_0} \sum_{n,m}^{N_0} \langle \exp[\mathrm{i}\boldsymbol{k} \cdot (\boldsymbol{R}_n - \boldsymbol{R}_m)] \rangle \tag{2.69}$$

which is independent of the system size (or N_0).

If the polymer solution is sufficiently dilute, the interference among different polymers can be neglected, so that eqn (2.69) is written as

$$g(\boldsymbol{k}) \equiv \frac{1}{N} \sum_{n,m}^{N} \langle \exp[\mathrm{i}\boldsymbol{k} \cdot (\boldsymbol{R}_n - \boldsymbol{R}_m)] \rangle \tag{2.70}$$

where N is, as before, the number of segments constituting the polymer. Since the direction of $\boldsymbol{R}_n - \boldsymbol{R}_m$ is random, eqn (2.70) can be rewritten as in eqn (2.32),

$$g(\boldsymbol{k}) = \frac{1}{N} \sum_{n,m} \left\langle \frac{\sin(|\boldsymbol{k}|\,|\boldsymbol{R}_n - \boldsymbol{R}_m|)}{|\boldsymbol{k}|\,|\boldsymbol{R}_n - \boldsymbol{R}_m|} \right\rangle. \tag{2.71}$$

The characteristic size of the polymer is obtained from $g(\boldsymbol{k})$ in the small $\boldsymbol{k}$ region. Expanding eqn (2.71) for small $\boldsymbol{k}$, we have

$$\begin{aligned} g(\boldsymbol{k}) &= \frac{1}{N}\sum_{n,m} \langle 1 - \tfrac{1}{6}\boldsymbol{k}^2(\boldsymbol{R}_n - \boldsymbol{R}_m)^2 + \dots \rangle \\ &= N(1 - \tfrac{1}{3}\boldsymbol{k}^2 R_g^2 + \dots) \end{aligned} \tag{2.72}$$

where R_g is defined by

$$R_g^2 \equiv \frac{1}{2N^2}\sum_{n,m=1}^{N} \langle (\boldsymbol{R}_n - \boldsymbol{R}_m)^2 \rangle \tag{2.73}$$

which represents the mean square length between all the pairs of the segments in the chain. The quantity R_g^2 is called the mean square radius of gyration since it can be rewritten as

$$R_g^2 = \frac{1}{N}\sum_{n=1}^{N} \langle (\boldsymbol{R}_n - \boldsymbol{R}_G)^2 \rangle \tag{2.74}$$

where $\boldsymbol{R}_G$ is the position of the centre of mass of the chain:

$$\boldsymbol{R}_G = \frac{1}{N}\sum_{n=1}^{N} \boldsymbol{R}_n. \tag{2.75}$$

The derivation of eqn (2.74) is straightforward: substitution of eqn (2.75) into eqn (2.74) gives

$$\begin{aligned} R_g^2 &= \frac{1}{N}\sum_{n} \langle \boldsymbol{R}_n^2 - 2\boldsymbol{R}_n \cdot \boldsymbol{R}_G + \boldsymbol{R}_G^2 \rangle \\ &= \frac{1}{N}\sum_{n} \left\langle \boldsymbol{R}_n^2 - 2\frac{\boldsymbol{R}_n}{N}\cdot\sum_{m} \boldsymbol{R}_m + \frac{1}{N^2}\sum_{m,i} \boldsymbol{R}_m \cdot \boldsymbol{R}_i \right\rangle \\ &= \left\langle \frac{1}{N}\sum_{n} \boldsymbol{R}_n^2 - \frac{1}{N^2}\sum_{n,m} \boldsymbol{R}_n \cdot \boldsymbol{R}_m \right\rangle \\ &= \frac{1}{2N^2}\sum_{n,m} \langle (\boldsymbol{R}_n - \boldsymbol{R}_m)^2 \rangle . \end{aligned} \tag{2.76}$$

In general, the size of the polymer is more appropriately represented by R_g than $\bar{R}$ (where $\bar{R}^2 = \langle \boldsymbol{R}^2 \rangle$) because R_g is well defined for branched polymers while $\bar{R}$ is not.

For a linear Gaussian polymer R_g is easily calculated. From eqns (2.42) and (2.73)

$$R_g^2 = \frac{1}{2N^2} \sum_{n=1}^{N} \sum_{m=1}^{N} |n-m|\, b^2 = \frac{1}{2N^2} \int_0^N \mathrm{d}n \int_0^N \mathrm{d}m\, |n-m|\, b^2$$

$$= \frac{1}{N^2} \int_0^N \mathrm{d}n \int_0^n \mathrm{d}m (n-m) b^2 = \tfrac{1}{6} N b^2. \tag{2.77}$$

Thus R_g is given by $\bar{R}/\sqrt{6}$ for the linear polymer.

Let us now calculate the structure factor for the Gaussian chain. Since the distribution of $\boldsymbol{R}_n - \boldsymbol{R}_m$ is Gaussian, $\langle \exp[\mathrm{i}\boldsymbol{k} \cdot (\boldsymbol{R}_n - \boldsymbol{R}_m)] \rangle$ is calculated using eqn (2.I.20) in the Appendix as

$$\langle \exp[\mathrm{i}\boldsymbol{k} \cdot (\boldsymbol{R}_n - \boldsymbol{R}_m)] \rangle = \left\langle \exp\left[\sum_{\alpha=x,y,z} \mathrm{i}k_\alpha (R_{n\alpha} - R_{m\alpha}) \right] \right\rangle$$

$$= \exp\left[-\tfrac{1}{2} \sum_{\alpha=x,y,z} k_\alpha^2 \langle (R_{m\alpha} - R_{n\alpha})^2 \rangle \right]. \tag{2.78}$$

Since $\langle (\boldsymbol{R}_{m\alpha} - \boldsymbol{R}_{n\alpha})^2 \rangle = |m-n|\, b^2/3$, we get

$$\langle \exp[\mathrm{i}\boldsymbol{k} \cdot (\boldsymbol{R}_n - \boldsymbol{R}_m)] \rangle = \exp\left[-\frac{b^2 \boldsymbol{k}^2}{6} |n-m| \right]. \tag{2.79}$$

Hence the structure factor is given by

$$g(\boldsymbol{k}) = \frac{1}{N} \int_0^N \mathrm{d}n \int_0^N \mathrm{d}m \exp\left[-\frac{b^2 \boldsymbol{k}^2}{6} |n-m| \right] = N f(\boldsymbol{k}^2 R_g^2) \tag{2.80}$$

where

$$f(x) = \frac{2}{x^2} (\mathrm{e}^{-x} - 1 + x) \tag{2.81}$$

which is called the Debye function. The asymptotic form of $g(\boldsymbol{k})$ is

$$g(\boldsymbol{k}) = \begin{cases} N(1 - \boldsymbol{k}^2 R_g^2/3) & \text{for} \quad |\boldsymbol{k}|\, R_g \ll 1 \\ 2N/\boldsymbol{k}^2 R_g^2 & \text{for} \quad |\boldsymbol{k}|\, R_g \gg 1. \end{cases} \tag{2.82}$$

For convenience of calculation, the Debye function is often approximated by

$$g(\boldsymbol{k}) = \frac{N}{1 + \boldsymbol{k}^2 R_g^2/2}. \tag{2.83}$$

The error of this equation is less than 15% for the entire region of k.

2.5 Excluded volume effect

2.5.1 Introduction

In the models of polymers considered in the previous sections, the interaction among the polymer segments is limited to within a few neighbours along the chain. In reality, however, segments distant along the chain do interact if they come close to each other in space. An obvious interaction is the steric effect: since the segment has finite volume, other segments cannot come into its own region. This interaction swells the polymer; the coil size of a chain with such an interaction is larger than that of the ideal chain which has no such interaction. Even when there are attractive forces, as long as the repulsive force dominates, the polymer will swell. This effect is called the excluded volume effect.

The excluded volume effect represents the effect of the interaction between segments which are far apart along the chain (see Fig. 2.7). Such an interaction is often called the 'long range interaction' in contrast to the 'short range interaction' which represents the interaction among a few neighbouring segments and is included in $\psi(\boldsymbol{r}_1, \ldots, \boldsymbol{r}_{1+n_c})$ in eqn (2.15). (Note that the terms 'long' and 'short' represent the distance along the chain, not the spatial separation.)

The excluded volume effect was first discussed by Kuhn,[14] and the modern development was initiated by Flory.[15,16] It had been recognized by these pioneers that the long range interaction changes the statistical property of the chain entirely. For example, $\langle \boldsymbol{R}^2 \rangle$ is no longer proportional to N but to a higher power of N

$$\langle \boldsymbol{R}^2 \rangle \propto N^{2\nu}. \tag{2.84}$$

The exponent ν is about 3/5, so that the excluded volume effect is very important for long chains.

Once the long range interaction is introduced, exact calculation becomes impossible. A great deal of work has been done on this

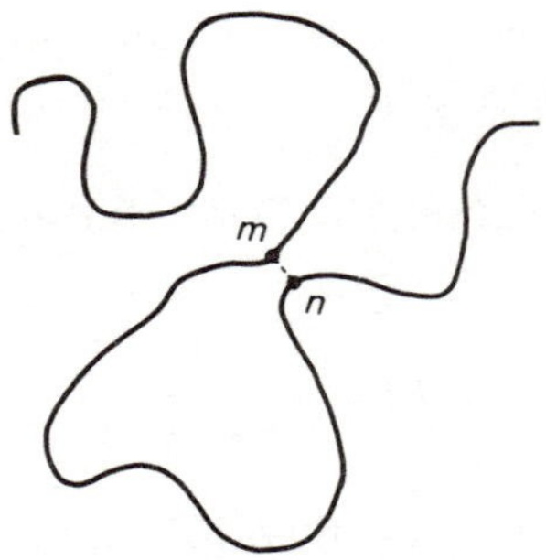

Fig. 2.7. Excluded volume interaction.

problem, and a detailed description is given in various literature.[4,6,7,17,18] Here we shall outline only a few typical approaches.

2.5.2 *Model of the excluded volume chain*

In real polymers, the nature of the long range interaction is quite complicated: the interaction will include steric effects, van der Waals attraction, and also may involve other specific interactions mediated by solvent molecules. However, as far as the property of large length scale is concerned, the detail of the interaction will not matter because the excluded volume effect is controlled by the interaction between distant parts of the chain. Thus the interaction between the polymer segments n and m can be expressed by a short range function

$$k_B T \tilde{v}(\boldsymbol{R}_n - \boldsymbol{R}_m) \tag{2.85}$$

which can usually be approximated even further to a delta function[19]

$$v k_B T \, \delta(\boldsymbol{R}_n - \boldsymbol{R}_m) \tag{2.86}$$

where v is the excluded volume and has the dimension of volume.†

The total interaction energy is thus written as

$$U_1 = \tfrac{1}{2} v k_B T \int_0^N \mathrm{d}n \int_0^N \mathrm{d}m \, \delta(\boldsymbol{R}_n - \boldsymbol{R}_m). \tag{2.87}$$

Using the local concentration of the segments

$$c(\boldsymbol{r}) = \sum_n \delta(\boldsymbol{r} - \boldsymbol{R}_n) = \int_0^N \mathrm{d}n \, \delta(\boldsymbol{r} - \boldsymbol{R}_n), \tag{2.88}$$

eqn (2.87) may be rewritten

$$U_1 = \int \mathrm{d}\boldsymbol{r} \tfrac{1}{2} v k_B T c(\boldsymbol{r})^2. \tag{2.89}$$

This expression indicates that eqn (2.87) is the first term in the virial expansion of the free energy with respect to the local concentration $c(\boldsymbol{r})$. Therefore the excluded volume parameter v can be regarded as the virial coefficient between the *segments*.

In principle the virial expansion can be continued to include higher

† The reader may be worried by the use of the delta function which diverges when $\boldsymbol{R}_n = \boldsymbol{R}_m$. It turns out, however, that all physically measurable quantities involve integrals of this potential and are well behaved. (The absolute entropy of the chain does diverge, but the entropy difference, which matters experimentally, is well behaved also.) This situation is familiar in quantum field theories and is discussed in great detail in the textbooks. The reader can be assured that this point causes no real difficulties.

order terms such as

$$U_1 = \int d\boldsymbol{r}[\tfrac{1}{2}vk_BTc(\boldsymbol{r})^2 + \tfrac{1}{6}wk_BTc(\boldsymbol{r})^3 + \ldots]. \tag{2.90}$$

However, the higher order terms may be neglected since the segment density inside the polymer coil is small: the segment density is estimated as

$$\bar{c} \simeq \frac{N}{\bar{R}^3} \propto N^{1-3\nu} = N^{-4/5} \quad (\text{when } \nu = 3/5) \tag{2.91}$$

which becomes very small for large N. Therefore the essential features of the excluded volume effect can be studied using the potential given by eqn (2.87).

For a given combination of polymer and solvent, v varies with temperature and can be zero at a certain temperature, called the Θ or Flory temperature. At the Θ temperature, the chain becomes nearly ideal.†

An appropriate expression for the temperature dependence of v may be obtained as follows. Suppose that the interaction between the segments is expressed by a potential energy $u(r)$ which depends only on their separation r. Then the second virial coefficient is evaluated by the standard formula for an imperfect gas[22]

$$v = \int d\boldsymbol{r}\left[1 - \exp\left(-\frac{u(r)}{k_BT}\right)\right]. \tag{2.92}$$

Usually $u(r)$ consists of a hard core potential $u_{\text{hard}}(r)$ and a weak attractive potential $u_{\text{attr}}(r)$ (see Fig. 2.8). In such a case, v is estimated as

$$v = \int d\boldsymbol{r}\left[1 - \exp\left(-\frac{u_{\text{hard}}(r)}{k_BT}\right)\left(1 - \frac{u_{\text{attr}}(r)}{k_BT}\right)\right] = A - \frac{B}{T} \tag{2.93}$$

where A and B are constants independent of temperature. For this model, the Θ temperature is defined as B/A, and eqn (2.93) may be written as[16]

$$v = v^{(0)}\left(1 - \frac{\Theta}{T}\right). \tag{2.94}$$

Now if the interaction (eqn (2.87)) is taken into account, the distribution function of $\boldsymbol{R}_n$ becomes

$$\Psi[\boldsymbol{R}_n] \propto \exp\left[-\frac{3}{2b^2}\int_0^N dn\left(\frac{\partial \boldsymbol{R}_n}{\partial n}\right)^2 - \tfrac{1}{2}v\int_0^N dn\int_0^N dm\,\delta(\boldsymbol{R}_n - \boldsymbol{R}_m)\right]. \tag{2.95}$$

† Even at the Θ temperature the chain is not ideal since there is a three-body collision term.[20,21] However the effect of the three-body collision is quite weak and gives only a logarithmic correction to $\langle \boldsymbol{R}^2 \rangle$.

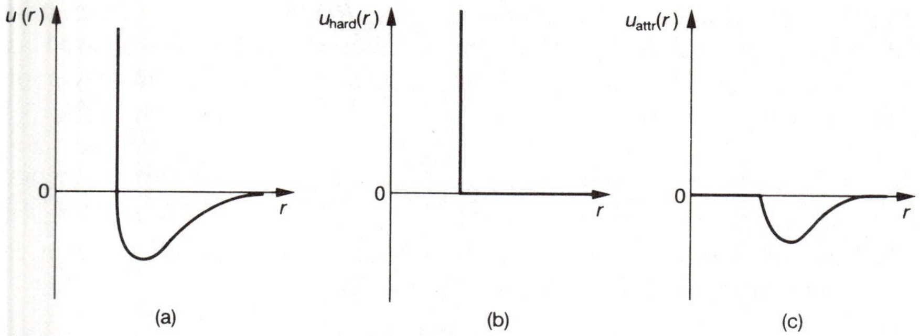

Fig. 2.8. A sketch of a potential. The potential can be decomposed into a hard core potential $u_{\text{hard}}(r)$ and a weak attractive potential $u_{\text{attr}}(r)$.

This model includes only two parameters; b, which represents the short range interaction, and v, which represents the long range interaction. The basic assumption of this two-parameter model is that there is a sharp distinction between the short range interaction and the long range interaction. Though the validity of this assumption can be questioned,[23] it is generally believed to be a correct starting point for the analysis of the excluded volume problem.

2.5.3 Theoretical approaches

We shall now discuss the statistical properties of the chain of eqn (2.95). We shall limit the discussion to the case of $v > 0$, i.e., the case of repulsive interation. The cases of $v = 0$ and $v < 0$ are discussed in refs 24 and 25.

A simple theory. The original idea of Flory[15] for calculating the size of a polymer is to consider the balance of two effects: a repulsive excluded volume interaction which tends to swell the polymer, and the elastic energy arising from the chain connectivity which tends to shrink the polymer. This idea can be put into a particularly simple form of theory.[26]

Consider the free energy of a chain whose end-to-end vector is fixed at $\boldsymbol{R}$. This is given by

$$A(\boldsymbol{R}) = -k_B T \ln \Phi(\boldsymbol{R}, N) + \text{terms independent of } \boldsymbol{R} \qquad (2.96)$$

because the equilibrium distribution function is proportional to $\exp(-A/k_B T)$. If $v = 0$, the free energy $A(\boldsymbol{R})$ is obtained from eqn (2.26) as

$$A(\boldsymbol{R}) = k_B T \frac{3\boldsymbol{R}^2}{2Nb^2} + \text{terms independent of } \boldsymbol{R}. \qquad (2.97)$$

To estimate the effect of the excluded volume interaction, we disregard the connectivity of the chain, and calculate the interaction energy of a 'segment gas' confined in a volume R^3 ($R = |\boldsymbol{R}|$). Since the concentration of the segment gas is $\bar{c} \simeq N/R^3$, the interaction energy is estimated as $vk_BT\bar{c}^2R^3$. (Here the numerical factor is disregarded at this level of approximation.) Thus

$$A(\boldsymbol{R}) \simeq k_BT\left(\frac{3R^2}{2Nb^2} + v\frac{N^2}{R^3}\right). \tag{2.98}$$

The average size of the polymer can be estimated from the value of R which minimizes A. From $\partial A/\partial R = 0$, we have

$$\bar{R} \simeq \sqrt{N}\, b\left(\frac{\sqrt{N}\, v}{b^3}\right)^{1/5} \propto N^{3/5}. \tag{2.99}$$

This gives the exponent $\nu = 3/5$ already quoted, which is close to the experimental value.[27,28]

Though the exponent ν is close to the experimental value, the prediction of this theory for other quantities turns out to be inadequate: for example, the expression for the free energy (eqn (2.98)) is not consistent with the distribution function of the end-to-end vector obtained by computer simulation.[29] Also it suffers from an unreasonable behaviour of the entropy.

Because of the very simple structure of the theory, the theory is generally regarded as a prototype of a mean field theory. It must, however, be realized that it is not a real mean field theory in that no mean field has been calculated. The mean field theory will be discussed later on.

Perturbation calculation. If the excluded volume v is small, the distribution function (2.95) can be expanded with respect to v.[30,31] Hence $\langle \boldsymbol{R}^2 \rangle$ can be calculated as a power series of v. The calculation of the first term is given in Appendix 2.III. Though straightforward, such a calculation becomes quite tedious.[32] The latest result[33] is

$$\langle R^2 \rangle = Nb^2(1 + \tfrac{4}{3}z - 2.075z^2 + 6.297z^3 - 25.057z^4 + 116.135z^5 - 594.717z^6 + \ldots) \tag{2.100}$$

where the expansion parameter is defined by

$$z = \left(\frac{3}{2\pi}\right)^{3/2} \frac{v\sqrt{N}}{b^3}. \tag{2.101}$$

Note that z is proportional to $\sqrt{N}$. Thus the perturbation expansion becomes useless for large N. This situation is entirely different from the virial expansion of the imperfect gas, where the expansion parameter is the average concentration $\bar{c}$, which is proportional to $N^{1-3\nu}$ and is very small in the case of polymer problems. The reason for this difference is that for polymers, the collision between any pair of segments affects the end-to-end vector seriously (the effect being of order 1), whilst in imperfect gases the effect is only of order of $1/N$. The perturbation calculation in the polymer problem is justified only when none of the polymer segments is likely to collide with the other segments. Thus what must be small is the average number of collisions taking place in the whole polymer coil:

$$Nv\bar{c} \simeq Nv\frac{N}{(\sqrt{N}\,b)^3} \simeq \frac{v\sqrt{N}}{b^3}, \tag{2.102}$$

which is z.

It has been shown that the series (2.100) is asymptotic, and suffers from an explosive increase in its coefficients which increase roughly like n^n.[34] However, by an appropriate resummation technique, useful information can be drawn about the asymptotic behaviour of large z.[35–37] In particular, the exponent ν is estimated as[38]

$$\nu = 0.588 \pm 0.001. \tag{2.103}$$

Uniform expansion model. The difficulty in the perturbation calculation can be improved by a simple scheme of calculation.[39] In this scheme, it is assumed that the expansion of the chain is represented by the expansion of the bond length, i.e., that the distribution function of $\boldsymbol{R}_n$ is well approximated by

$$\Psi'[\boldsymbol{R}_n] \propto \exp\left[-\frac{3}{2b'^2}\int_0^N \mathrm{d}n\left(\frac{\partial \boldsymbol{R}_n}{\partial n}\right)^2\right] \tag{2.104}$$

where b' is the expanded bond length to be determined. Let $\langle \ldots \rangle'$ be the average for this distribution function. The average $\langle \boldsymbol{R}^2 \rangle$ for the distribution function (2.95) is written as

$$\langle \boldsymbol{R}^2 \rangle = \frac{\langle \exp(-B[\boldsymbol{R}_n])(\boldsymbol{R}_N - \boldsymbol{R}_0)^2 \rangle'}{\langle \exp(-B[\boldsymbol{R}_n]) \rangle'} \tag{2.105}$$

where

$$B[\boldsymbol{R}_n] = \tfrac{3}{2}(b^{-2} - b'^{-2})\int_0^N \mathrm{d}n\left(\frac{\partial \boldsymbol{R}_n}{\partial n}\right)^2 + \tfrac{1}{2}v\int_0^N \mathrm{d}n\int_0^N \mathrm{d}m\,\delta(\boldsymbol{R}_n - \boldsymbol{R}_m). \tag{2.106}$$

Now if eqn (2.104) is a good approximation to eqn (2.95), $B[\boldsymbol{R}_n]$ can be regarded as small. Hence eqn (2.105) is evaluated as

$$\langle \boldsymbol{R}^2 \rangle = \frac{\langle (1 - B[\boldsymbol{R}_n])(\boldsymbol{R}_N - \boldsymbol{R}_0)^2 \rangle'}{\langle 1 - B[\boldsymbol{R}_n] \rangle'}. \tag{2.107}$$

The average is evaluated straightforwardly (see Appendix 2.III) as

$$\langle \boldsymbol{R}^2 \rangle = \langle (\boldsymbol{R}_N - \boldsymbol{R}_0)^2 \rangle' - \langle B[\boldsymbol{R}_n](\boldsymbol{R}_N - \boldsymbol{R}_0)^2 \rangle' + \langle (\boldsymbol{R}_N - \boldsymbol{R}_0)^2 \rangle' \langle B[\boldsymbol{R}_n] \rangle'$$
$$= Nb'^2 + \underline{Nb'^2\left(\frac{b'^2}{b^2} - 1 - \frac{4}{3}\frac{b^3}{b'^3} z\right)} \tag{2.108}$$

where z is a parameter defined by eqn (2.101). Now if eqn (2.104) is the best approximation of eqn (2.95), the first-order correction (the underlined part) must vanish. This condition gives

$$\frac{b'^2}{b^2} - 1 - \frac{4}{3}\frac{b^3}{b'^3} z = 0. \tag{2.109}$$

Let us define a swelling coefficient α:

$$\alpha^2 \equiv \frac{\langle \boldsymbol{R}^2 \rangle}{Nb^2}. \tag{2.110}$$

Equation (2.108) gives $\alpha = b'/b$, so that eqn (2.109) is written as

$$\alpha^5 - \alpha^3 = \tfrac{4}{3} z. \tag{2.111}$$

Equations (2.110) and (2.111) determine $\langle \boldsymbol{R}^2 \rangle$. For small z, this theory agrees with the result of the first-order perturbation expansion (eqn (2.100)). On the other hand, for large z the theory gives $\bar{R} \propto N^{3/5}$ in agreement with eqn (2.99). Thus the theory gives an interpolation between the two cases.

A warning must be given about the assumption involved in this theory. Although eqn (2.104) gives a good approximation for $\langle \boldsymbol{R}^2 \rangle$, it gives erroneous predictions for other quantities; for example eqn (2.104) wrongly predicts that the distribution of $\boldsymbol{R}$ is Gaussian. In general the optimum choice of b' depends on the quantity under consideration: to calculate a certain quantity $A[\boldsymbol{R}_n]$, we expand $\langle A[\boldsymbol{R}_n] \rangle$ as

$$\langle A[\boldsymbol{R}_n] \rangle = \langle A[\boldsymbol{R}_n] \rangle' \underline{- \langle B[\boldsymbol{R}_n] A[\boldsymbol{R}_n] \rangle' + \langle B[\boldsymbol{R}_n] \rangle' \langle A[\boldsymbol{R}_n] \rangle'} \tag{2.112}$$

and choose the parameter b' so that the underlined term vanishes. This prescription works well. For example, the distribution function of the end-to-end vector is calculated as[39]

$$\Phi(\boldsymbol{R}, N) \propto \begin{cases} \exp\left[-\dfrac{1}{2}\left(\dfrac{R}{\bar{R}}\right)^2\right] & \text{for} \quad R \lesssim \bar{R}, \\ \exp\left[-\left(\dfrac{2}{5}\right)^{3/2} \dfrac{3}{\pi}\left(\dfrac{R}{\bar{R}}\right)^{5/2}\right] & \text{for} \quad R \gtrsim \bar{R}. \end{cases} \tag{2.113}$$

This function has been found by Domb *et al.*[29] in numerical simulation of the problem on a lattice, and has also been discussed theoretically by Fisher.[40]

Mean field theory. The simple theory given in the first part of this section can be improved by mean field calculation.[41] Let us consider an ensemble of chains whose ends are fixed in space, one at the origin and the other at point $\boldsymbol{R}$. Let $\bar{c}(\boldsymbol{r})$ be the average density of the segment:

$$\bar{c}(\boldsymbol{r}) = \int_0^N \mathrm{d}n \langle \delta(\boldsymbol{r} - \boldsymbol{R}_n) \rangle . \tag{2.114}$$

Then the segment at $\boldsymbol{r}$ feels the mean field potential $\bar{c}(\boldsymbol{r}) v k_B T$, whence the statistical distribution of the polymer becomes

$$\Psi_{\mathrm{MF}}[\boldsymbol{R}_n] \propto \exp\left[-\frac{3}{2b^2} \int_0^N \mathrm{d}n \left(\frac{\partial \boldsymbol{R}_n}{\partial n} \right)^2 - v \int_0^N \mathrm{d}n \bar{c}(\boldsymbol{R}_n) \right] . \tag{2.115}$$

For the distribution function, the Green function $G(\boldsymbol{R}, 0, n)$ is calculated by (see eqn (2.53))

$$\left[\frac{\partial}{\partial n} - \frac{b^2}{6} \frac{\partial^2}{\partial \boldsymbol{R}^2} + v \bar{c}(\boldsymbol{R}) \right] G(\boldsymbol{R}, 0, n) = \delta(\boldsymbol{R})\, \delta(N). \tag{2.116}$$

Given $G(\boldsymbol{R}, 0, n)$, the segment density $\bar{c}(\boldsymbol{r})$ is calculated by

$$\bar{c}(\boldsymbol{r}) = \frac{1}{G(\boldsymbol{R}, 0, N)} \int_0^N \mathrm{d}n G(\boldsymbol{R}, \boldsymbol{r}, N - n) G(\boldsymbol{r}, 0, n). \tag{2.117}$$

Equations (2.116) and (2.117) give a closed equation for $G(\boldsymbol{R}, 0, N)$. The detailed calculation within this theory is involved, but it predicts many interesting features of the excluded volume chain. For example, the structure factor $g(\boldsymbol{k})$ at high $\boldsymbol{k}$ region is shown to be

$$g(\boldsymbol{k}) \propto k^{-5/3} \quad \text{for} \quad kR_g \gg 1 \tag{2.118}$$

which can also be obtained from the uniform expansion model. Fluctuations have been added to this model by Kosmas and Freed,[42] but the essential features of the result turned out to be the same.

Renormalization group theory. The mean field theory neglects the fact that the collisions between the segments in the polymer are strongly correlated: a collision of any pair of segments is likely to induce other collisions since the segments are connected. To take into account such correlation, the renormalization group technique, developed in the study of the critical phenomena, has been applied to the polymer problem. This method, originally invented by Wilson,[43] was first applied to polymer problems by de Gennes[44] and des Cloizeaux.[45] A particularly clear result is given by Wilson and Fisher[46] based on the observation that in a space of higher dimensionality, the effect of correlations becomes weak and the simple perturbation expansion becomes applicable. For example, in the excluded volume problem if the polymer is embedded in d-dimensional Euclidean space, the expansion parameter z is $N^2 v/(\sqrt{N}\, b)^d \propto N^{(4-d)/2}$, which becomes small for large N if $d > 4$. (The dimension 4 is called the critical dimension.) It is thus possible to develop an expansion scheme regarding $\varepsilon = 4 - d$ as an expansion parameter. Such a scheme gives ν as

$$\nu = \tfrac{1}{2}\left(1 + \tfrac{1}{8}\varepsilon + \tfrac{15}{256}\varepsilon^2 \ldots\right). \tag{2.119}$$

For $d = 3$ this gives $\nu = 0.592$, which compares well with the more exact result[36,38] $\nu = 0.588 \pm 0.001$. In any case, the deviation in the indices from 0.6 is quite small.

The actual calculation method of the renormalization group theory is quite complex and has many variations. Details can be found in references 18, 45, 47, and 48.

2.6 Scaling

Though the renormalization group method is highly sophisticated, certain conclusions derived from the theory are easy to understand and quite powerful in understanding the nature of the excluded volume chain.[17]

The basis of the renormalization group theory is to study how physical quantities change when the basic units of the physics are changed. Before explaining this idea for the excluded volume chain, let us first consider the Gaussian chain.

As was explained in Section 2.2, the statistical property of the Gaussian chain does not depend on the local structure of the chain. Therefore instead of the original Gaussian chain consisting of N segments of bond length b, we can start from a new Gaussian chain which consists of $N' = N/\lambda$ segments with the bond length $\sqrt{\lambda}\, b$ (see Fig. 2.9). The transformation from the old chain to the new chain is the change in the parameter

$$N \to N/\lambda, \qquad b \to b\sqrt{\lambda}. \tag{2.120}$$

If one knows how a physical quantity changes under this transformation, one can draw conclusions about the dependence of the physical quantity on the parameters N and b.

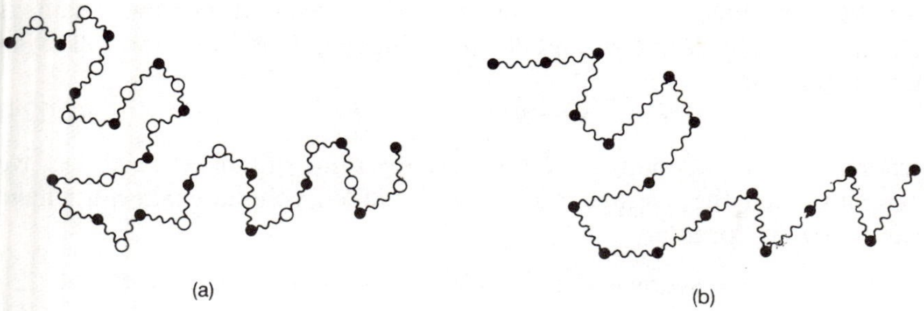

Fig. 2.9. (*a*) Original chain and (*b*) new chain, in which $\lambda = 2$.

As an example, consider the size of the Gaussian chain. Various length can be defined to characterize it. For example:

(i) Root of the mean square of the end-to-end distance, $\bar{R} = \langle \boldsymbol{R}^2 \rangle^{1/2}$

(ii) Root mean square of radius of gyration, $R_g = \langle \sum_n (\boldsymbol{R}_n - \boldsymbol{R}_G)^2 / N \rangle^{1/2}$

(iii) Mean end-to-end distance $\langle |\boldsymbol{R}| \rangle$.

(iv) The average of the longest distance connecting two beads in the chain, $\langle \max_{n,m} |\boldsymbol{R}_n - \boldsymbol{R}_m| \rangle$.

The first three quantities are easily calculated, but the last is not. However, without doing any calculation, one can show that all these quantities are proportional to $\sqrt{N}\, b$. This can be shown by the following argument: the average size of the polymer, however it is defined, has the dimension of length and must be written as

$$\text{average size} = F(N)b. \qquad (2.121)$$

The size of the polymer must be invariant under the transformation (2.120), i.e.,

$$F(N)b = F(N/\lambda)\sqrt{\lambda}\, b \qquad (2.122)$$

which is satisfied only when

$$F(N)b = \text{numerical constant} \times \sqrt{N}\, b. \qquad (2.123)$$

Thus the distinction among the various lengths is only in the numerical constant.

The scaling argument is developed quite generally in statistical mechanics[22,49,50] and indeed historically was the source of the renormalization group theory. Suppose therefore that a similar property exists for the excluded volume chain: i.e. the excluded volume chain is characterized by N and b, and its size is independent under the following transformation:

$$N \to N/\lambda, \qquad b \to b\lambda^{\nu} \tag{2.124}$$

where ν is the exponent in $\bar{R} \propto N^{\nu}$. By the same argument as above, we can show that the average size of the excluded volume chain must have the following form:

$$\text{average size} = \text{numerical constant } N^{\nu}b. \tag{2.125}$$

Equation (2.125) indicates that there is only one length-scale to characterize the macroscopic size of the polymer. In particular, the difference between $\bar{R}$ and R_g is only in the numerical factor

$$R_g = \text{numerical factor} \times \bar{R} \simeq N^{\nu}b. \tag{2.126}$$

According to the renormalization group calculation,[51,52] the numerical factor is about 0.406 which is close to $1/\sqrt{6} = 0.40825$.

In general, under the transformation (2.124), the physical quantity A changes as

$$A \to \lambda^{x} A. \tag{2.127}$$

The parameter x depends on the nature of A and can be inferred by physical argument. From this property much information can be obtained on the nature of the polymer chain.

As an example, consider the structure factor $g(\boldsymbol{k})$ of an excluded volume chain. From the dimensional analysis, $g(\boldsymbol{k})$ must be written as

$$g(\boldsymbol{k}) = F(kb, N). \tag{2.128}$$

Under the transformation, $g(\boldsymbol{k})$ changes from $g(\boldsymbol{k})$ to $g(\boldsymbol{k})/\lambda$ since $g(\boldsymbol{k})$ is proportional to the number of scattering units N. Thus the function F must satisfy

$$F(kb\lambda^{\nu}, N/\lambda) = \frac{1}{\lambda} F(kb, N). \tag{2.129}$$

For this to hold for arbitrary λ, $g(\boldsymbol{k})$ must have the following form:†

$$g(\boldsymbol{k}) = NF(kbN^{\nu}). \tag{2.130}$$

Since $N^{\nu}b \propto R_g$, this equation may be rewritten

$$g(\boldsymbol{k}) = NF(kR_g). \tag{2.131}$$

This equation determines the general functional form of the scattering

† Here a common symbol F is used to denote various functions. The functional form of F in eqn (2.131) is not the same as in eqn (2.130).

function. Particularly useful information is obtained from eqn (2.131) for the high k region. If $kR_g \gg 1$, $g(\boldsymbol{k})$ should be independent of N. This happens only when $g(\boldsymbol{k})$ is written as

$$g(\boldsymbol{k}) = \text{const}\, N(kN^{\nu})^{-1/\nu} \propto k^{-1/\nu}. \tag{2.132}$$

If $\nu = 3/5$ this gives $g(\boldsymbol{k}) \propto k^{-5/3}$. This k dependence is consistent with the result of the mean field theory. Calculations of $g(\boldsymbol{k})$ by the renormalization group method are given in refs 51 and 52.

The scaling concept has been extensively applied to various polymer problems. A variety of beautiful applications are described in the book by de Gennes.[17] We shall see them again in later chapters.

Appendix 2.I Gaussian distribution functions

2.I.1 Gaussian distribution for a single variable

Here we summarize some useful properties of the Gaussian distribution function. First we consider the case of a single variable

$$\Psi(x) = (A/2\pi)^{1/2} \exp[-\tfrac{1}{2}A(x-B)^2]. \tag{2.I.1}$$

The well-known formula for the Gaussian integral is

$$\int_{-\infty}^{\infty} \mathrm{d}x \exp(-ax^2 + bx) = (\pi/a)^{1/2} \exp\left(\frac{b^2}{4a}\right) \tag{2.I.2}$$

where a is a positive constant and b is an arbitrary complex variable. When b is real, eqn (2.I.2) may be written

$$\int_{-\infty}^{\infty} \mathrm{d}x \exp(-ax^2 + bx) = (\pi/a)^{1/2} \exp[\max_x(-ax^2 + bx)] \tag{2.I.3}$$

where $\max_x(\ldots)$ means the maximum value of the expression in the parenthesis when x is varied from $-\infty$ to $+\infty$.

Let $\langle \ldots \rangle$ be the average of the distribution function of (2.I.1),

$$\langle \ldots \rangle \equiv \int_{-\infty}^{\infty} \mathrm{d}x \ldots \Psi(x). \tag{2.I.4}$$

Equation (2.I.3) then gives

$$\langle \exp(\xi x) \rangle = \exp[\max_x(\xi x - \tfrac{1}{2}A(x-B)^2)] = \exp\left(\xi B + \frac{\xi^2}{2A}\right). \tag{2.I.5}$$

From eqn (2.I.5), the first and the second moment are calculated as

$$\langle x \rangle = \frac{\partial}{\partial \xi} \langle \exp(\xi x) \rangle |_{\xi=0} = B \tag{2.I.6}$$

and

$$\langle x^2 \rangle = \frac{\partial^2}{\partial \xi^2} \langle \exp(\xi x) \rangle |_{\xi=0} = B^2 + \frac{1}{A} \tag{2.I.7}$$

whence

$$A = [\langle x^2 \rangle - \langle x \rangle^2]^{-1} = \langle (x - \langle x \rangle)^2 \rangle^{-1} \quad \text{and} \quad B = \langle x \rangle.$$

The Gaussian distribution is thus completely specified by the mean value $\langle x \rangle$ and the variance $\langle (x - \langle x \rangle)^2 \rangle \equiv \langle \Delta x^2 \rangle$, and eqn (2.I.1) may be written as

$$\Psi(x) = (2\pi \langle \Delta x^2 \rangle)^{-1/2} \exp\left(- \frac{(x - \langle x \rangle)^2}{2 \langle \Delta x^2 \rangle} \right). \tag{2.I.8}$$

2.I.2 Gaussian distribution for many variables

The Gaussian distribution for a set of real variables $x_1, x_2, \ldots, x_N = \{x\}$ is defined as

$$\Psi(x_1, x_2, \ldots, x_N) = C \exp\left[-\frac{1}{2} \sum_{n,m} A_{nm}(x_n - B_n)(x_m - B_m) \right] \tag{2.I.9}$$

where A_{nm} is a symmetric positive definite matrix, i.e.,

$$A_{nm} = A_{mn} \quad \text{and} \quad \sum_{n,m} A_{nm} x_n x_m \geqslant 0 \quad \text{for all } x_n, \tag{2.I.10}$$

and C is a normalization constant given by

$$C = (\det[A_{nm}])^{1/2} (2\pi)^{-N/2}. \tag{2.I.11}$$

Using the coordinate transformation $x'_n = x_n - B_n$, eqn (2.I.9) is transformed to

$$\Psi(x_1, x_2, \ldots, x_N) = C \exp\left[-\frac{1}{2} \sum_{n,m} A_{nm} x_n x_m \right]. \tag{2.I.12}$$

Hence we shall consider only this form.

For the distribution (2.I.12), the generalization of the formula (2.I.5) is

$$\left\langle \exp\left[\sum_n \xi_n x_n \right] \right\rangle = \exp\left[\frac{1}{2} \sum_{n,m} (A^{-1})_{nm} \xi_n \xi_m \right] \tag{2.I.13}$$

where $(A^{-1})_{nm}$ is the inverse of the matrix of A_{nm},

$$\sum_m (A^{-1})_{nm} A_{mk} = \delta_{nk}. \tag{2.I.14}$$

Proof: Eqn (2.I.3) gives

$$\int dx_1 \exp\Big[-\sum_{n,m} \tfrac{1}{2}A_{nm}x_n x_m + \sum_n \xi_n x_n\Big]$$
$$= \sqrt{2\pi/A_{11}} \exp\Big[\max_{x_1}\Big(-\frac{1}{2}\sum_{n,m} A_{nm}x_n x_m + \sum_n \xi_n x_n\Big)\Big]. \quad (2.I.15)$$

The exponent on the right-hand side of eqn (2.I.15) is a quadratic function of x_2. Thus the integral over x_2 is repeated using eqn (2.I.3). Repeating this process for x_3, x_4, . . . , we get

$$\Big\langle \exp\Big[\sum_n \xi_n x_n\Big]\Big\rangle$$
$$= \text{const} \exp\Big[\max_{x_1,\ldots,x_N}\Big(-\frac{1}{2}\sum_{n,m} A_{nm}x_n x_m + \sum_n \xi_n x_n\Big)\Big]. \quad (2.I.16)$$

The maximum is obtained at $x_n = \sum_m (A^{-1})_{nm}\xi_m$, and

$$\max_{x_1,\ldots,x_N}\Big(-\frac{1}{2}\sum_{n,m} A_{nm}x_n x_m + \sum_n \xi_n x_n\Big) = \frac{1}{2}\sum_{n,m}(A^{-1})_{nm}\xi_n\xi_m. \quad (2.I.17)$$

The constant in eqn (2.I.16) must be 1 since eqn (2.I.16) must hold for $\xi_1 = \xi_2 = \cdots = \xi_N = 0$. Thus eqn (2.I.13) is proved.

Since the moment $\langle x_n x_m\rangle$ is calculated as

$$\langle x_n x_m\rangle = \Big[\frac{\partial^2}{\partial\xi_n\,\partial\xi_m}\Big\langle\exp\Big[\sum_j \xi_j x_j\Big]\Big\rangle\Big]_{\{\xi_j\}=0} \quad (2.I.18)$$

it follows from eqns (2.I.13) and (2.I.18), that

$$\langle x_n x_m\rangle = (A^{-1})_{nm}. \quad (2.I.19)$$

Equation (2.I.13) is thus rewritten as

$$\Big\langle\exp\Big(\sum_n \xi_n x_n\Big)\Big\rangle = \exp\Big[\frac{1}{2}\sum_{n,m}\xi_n\xi_m\langle x_n x_m\rangle\Big]. \quad (2.I.20)$$

By straightforward calculation, it can be shown from eqn (2.I.20) that

$$\langle x_n x_m x_k x_l\rangle = \frac{\partial^4}{\partial\xi_n\,\partial\xi_m\,\partial\xi_k\,\partial\xi_l}\exp\Big[\frac{1}{2}\sum_{i,j}\xi_i\xi_j\langle x_i x_j\rangle\Big]\Big|_{\{\xi_j\}=0}$$
$$= \langle x_n x_m\rangle\langle x_k x_l\rangle + \langle x_n x_k\rangle\langle x_m x_l\rangle + \langle x_n x_l\rangle\langle x_m x_k\rangle. \quad (2.I.21)$$

In general we have the following formula (Wick's theorem):

$$\langle x_{n_1}x_{n_2}\ldots x_{n_{2p}}\rangle = \sum_{\text{all pairing}}\langle x_{m_1}x_{m_2}\rangle\langle x_{m_3}x_{m_4}\rangle\ldots\langle x_{m_{2p-1}}x_{m_{2p}}\rangle \quad (2.I.22)$$

where $\langle m_1, m_2, \ldots, m_{2p}\rangle$ stands for the permutation of $\langle n_1, n_2, \ldots, n_{2p}\rangle$ and the summation is taken over all possible pairings.

An important property of the Gaussian distribution is that if the distribution of x_n is Gaussian, any linear combination of x_n,

$$X = \sum_n a_n x_n, \tag{2.I.23}$$

obeys the Gaussian distribution, i.e.,

$$\Psi(X) = [2\pi\langle X^2\rangle]^{-1/2} \exp\left[-\frac{X^2}{2\langle X^2\rangle}\right] \tag{2.I.24}$$

where

$$\langle X^2\rangle = \sum_{n,m} a_n a_m \langle x_n x_m\rangle. \tag{2.I.25}$$

The proof is straightforward. By definition

$$\Psi(X) = \int \prod_{k=1}^{N} \mathrm{d}x_k \Psi(x_1, x_2, \ldots, x_N)\, \delta\Big(X - \sum_n a_n x_n\Big).$$

After the integral over x_1, which is carried out easily because of the delta function, the integrand becomes a Gaussian function of $x_2, x_3, \ldots, x_N$ and X. Successive integrations over $x_2, x_3, \ldots, x_N$ are done using eqn (2.I.2) and give a Gaussian function of X, which is eqn (2.I.24).

2.I.3 Gaussian distribution for complex variables

For the complex variables $z_1, z_2, \ldots, z_N$, we can also define the Gaussian distribution function

$$\Psi(z_1, z_2, \ldots, z_N) = C \exp\left[-\sum_{n,m} A_{nm} z_n^* z_m\right] \tag{2.I.26}$$

where the matrix A_{nm} is assumed Hermitian with positive eigenvalues,

$$A_{nm}^* = A_{mn} \quad \text{and} \quad \sum_{n,m} A_{nm} z_n^* z_m \geqslant 0. \tag{2.I.27}$$

The integral over z_n is defined to be made over the entire surface of the complex plane; i.e., if x and y denote the real part and the imaginary part of z respectively, then

$$\int \mathrm{d}z \cdots \equiv \int_{-\infty}^{+\infty} \mathrm{d}x \int_{-\infty}^{+\infty} \mathrm{d}y \ldots. \tag{2.I.28}$$

The average of a function $F(z_1, z_2, \ldots, z_N)$ is defined as

$$\langle F(z_1, z_2, \ldots)\rangle = \int_{-\infty}^{+\infty} \prod_n \mathrm{d}x_n \int_{-\infty}^{+\infty} \prod_n \mathrm{d}y_n F(z_1, z_2, \ldots)\Psi(z_1, z_2, \ldots).$$

The generalization of eqn (2.I.3) to complex variables is, for a positive A and a complex ξ,

$$\int dz \exp[-Azz^* + z\xi^* + z^*\xi] = \frac{\pi}{A}\exp[\max_z(-Azz^* + z\xi^* + z^*\xi)] \tag{2.I.29}$$

where $\max_z(\ldots)$ denotes the maximum value when z is varied over the entire complex plane. Equation (2.I.29) can be proved by direct calculation. Using eqn (2.I.29), one can prove:

$$\text{(i)}\quad \left\langle \exp\left[\sum_n (\xi_n z_n^* + \xi_n^* z_n)\right]\right\rangle = \exp\left[\sum_{n,m} \xi_n^* \xi_m \langle z_n z_m^* \rangle\right] \tag{2.I.30}$$

and

$$\text{(ii)}\quad \langle z_n^* z_m \rangle = (A^{-1})_{nm}. \tag{2.I.31}$$

2.I.4 *Functional integral*

If the subscript n of x_n is regarded as a continuous variable, the set of points $(x_1, x_2, \ldots, x_N)$ represents a continuous function, and the integral over the set $(x_1, x_2, \ldots, x_N)$ reduces to the integration over all the functions, and is called the 'functional integral'. It is denoted by the symbol δx_n; i.e.,

$$\int \prod_n dx_n \xrightarrow[\text{continuous limit}]{} \int \delta x_n.$$

The formula in the functional integral is obtained by taking the continuous limit of the discrete variables. For example, as a generalization of eqn (2.I.16), we have

$$\int \delta x_n \exp\left[-\frac{1}{2}\int_0^N dn \int_0^N dm A_{nm} x_n x_m + \int_0^N dn \xi_n x_n\right]$$

$$\propto \exp\left[\max\left(-\frac{1}{2}\int_0^N dn \int_0^N dm A_{nm} x_n x_m + \int_0^N dn \xi_n x_n\right)\right] \tag{2.I.32}$$

or

$$\left\langle \exp\left[\int_0^N dn \xi_n x_n\right]\right\rangle = \exp\left[\frac{1}{2}\int_0^N dn \int_0^N dm (A^{-1})_{nm} \xi_n \xi_m\right] \tag{2.I.33}$$

where

$$\int_0^N dm (A^{-1})_{nm} A_{mj} = \delta(n-j). \tag{2.I.34}$$

Appendix 2.II Differential equation for $G(\boldsymbol{R}, \boldsymbol{R}'; N)$

For $U_e = 0$, $G(\boldsymbol{R}, \boldsymbol{R}'; N)$ becomes the probability distribution of the end-to-end vector and is given by

$$G_0(\boldsymbol{R} - \boldsymbol{R}'; N) = \left(\frac{2\pi N b^2}{3}\right)^{-3/2} \exp\left(-\frac{3(\boldsymbol{R} - \boldsymbol{R}')^2}{2Nb^2}\right)\Theta(N) \quad (2.\text{II}.1)$$

where

$$\Theta(N) = \begin{cases} 1, & N > 0, \\ 0, & N < 0. \end{cases} \quad (2.\text{II}.2)$$

Using the Fourier transform (see eqn (2.25))

$$G_0(\boldsymbol{R} - \boldsymbol{R}'; N) = \int \frac{\mathrm{d}\boldsymbol{k}}{(2\pi)^3} \exp(-\mathrm{i}\boldsymbol{k} \cdot (\boldsymbol{R} - \boldsymbol{R}'))\exp\left(-\frac{Nb^2\boldsymbol{k}^2}{6}\right)\Theta(N), \quad (2.\text{II}.3)$$

one can easily check

$$\frac{\partial}{\partial N} G_0(\boldsymbol{R} - \boldsymbol{R}'; N) = \int \frac{\mathrm{d}\boldsymbol{k}}{(2\pi)^3} \exp(-\mathrm{i}\boldsymbol{k} \cdot (\boldsymbol{R} - \boldsymbol{R}'))\left[-\frac{b^2\boldsymbol{k}^2}{6} \times \exp\left(-\frac{Nb^2\boldsymbol{k}^2}{6}\right)\Theta(N) + \delta(N)\right]$$

$$= \frac{b^2}{6}\frac{\partial^2}{\partial \boldsymbol{R}^2} G_0(\boldsymbol{R} - \boldsymbol{R}'; N) + \delta(\boldsymbol{R} - \boldsymbol{R}')\,\delta(N). \quad (2.\text{II}.4)$$

To derive an equation for $G(\boldsymbol{R}, \boldsymbol{R}'; N)$ for $U_e \neq 0$, we use eqn (2.49) for small ΔN:

$$G(\boldsymbol{R}, \boldsymbol{R}'; N + \Delta N) = \int \mathrm{d}\boldsymbol{R}'' G(\boldsymbol{R}, \boldsymbol{R}''; \Delta N) G(\boldsymbol{R}'', \boldsymbol{R}'; N). \quad (2.\text{II}.5)$$

For n between N and $N + \Delta N$, $\boldsymbol{R}_n$ is not far apart from $\boldsymbol{R}$. Thus if $U_e(\boldsymbol{r})$ is a smooth function of $\boldsymbol{r}$, the energy is approximated as

$$\int_N^{N+\Delta N} \mathrm{d}n U_e(\boldsymbol{R}_n) \simeq \Delta N U_e(\boldsymbol{R}). \quad (2.\text{II}.6)$$

Then $G(\boldsymbol{R}, \boldsymbol{R}''; \Delta N)$ is easily obtained from eqn (2.46),

$$G(\boldsymbol{R}, \boldsymbol{R}''; \Delta N) = \exp\left(\frac{-1}{k_B T} \Delta N U_e(\boldsymbol{R})\right) G_0(\boldsymbol{R} - \boldsymbol{R}''; \Delta N). \quad (2.\text{II}.7)$$

From eqns (2.II.5) and (2.II.7),

$$G(\boldsymbol{R}, \boldsymbol{R}'; N + \Delta N) = \exp\left(-\frac{1}{k_B T} \Delta N U_e(\boldsymbol{R})\right) \times \int \mathrm{d}\boldsymbol{R}'' G_0(\boldsymbol{R} - \boldsymbol{R}''; \Delta N) G(\boldsymbol{R}'', \boldsymbol{R}'; N). \quad (2.\text{II}.8)$$

For small ΔN, $G_0(\boldsymbol{R}-\boldsymbol{R}''; \Delta N)$ has a sharp peak at $\boldsymbol{R}=\boldsymbol{R}'$. Therefore the integral in eqn (2.II.8) can be evaluated by expanding $G(\boldsymbol{R}'', \boldsymbol{R}'; N)$ with respect to $\boldsymbol{r} \equiv \boldsymbol{R}-\boldsymbol{R}''$:

$$
\begin{aligned}
I &= \int \mathrm{d}\boldsymbol{R}'' G_0(\boldsymbol{R}-\boldsymbol{R}''; \Delta N) G(\boldsymbol{R}'', \boldsymbol{R}'; N) \\
&= \int \mathrm{d}\boldsymbol{r} G_0(\boldsymbol{r}; \Delta N) G(\boldsymbol{R}-\boldsymbol{r}, \boldsymbol{R}'; N) \\
&= \int \mathrm{d}\boldsymbol{r} G_0(\boldsymbol{r}; \Delta N)\left(1 - r_\alpha \frac{\partial}{\partial R_\alpha} + \tfrac{1}{2} r_\alpha r_\beta \frac{\partial^2}{\partial R_\alpha \, \partial R_\beta}\right) G(\boldsymbol{R}, \boldsymbol{R}'; N). \qquad (2.II.9)
\end{aligned}
$$

Since

$$
\int \mathrm{d}\boldsymbol{r} G_0(\boldsymbol{r}; \Delta N) r_\alpha = 0, \qquad \int \mathrm{d}\boldsymbol{r} G_0(\boldsymbol{r}; \Delta N) r_\alpha r_\beta = \Delta N \frac{b^2}{3} \delta_{\alpha\beta}, \qquad (2.II.10)
$$

the integral becomes

$$
I = \left(1 + \tfrac{1}{6} \Delta N b^2 \frac{\partial^2}{\partial \boldsymbol{R}^2}\right) G(\boldsymbol{R}, \boldsymbol{R}'; N). \qquad (2.II.11)
$$

Thus to the order of ΔN, eqn (2.II.8) is written as

$$
\begin{aligned}
\left(1 + \Delta N \frac{\partial}{\partial N}\right) G(\boldsymbol{R}, \boldsymbol{R}'; N) &= \left(1 - \frac{1}{k_B T} \Delta N U_e(\boldsymbol{R})\right) \\
&\quad \times \left(1 + \tfrac{1}{6} \Delta N b^2 \frac{\partial^2}{\partial \boldsymbol{R}^2}\right) G(\boldsymbol{R}, \boldsymbol{R}'; N). \qquad (2.II.12)
\end{aligned}
$$

Comparing the terms of order ΔN, we have

$$
\left(\frac{\partial}{\partial N} - \frac{b^2}{6} \frac{\partial^2}{\partial \boldsymbol{R}^2} + \frac{1}{k_B T} U_e(\boldsymbol{R})\right) G(\boldsymbol{R}, \boldsymbol{R}'; N) = 0. \qquad (2.II.13)
$$

This equation holds for $N > 0$. To account for the singularity at $N = 0$, we use eqn (2.II.7) for small N, i.e.,

$$
G(\boldsymbol{R}, \boldsymbol{R}'; N) = \exp\left(\frac{-1}{k_B T} N U_e(\boldsymbol{R})\right) G_0(\boldsymbol{R}-\boldsymbol{R}'; N) \qquad (2.II.14)
$$

which gives $\delta(\boldsymbol{R}-\boldsymbol{R}')\,\delta(N)$ on the right-hand side of eqn (2.53).

Appendix 2.III Perturbation calculation for the excluded volume effect

Let $\langle \ldots \rangle_0$ denote the average for the distribution function of the ideal chain, then for the excluded volume chain

$$
\langle \boldsymbol{R}^2 \rangle = \frac{\langle \exp(-U_1)(\boldsymbol{R}_N - \boldsymbol{R}_0)^2 \rangle_0}{\langle \exp(-U_1) \rangle_0}, \qquad (2.III.1)
$$

where

$$U_1 = v \int_0^N dm \int_0^m dn\, \delta(\boldsymbol{R}_n - \boldsymbol{R}_m) \tag{2.III.2}$$

(here $k_B T$ is put equal to unity). Expanding eqn (2.III.1) to the first order in U_1, we have

$$\begin{aligned}
\langle \boldsymbol{R}^2 \rangle &= \frac{\langle (1 - U_1)(\boldsymbol{R}_N - \boldsymbol{R}_0)^2 \rangle_0}{\langle 1 - U_1 \rangle_0} \\
&= \langle (\boldsymbol{R}_N - \boldsymbol{R}_0)^2 \rangle_0 (1 + \langle U_1 \rangle_0) - \langle U_1 (\boldsymbol{R}_N - \boldsymbol{R}_0)^2 \rangle_0 \\
&= Nb^2 \Big(1 + v \int_0^N dm \int_0^m dn \langle \delta(\boldsymbol{R}_n - \boldsymbol{R}_m) \rangle_0 \Big) \\
&\quad - v \int_0^N dm \int_0^m dn \langle \delta(\boldsymbol{R}_n - \boldsymbol{R}_m)(\boldsymbol{R}_N - \boldsymbol{R}_0)^2 \rangle_0 .
\end{aligned} \tag{2.III.3}$$

First we calculate

$$I = \langle \delta(\boldsymbol{R}_n - \boldsymbol{R}_m)(\boldsymbol{R}_N - \boldsymbol{R}_0)^2 \rangle_0 , \tag{2.III.4}$$

which is written, using the Green function $G_0(\boldsymbol{R}, \boldsymbol{R}', N)$ in free space (see eqns (2.50) and (2.51)), as

$$\begin{aligned}
I &= \int d\boldsymbol{R}_N \, d\boldsymbol{R}_m \, d\boldsymbol{R}_n G_0(\boldsymbol{R}_N - \boldsymbol{R}_m; N - m) \\
&\quad \times G_0(\boldsymbol{R}_m - \boldsymbol{R}_n; m - n) G_0(\boldsymbol{R}_n - \boldsymbol{R}_0; n)\, \delta(\boldsymbol{R}_n - \boldsymbol{R}_m)(\boldsymbol{R}_N - \boldsymbol{R}_0)^2 \\
&= \int d\boldsymbol{R}_N \, d\boldsymbol{R}_n G_0(\boldsymbol{R}_N - \boldsymbol{R}_n; N - m) G_0(0; m - n) \\
&\quad \times G_0(\boldsymbol{R}_n - \boldsymbol{R}_0; n)(\boldsymbol{R}_N - \boldsymbol{R}_0)^2
\end{aligned} \tag{2.III.5}$$

To do the integral, we write $\boldsymbol{R}_N - \boldsymbol{R}_0$ as $(\boldsymbol{R}_N - \boldsymbol{R}_n) - (\boldsymbol{R}_n - \boldsymbol{R}_0)$ and integrate first over $\boldsymbol{R}_N$ and then over $\boldsymbol{R}_n$. The result is

$$I = G_0(0; m - n)(N - m + n) b^2 . \tag{2.III.6}$$

Thus

$$\langle U_1 (\boldsymbol{R}_N - \boldsymbol{R}_0)^2 \rangle_0 = v \int_0^N dm \int_0^m dn (N - m + n) b^2 G_0(0; m - n). \tag{2.III.7}$$

Similarly

$$\langle U_1 \rangle = v \int_0^N dm \int_0^m dn G_0(0; m - n). \tag{2.III.8}$$

The integrals in eqns (2.III.7) and (2.III.8) diverge, but the divergent terms cancel with each other in $\langle \boldsymbol{R}^2 \rangle$. In fact, from eqns (2.III.3), (2.III.7), and (2.III.8), we have

$$\langle \boldsymbol{R}^2 \rangle = Nb^2\Big(1 + v\int_0^N \mathrm{d}m \int_0^m \mathrm{d}n G_0(0; m-n)(m-n)\Big)$$

$$= Nb^2\Big(1 + v\int_0^N \mathrm{d}m \int_0^m \mathrm{d}n\Big(\frac{3}{2\pi(m-n)b^2}\Big)^{3/2}(m-n)\Big), \quad (2.\mathrm{III}.9)$$

which converges and gives the first term in eqn (2.100).

Next we derive eqn (2.108). The integral we now have to evaluate is

$$J = \left\langle \left(\frac{\partial \boldsymbol{R}_n}{\partial n}\right)^2 (\boldsymbol{R}_N - \boldsymbol{R}_0)^2 \right\rangle'. \quad (2.\mathrm{III}.10)$$

To calculate this we replace $\partial \boldsymbol{R}_n/\partial n$ by $\boldsymbol{R}_n - \boldsymbol{R}_{n-1}$, and rewrite $\boldsymbol{R}_N - \boldsymbol{R}_0$ as

$$\boldsymbol{R}_N - \boldsymbol{R}_0 = (\boldsymbol{R}_N - \boldsymbol{R}_n) + (\boldsymbol{R}_n - \boldsymbol{R}_{n-1}) + (\boldsymbol{R}_{n-1} - \boldsymbol{R}_0). \quad (2.\mathrm{III}.11)$$

Since there is no correlation among $\boldsymbol{R}_N - \boldsymbol{R}_n$, $\boldsymbol{R}_n - \boldsymbol{R}_{n-1}$, and $\boldsymbol{R}_{n-1} - \boldsymbol{R}_0$, eqn (2.III.10) becomes

$$J = \langle(\boldsymbol{R}_n - \boldsymbol{R}_{n-1})^2\rangle'\langle(\boldsymbol{R}_N - \boldsymbol{R}_n)^2\rangle' + \langle(\boldsymbol{R}_n - \boldsymbol{R}_{n-1})^4\rangle' + \langle(\boldsymbol{R}_n - \boldsymbol{R}_{n-1})^2\rangle'\langle(\boldsymbol{R}_{n-1} - \boldsymbol{R}_0)^2\rangle'. \quad (2.\mathrm{III}.12)$$

Since $\langle(\boldsymbol{R}_m - \boldsymbol{R}_n)^2\rangle' = (m-n)b'^2$ and $\langle(\boldsymbol{R}_m - \boldsymbol{R}_n)^4\rangle' = (5/3)(m-n)^2b^4$, eqn (2.III.12) becomes

$$J = Nb'^4 + \tfrac{2}{3}b'^4, \quad (2.\mathrm{III}.13)$$

which gives eqn (2.108).

References

1. Flory, P. J., *Statistical Mechanics of Chain Molecules*. Interscience, New York (1969).
2. Volkenstein, M. V., *Configurational Statistics of Polymeric Chains*. Interscience, New York (1963).
3. Birshtein, T. M., and Ptitsyn, O. B., *Conformations of Macromolecules*. Interscience, New York, (1966).
4. Yamakawa, H., *Modern Theory of Polymer Solutions*. Harper & Row, New York (1971).
5. Feller, W., *An Introduction to Probability Theory and its Applications* (3rd edn). Wiley, New York, (1968).

6. Edwards, S. F., The configurations and dynamics of polymer chains, in *Molecular Fluids* (eds. R. Balian and G. Weill). Gordon & Breach, London (1976).
7. Freed, K. F., *Adv. Chem. Phys.* **22,** 1 (1972).
8. Edwards, S. F., and Freed, K. F., *J. Phys. A* **2,** 145 (1969).
9. Helfand, G., in *Polymer Compatibility and Incompatibility* (ed. K. Solc), p. 143. Harwood Academic Publishers (1982).
10. Napper, D. H., *Polymeric Stabilization of Colloidal Dispersions*. Academic Press, London (1983).
11. Chu, B., *Laser Light Scattering*. Academic Press, New York (1974).
12. Berne, B. J., and Pecora, R., *Dynamic Light Scattering*. Wiley, New York (1976).
13. Marshall, W. and Lovesey, S. W., *Theory of Thermal Neutron Scattering*. Oxford Univ. Press (1971).
14. Kuhn, W., *Kolloid Z.* **68,** 2 (1934).
15. Flory, P. J., *J. Chem. Phys.* **17,** 303 (1949).
16. Flory, P. J., *Principles of Polymer Chemistry*. Cornell Univ. Press, Ithaca, (1953).
17. de Gennes, P. G., *Scaling Concepts in Polymer Physics*. Cornell Univ. Press, Ithaca (1979).
18. Oono, Y., *Adv. Chem Phys.* **61,** 301 (1985).
19. Stockmayer, W. H., *Macromol. Chem.* **35,** 54 (1960). The potential of the delta function was first used by Zimm, B. H., Stockmayer, W. H., and Fixman, M., *J. Chem. Phys.* **21,** 1716 (1953).
20. Stephen, M. J., *Phys. Lett.* **53A,** 363 (1975).
21. de Gennes, P. G., *J. Physique* **36,** L55 (1975); **39,** L299 (1978).
22. See for example, Reichl, L. E., *A Modern Course in Statistical Physics*, Chap. 11, Univ. of Texas Press, Austin (1980).
23. See Ref. 4, Section 39. A recent discussion on this problem can be found in Tanaka, G., *Macromolecules* **13,** 1513 (1980) and ref. 18.
24. Lifshitz, I. M., Grosberg, A. Y., and Khokhlov, A. R., *Rev. Mod. Phys.* **50,** 683 (1978).
25. Williams, C., Brochard, F., and Frisch, H. L., *Ann. Rev. Phys. Chem.* **32,** 433 (1981).
26. Fisher, M. E., *J. Phys. Soc. Jpn* **26** (Suppl.), 44 (1969).
27. See Cotton, J. P., *J. Physique Lett.* **L231** (1980), which summarizes experimental results obtained by Miyaki, Y., Einaga, Y., and Fujita, H., *Macromolecules* **11,** 1180 (1978); Fukuda, M., Fukutomi, M., and Hashimoto, T., *J. Polym. Sci.* **12,** 87 (1974); Yamamoto, A., Fujii, M., Tanaka, G., and Yamakawa, H., *Polymer J.* **2,** 799 (1971).
28. Wall, F. T., Windwer, S., and Gans, P. J., Monte Carlo methods applied to configurations of flexible polymer molecules, *J. Chem. Phys.* **38,** 2220; 2228 (1963).
29. Domb, C., Gillis, J., and Wilmers, G., *Proc. Phys. Soc.* **85,** 625 (1965).
30. Teramoto, E., *Busseiron Kenkyu* **39,** 1 (1951).
31. Fixman, M., *J. Chem. Phys.* **23,** 1656 (1955).
32. Yamakawa, H. and Tanaka, G., *J. Chem. Phys.* **47,** 3991 (1967).
33. Muthukumar, M., and Nickel, B. G., *J. Chem. Phys.*, **80,** 5839 (1984).
34. Edwards, S. F., *J. Phys. A* **8,** 1171 (1975). See also Oono, Y., *J. Phys. Soc. Jpn.* **39,** 25 (1975); and **41,** 787 (1976).

35. Domb, C., *Adv. Chem. Phys.* **15,** 229 (1969).
36. McKenzie, D. S., *Phys. Rep.* **27C,** 35 (1976).
37. Brézin, E., Le Guillou, J. C., and Zinn-Justin, J., *Phys. Rev.* **D15,** 1544, 1558 (1977).
38. Le Guillou, J. C., and Zinn-Justin, J., *Phys. Rev. Lett.* **39,** 95 (1977).
39. Edwards, S. F., and Singh, P., *J. Chem. Soc. Faraday Trans. II* **75,** 1001 (1979).
40. Fisher, M. E., *J. Chem. Phys.* **44,** 616 (1966).
41. Edwards, S. F., *Proc. Phys. Soc.* **85,** 613 (1965).
42. Kosmas, M. K., and Freed, K. F., *J. Chem. Phys.* **68,** 4878 (1978).
43. Wilson, K. G., *Phys. Rev.* **B4,** 3184 (1971); Wilson, K. G., and Kogut, J. B., *Phys. Rep.* **C12,** 75 (1974).
44. de Gennes, P. G., *Phys. Lett.* **38A,** 339 (1972).
45. des Cloizeaux, J., *Phys. Rev.* **A10,** 1665 (1974); *J. Physique* **42,** 635 (1981).
46. Wilson, K. G., and Fisher, M. E., *Phys. Rev. Lett.* **28,** 240 (1972).
47. Burch, D. J., and Moore, M. A., *J. Phys.* **A9,** 435 (1976).
48. Witten, T. A. Jr. and Schäfer, L., *J. Chem. Phys.* **74,** 2582 (1981).
49. Widom, B., *J. Chem. Phys.* **43,** 3898 (1965).
50. Kadanoff, L., *Physics* **A2,** 263 (1966).
51. Witten, T., Jr., *J. Chem. Phys.* **76,** 3300 (1982).
52. Ohta, T., Oono, Y., and Freed, K. F., *Phys. Rev.* **A25,** 2801 (1982).

3

BROWNIAN MOTION

3.1 Introduction

Polymers in solutions incessantly change both their shape and position randomly by thermal agitation. This Brownian motion dominates various time-dependent phenomena in polymer solutions such as viscoelasticity, diffusion, birefringence, and dynamic light scattering, which are to be discussed in subsequent chapters. In this chapter, we study the basic theory of Brownian motion. Since the general aspects of the theory of Brownian motion have already been discussed in many articles,[1–3] we shall limit the discussion to topics which will be useful in the application to polymer solutions and suspensions.

In principle, Brownian motion can be discussed starting from the dynamical equation of motion of the Brownian particle and the fluid molecules. However, this microscopic approach is not useful for calculating the various dynamical quantities we are interested in. Here we take a phenomenological approach, regarding Brownian motion as a kind of stochastic process, and construct a phenomenological equation describing Brownian motion based on known macroscopic laws. This approach, originated by Einstein,[4] is limited by several conditions, such as that the time-scale and the length-scale under consideration are much longer than those characteristic of solvent molecules, and that a linear relation holds between fluxes and forces. For polymer solutions and suspensions, these conditions are normally satisfied without any problems, and the theory we shall describe in this chapter can be regarded as a general base for describing their dynamics.

The phenomenological equation for Brownian motion has two seemingly different, but essentially the same, forms—the Smoluchowski equation and the Langevin equation. The Smoluchowski equation is derived from the generalization of the diffusion equation and has a clear relevance to the thermodynamics of irreversible processes. The Langevin equation, on the other hand, has no direct relationship to thermodynamics, but is capable of describing wider classes of stochastic processes.[2] We shall first study the Smoluchowski equation, and then consider its equivalence to the Langevin equation.

3.2 The Smoluchowski equation

3.2.1 Diffusion of particles

The phenomena in which the effect of Brownian motion appears most clearly is diffusion: small particles placed at a certain point will spread out in time. For the sake of simplicity we will consider one-dimensional diffusion. Let $c(x, t)$ be the concentration at x and t. The process of diffusion is phenomenologically described by Fick's law, which says that if the concentration is not uniform, there is a flux $j(x, t)$ which is proportional to the spatial gradient of the concentration, i.e.,

$$j(x, t) = -D\frac{\partial c}{\partial x} \tag{3.1}$$

where the constant D is called the diffusion constant.

The microscopic origin of the flux (eqn (3.1)) is the random motion of the particles: if the concentration is not uniform, the number of particles which happen to flow from the higher concentration region to the lower concentration region is larger than the number of particles flowing in the other direction (see Fig. 3.1). This imbalance in the number of flowing particles gives (to the first order in the concentration gradient) eqn (3.1) for the flux. Note that the flux comes entirely from fluctuations in the

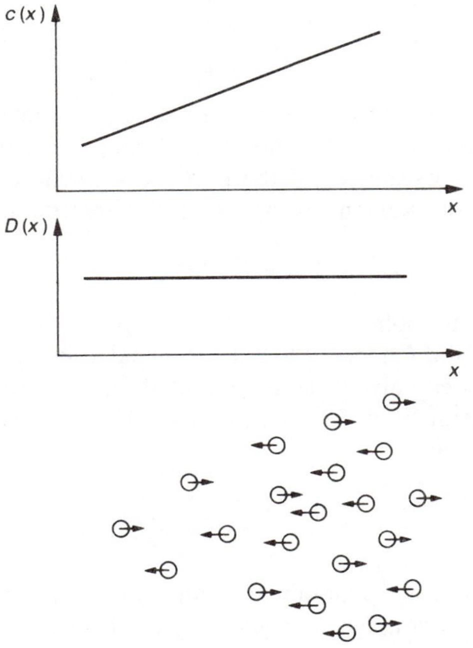

Fig. 3.1. Microscopic explanation for Fick's law. Suppose that the particle concentration $c(x)$ is not uniform. If the particles move randomly as shown by the arrows, there is a net flux of particles flowing from the higher concentration region to the lower concentration region. Here the diffusion constant of the particle, which determines the average length of the arrows, is assumed to be constant.

velocity: the average velocity of individual particles is zero as their motion is independent of each other.

Equation (3.1), together with the continuity equation

$$\frac{\partial c}{\partial t} = -\frac{\partial j}{\partial x}, \tag{3.2}$$

gives the well-known diffusion equation for constant D,

$$\frac{\partial c}{\partial t} = D\frac{\partial^2 c}{\partial x^2}. \tag{3.3}$$

If there is an external potential $U(x)$, Fick's law must be modified. The potential $U(x)$ exerts a force

$$F = -\frac{\partial U}{\partial x} \tag{3.4}$$

on the particle, and gives a non-vanishing average velocity v, which in the usual condition of weak force, is linear in F so that

$$v = -\frac{1}{\zeta}\frac{\partial U}{\partial x}. \tag{3.5}$$

The constant ζ is called the friction constant and its inverse $1/\zeta$ is called the mobility. If the particle is sufficiently large, ζ can be obtained from hydrodynamics. For example, if the particle is a sphere of radius a, and the viscosity of the solvent is η_s, then the hydrodynamic calculation indicates[5]

$$\zeta = 6\pi\eta_s a. \tag{3.6}$$

If the particle is not spherical, the formula for the velocity is not simple (see Section 3.8 and Chapter 8), but the linear relationship between the force and the velocity always holds provided that the force is weak.

The average velocity of the particle gives an additional flux cv, so that the total flux will be

$$j = -D\frac{\partial c}{\partial x} - \frac{c}{\zeta}\frac{\partial U}{\partial x}. \tag{3.7}$$

An important relation is obtained from eqn (3.7). In the equilibrium state, the concentration $c(x, t)$ is given by the Boltzmann distribution

$$c_{\mathrm{eq}}(x) \propto \exp(-U(x)/k_B T) \tag{3.8}$$

for which the flux must vanish:

$$-D\frac{\partial}{\partial x}c_{\mathrm{eq}} - \frac{1}{\zeta}c_{\mathrm{eq}}\frac{\partial U}{\partial x} = 0. \tag{3.9}$$

From eqns (3.8) and (3.9), it follows that in equilibrium

$$D = \frac{k_B T}{\zeta}. \tag{3.10}$$

This relation is called the Einstein relation.

The Einstein relation states that the quantity D which characterizes the thermal motion is related to the quantity ζ which specifies the response to the external force. We shall show later that the Einstein relation is a special case of a more general theorem, called the fluctuation dissipation theorem, which states that the characteristics of the spontaneous thermal flucutation are related to the characteristics of the response of the system to an external field.

Using eqn (3.10), eqn (3.7) can be written as

$$j = -\frac{1}{\zeta}\left(k_B T \frac{\partial c}{\partial x} + c \frac{\partial U}{\partial x}\right). \tag{3.11}$$

Hence the diffusion equation becomes

$$\frac{\partial c}{\partial t} = \frac{\partial}{\partial x}\frac{1}{\zeta}\left(k_B T \frac{\partial c}{\partial x} + c \frac{\partial U}{\partial x}\right). \tag{3.12}$$

This equation is called the Smoluchowski equation.

The above argument for deriving the Smoluchowski equation can be summarized in a more formal way. Equation (3.11) can be rewritten as

$$j = -\frac{1}{\zeta} c \frac{\partial}{\partial x}(k_B T \ln c + U), \tag{3.13}$$

which has a thermodynamic significance. The quantity $U(x) + k_B T \ln c$ is the chemical potential of noninteracting particles of concentration c. Thus eqn (3.13) states that the flux is proportional to the spatial gradient of the chemical potential. This is a natural generalization of Fick's law because when the external field is nonzero, what must be constant in the equilibrium state is not the concentration, but the chemical potential.

Now if we define the flux velocity by $v_f \equiv j/c$, eqn (3.13) gives

$$v_f = -\frac{1}{\zeta}\frac{\partial}{\partial x}(k_B T \ln c + U) \tag{3.14}$$

which is quite similar to eqn (3.5). The only difference is that the potential $U(x)$ is now replaced by the chemical potential $U(x) + k_B T \ln c(x, t)$. If this replacement is made, the Smoluchowski equation is derived from the usual continuity equation

$$\frac{\partial c}{\partial t} = -\frac{\partial}{\partial x}(c v_f). \tag{3.15}$$

Thus the Smoluchowski equation can be formally derived from the macroscopic equations (eqns (3.14) and (3.15)), where the fluctuations have been dealt with by invoking the thermodynamic variable—the chemical potential $U(x) + k_B T \ln c(x, t)$—instead of the potential $U(x)$.

3.2.2 Diffusion in phase space

So far we have been considering the diffusion of concentration. The same argument will hold for the probability distribution function $\Psi(x, t)$ that a particular particle is found at point x at time t since the distinction between $c(x, t)$ and $\Psi(x, t)$ is, for non-interacting particles, only in the fact that Ψ is normalized. Thus the evolution equation for the probability $\Psi(x, t)$ is written as

$$\frac{\partial \Psi}{\partial t} = \frac{\partial}{\partial x} \frac{1}{\zeta} \left(k_B T \frac{\partial \Psi}{\partial x} + \frac{\partial U}{\partial x} \Psi \right) \tag{3.16}$$

which will also be termed the Smoluchowski equation. (In some literature this equation is called the Fokker–Planck equation, or the generalized diffusion equation, but the original Fokker–Planck equation was for diffusion in velocity space, so we do not employ the term here.)

Having seen the basic principle, it is easy to derive the Smoluchowski equation for a system which has many degrees of freedom. Let $x_1, x_2, \ldots, x_N \equiv \{x\}$ be the set of dynamical variables describing the state of Brownian particles. To construct the Smoluchowski equation, we have to know first the relation between the average velocity v_n and the force $F_n = -\partial U / \partial x_n$. Such a relation is generally written as

$$v_n = \sum_m L_{nm}(\{x\}) F_m. \tag{3.17}$$

The coefficients L_{nm} are called the mobility matrix, and may be obtained using hydrodynamics. It can be proved that L_{nm} is a symmetric positive definite matrix:

$$L_{nm} = L_{mn} \quad \text{and} \quad \sum_{n,m} F_n F_m L_{nm} \geqslant 0 \quad \text{for all } F_n. \tag{3.18}$$

Given the mobility matrix, the Smoluchowski equation is obtained from the continuity equation

$$\frac{\partial \Psi}{\partial t} = -\sum_n \frac{\partial}{\partial x_n} (v_{fn} \Psi) \tag{3.19}$$

with the flux velocity being given by

$$v_{fn} = -\sum_m L_{nm} \frac{\partial}{\partial x_m} (k_B T \ln \Psi + U). \tag{3.20}$$

Hence

$$\frac{\partial \Psi}{\partial t} = \sum_{n,m} \frac{\partial}{\partial x_n} L_{nm}\left(k_B T \frac{\partial \Psi}{\partial x_m} + \frac{\partial U}{\partial x_m} \Psi\right) \tag{3.21}$$

which is the basic equation for the dynamics of polymer solutions and suspensions.

The Smoluchowski equation may be regarded as a phenomenological tool for describing the fluctuation of physical quantities, and can be applied to more general situations: for example the equation can be used to describe the fluctuation of thermodynamic variables (such as concentration). In such cases, the potential $U(x)$ must be regarded as the free energy which determines the equilibrium distribution of those variables, and the relation (3.17) must be replaced by phenomenological kinetic equations. We shall see such applications in Chapter 5.

3.2.3 *Irreversibility of the Smoluchowski equation*

An important property of the Smoluchowski equation is that if $U(\{x\})$ is independent of time and if there is no flux at the boundary, the distribution function Ψ always approaches Ψ equilibrium (Ψ_{eq});

$$\Psi_{\text{eq}} = \exp(-U(\{x\})/k_B T) \Big/ \int \mathrm{d}\{x\} \exp(-U(\{x\})/k_B T). \tag{3.22}$$

To prove this we consider a functional[6]

$$\mathscr{A}[\Psi] \equiv \int \mathrm{d}\{x\} \Psi (k_B T \ln \Psi + U) \tag{3.23}$$

for the general solution Ψ of eqn (3.21). The time derivative of $\mathscr{A}$ is calculated as

$$\frac{\mathrm{d}}{\mathrm{d}t} \mathscr{A} = \int \mathrm{d}\{x\} \left[\frac{\partial \Psi}{\partial t} (k_B T \ln \Psi + U) + k_B T \frac{\partial \Psi}{\partial t} \right]. \tag{3.24}$$

Using eqn (3.21) and the integral by parts (the integral at the boundary being zero by the assumption), we have

$$\frac{\mathrm{d}\mathscr{A}}{\mathrm{d}t} = -\int \mathrm{d}\{x\} \Psi \sum_{n,m} L_{nm} \left[\frac{\partial}{\partial x_n} (k_B T \ln \Psi + U) \right] \left[\frac{\partial}{\partial x_m} (k_B T \ln \Psi + U) \right]$$

$$= -\int \mathrm{d}\{x\} \Psi \sum_{n,m} L_{nm} \left[k_B T \frac{\partial}{\partial x_n} \ln (\Psi/\Psi_{\text{eq}}) \right] \left[k_B T \frac{\partial}{\partial x_m} \ln (\Psi/\Psi_{\text{eq}}) \right] \tag{3.25}$$

which is negative according to eqn (3.18) unless Ψ is identical to Ψ_{eq}. As $\mathscr{A}$ will eventually reach the minimum value, Ψ also reaches Ψ_{eq} after a sufficiently long time.

At equilibrium, $\mathscr{A}[\Psi]$ becomes the free energy defined in the equilibrium statistical mechanics:

$$\mathscr{A}[\Psi_{\text{eq}}] = -k_B T \ln\left(\int \mathrm{d}\{x\} \exp(-U/k_B T)\right). \tag{3.26}$$

When the system is not in equilibrium, $\mathscr{A}[\Psi]$ is larger than $\mathscr{A}[\Psi_{\text{eq}}]$ and the difference $\mathscr{A}[\Psi] - \mathscr{A}[\Psi_{\text{eq}}]$ represents how far the system is away from the equilibrium state. We shall call $\mathscr{A}$ the dynamical free energy. This quantity plays an important role in the subsequent discussions.

The approach to the equilibrium state is also shown directly by the eigenfunction expansion of the distribution function. This is discussed in Appendix 3.I. More detailed discussion on the Smoluchowski equation is given in ref. 7.

3.3 The Langevin equation

An alternative description of Brownian motion is to study the equation of motion of the Brownian particle writing the random force $f(t)$ explicitly in the Langevin form:

$$\zeta \frac{\mathrm{d}x}{\mathrm{d}t} = -\frac{\partial U}{\partial x} + f(t). \tag{3.27}$$

Physically, the random force $f(t)$ represents the sum of the forces due to the incessant collision of the fluid molecules with the Brownian particle. As we cannot know the precise time-dependence of such a force, we regard it as a stochastic variable, and assume a plausible distribution $\Psi[f(t)]$ for it. In this scheme, the average of a physical quantity $A(x(t))$ is calculated, in principle, by the following procedure. First eqn (3.27) is solved for any given $f(t)$, and $A(x(t))$ is expressed by $f(t)$. The average is then taken with respect to $f(t)$ for the given distribution function $\Psi[f(t)]$.

Though various distributions can be conceived for $f(t)$ depending on physical modelling, here we shall consider only the process which is equivalent to that described by the Smoluchowski equation. It is shown in Appendix 3.II that if the distribution of $f(t)$ is assumed to be Gaussian characterized by the moment

$$\langle f(t) \rangle = 0, \qquad \langle f(t) f(t') \rangle = 2\zeta k_B T\, \delta(t - t'), \tag{3.28}$$

i.e., if the functional probability distribution of $f(t)$ is

$$\Psi[f(t)] \propto \exp\left(-\frac{1}{4\zeta k_B T}\int \mathrm{d}t f(t)^2\right) \tag{3.29}$$

then the distribution of $x(t)$ determined by eqn (3.27) satisfies the Smoluchowski equation. Here we check this for a simple special case.

Consider the Brownian motion of a free particle ($U=0$) for which the Langevin equation reads

$$\zeta \frac{dx}{dt} = f(t). \tag{3.30}$$

If the particle was at x' at time $t=0$, its position at time t is given by

$$x(t) = x' + \frac{1}{\zeta}\int_0^t dt' f(t'). \tag{3.31}$$

Equation (3.31) indicates that $x(t)-x'$ is a linear combination of Gaussian random variables $f(t)$. Therefore, according to the theorem in Appendix 2.I, the distribution of $x(t)$ must be Gaussian. Hence the probability distribution of $x(t)$ is written as

$$\Psi(x, t) = (2\pi B)^{-1/2} \exp\left(-\frac{(x-A)^2}{2B}\right) \tag{3.32}$$

where

$$A = \langle x(t)\rangle, \qquad B = \langle (x(t)-A)^2\rangle. \tag{3.33}$$

From eqns (3.28) and (3.31), these moments are easily calculated:

$$A = x' + \frac{1}{\zeta}\int_0^t dt' \langle f(t')\rangle = x'$$

and

$$B = \left\langle \left(\frac{1}{\zeta}\int_0^t dt' f(t')\right)\left(\frac{1}{\zeta}\int_0^t dt'' f(t'')\right)\right\rangle$$

$$= \frac{1}{\zeta^2}\int_0^t dt' \int_0^t dt'' \langle f(t') f(t'')\rangle \tag{3.34}$$

$$= \frac{2k_B T}{\zeta}\int_0^t dt' \int_0^t dt''\, \delta(t'-t'') = \frac{2k_B T}{\zeta} t \tag{3.35}$$

or by the Einstein relation (3.10)

$$B = 2Dt.$$

Thus

$$\Psi(x, t) = (4\pi Dt)^{-1/2} \exp\left(-\frac{(x-x')^2}{4Dt}\right). \tag{3.36}$$

This is the solution of the Smoluchowski equation

$$\frac{\partial \Psi}{\partial t} = D\frac{\partial^2}{\partial x^2}\Psi. \tag{3.37}$$

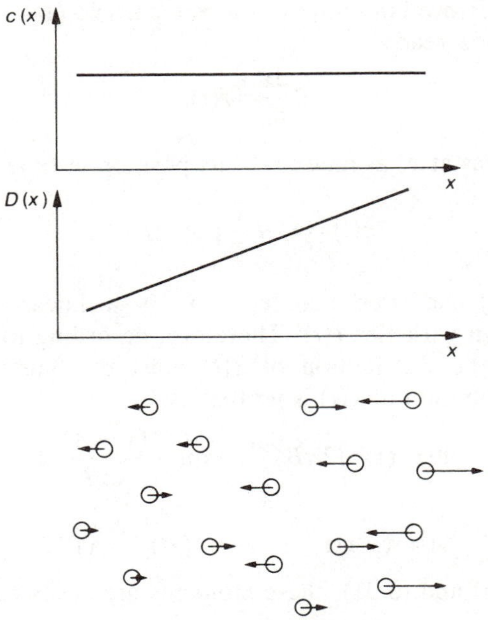

Fig. 3.2. Explanation for the term $\partial D/\partial x$ in eqn (3.38). Consider the equilibrium state for the case of $U=0$ and $\partial D/\partial x>0$. The random force $f(t)$ in eqn (3.38) causes displacement of individual particles as shown by the arrows: the particles on the right are more mobile than those on the left. This creates a net flux of particles toward the negative direction. To compensate for this, the term $\partial D/\partial x$ is needed.

The Langevin equation (3.27) is equivalent to the Smoluchowski equation only when the diffusion constant $D\equiv k_BT/\zeta$ is independent of x. If D depends on x, an additional term must be added to eqn (3.27)[2]†

$$\zeta\frac{\mathrm{d}x}{\mathrm{d}t}=-\frac{\partial U}{\partial x}+f(t)+\frac{\zeta}{2}\frac{\partial D}{\partial x}. \tag{3.38}$$

† In some literature, the added term is $\zeta\,\partial D/\partial x$. This difference comes from the difference in the definition of the integral over the random force which is not properly defined in the conventional Stieltjes integration. Here we have followed the definition given in ref. 2. Such complication can be avoided if one uses the Langevin equation written in a difference form in a finite time interval Δt:[9,10]

$$x(t+\Delta t)=x(t)-\frac{1}{\zeta}\frac{\partial U}{\partial x}\Delta t+\sqrt{D}\,(x(t))g(t)+\frac{\partial D}{\partial x}\Delta t$$

where $g(t)$ is a random variable satisfying

$$\langle g(t)\rangle=0,\qquad \langle g(t)g(t')\rangle=2\Delta t\,\delta_{tt'}.$$

See ref. 7 for more detailed discussion.

The derivation is given in Appendix 3.II. Physically, the term $\partial D/\partial x$ is needed to compensate for the flux caused by the random force which is dependent on particle position (see Fig. 3.2).

The Langevin equation corresponding to the Smoluchowski equation in multidimensional phase space (eqn (3.21)) is given by[2]

$$\frac{\mathrm{d}}{\mathrm{d}t}x_n = \sum_m L_{nm}\left(-\frac{\partial U}{\partial x_m} + f_m(t)\right) + \tfrac{1}{2}k_B T \sum_m \frac{\partial}{\partial x_m} L_{nm}. \tag{3.39}$$

The distribution of the random force is Gaussian, characterized by the moment

$$\langle f_n(t)\rangle = 0, \tag{3.40}$$

$$\langle f_n(t) f_m(t')\rangle = 2(L^{-1})_{nm} k_B T\, \delta(t-t'), \tag{3.41}$$

where $(L^{-1})_{nm}$ denotes the inverse matrix of L_{nm}

$$\sum_m L_{nm}(L^{-1})_{mk} = \delta_{nk}. \tag{3.42}$$

The Langevin equation (3.39) represents the same motion as the Smoluchowski equation (3.21). However, each of the equations has advantages and disadvantages in solving our problems; we shall therefore use both equations interchangeably.

3.4 Time correlation function and response function

3.4.1 *Time correlation function*

An important quantity characterizing the Brownian motion is the time correlation function, which is operationally defined in the following way. Suppose we measure a physical quantity A of a system of Brownian particles for many samples in the equilibrium state. Let $A(t)$ be the measured values of A at time t. Usually $A(t)$ looks like a noise pattern as shown in Fig. 3.3*a*. The time correlation function $C_{AA}(t)$ is defined as the average of the product $A(t)A(0)$ over many measurements:

$$C_{AA}(t) = \langle A(t)A(0)\rangle. \tag{3.43}$$

Typical behaviour of $C_{AA}(t)$ is shown in Fig. 3.3*b*: at $t=0$, $C_{AA}(0)$ is positive and is equal to the mean square of A in the equilibrium state, $\langle A^2\rangle$. As time passes, $C_{AA}(t)$ usually decreases with time since the value of $A(t)$ becomes uncorrelated to that at $t=0$. After a sufficiently long time, the correlation between $A(t)$ and $A(0)$ vanishes completely, and $C_{AA}(t)$ becomes equal to $\langle A(t)\rangle\langle A(0)\rangle = \langle A\rangle^2$. The characteristic time with which $C_{AA}(t)$ approaches the asymptotic value is called the correlation time.

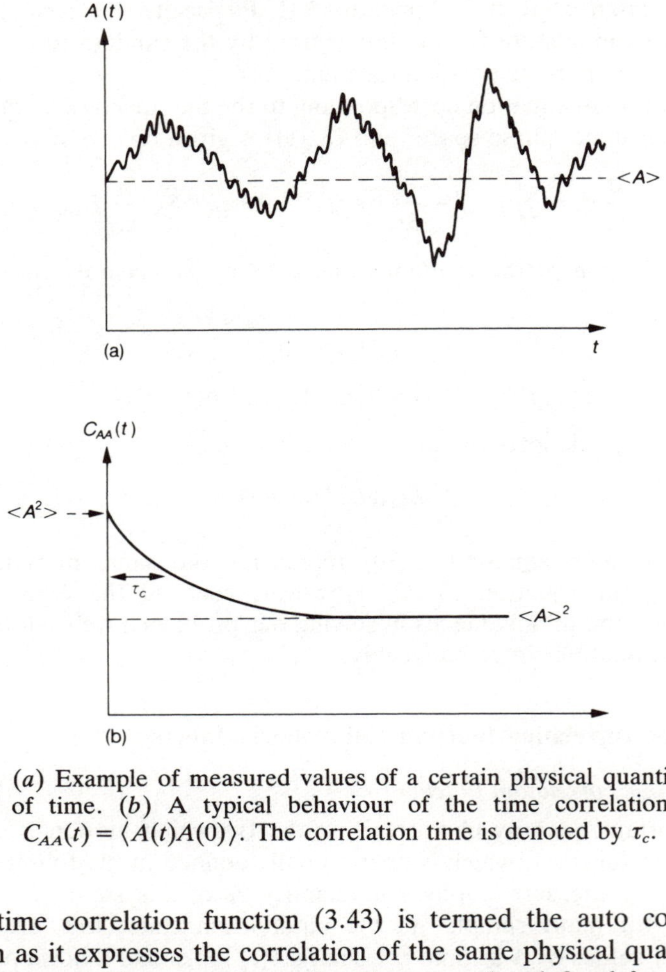

Fig. 3.3. (*a*) Example of measured values of a certain physical quantity A as a function of time. (*b*) A typical behaviour of the time correlation function $C_{AA}(t) = \langle A(t)A(0)\rangle$. The correlation time is denoted by τ_c.

The time correlation function (3.43) is termed the auto correlation function as it expresses the correlation of the same physical quantities at different times. Time correlation functions are also defined for different physical quantities,

$$C_{AB}(t) = \langle A(t)B(0)\rangle \tag{3.44}$$

which is called the cross correlation function.

3.4.2 *Microscopic expression for the time correlation function*

Given the Smoluchowski equation, time correlation functions can be calculated. For the sake of simplicity we use x to denote the whole set of coordinates $x_1, x_2, \ldots, x_N$ appearing in the Smoluchowski equation. Let $G(x, x'; t)$ be the probability that the system which was in the state x' at time $t = 0$ is in the state x at time t. Clearly such probability is obtained from the Smoluchowski equation

$$\frac{\partial}{\partial t} G(x, x'; t) = \sum_{n,m} \frac{\partial}{\partial x_n} L_{nm}\left(k_B T \frac{\partial G}{\partial x_m} + \frac{\partial U}{\partial x_m} G\right) \tag{3.45}$$

and the initial condition

$$G(x, x'; t=0) = \delta(x - x') \equiv \prod_n \delta(x_n - x'_n). \tag{3.46}$$

Given $G(x, x'; t)$, the time correlation function is evaluated by

$$\langle A(t)B(0)\rangle = \int \mathrm{d}x \int \mathrm{d}x' A(x)B(x')G(x, x'; t)\Psi_{\mathrm{eq}}(x') \tag{3.47}$$

where $A(x)$ and $B(x)$ denotes the value of the physical quantity A and B when the system is in the state x.† The meaning of eqn (3.47) is clear. If the system is in the state x' at time $t = 0$ and in the state x at time t, the measured value of A and B are $A(x)$ and $B(x')$, respectively. The probability that this happens at equilibrium is $G(x, x'; t)\Psi_{\mathrm{eq}}(x')$. Averaging $A(x)B(x')$ for this probability, we get eqn (3.47).

Though eqn (3.47) gives a general method for calculating the time correlation function, it is not easy to carry out this procedure since $G(x, x'; t)$ is difficult to obtain. Usually, more convenient methods are available, which will be demonstrated in subsequent sections. However, the initial slope of the time correlation function can be calculated directly from eqn (3.47). The time derivative of eqn (3.47) is calculated as

$$\begin{aligned}\frac{\mathrm{d}}{\mathrm{d}t}\langle A(t)B(0)\rangle &= \int \mathrm{d}x \int \mathrm{d}x' A(x) \frac{\partial G}{\partial t} B(x')\Psi_{\mathrm{eq}}(x') \\ &= \int \mathrm{d}x \int \mathrm{d}x' A(x) \sum_{n,m} \frac{\partial}{\partial x_n} L_{nm}\left(k_B T \frac{\partial G}{\partial x_m} + G \frac{\partial U}{\partial x_m}\right) \\ &\quad \times B(x')\Psi_{\mathrm{eq}}(x').\end{aligned}$$

At $t = 0$, $G(x, x'; t)$ becomes $\delta(x - x')$, hence

$$\frac{\mathrm{d}}{\mathrm{d}t}\langle A(t)B(0)\rangle\bigg|_{t=0} = \int \mathrm{d}x A(x) \sum_{n,m} \frac{\partial}{\partial x_n} L_{nm}\left[k_B T \frac{\partial}{\partial x_m}(B(x)\Psi_{\mathrm{eq}}(x)) + B(x)\Psi_{\mathrm{eq}}(x)\frac{\partial U}{\partial x_m}\right]. \tag{3.48}$$

† Note that here two functions, $A(t)$ which represents the measured value of the physical quantity at time t, and $A(x)$ which represents the dependence of the physical quantity on the dynamical variables x, are distinguished by their arguments. To avoid confusion, it may be preferable to denote the latter as $\tilde{A}(x)$, then $A(t)$ can be written as $\tilde{A}(x(t))$. However, since this leads to over cumbersome nomenclature, we do not use it here.

Using integration by parts and eqn (3.22), we have

$$\left.\frac{\mathrm{d}}{\mathrm{d}t}\langle A(t)B(0)\rangle\right|_{t=0} = -k_B T\int \mathrm{d}x \sum_{n,m} \Psi_{\mathrm{eq}}(x)\frac{\partial A(x)}{\partial x_n} L_{nm}\frac{\partial B(x)}{\partial x_m}$$

$$= -k_B T\sum_{n,m}\left\langle \frac{\partial A}{\partial x_n} L_{nm}\frac{\partial B}{\partial x_m}\right\rangle. \tag{3.49}$$

The average in the final expression is for the equilibrium distribution function $\Psi_{\mathrm{eq}}(x)$. The initial decay rate $\Gamma^{(0)}$ defined by

$$\Gamma^{(0)} = -\left.\frac{\mathrm{d}}{\mathrm{d}t}\langle A(t)B(0)\rangle\right|_{t=0} /(\langle AB\rangle - \langle A\rangle\langle B\rangle) \tag{3.50}$$

is thus given by

$$\Gamma^{(0)} = \frac{k_B T}{\langle AB\rangle - \langle A\rangle\langle B\rangle}\sum_{n,m}\left\langle \frac{\partial A}{\partial x_n} L_{nm}\frac{\partial B}{\partial x_m}\right\rangle. \tag{3.51}$$

3.4.3 Fluctuation dissipation theorem

Consider a time-dependent external field $h(t)$ (magnetic field, electric field, or velocity gradient field) applied to a system in equilibrium. In general, the field perturbs the system, and changes the average values of physical quantities from those in the equilibrium state. If the field is weak, the change in any physical quantity is a linear functional of the field, and is written as

$$\langle A(t)\rangle_h - \langle A\rangle_0 = \int_{-\infty}^{t} \mathrm{d}t'\mu(t-t')h(t'), \tag{3.52}$$

where $\langle A(t)\rangle_h$ denotes the value of A at time t when the field is applied and $\langle A\rangle_0$ the equilibrium value of A in the absence of the field. The function $\mu(t)$ is called the response function.

In many cases, the effect of the field on the system is expressed by a potential such as

$$U_{\mathrm{ext}}(x, t) = -h(t)B(x). \tag{3.53}$$

The quantity $B(x)$ is said to be conjugate to the field $h(t)$. In such cases, the response function is related to the time correlation function by

$$\mu(t) = -\frac{1}{k_B T}\frac{\mathrm{d}}{\mathrm{d}t}C_{AB}(t). \tag{3.54}$$

This theorem is called the fluctuation dissipation theorem.[3,7]

To prove eqn (3.54), we consider the situation that a constant field h is applied for a long time until the system reaches equilibrium, and that the

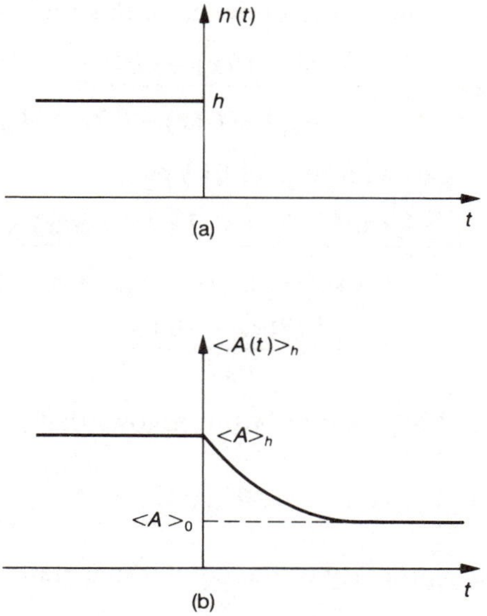

Fig. 3.4. Relaxation function. When a constant external field is switched off at $t=0$, the average of a physical quantity A relaxes from the equilibrium value in the presence of the field, $\langle A\rangle_h$, to that in the absence of the field $\langle A\rangle_0$. The time-dependence for $t>0$ is described by the relaxation function $\alpha(t)$ as $\langle A\rangle_0+\alpha(t)h$.

field is then switched off at $t=0$ (see Fig. 3.4). In such a case, the average value of A will change as shown in Fig. 3.4*b*.

$$\langle A(t)\rangle_h=\alpha(t)h+\langle A\rangle_0 \tag{3.55}$$

The function $\alpha(t)$, called the relaxation function, is expressed by the response function $\mu(t)$ as

$$\alpha(t)=\int_{-\infty}^{0}\mathrm{d}t'\mu(t-t')=\int_{t}^{\infty}\mathrm{d}t'\mu(t'). \tag{3.56}$$

Now $\langle A(t)\rangle_h$ can be calculated if the distribution function $\Psi(x,t)$ is known:

$$\langle A(t)\rangle_h=\int\mathrm{d}xA(x)\Psi(x,t). \tag{3.57}$$

Since there is no external field for $t>0$, $\Psi(x,t)$ is related to $\Psi(x,t=0)$ by the Green function $G(x,x';t)$ *in the absence* of the field

$$\Psi(x,t)=\int\mathrm{d}x'G(x,x';t)\Psi(x,0). \tag{3.58}$$

Since Ψ is at equilibrium in the presence of the field at time $t = 0$:

$$\Psi(x, 0) = \frac{\exp[-(U(x) - hB(x))/k_B T]}{\int dx \exp[-(U(x) - hB(x))/k_B T]} \tag{3.59}$$

which can be expanded with respect to h as

$$\Psi(x, 0) = \frac{\exp(-U(x)/k_B T)(1 + hB(x)/k_B T)}{\left[\int dx \exp(-U(x)/k_B T)\right](1 + h\langle B\rangle_0/k_B T)}$$
$$= \Psi_{\text{eq}}\left(1 + \frac{h[B(x) - \langle B\rangle_0]}{k_B T}\right). \tag{3.60}$$

From eqns (3.57), (3.58), and (3.60), it follows that

$$\langle A(t)\rangle_h = \int dx \int dx' A(x) G(x, x'; t) \Psi_{\text{eq}}(x')\left(1 + \frac{h[B(x') - \langle B\rangle_0]}{k_B T}\right). \tag{3.61}$$

Using the stationary property of the equilibrium state,

$$\Psi_{\text{eq}}(x) = \int dx' G(x, x'; t) \Psi_{\text{eq}}(x'), \tag{3.62}$$

and the definition of the time correlation function, we get

$$\langle A(t)\rangle_h = \langle A\rangle_0 + \frac{h}{k_B T}\int dx \int dx' A(x) G(x, x'; t)(B(x') - \langle B\rangle_0)\Psi_{\text{eq}}(x')$$
$$= \langle A\rangle_0 + \frac{h}{k_B T}[\langle A(t)B(0)\rangle_0 - \langle A\rangle_0\langle B\rangle_0]. \tag{3.63}$$

From eqns (3.63) and (3.55), it follows that

$$\alpha(t) = \frac{1}{k_B T}(C_{AB}(t) - \langle A\rangle_0\langle B\rangle_0) \tag{3.64}$$

which leads to eqn (3.54) after differentiation with respect to t.

Note that in the above proof the explicit form of the time evolution equation for Ψ is not used. Therefore the proof applies to a pure dynamical system which is described by the Liouville equation. The fluctuation dissipation theorem holds quite generally in physical systems near equilibrium.

In the case of $A = B$, the fluctuation dissipation theorem can be stated in a more convenient form. Let us define the growth function $\beta(t)$ as the response to the sudden application of a step field (see Fig. 3.5):

$$\langle A(t)\rangle_h - \langle A\rangle_0 = \beta(t)h \tag{3.65}$$

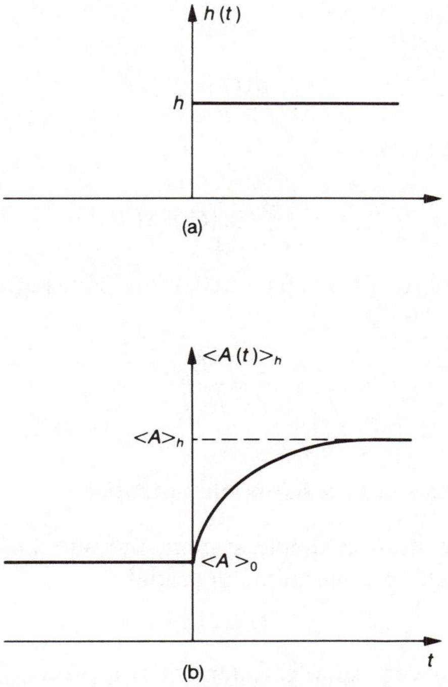

Fig. 3.5. Growth function. When a stepwise external field is applied (a), the average of a physical quantity A changes as shown in (b). The time evolution is described by the growth function $\beta(t)$ as $\langle A \rangle_0 + \beta(t)h$.

From eqn (3.52) we have

$$\beta(t) = \int_0^t dt' \mu(t'). \tag{3.66}$$

Using eqn (3.54),

$$\beta(t) = \frac{1}{k_B T}(C_{AA}(0) - C_{AA}(t)) = \frac{1}{2k_B T}(\langle A(t)^2 \rangle + \langle A(0)^2 \rangle - 2\langle A(t)A(0) \rangle)$$
$$= \frac{1}{2k_B T}\langle (A(t) - A(0))^2 \rangle. \tag{3.67}$$

A simple application of eqn (3.67) is the Einstein relation. Let us consider that A denotes the x component of the position vector of the Brownian particle. The field conjugate to x is the external force F_{ext}. When F_{ext} is applied, the particle begins to move with the constant velocity F_{ext}/ζ. Thus

$$\beta(t)=\frac{t}{\zeta}. \tag{3.68}$$

Hence eqn (3.67) becomes

$$\frac{t}{\zeta}=\frac{1}{2k_BT}\langle(x(t)-x(0))^2\rangle. \tag{3.69}$$

Since, by definition of D, $\langle(x(t)-x(0))^2\rangle$ is $2Dt$, eqn (3.69) leads to the Einstein relation

$$\frac{1}{\zeta}=\frac{D}{k_BT}. \tag{3.70}$$

3.5 Brownian motion in a harmonic potential

In this section we study a simple system, the one-dimensional Brownian motion of a particle in a harmonic potential:

$$U=\tfrac{1}{2}kx^2. \tag{3.71}$$

Although this is a very simple system, it is a prototype of the problems which we shall discuss later on.

3.5.1 Smoluchowski equation

Let us first calculate the time correlation function $\langle x(t)x(0)\rangle$ using the Smoluchowski equation. In principle this is obtained from eqn (3.47)

$$\langle x(t)x(0)\rangle=\int \mathrm{d}x\int \mathrm{d}x'\,xx'G(x,x';t)\Psi_{\text{eq}}(x') \tag{3.72}$$

where

$$\Psi_{\text{eq}}(x)=\left(\frac{2\pi k_BT}{k}\right)^{-1/2}\exp\left(-\frac{kx^2}{2k_BT}\right) \tag{3.73}$$

and the Green function $G(x,x';t)$ is to be determined from the Smoluchowski equation

$$\frac{\partial}{\partial t}G(x,x';t)=\frac{1}{\zeta}\frac{\partial}{\partial x}\left(k_BT\frac{\partial G}{\partial x}+kxG\right) \tag{3.74}$$

under the initial condition

$$G(x,x';0)=\delta(x-x'). \tag{3.75}$$

However, we can show that $\langle x(t)x(0)\rangle$ can be calculated without complete knowledge of $G(x, x'; t)$. Using eqn (3.74), the time derivative of eqn (3.72) is calculated as

$$\frac{\partial}{\partial t}\langle x(t)x(0)\rangle = \int \mathrm{d}x \int \mathrm{d}x' xx' \left[\frac{1}{\zeta}\frac{\partial}{\partial x}\left(k_B T \frac{\partial G}{\partial x} + kxG\right)\right]\Psi_{\mathrm{eq}}(x'). \quad (3.76)$$

The right-hand side is rewritten by integral by parts as

$$\begin{aligned}\frac{\partial}{\partial t}\langle x(t)x(0)\rangle &= \int \mathrm{d}x \int \mathrm{d}x' G\Psi_{\mathrm{eq}}(x')\left[\frac{k_B T}{\zeta}\frac{\partial}{\partial x}\frac{\partial}{\partial x}(xx') - \frac{kx}{\zeta}\frac{\partial}{\partial x}(xx')\right]\\ &= -\frac{k}{\zeta}\int \mathrm{d}x \int \mathrm{d}x' xx' G(x, x'; t)\Psi_{\mathrm{eq}}(x')\\ &= -\frac{1}{\tau}\langle x(t)x(0)\rangle \end{aligned} \quad (3.77)$$

where

$$\tau = \zeta / k. \quad (3.78)$$

The initial condition for the differential equation is obtained from

$$\langle x(0)^2\rangle = \int \mathrm{d}x x^2 \Psi_{\mathrm{eq}}(x) = \frac{k_B T}{k}. \quad (3.79)$$

Hence

$$\langle x(t)x(0)\rangle = \frac{k_B T}{k}\exp(-t/\tau). \quad (3.80)$$

3.5.2 Langevin equation

The same result as eqn (3.80) is obtained from the Langevin equation. For the potential (eqn (3.71)), the Langevin equation is written as

$$\zeta \frac{\mathrm{d}x}{\mathrm{d}t} = -kx + f(t) \quad (3.81)$$

with

$$\langle f(t)\rangle = 0, \qquad \langle f(t)f(t')\rangle = 2\zeta k_B T\,\delta(t-t'). \quad (3.82)$$

To calculate $\langle x(t)x(0)\rangle$, we first express $x(t)$ in terms of $f(t')$:

$$x(t) = \frac{1}{\zeta}\int_{-\infty}^{t} \mathrm{d}t' \exp(-(t-t')/\tau) f(t'). \quad (3.83)$$

Hence

$$\langle x(t)x(0)\rangle = \frac{1}{\zeta^2}\int_{-\infty}^{t} dt_1 \int_{-\infty}^{0} dt_2 \exp[-(t-t_1-t_2)/\tau]\langle f(t_1)f(t_2)\rangle$$
$$= \frac{1}{\zeta^2}\int_{-\infty}^{t} dt_1 \int_{-\infty}^{0} dt_2 \exp[-(t-t_1-t_2)/\tau] 2\zeta k_B T\,\delta(t_1-t_2)$$
$$= \frac{2k_BT}{\zeta}\int_{-\infty}^{0} dt_2 \exp(-(t-2t_2)/\tau) = \frac{k_BT}{k}\exp(-t/\tau) \qquad (3.84)$$

which agrees with eqn (3.80).

Finally we derive an explicit form of the Green function $G(x, x'; t)$. Again we use the same argument as used in deriving eqn (3.36). Since $x(t)$ is a linear combination of $f(t)$, the distribution of x is Gaussian, which is generally written as

$$G(x, x'; t) = [2\pi B(t)]^{-1/2} \exp\left[-\frac{(x-A(t))^2}{2B(t)}\right] \qquad (3.85)$$

where

$$A(t) = \langle x(t)\rangle, \; B(t) = \langle (x(t)-A(t))^2\rangle. \qquad (3.86)$$

To calculate these quantities, we solve the Langevin equation under the initial condition $x(0) = x'$:

$$x(t) = x'\exp(-t/\tau) + \frac{1}{\zeta}\int_0^t dt' \exp(-(t-t')/\tau) f(t'). \qquad (3.87)$$

Using eqn (3.82), $A(t)$ and $B(t)$ are calculated from eqns (3.86) and (3.87):

$$A(t) = x'\exp(-t/\tau) \qquad (3.88)$$

and

$$B(t) = \frac{1}{\zeta^2}\int_0^t dt_1 \int_0^t dt_2 \exp(-(t-t_1+t-t_2)/\tau)\langle f(t_1)f(t_2)\rangle$$
$$= \frac{1}{\zeta^2}\int_0^t dt_1 \exp[-2(t-t_1)/\tau]\cdot 2\zeta k_B T$$
$$= \frac{k_BT}{k}[1-\exp(-2t/\tau)]. \qquad (3.89)$$

Hence the Green function is given by

$$G(x, x'; t) = \left[\frac{2\pi k_B T}{k}(1 - \exp(-2t/\tau))\right]^{-1/2} \times \exp\left[-\frac{k[x - x'\exp(-t/\tau)]^2}{2k_B T[1 - \exp(-2t/\tau)]}\right]. \quad (3.90)$$

Consider the two limiting cases:

(i) When t is small, $t \ll \tau$, $G(x, x'; t)$ is the same as free diffusion:

$$G(x, x'; t) = (4\pi D t)^{-1/2} \exp\left[-\frac{(x - x')^2}{4Dt}\right] \quad (3.91)$$

(ii) When t is large, $t \gg \tau$, $G(x, x'; t)$ becomes the Boltzmann distribution,

$$G(x, x'; t) = (k/2\pi k_B T)^{1/2} \exp\left[-\frac{k}{2k_B T}x^2\right], \quad (3.92)$$

in accordance with the second law of thermodynamics.

Using Mehler's formula[11] for the Hermite polynomial:

$$\sum_{p=1}^{\infty} \frac{H_p(\xi)H_p(\eta)}{2^p p!} s^p = (1 - s^2)\exp\left[\xi^2 - \frac{(\xi - \eta s)^2}{1 - s^2}\right] \quad (3.93)$$

with $H_p(\xi)$ defined by

$$H_p(\xi) = (-1)^p \exp(\xi^2)\frac{\mathrm{d}^p}{\mathrm{d}\xi^p}\exp(-\xi^2), \quad (3.94)$$

eqn (3.90) can be written in the eigenfunction expansion form:

$$G(x, x'; t) = \sum_{p=0}^{\infty} \exp(-\lambda_p t)\psi_p(x)\psi_p(x')\Psi_{\text{eq}}(x) \quad (3.95)$$

where

$$\lambda_p = p/\tau; \qquad \psi_p(x) = (2^p p!)^{-1/2} H_p(x/\gamma), \quad \text{with} \quad \gamma = (2k_B T/k)^{1/2}. \quad (3.96)$$

3.6 Interacting Brownian particles

Having seen the basic tools for describing Brownian motion, we now consider more realistic situations. Suppose that a collection of spherical Brownian particles, all having equal size, are suspended in a fluid and interacting with each other. Such systems are often found in colloidal suspensions.[12] As we shall discuss in the next section, the study of this system is the basis for the general theory of polymer solutions and suspensions.

To obtain the Smoluchowski equation for such a system, we first calculate the mobility matrix. Let $\boldsymbol{R}_1, \boldsymbol{R}_2, \ldots, \boldsymbol{R}_N \equiv \{\boldsymbol{R}\}$ be the positions of the spheres and $\boldsymbol{F}_1, \boldsymbol{F}_2, \ldots, \boldsymbol{F}_N$ be the forces acting on them. We assume that there are no external torques acting on the particles. Then the velocities of the particles are written as†

$$\boldsymbol{V}_n = \sum_m \boldsymbol{H}_{nm} \cdot \boldsymbol{F}_m \tag{3.97}$$

which defines the mobility matrix $\boldsymbol{H}_{nm}$. (Note that $\boldsymbol{V}_n$ and $\boldsymbol{F}_n$ are vectors and therefore each component of the mobility matrix $\boldsymbol{H}_{nm}$ is a tensor.)

In a very dilute suspension, the velocity of a particle is determined only by the force acting on it, and the mobility matrix becomes

$$\boldsymbol{H}_{nm} = \frac{\boldsymbol{I}\delta_{nm}}{\zeta}, \tag{3.98}$$

where $\zeta = 6\pi\eta_s a$ is the friction constant of the particle, and $\boldsymbol{I}$ is the unit tensor ($I_{\alpha\beta} = \delta_{\alpha\beta}$). In general, however, the velocities of particles depend on the forces acting on their surrounding particles, because the force acting on a certain particle causes the fluid motion around it and affects the velocity of the other particles (see Fig. 3.6). This interaction, mediated by the motion of the solvent fluid, is called the hydrodynamic interaction. As a result of the hydrodynamic interaction, the off-diagonal components of the mobility matrix become nonzero.

To calculate the particle velocities $\boldsymbol{V}_n$ ($n = 1, 2, \ldots N$) we have to know the fluid velocity $\boldsymbol{v}(\boldsymbol{r})$ created by the external forces acting on the particles. In the usual condition of Brownian motion, the relevant hydrodynamic equation of motion is that of the low Reynolds number

† This form of the mobility matrix is correct only for spherical particles. In general, particles of finite shape have both translational and the rotational degrees of freedom, and the general form of the mobility matrix is written as

$$\boldsymbol{V}_n = \sum_m \boldsymbol{H}_{nm}^{(TT)} \cdot \boldsymbol{F}_m + \boldsymbol{H}_{nm}^{(TR)} \cdot \boldsymbol{N}_m,$$

$$\boldsymbol{\omega}_n = \sum_m \boldsymbol{H}_{nm}^{(RT)} \cdot \boldsymbol{F}_m + \boldsymbol{H}_{nm}^{(RR)} \cdot \boldsymbol{N}_m,$$

where $\boldsymbol{\omega}_m$ is the angular velocity of the particle m, and $\boldsymbol{N}_m$ is the external torque acting on it. The coefficients $\boldsymbol{H}_{nm}^{(TT)}$, $\boldsymbol{H}_{nm}^{(TR)}$, etc, depend on both the spatial arrangement of the particles and their orientation. However, in the case of spherical particles, $\boldsymbol{H}_{nm}^{(TT)}$ is independent of the orientation, so that if $\boldsymbol{N}_m = 0$, the rotational motion need not be considered explicitly.

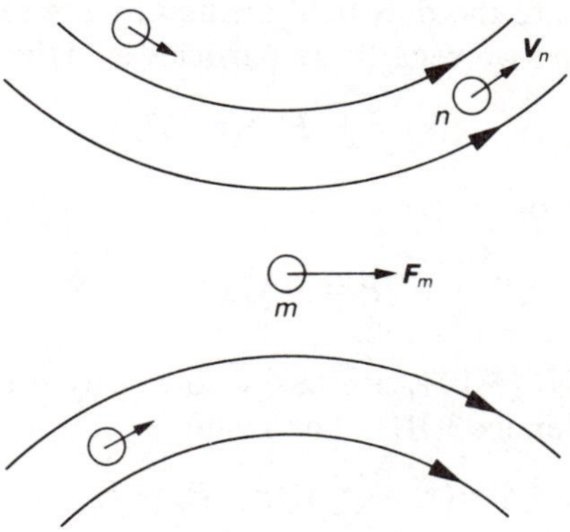

Fig. 3.6. The hydrodynamic interaction. The force acting on the particle m creates a velocity field and causes the motion of other particles.

hydrodynamics,[5,13] which assumes:

(i) The fluid is incompressible

$$\frac{\partial}{\partial r_\alpha} v_\alpha = 0. \tag{3.99}$$

(In this book Greek indices $\alpha, \beta, \mu, \nu \ldots$ are used to indicate the components of vectors and tensors, and the summation convention is used over the repeated indices.)

(ii) The inertia force of the fluid is negligibly small, so that if $\sigma_{\alpha\beta}(\boldsymbol{r})$ and $g_\alpha(\boldsymbol{r})$ denote the stress tensor and the external force acting on unit volume of the fluid, respectively,

$$\frac{\partial}{\partial r_\beta} \sigma_{\alpha\beta} = -g_\alpha(\boldsymbol{r}). \tag{3.100}$$

The stress in the fluid is written as

$$\sigma_{\alpha\beta} = \eta_s\left(\frac{\partial v_\beta}{\partial r_\alpha} + \frac{\partial v_\alpha}{\partial r_\beta}\right) + P\delta_{\alpha\beta} \tag{3.101}$$

(P being the pressure). From eqns (3.99)–(3.101) it follows that

$$\eta_s \frac{\partial^2}{\partial r_\beta^2} v_\alpha + \frac{\partial}{\partial r_\alpha} P = -g_\alpha. \tag{3.102}$$

Equations (3.99) and (3.102) are called the Stokes approximation, and are the basis of the hydrodynamic interactions in suspensions and polymer solutions.

Let us now calculate the flow field created by the external forces acting on the particles. First we regard the particles as points, and write $\boldsymbol{g}(\boldsymbol{r})$ as

$$\boldsymbol{g}(\boldsymbol{r}) = \sum_n \boldsymbol{F}_n\, \delta(\boldsymbol{r} - \boldsymbol{R}_n), \tag{3.103}$$

then eqn (3.102) reads

$$\eta_s \nabla^2 \boldsymbol{v} + \nabla P = -\sum_n \boldsymbol{F}_n\, \delta(\boldsymbol{r} - \boldsymbol{R}_n). \tag{3.104}$$

Equations (3.99) and (3.104) are easily solved by the use of the Fourier transform (see Appendix 3.III). The result is

$$\boldsymbol{v}(\boldsymbol{r}) = \sum_n \boldsymbol{H}(\boldsymbol{r} - \boldsymbol{R}_n) \cdot \boldsymbol{F}_n \tag{3.105}$$

with

$$\boldsymbol{H}(\boldsymbol{r}) = \frac{1}{8\pi\eta_s r}(\boldsymbol{I} + \hat{\boldsymbol{r}}\hat{\boldsymbol{r}}) \tag{3.106}$$

where $\hat{\boldsymbol{r}}$ denotes the unit vector parallel to $\boldsymbol{r}$. The tensor $\boldsymbol{H}(\boldsymbol{r})$ is called the Oseen tensor.

Since the particles move with the same velocity as the fluid, their velocities are given by

$$\boldsymbol{V}_n = \boldsymbol{v}(\boldsymbol{R}_n) = \sum_m \boldsymbol{H}(\boldsymbol{R}_n - \boldsymbol{R}_m) \cdot \boldsymbol{F}_m. \tag{3.107}$$

Thus

$$\boldsymbol{H}_{nm} = \boldsymbol{H}(\boldsymbol{R}_n - \boldsymbol{R}_m). \tag{3.108}$$

Unfortunately, eqn (3.108) is not appropriate since $\boldsymbol{H}_{nn} = \boldsymbol{H}(0)$ is infinite. This failure comes from the approximation that the particles are regarded as points. If we start from a collection of particles with finite size, this difficulty does not arise. Unfortunately, for a collection of particles with finite sizes the solution of Stokes equation is obtained only in the form of a perturbation expansion,[13–16] which is not easy to handle.

A simple approximation commonly adopted in the theory of polymer solutions is to use $\boldsymbol{I}/\zeta$ for $\boldsymbol{H}_{nn}$, i.e.,

$$\boldsymbol{H}_{nn} = \frac{\boldsymbol{I}}{\zeta}; \qquad \boldsymbol{H}_{nm} = \boldsymbol{H}(\boldsymbol{R}_n - \boldsymbol{R}_m), \qquad n \neq m. \tag{3.109}$$

With this definition of the mobility matrix, the general form of the Smoluchowski equation is written as

$$\frac{\partial \Psi}{\partial t} = \sum_{n,m} \frac{\partial}{\partial \boldsymbol{R}_n} \cdot \boldsymbol{H}_{nm} \cdot \left(k_B T \frac{\partial \Psi}{\partial \boldsymbol{R}_m} + \frac{\partial U}{\partial \boldsymbol{R}_m} \Psi \right). \tag{3.110}$$

This is the basic equation describing the interacting Brownian particles.

3.7 Microscopic basis of viscoelasticity

3.7.1 Introduction

Colloidal suspensions and polymer solutions have interesting mechanical properties. In general these materials have both viscosity and elasticity and hence are called viscoelastic. Colloidal suspensions show curious nonlinear hysteresis effects called thixotropy, rheopexy, and dilatancy. These unusual flow behaviours are the central problems of rheology.[17–19] A fundamental question in rheology is how those phenomena can be understood from the microscopic characteristic of the materials, i.e., their structure and the type of interaction. In later chapters, we shall discuss this in detail for polymeric liquids. In this section, we shall give a general base for developing microscopic theory for the mechanical properties of suspensions and polymer solutions.

The theory presented here is based on the classical work of Kirkwood,[20] who summarized the earlier works of Burgers,[21] Kuhn,[22] and Kramers[23] and established how to take Brownian motion into account, how to include the effect of the macroscopic flow, and how to calculate the stress tensor.[24] Though the original theory of Kirkwood was for dilute polymer solutions, the theory equally applies to concentrated polymer solutions and suspensions.

The basis of the Kirkwood theory is to assume that the microscopic dynamics is described by a system of spherical Brownian particles of equal size interacting via the potential $U(\{\boldsymbol{R}\})$ and the hydrodynamic interaction. Each Brownian particle is called a bead. At first sight this model may appear to be quite specific, but it actually represents a quite wide class of systems. For example:

(i) Polymer solutions: A polymer in solution can be modeled as a collection of beads connected sequentially (see Fig. 3.7*a*). The connec-

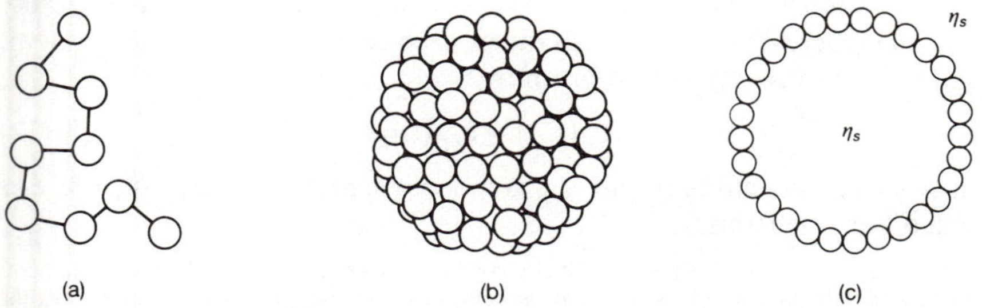

Fig. 3.7. Systems described by interacting beads. (*a*) Polymer (the freely jointed model). (*b*) Solid particle of an arbitrary shape. In this model, it is convenient to assume that the inside of the particle is filled with the same fluid as that outside, as shown in (*c*).

tivity of the chain and the excluded volume interaction are included in the potential $U(\{\boldsymbol{R}_n\})$.

(ii) Suspensions of solid particles of arbitrary shape. A solid particle in suspension can be modelled as a collection of beads fixed on the surface the solid particle (see Fig. 3.7*b*). In this model, the bead m represents a discretized surface element of the particle, and when the size of the beads is taken to be infinitesimal, the formulation becomes equivalent to that obtained by hydrodynamical considerations.[25]†

To apply the theory to such general systems, we have to consider a system with rigid constraints. However, in this section we shall first consider the case in which there are no rigid constraints, i.e., the force $\partial U/\partial \boldsymbol{R}_m$ is finite and well behaved.

3.7.2 Constitutive equation

The mechanical property of a homogeneous material is expressed by the constitutive equation which relates the stress tensor $\sigma_{\alpha\beta}$ to the velocity gradient tensor $\kappa_{\alpha\beta}$, where

$$\kappa_{\alpha\beta}(\boldsymbol{r}, t) = \frac{\partial \bar{v}_\alpha(\boldsymbol{r}, t)}{\partial r_\beta} \tag{3.111}$$

and $\bar{\boldsymbol{v}}(\boldsymbol{r}, t)$ is the macroscopic velocity field.

Generally, both $\sigma_{\alpha\beta}$ and $\kappa_{\alpha\beta}$ depend on position and time. However, in considering the constitutive equation, the positional dependence of $\kappa_{\alpha\beta}$ can be eliminated because the stress at a certain fluid element depends only on the previous values of the velocity gradient evaluated at that fluid element (principle of locality). This is a consequence of the length-scales of the particles being much smaller than the macroscopic length: although the stress depends, strictly speaking, not only on $\kappa_{\alpha\beta}$ but also on its spatial derivative $\partial\kappa_{\alpha\beta}/\partial r_\gamma$, the effect is of the order of (size of the particle)/(macroscopic length), which is usually negligibly small. Therefore to study the constitutive equation, we may assume without loss of generality that the macroscopic velocity gradient is constant throughout the system, and that the velocity field is given by

$$\bar{v}_\alpha(\boldsymbol{r}, t) = \kappa_{\alpha\beta}(t) r_\beta. \tag{3.112}$$

In this case, the stress tensor $\sigma_{\alpha\beta}$ is independent of position. Such flow is called homogeneous.

† In this modelling of solid particles, it is convenient to assume that the frictional elements are placed only on the surface and that the interior of the particle is filled with fluid of the same viscosity as that of the outside fluid (see Fig. 3.7*c*). Provided that the shell undergoes only translation and rotation, the interior fluid behaves like a solid (the strain rate inside the fluid is zero). In such a case the stress is zero inside the particle and there is a discontinuity in the stress at the surface of the particle, i.e., if $\boldsymbol{F}$ is the surface force density and $\boldsymbol{n}$ the unit vector normal to the surface then $\sigma_{\alpha\beta} n_\beta = F_\alpha$ at the surface.

The stress tensor $\sigma_{\alpha\beta}$ can be expressed as the sum of an isotropic tensor $\bar{\sigma}\delta_{\alpha\beta}$ and an anisotropic tensor $\sigma^{(a)}_{\alpha\beta}$ whose trace is zero:

$$\sigma_{\alpha\beta} = \bar{\sigma}\delta_{\alpha\beta} + \sigma^{(a)}_{\alpha\beta} \tag{3.113}$$

with

$$\bar{\sigma} = \tfrac{1}{3}\sigma_{\alpha\alpha}, \qquad \sigma^{(a)}_{\alpha\alpha} = 0. \tag{3.114}$$

In an incompressible fluid for which

$$\frac{\partial}{\partial r_\alpha}\bar{v}_\alpha = \kappa_{\alpha\alpha} = 0 \tag{3.115}$$

the isotropic part is determined by the external conditions and is irrelevant in the discussion of the constitutive equation. In this book we shall consider only such fluids, and neglect the difference in the isotropic part of the stress tensor. Thus two stress tensors $\sigma^{(A)}_{\alpha\beta}$ and $\sigma^{(B)}_{\alpha\beta}$ are regarded as equal if their difference, $\sigma^{(A)}_{\alpha\beta} - \sigma^{(B)}_{\alpha\beta}$, is an isotropic tensor.

3.7.3 The Smoluchowski equation for a system in macroscopic flow

Now our aim is to calculate the stress for given history of macroscopic velocity gradient. First we consider the form of the Smoluchowski equation for a given *macroscopic* velocity field $\bar{\boldsymbol{v}}(\boldsymbol{r}, t) = \boldsymbol{\kappa}(t)\cdot\boldsymbol{r}$. To do this we have to know the *microscopic* velocity field $\boldsymbol{v}(\boldsymbol{r}, t)$. This is obtained from the conditions: (i) $\boldsymbol{v}(\boldsymbol{r}, t)$ is a solution of eqn (3.104), and (ii) the average of $\boldsymbol{v}(\boldsymbol{r}, t)$ is the macroscopic field, i.e.,

$$\bar{\boldsymbol{v}}(\boldsymbol{r}, t) = \langle \boldsymbol{v}(\boldsymbol{r}, t)\rangle \tag{3.116}$$

where $\langle \ldots \rangle$ means the configurational average of the beads,

$$\langle \ldots \rangle \equiv \int \mathrm{d}\{\boldsymbol{R}\}\Psi(\{\boldsymbol{R}\}; t)\ldots \tag{3.117}$$

Though at first sight it may seem difficult to find the velocity field $\bar{\boldsymbol{v}}(\boldsymbol{r}, t)$, the answer is simple:

$$\boldsymbol{v}(\boldsymbol{r}, t) = \boldsymbol{\kappa}(t)\cdot\boldsymbol{r} + \sum_m \boldsymbol{H}(\boldsymbol{r} - \boldsymbol{R}_m)\cdot\boldsymbol{F}_m. \tag{3.118}$$

For this flow, the first condition is obviously satisfied. That the second condition is satisfied is seen by the symmetry argument: in the homogeneous flow, the average

$$\bar{\boldsymbol{v}}'(\boldsymbol{r}, t) \equiv \left\langle \sum_m \boldsymbol{H}(\boldsymbol{r} - \boldsymbol{R}_m)\cdot\boldsymbol{F}_m \right\rangle \tag{3.119}$$

must be a constant vector which can depend only on the tensor $\boldsymbol{\kappa}$.† Since one cannot construct a vector from a single tensor $\boldsymbol{\kappa}$, $\boldsymbol{v}'(\boldsymbol{r}, t)$ must be zero.‡

Given the microscopic velocity field $\boldsymbol{v}(\boldsymbol{r}, t)$, the velocity of the beads is immediately obtained:

$$\boldsymbol{V}_m = \boldsymbol{v}(\boldsymbol{R}_m; t) = \boldsymbol{\kappa}(t) \cdot \boldsymbol{R}_m + \sum_n \boldsymbol{H}_{mn} \cdot \boldsymbol{F}_n \tag{3.120}$$

so that the Smoluchowski equation becomes

$$\frac{\partial}{\partial t}\Psi = \sum_{m,n} \frac{\partial}{\partial \boldsymbol{R}_m} \cdot \boldsymbol{H}_{mn} \cdot \left(k_B T \frac{\partial \Psi}{\partial \boldsymbol{R}_n} + \Psi \frac{\partial U}{\partial \boldsymbol{R}_n}\right) - \sum_m \frac{\partial}{\partial \boldsymbol{R}_m} \cdot \boldsymbol{\kappa}(t) \cdot \boldsymbol{R}_m \Psi. \tag{3.121}$$

This equation describes the change in the distribution function of a system under macroscopic velocity gradient.

3.7.4 *Expression for the stress tensor*

Next we consider how to calculate the stress tensor, say the component $\sigma_{\alpha z}(\alpha = x, y, z)$. To this end we consider a region of volume V in the

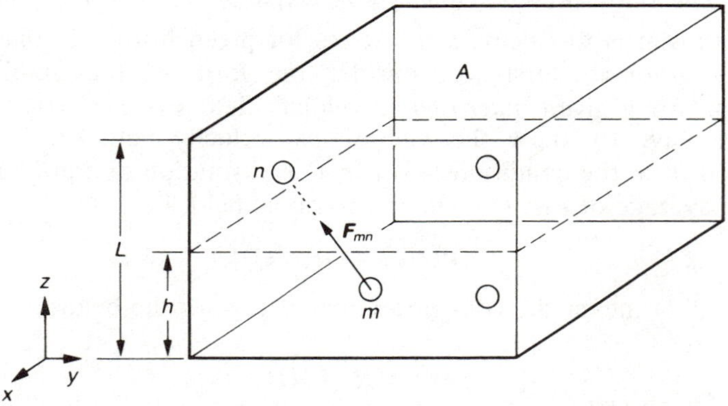

Fig. 3.8. Microscopic definition of the stress tensor. The stress component $\sigma_{\alpha z}(\alpha = x, y, z)$ is the α component of the force (per area) that the material above the plane (denoted by the dashed line) exerts on the material below the plane. The force $\boldsymbol{S}$ consists of two parts $\boldsymbol{S}^{(s)}$ and $\boldsymbol{S}^{(p)}$. $\boldsymbol{S}^{(s)}$ is the force acting through the solvent at the plane, and $\boldsymbol{S}^{(p)}$ is the sum of the forces that the upper beads exert on the lower beads via a potential.

† Strictly speaking a more careful analysis is needed to justify this argument since the sum in eqn (3.119) is ill converged, and depends on the shape of the outer boundary. However, as in the case of dieletrics, the contribution from the outer boundary can be included in the imposed field $\boldsymbol{\kappa}(t)$, so that in practice the contribution from the outer boundary can be neglected. (See the discussion in ref. 26.)

‡ If there is an external force field (such as gravitational force) $\bar{\boldsymbol{v}}'(\boldsymbol{r}, t)$ does not vanish. In such a case, the problem must be treated by a multi-fluid model,[27] which is not considered in this book.

homogenous flow, and divide it by a hypothetical plane which is perpendicular to the z-axis (see Fig. 3.8). By the definition of the stress tensor, $\sigma_{\alpha z}$ is given by the force S_α which the upper part exerts on the lower part through the plane:

$$\sigma_{\alpha z} = \langle S_\alpha \rangle / A \tag{3.122}$$

where A is the area of the plane.

Now S_α consists of two parts, the force $S_\alpha^{(s)}$ which acts through the solvent fluid and the force $S_\alpha^{(p)}$ which acts directly between the beads. The former is written as

$$S_\alpha^{(s)}(h) = \int_{r \in V} \mathrm{d}\boldsymbol{r} \left[\eta_s \left(\frac{\partial v_\alpha}{\partial r_z} + \frac{\partial v_z}{\partial r_\alpha} \right) + P\delta_{\alpha z} \right] \delta(r_z - h), \tag{3.123}$$

where h is the position of the plane defined in Fig. 3.8, and the integral is carried out in the region V. The force $S_\alpha^{(p)}$ is given by the force $\boldsymbol{F}_{mn}$ which the bead n exerts on the bead m:†

$$S_\alpha^{(p)}(h) = \sum_{m,n} F_{mn\alpha} \Theta(h - R_{mz}) \Theta(R_{nz} - h), \tag{3.124}$$

where

$$\Theta(x) = \begin{cases} 1 & \text{for} \quad x > 0, \\ 0 & \text{for} \quad x < 0. \end{cases} \tag{3.125}$$

The two Θ functions in eqn (3.124) restrict the bead n to the upper part and the bead m to the lower part.

In the homogeneous flow, the average $\langle S_\alpha(h) \rangle$ is independent of h, so that eqn (3.122) can be written as

$$\sigma_{\alpha z} = \frac{1}{AL} \int_0^L \mathrm{d}h \langle S_\alpha(h) \rangle = \frac{1}{AL} \int_0^L \mathrm{d}h \langle S_\alpha^{(s)}(h) + S_\alpha^{(p)}(h) \rangle \tag{3.126}$$

The integral for $S_\alpha^{(s)}$ is

$$\begin{aligned} X_\alpha^{(s)} &= \int_0^L \mathrm{d}h \int_{r \in V} \mathrm{d}\boldsymbol{r} \left[\eta_s \left(\frac{\partial}{\partial r_z} \langle v_\alpha \rangle + \frac{\partial}{\partial r_\alpha} \langle v_z \rangle \right) + \langle P \rangle \delta_{\alpha z} \right] \delta(r_z - h) \\ &= \int_{r \in V} \mathrm{d}\boldsymbol{r} \left[\eta_s \left(\frac{\partial}{\partial r_z} \langle v_\alpha \rangle + \frac{\partial}{\partial r_\alpha} \langle v_z \rangle \right) + \langle P \rangle \delta_{\alpha z} \right]. \end{aligned} \tag{3.127}$$

† Note that here the size of the beads is taken to be infinitely small. Since rigid bodies of finite size are expressed by the model shown in Fig. 3.7*b*, this does not impose any limitation on the applicability of the present formulation.

By eqn (3.116),

$$X_\alpha^{(s)} = \int_{r\in V} d\boldsymbol{r}\left[\eta_s\left(\frac{\partial}{\partial r_z}(\kappa_{\alpha\mu}r_\mu) + \frac{\partial}{\partial r_\alpha}(\kappa_{z\mu}r_\mu)\right) + \langle P\rangle\delta_{\alpha z}\right]$$

$$= V[\eta_s(\kappa_{\alpha z} + \kappa_{z\alpha}) + \langle P\rangle\delta_{\alpha z}]. \tag{3.128}$$

The integral for $S_\alpha^{(p)}$ is rewritten as

$$X_\alpha^{(p)} \equiv \left\langle\int_0^L dh S_\alpha^{(p)}(h)\right\rangle = \sum_{n,m}\left\langle F_{mn\alpha}\int_0^L dh\Theta(h - R_{mz})\Theta(R_{nz} - h)\right\rangle$$

$$= \left\langle\sum_{m,n} F_{mn\alpha}(R_{nz} - R_{mz})\Theta(R_{nz} - R_{mz})\right\rangle. \tag{3.129}$$

Exchanging m and n, and using Newton's third law, $\boldsymbol{F}_{mn} = -\boldsymbol{F}_{nm}$, we have

$$X_\alpha^{(p)} = \left\langle\tfrac{1}{2}\sum_{m,n}[F_{mn\alpha}(R_{nz} - R_{mz})\Theta(R_{nz} - R_{mz})\right.$$

$$\left. + F_{nm\alpha}(R_{mz} - R_{nz})\Theta(R_{mz} - R_{nz})]\right\rangle$$

$$= \left\langle\tfrac{1}{2}\sum_{m,n} F_{mn\alpha}(R_{nz} - R_{mz})[\Theta(R_{nz} - R_{mz}) + \Theta(R_{mz} - R_{nz})]\right\rangle$$

$$= \left\langle -\tfrac{1}{2}\sum_{m,n} F_{mn\alpha}(R_{mz} - R_{nz})\right\rangle. \tag{3.130}$$

Let $\boldsymbol{F}_m$ be the sum of the nonhydrodynamic forces acting on the bead m:

$$\boldsymbol{F}_m = \sum_n \boldsymbol{F}_{mn}. \tag{3.131}$$

Then eqn (3.130) is rewritten as

$$X_\alpha^{(p)} = \left\langle -\tfrac{1}{2}\sum_{m,n} F_{mn\alpha}R_{mz} + \tfrac{1}{2}\sum_{m,n} F_{mn\alpha}R_{nz}\right\rangle$$

$$= -\left\langle\sum_m F_{m\alpha}R_{mz}\right\rangle. \tag{3.132}$$

From eqns (3.126), (3.128), and (3.132), we finally obtain the stress tensor

$$\sigma_{\alpha\beta} = \eta_s(\kappa_{\alpha\beta} + \kappa_{\beta\alpha}) + \langle P\rangle\delta_{\alpha\beta} - \frac{1}{V}\sum_m\langle F_{m\alpha}R_{m\beta}\rangle. \tag{3.133}$$

The first term in eqn (3.133) represents the stress in the absence of the beads, and the last term,

$$\sigma^{(p)}_{\alpha\beta} = -\frac{1}{V}\sum_m \left\langle F_{m\alpha} R_{m\beta} \right\rangle, \tag{3.134}$$

denotes the extra stress due to the beads.

It is important to notice that $\boldsymbol{F}_m$ is given by

$$\boldsymbol{F}_m = -\frac{\partial}{\partial \boldsymbol{R}_m}(k_B T \ln \Psi + U) \tag{3.135}$$

and includes the thermodynamic force $\partial k_B T \ln \Psi/\partial \boldsymbol{R}_m$. The necessity of such a term can be understood by studying a special situation. Consider for example that the suspension is in equilibrium under the gravitational field $U = -gz$. If the term $k_B T\, \partial \ln \Psi/\partial \boldsymbol{R}_m$ is not included, the stress tensor is not isotropic, which contradicts the condition that the stress must be isotropic in a liquid in equilibrium.

Equations (3.121) and (3.133) together may be regarded as a constitutive equation: for a given macroscopic velocity gradient $\kappa_{\alpha\beta}(t)$, the distribution function Ψ is obtained from eqn (3.121) and the stress is calculated using eqn (3.133).

3.7.5 *Principle of virtual work*

The stress formula (3.134) can be put in the form equivalent to the principle of virtual work.[6] Consider a virtual deformation which displaces the point r_α on the material to $r_\alpha + \delta\varepsilon_{\alpha\beta} r_\beta$ in a very short time δt. In the limit of $\delta t \to 0$, the velocity gradient $\kappa_{\alpha\beta} = \delta\varepsilon_{\alpha\beta}/\delta t$ becomes very large, so that the time evolution of Ψ in the time interval δt is dominated by $\kappa_{\alpha\beta}$,†

$$\frac{\partial \Psi}{\partial t} = -\sum_m \frac{\partial}{\partial \boldsymbol{R}_m} \cdot \boldsymbol{\kappa} \cdot \boldsymbol{R}_m \Psi. \tag{3.136}$$

Hence the change in Ψ by the hypothetical deformation is

$$\delta\Psi = \frac{\partial \Psi}{\partial t}\delta t = -\sum_m \frac{\partial}{\partial \boldsymbol{R}_m} \cdot (\delta\boldsymbol{\varepsilon} \cdot \boldsymbol{R}_m \Psi). \tag{3.137}$$

Let $\mathscr{A}$ be the dynamical free energy of the system:

$$\mathscr{A} = \int \mathrm{d}\{\boldsymbol{R}\}\Psi(k_B T \ln \Psi + U)$$

It can be shown that the stress $\sigma^{(p)}_{\alpha\beta}$ due to the Brownian particles is

† This statement is not true if there are rigid constraints which produce infinite forces. See the discussion in the next section.

related to the change in the dynamical free energy by

$$\delta\mathscr{A} = \sigma^{(\mathrm{p})}_{\alpha\beta}\delta\varepsilon_{\alpha\beta}V. \tag{3.138}$$

The proof is straightforward. The change in the dynamical free energy is

$$\delta\mathscr{A} = \int \mathrm{d}\{\boldsymbol{R}\}(k_BT\delta\Psi \ln\Psi + k_BT\delta\Psi + U\delta\Psi). \tag{3.139}$$

Substituting eqn (3.137) and using the integration by parts, we have

$$\begin{aligned}\delta\mathscr{A} &= \int \mathrm{d}\{\boldsymbol{R}\}\sum_m -\frac{\partial}{\partial R_{m\alpha}}(\delta\varepsilon_{\alpha\beta}R_{m\beta}\Psi)(k_BT\ln\Psi + k_BT + U)\\ &= \int \mathrm{d}\{\boldsymbol{R}\}\sum_m \delta\varepsilon_{\alpha\beta}R_{m\beta}\Psi\frac{\partial}{\partial R_{m\alpha}}(k_BT\ln\Psi + U)\\ &= -\delta\varepsilon_{\alpha\beta}\sum_m\langle R_{m\beta}F_{m\alpha}\rangle\\ &= \delta\varepsilon_{\alpha\beta}\sigma^{(\mathrm{p})}_{\alpha\beta}V.\end{aligned} \tag{3.140}$$

Equation (3.138) represents the principle of virtual work.

3.8 Systems with rigid constraints

3.8.1 Introduction

We now consider the case where the beads are subject to rigid constraints. This is necessary to deal with the problems of suspensions of a rigid body, or polymers with rigid constraints (such as the rodlike polymer, or the freely jointed model), but the reader who is interested only in flexible polymers can omit this section.

Here we shall consider only the *holonomic* constraints† which can be expressed as equations connecting the coordinates of the beads:

$$C_p(\{\boldsymbol{R}\}) = 0, \qquad p = 1, 2, \ldots, N_c. \tag{3.141}$$

Examples of systems which have such constraints are:

(i) the freely jointed model (Fig. 3.7*a*), in which the beads are successively connected at constant distance b, so that

$$(\boldsymbol{R}_n - \boldsymbol{R}_{n-1})^2 - b^2 = 0, \qquad n = 1, 2, \ldots, N. \tag{3.142}$$

(ii) the rigid body model (Fig. 3.7*b*), in which the mutual distance between the beads is fixed.

When the constraints are introduced, the force $\boldsymbol{F}_m$ is no longer a known function of $\{\boldsymbol{R}\}$ and must be determined by the equation of motion. This

† The topological constraints that the chains cannot cross each other do not belong to this class.

situation is familiar in classical mechanics,[28] where the forces are determined using the condition that the solution of the equation of motion must satisfy the constraints. The same rule applies to the present system except that in our problem, the equation of motion is not Newton's equation, but the hydrodynamic relation

$$\boldsymbol{V}_m = \boldsymbol{\kappa} \cdot \boldsymbol{R}_m + \sum_n \boldsymbol{H}_{mn} \cdot \boldsymbol{F}_n. \tag{3.143}$$

From a practical viewpoint, there are two ways of doing this.

One is to introduce generalized coordinates which are independent of each other, and specify the configuration of the beads uniquely.[29] This method is suitable when the positions of the beads are expressed explicitly as a function of such coordinates. For example, rigid body problems are conveniently handled by this method. In this example, the generalized coordinates will stand for the three components of the position vector of the centre of mass, and the three Euler angles specifying the orientation of the rigid body. However it is impractical to apply this method to the freely jointed model.

Another method is to use the Lagrangian multipliers for the constraints. This method is complementary to the first, and indeed has been successfully used for the freely jointed model (and semiflexible polymer models[30]).

Here we shall describe the formal part of the methods,[29,30] leaving the detail of calculation to the literature.[30–35]

3.8.2 *The method of generalized coordinates*

Let $\{Q\} \equiv Q_1, Q_2, \ldots, Q_{N_f}$ be the set of generalized coordinates. The position $\boldsymbol{R}_m$ is expressed as a function of $\{Q\}$ as

$$\boldsymbol{R}_m = \boldsymbol{R}_m(\{Q\}), \qquad m = 1, 2, \ldots, N. \tag{3.144}$$

If the velocity of the generalized coordinate is V_a, the velocity of the particle is given as

$$\boldsymbol{V}_m = \sum_{a=1}^{N_f} \frac{\partial \boldsymbol{R}_m}{\partial Q_a} V_a. \tag{3.145}$$

In this section we shall employ the summation convention, and write eqn (3.145) as

$$\boldsymbol{V}_m = \frac{\partial \boldsymbol{R}_m}{\partial Q_a} V_a. \tag{3.146}$$

To obtain $\boldsymbol{F}_m$, we use the principle of virtual work.[28] Let us consider the work necessary to change Q_a by δQ_a, which is

$$\delta(U + k_B T \ln \Psi) = \left[\frac{\partial}{\partial Q_a}(U + k_B T \ln \Psi)\right]\delta Q_a. \tag{3.147}$$

Alternatively, the work can also be calculated using the force $\boldsymbol{F}_m$ and the displacement $\delta \boldsymbol{R}_m$ caused by the change in Q_a, i.e.

$$\delta(U + k_B T \ln \Psi) = -\boldsymbol{F}_m \cdot \delta \boldsymbol{R}_m \tag{3.148}$$

where

$$\delta \boldsymbol{R}_m = \frac{\partial \boldsymbol{R}_m}{\partial Q_a}\delta Q_a. \tag{3.149}$$

From eqns (3.147)–(3.149) we have

$$\boldsymbol{F}_m \cdot \frac{\partial \boldsymbol{R}_m}{\partial Q_a} = -\frac{\partial}{\partial Q_a}(k_B T \ln \Psi + U). \tag{3.150}$$

Equations (3.143), (3.145), and (3.150) determine V_a and $\boldsymbol{F}_m$.

To obtain $\boldsymbol{F}_m$ and V_a explicitly, we first solve eqn (3.143) for $\boldsymbol{F}_n$:

$$\begin{aligned}\boldsymbol{F}_n &= (\boldsymbol{H}^{-1})_{nm} \cdot (\boldsymbol{V}_m - \boldsymbol{\kappa} \cdot \boldsymbol{R}_m)\\ &= (\boldsymbol{H}^{-1})_{nm} \cdot \left(\frac{\partial \boldsymbol{R}_m}{\partial Q_a} V_a - \boldsymbol{\kappa} \cdot \boldsymbol{R}_m\right)\end{aligned} \tag{3.151}$$

where $(\boldsymbol{H}^{-1})_{nm}$ is the inverse of $\boldsymbol{H}_{nm}$:

$$(\boldsymbol{H}^{-1})_{nm} \cdot \boldsymbol{H}_{mk} = \delta_{nk}\boldsymbol{I}. \tag{3.152}$$

Substituting this into eqn (3.150), we have

$$\frac{\partial \boldsymbol{R}_n}{\partial Q_a} \cdot (\boldsymbol{H}^{-1})_{nm} \cdot \left[\frac{\partial \boldsymbol{R}_m}{\partial Q_b} V_b - \boldsymbol{\kappa} \cdot \boldsymbol{R}_m\right] = -\frac{\partial}{\partial Q_a}(U + k_B T \ln \Psi). \tag{3.153}$$

Putting

$$(h^{-1})_{ab} = \frac{\partial \boldsymbol{R}_n}{\partial Q_a} \cdot (\boldsymbol{H}^{-1})_{nm} \cdot \frac{\partial \boldsymbol{R}_m}{\partial Q_b}, \tag{3.154}$$

define

$$F_a^{(\mathrm{E})} = -\frac{\partial}{\partial Q_a}(U + k_B T \ln \Psi), \tag{3.155}$$

and

$$V_a^{(\mathrm{V})} = h_{ab}\frac{\partial \boldsymbol{R}_n}{\partial Q_b} \cdot (\boldsymbol{H}^{-1})_{nm} \cdot \boldsymbol{\kappa} \cdot \boldsymbol{R}_m, \tag{3.156}$$

we can rewrite eqn (3.153) as

$$(h^{-1})_{ab}(V_b - V_b^{(V)}) = F_a^{(E)} \tag{3.157}$$

which can be solved using h_{ab}, the inverse of $(h^{-1})_{ab}$, giving

$$\begin{aligned} V_a &= V_a^{(V)} + h_{ab}F_b^{(E)} \\ &= -h_{ab}\frac{\partial}{\partial Q_b}(U + k_B T \ln \Psi) + V_a^{(V)}. \end{aligned} \tag{3.158}$$

Hence

$$\boldsymbol{F}_n = (\boldsymbol{H}^{-1})_{nm} \cdot \left(\frac{\partial \boldsymbol{R}_m}{\partial Q_a} h_{ab}F_b^{(E)} + \frac{\partial \boldsymbol{R}_m}{\partial Q_a} V_a^{(V)} - \boldsymbol{\kappa} \cdot \boldsymbol{R}_m\right). \tag{3.159}$$

In the generalized coordinate space, the conservation equation is written as[36]

$$\frac{\partial \Psi}{\partial t} = -\frac{1}{\sqrt{g}} \sum_{a=1}^{N_f} \frac{\partial}{\partial Q_a} [\sqrt{g}\, V_a \Psi] \tag{3.160}$$

where g is the determinant of the matrix g_{ab} defined by

$$g_{ab} \equiv \frac{\partial \boldsymbol{R}_m}{\partial Q_a} \cdot \frac{\partial \boldsymbol{R}_m}{\partial Q_b}. \tag{3.161}$$

Thus the diffusion equation is obtained as†

$$\frac{\partial \Psi}{\partial t} = \frac{1}{\sqrt{g}} \frac{\partial}{\partial Q_a} \sqrt{g}\left[h_{ab}\left(k_B T \frac{\partial \Psi}{\partial Q_b} + \frac{\partial U}{\partial Q_b}\Psi\right) - V_a^{(V)}\Psi\right]. \tag{3.162}$$

This equation was first given by Kirkwood.[20]

3.8.3 *The method of Lagrangian multipliers*

The constraints can be treated by the alternative method of Lagrangian multipliers.[30] Here $\{\boldsymbol{R}\}$ are regarded as independent variables, and the constraining forces are explicitly added to the right-hand side of eqn (3.135):

$$\boldsymbol{F}_m = -\frac{\partial}{\partial \boldsymbol{R}_m}(k_B T \ln \Psi + U) + \lambda_p \frac{\partial C_p}{\partial \boldsymbol{R}_m}. \tag{3.163}$$

† The factor $\sqrt{g}$ appears naturally as a result of the coordinate transformation, but it is unnecessary from the viewpoint of the general theory of Brownian motion. Indeed, in terms of $\tilde{\Psi} = \Psi\sqrt{g}$ and $\tilde{U} = U - k_B T \ln \sqrt{g}$, eqn (3.162) can be written in the form of eqn (3.21)

$$\frac{\partial \tilde{\Psi}}{\partial t} = \frac{\partial}{\partial Q_a}\left[h_{ab}\left(k_B T \frac{\partial \tilde{\Psi}}{\partial Q_b} + \frac{\partial \tilde{U}}{\partial Q_b}\tilde{\Psi}\right) - V_a^{(V)}\tilde{\Psi}\right],$$

which includes no $\sqrt{g}$ factor. Notice that the theory here can be presented without using the Riemannian geometry.

The unknown parameters λ_p are chosen such that the velocity $\boldsymbol{V}_m$ determined by eqn (3.143) satisfies.

$$\frac{\partial C_p}{\partial \boldsymbol{R}_m} \cdot \boldsymbol{V}_m = 0. \tag{3.164}$$

Using eqns (3.143), (3.163), and (3.164),

$$\lambda_p = (\tilde{h}^{-1})_{pq} \frac{\partial C_q}{\partial \boldsymbol{R}_n} \cdot \boldsymbol{H}_{nm} \cdot \frac{\partial}{\partial \boldsymbol{R}_m} [k_B T \ln \Psi + U] - (\tilde{h}^{-1})_{pq} \frac{\partial C_q}{\partial \boldsymbol{R}_n} \cdot \boldsymbol{\kappa} \cdot \boldsymbol{R}_n \tag{3.165}$$

where $(\tilde{h}^{-1})_{pq}$ is the inverse of the matrix

$$\tilde{h}_{pq} \equiv \frac{\partial C_p}{\partial \boldsymbol{R}_n} \cdot \boldsymbol{H}_{nm} \cdot \frac{\partial C_q}{\partial \boldsymbol{R}_m}. \tag{3.166}$$

The Smoluchowski equation is obtained if eqn (3.163) is substituted into the continuity equation:

$$\begin{aligned}\frac{\partial \Psi}{\partial t} &= -\frac{\partial}{\partial \boldsymbol{R}_m} \cdot (\boldsymbol{V}_m \Psi) \\ &= \frac{\partial}{\partial \boldsymbol{R}_n} \cdot \boldsymbol{H}_{nm} \cdot \Psi \left(\frac{\partial}{\partial \boldsymbol{R}_m} - \frac{\partial C_p}{\partial \boldsymbol{R}_m} (\tilde{h}^{-1})_{pq} \frac{\partial C_q}{\partial \boldsymbol{R}_k} \cdot \boldsymbol{H}_{ki} \cdot \frac{\partial}{\partial \boldsymbol{R}_i} \right) (k_B T \ln \Psi + U) \\ &\quad - \frac{\partial}{\partial \boldsymbol{R}_m} \cdot \left(\boldsymbol{\kappa} \cdot \boldsymbol{R}_m - \boldsymbol{H}_{mk} \cdot \frac{\partial C_p}{\partial \boldsymbol{R}_k} (\tilde{h}^{-1})_{pq} \frac{\partial C_q}{\partial \boldsymbol{R}_n} \cdot \boldsymbol{\kappa} \cdot \boldsymbol{R}_n \right).\end{aligned} \tag{3.167}$$

3.8.4 *Elastic stress and viscous stress*

If the distribution function is obtained from eqns (3.162) or (3.167), the stress can be calculated from eqn (3.134) in which $\boldsymbol{F}_m$ is now given by either eqn (3.159) or (3.163). In either expression, the force $\boldsymbol{F}_m$ consists of two terms, one independent of $\boldsymbol{\kappa}$ and the other proportional to $\boldsymbol{\kappa}$:

$$F_{m\alpha} = F_{m\alpha}^{(\mathrm{E})} + \Xi_{m\alpha\beta\gamma} \kappa_{\beta\gamma}. \tag{3.168}$$

For example in the generalized coordinate representation, eqns (3.159), (3.155), and (3.156) give

$$\begin{aligned}\boldsymbol{F}_m^{(\mathrm{E})} &= (\boldsymbol{H}^{-1})_{mn} \cdot \frac{\partial \boldsymbol{R}_n}{\partial Q_a} h_{ab} F_b^{(\mathrm{E})} \\ &= -(\boldsymbol{H}^{-1})_{mn} \cdot \frac{\partial \boldsymbol{R}_n}{\partial Q_a} h_{ab} \frac{\partial}{\partial Q_b} (U + k_B T \ln \Psi)\end{aligned} \tag{3.169}$$

and

$$\Xi_m : \boldsymbol{\kappa} = (\boldsymbol{H}^{-1})_{mn} \cdot \left(\frac{\partial \boldsymbol{R}_n}{\partial Q_a} V_a^{(V)} - \boldsymbol{\kappa} \cdot \boldsymbol{R}_n \right)$$

$$= (\boldsymbol{H}^{-1})_{mn} \cdot \left(\frac{\partial \boldsymbol{R}_n}{\partial Q_a} h_{ab} \frac{\partial \boldsymbol{R}_i}{\partial Q_b} \cdot (\boldsymbol{H}^{-1})_{ij} \cdot \boldsymbol{\kappa} \cdot \boldsymbol{R}_j - \boldsymbol{\kappa} \cdot \boldsymbol{R}_n \right). \quad (3.170)$$

The force $\boldsymbol{F}_m^{(E)}$ represents the force due to the potential, and $\Xi_m : \boldsymbol{\kappa}$ represents the effect of the constraining force. Hence the excess stress due to the Brownian particles is generally written as

$$\sigma_{\alpha\beta}^{(p)} = \sigma_{\alpha\beta}^{(E)} + \sigma_{\alpha\beta}^{(V)} \quad (3.171)$$

where

$$\sigma_{\alpha\beta}^{(E)} = -\frac{1}{V} \sum_{m=1}^{N} \langle F_{m\alpha}^{(E)} R_{m\beta} \rangle, \quad (3.172)$$

$$\sigma_{\alpha\beta}^{(V)} = -\frac{1}{V} \sum_{m=1}^{N} \langle \Xi_{m\alpha\mu\nu} R_{m\beta} \rangle \kappa_{\mu\nu}(t). \quad (3.173)$$

We shall call $\boldsymbol{\sigma}^{(E)}$ the elastic stress and $\boldsymbol{\sigma}^{(V)}$ the viscous stress. The viscous stress is proportional to the current velocity gradient $\boldsymbol{\kappa}(t)$ and can be written as

$$\sigma_{\alpha\beta}^{(V)}(t) = \eta_{\alpha\beta\mu\nu}^{(V)} \kappa_{\mu\nu}(t), \quad (3.174)$$

while the elastic stress does not include $\boldsymbol{\kappa}(t)$ explicitly.

Phenomenologically, the viscous stress is the stress which vanishes instantaneously when the flow is stopped. On the other hand the elastic stress does not vanish until the system is in equilibrium. The elastic stress is dominant in concentrated polymer solutions, while viscous stress often dominates in the suspensions of larger particles for which the Brownian motion is not effective. Whichever stress dominates, the rheological properties can be quite complex since both $\sigma_{\alpha\beta}^{(E)}$ and $\eta_{\alpha\beta\mu\nu}^{(V)}$ are functions of the configuration of the beads and therefore depend on the previous values of the velocity gradient. Note that the viscous stress $\boldsymbol{\sigma}^{(V)}$ only appears in the system with rigid constraints.†

The actual formula for the stress tensor is complicated. However, a neat expression is obtained in the form of the principle of virtual work. By a straightforward calculation, it can be shown[6]

† It must be mentioned that the distinction between the elastic stress and the viscous stress is a matter of time-scale. The elastic stress which has very short relaxation times, cannot be distinguished from the viscous stress. A discussion from a phenomenological viewpoint is given by J. D. Goddard, *J. Non-Newtonian Fluid Mech.* **14,** 41 (1984).

(i) The viscous stress $\sigma^{(V)}_{\alpha\beta}$ or the coefficient $\eta^{(V)}_{\alpha\beta\mu\nu}$ are related to the energy dissipation function

$$W(\{V\}) = \sum_{m,n} (\boldsymbol{V}_m - \boldsymbol{\kappa} \cdot \boldsymbol{R}_m) \cdot (\boldsymbol{H}^{-1})_{mn} \cdot (\boldsymbol{V}_n - \boldsymbol{\kappa} \cdot \boldsymbol{R}_n) \quad (3.175)$$

as

$$\langle \text{Mini } W \rangle = \eta^{(V)}_{\alpha\beta\mu\nu} \kappa_{\alpha\beta} \kappa_{\mu\nu} V \quad (3.176)$$

Note that Mini W does not vanish since $\boldsymbol{V}_m$ must satisfy eqn (3.145) (or eqn (3.164)) for the system with constraints.

(ii) The elastic stress is related to the change in the dynamical free energy $\mathscr{A}$ caused by the instantaneous deformation $\delta\varepsilon_{\alpha\beta} = \kappa_{\alpha\beta}\delta t$:

$$\delta\mathscr{A} = \sigma^{(E)}_{\alpha\beta} \delta\varepsilon_{\alpha\beta} V, \quad (3.177)$$

where

$$\delta\mathscr{A} = \int \mathrm{d}\{\boldsymbol{R}\}(k_B T \delta\Psi \ln \Psi + k_B T \delta\Psi + U \delta\Psi). \quad (3.178)$$

Here $\delta\Psi$ is given by

$$\delta\Psi = -\frac{\partial}{\partial \boldsymbol{R}_m} \cdot (\boldsymbol{V}_m \delta t \Psi) \quad (3.179)$$

where $\boldsymbol{V}_m$ is the velocity which minimizes eqn (3.175).

3.8.5 Variational formulation

The theory described above can be formulated in the form of a variational principle[6] which is similar to the Lagrangian formulation of classical mechanics. The advantage of this formulation is that it is independent of the coordinate system, and allows a great flexibility in choosing coordinates.

Let us regard $\boldsymbol{V}_m$ as a function of $\{\boldsymbol{R}\}$, and consider the functional defined by

$$K = \tfrac{1}{2} W + \dot{\mathscr{A}} \quad (3.180)$$

with

$$W \equiv \int \mathrm{d}\{\boldsymbol{R}\} \Psi \sum_{m,n} (\boldsymbol{V}_m - \boldsymbol{\kappa} \cdot \boldsymbol{R}_m) \cdot (\boldsymbol{H}^{-1})_{mn} \cdot (\boldsymbol{V}_n - \boldsymbol{\kappa} \cdot \boldsymbol{R}_n) \quad (3.181)$$

and

$$\dot{\mathscr{A}} \equiv \int \mathrm{d}\{\boldsymbol{R}\}(k_B T \dot{\Psi} \ln \Psi + k_B T \dot{\Psi} + \dot{\Psi} U), \quad (3.182)$$

where $\dot{\Psi}$ is defined by

$$\dot{\Psi} \equiv -\sum_m \frac{\partial}{\partial \boldsymbol{R}_m} \cdot (\boldsymbol{V}_m \Psi). \quad (3.183)$$

It is then shown that:

(i) The minimization of K for all variations of V_m subject to the constraints (3.145) (or (3.164)) determines the time evolution of Ψ to be

$$\frac{\partial}{\partial t}\Psi(\{\boldsymbol{R}\};t)=\dot{\Psi}(\{\boldsymbol{R}\}). \tag{3.184}$$

(ii) The stress is given by

$$\sigma_{\alpha\beta}=\frac{\partial \text{ Mini } K}{\partial \kappa_{\alpha\beta}}+\eta_s(\kappa_{\alpha\beta}+\kappa_{\beta\alpha})+P\delta_{\alpha\beta} \tag{3.185}$$

Some applications of this principle are given in ref. 6.

In closing this section it is worthwhile to stress again that no condition has been imposed on the concentration of the particles. Therefore the theory will apply to concentrated suspensions as well as dilute suspensions.

Appendix 3.I Eigenfunctions of the Smoluchowski equations

In this Appendix we discuss the eigenfunction expansion of the Smoluchowski equation (3.21). For the sake of simplicity, we represent the set of variables $\{x\}\equiv(x_1,\ldots,x_N)$ by x, and write the Smoluchowski equation as

$$\frac{\partial\Psi}{\partial t}=-\Gamma(x)\Psi(x,t) \tag{3.I.1}$$

where Γ is a linear differential operator:

$$\Gamma(x)\Psi=-\sum_{n,m}\frac{\partial}{\partial x_n}L_{nm}\left(k_BT\frac{\partial\Psi}{\partial x_m}+\Psi\frac{\partial U}{\partial x_m}\right). \tag{3.I.2}$$

The conjugate of the operator $\Gamma(x)$ is denoted by $\Gamma^+(x)$: for any $\Psi(x)$ and $\Phi(x)$

$$\int dx\Psi(x)(\Gamma(x)\Phi(x))=\int dx(\Gamma^+(x)\Psi(x))\Phi(x). \tag{3.I.3}$$

From eqns (3.I.2) and (3.I.3), $\Gamma^+(x)$ is obtained as

$$\Gamma^+(x)=-\sum_{n,m}\left(k_BT\frac{\partial}{\partial x_n}-\frac{\partial U}{\partial x_n}\right)L_{nm}\frac{\partial}{\partial x_m}. \tag{3.I.4}$$

Let $\Psi_p(x)$ be the right-hand eigenfunctions and $\psi_p(x)$ be the left-hand eigenfunctions:

$$\Gamma(x)\Psi_p(x)=\lambda_p\Psi_p(x) \tag{3.I.5}$$

and

$$\Gamma^+(x)\psi_p(x) = \lambda_p \psi_p(x). \tag{3.I.6}$$

The eigenfunctions are chosen to be orthonormal so that

$$\int dx \Psi_p(x)\psi_q(x) = \delta_{pq}. \tag{3.I.7}$$

It is easy to prove by direct substitution that the right-hand and left-hand eigenfunctions are related by

$$\Psi_p(x) = \Psi_{eq}(x)\psi_p(x) \tag{3.I.8}$$

Hence eqn (3.I.7) can be written as

$$\int dx \Psi_{eq}\psi_p\psi_q = \langle \psi_p\psi_q \rangle_{eq} = \delta_{pq} \tag{3.I.9}$$

where $\langle \ldots \rangle_{eq}$ denotes the equilibrium average.

The equilibrium distribution function Ψ_{eq} is an eigenfunction with eigenvalue 0, which will be denoted by the suffix $p = 0$, so that $\psi_0 = 1$. All the other eigenvalues are positive. To show this we multiply both sides of eqn (3.I.5) by $\psi_p(x)$ and integrate over x:

$$\lambda_p \int dx \Psi_{eq}\psi_p^2 = \int dx \psi_p \Gamma(\Psi_{eq}\psi_p). \tag{3.I.10}$$

Using eqn (3.I.2) and the integral by parts, the right-hand side is rewritten as

$$\text{rhs} = k_B T \int dx \sum_{n,m} L_{nm}\Psi_{eq} \frac{\partial \psi_p}{\partial x_n}\frac{\partial \psi_p}{\partial x_m} \tag{3.I.11}$$

which is positive due to eqn (3.18). Therefore all λ_p are positive except $\lambda_0 = 0$.

Now the distribution function $\Psi(x, t)$ can be expanded by the eigenfunctions as

$$\Psi(x, t) = \sum_p a_p(t)\psi_p(x)\Psi_{eq}(x) \tag{3.I.12}$$

where by eqn (3.I.9)

$$a_p(t) = \int dx \psi_p(x)\Psi(x, t). \tag{3.I.13}$$

Since $\psi_0 = 1$, and $\Psi(x, t)$ satisfies the normalization condition,

$$a_0 = 1. \tag{3.I.14}$$

From eqns (3.I.1), (3.I.5), and (3.I.12), $a_p(t)$ satisfies

$$\frac{\mathrm{d}}{\mathrm{d}t} a_p(t) = -\lambda_p a_p(t) \tag{3.I.15}$$

which gives

$$a_p(t) = a_p(0)\exp(-\lambda_p t). \tag{3.I.16}$$

Thus

$$\Psi(x, t) = \sum_p a_p(0)\exp(-\lambda_p t)\psi_p(x)\Psi_{\mathrm{eq}}(x) \tag{3.I.17}$$

$$= \Psi_{\mathrm{eq}}(x) + \underline{\sum_{p>0} a_p(0)\exp(-\lambda_p t)\psi_p(x)\Psi_{\mathrm{eq}}(x)}. \tag{3.I.18}$$

After a long time, the underlined terms become very small, and $\Psi(x, t)$ becomes the equilibrium distribution function.

As a special case of formula (3.I.18), the Green function $G(x, x'; t)$ which satisfies the initial condition (3.46) is obtained as

$$G(x, x'; t) = \sum_p \exp(-\lambda_p t)\psi_p(x)\psi_p(x')\Psi_{\mathrm{eq}}(x). \tag{3.I.19}$$

Appendix 3.II Relationship between the Langevin equation and the Smoluchowski equation

Here, we shall show that the probability distribution of the solution of the Langevin equation (3.27) satisfies the Smoluchowski equation (3.16). First we consider the case when ζ is independent of x. We write the Langevin equation (3.27) as

$$\frac{\mathrm{d}x}{\mathrm{d}t} = V(x) + \sigma g(t) \tag{3.II.1}$$

where

$$V(x) = -\frac{1}{\zeta}\frac{\partial U(x)}{\partial x}, \quad \sigma = \left(\frac{k_B T}{\zeta}\right)^{1/2} = D^{1/2}, \tag{3.II.2}$$

and $g(t)$ is a random variable satisfying

$$\langle g(t)\rangle = 0, \quad \langle g(t)g(t')\rangle = 2\delta(t - t'). \tag{3.II.3}$$

Suppose that the particle was at x at time t. The displacement ξ of the particle in a very short time Δt is easily obtained from eqn (3.II.1) since

$V(x)$ can be regarded as constant in a short time interval:

$$\xi = V(x)\Delta t + \sigma \int_t^{t+\Delta t} \mathrm{d}t_1 g(t_1). \tag{3.II.4}$$

Since ξ is a linear combination of the Gaussian random variable $g(t)$, its probability distribution $\phi(\xi, \Delta t; x)$ is also Gaussian, the moments of which are obtained from eqn (3.II.3) and (3.II.4):

$$\langle \xi \rangle = V(x)\Delta t, \tag{3.II.5}$$

$$\langle (\xi - \langle \xi \rangle)^2 \rangle = \sigma^2 \int_t^{t+\Delta t} \mathrm{d}t_1 \int_t^{t+\Delta t} \mathrm{d}t_2 \langle g(t_1)g(t_2) \rangle = 2\sigma^2 \Delta t = 2D\Delta t. \tag{3.II.6}$$

Hence

$$\phi(\xi, \Delta t; x) = (4\pi D\Delta t)^{-1/2} \exp\left[-\frac{(\xi - V(x)\Delta t)^2}{4D\Delta t}\right]. \tag{3.II.7}$$

If the probability that the particle is at x at time t is $\Psi(x, t)$, then the probability that it is at x at time $t + \Delta t$ is given by

$$\begin{aligned}\Psi(x, t + \Delta t) &= \int \mathrm{d}\xi \int \mathrm{d}x' \delta(x - x' - \xi)\phi(\xi, \Delta t; x')\Psi(x', t) \\ &= \int \mathrm{d}\xi (4\pi D\Delta t)^{-1/2} \exp\left[-\frac{(\xi - V(x-\xi)\Delta t)^2}{4D\Delta t}\right]\Psi(x - \xi, t).\end{aligned} \tag{3.II.8}$$

Since the integrand has a sharp peak at $\xi = 0$, the integral is evaluated by expanding $V(x - \xi)$ and $\Psi(x - \xi, t)$ with respect to ξ:

$$\Psi(x, t + \Delta t) = \int \mathrm{d}\xi (4\pi D\Delta t)^{-1/2} \exp\left[-\frac{\left(\left(1 + \frac{\mathrm{d}V}{\mathrm{d}x}\Delta t\right)\xi - V\Delta t\right)^2}{4D\Delta t}\right] \times \left(1 - \xi\frac{\partial}{\partial x} + \tfrac{1}{2}\xi^2 \frac{\partial^2}{\partial x^2}\right)\Psi(x, t).$$

Neglecting the terms of order Δt^2 and higher, we get

$$\Psi(x, t + \Delta t) = \left(1 - \frac{\mathrm{d}V}{\mathrm{d}x}\Delta t\right)\Psi - \Delta t V \frac{\partial \Psi}{\partial x} + D\Delta t \frac{\partial^2 \Psi}{\partial x^2}. \tag{3.II.9}$$

Collecting terms of order Δt, we get

$$\frac{\partial \Psi}{\partial t} = -\frac{\partial}{\partial x}(V(x)\Psi(x, t)) + D\frac{\partial^2 \Psi}{\partial x^2} = D\frac{\partial^2 \Psi}{\partial x^2} + \frac{1}{\zeta}\frac{\partial}{\partial x}\left(\frac{\partial U}{\partial x}\Psi\right). \tag{3.II.10}$$

This agrees with eqn (3.16)

Next we consider the case when ζ depends on x. The Langevin equation (3.38) gives

$$\frac{dx}{dt} = V(x) + \sigma(x)g(t) + \sigma(x)\frac{d\sigma}{dx}, \tag{3.II.11}$$

where $\sigma(x)^2 = k_B T/\zeta(x)$. Let $X(t)$ be defined by

$$X(t) = \int^{x(t)} dx' \frac{1}{\sigma(x')} \tag{3.II.12}$$

The Langevin equation for X is then obtained from eqn (3.II.11)

$$\frac{dX}{dt} = \tilde{V}(X) + g(t) \tag{3.II.13}$$

with

$$\tilde{V}(X) = \frac{V}{\sigma} + \frac{d\sigma}{dx}. \tag{3.II.14}$$

Equation (3.II.13) is the Langevin equation studied earlier. Thus, the probability distribution function for X satisfies

$$\frac{\partial \tilde{\Psi}}{\partial t} = \frac{\partial}{\partial X}\left(\frac{\partial \tilde{\Psi}}{\partial X} - \tilde{V}\tilde{\Psi}\right). \tag{3.II.15}$$

From eqn (3.II.12), it follows that

$$\tilde{\Psi}(X, t) = \sigma(x)\Psi(x, t) \tag{3.II.16}$$

and

$$\frac{\partial}{\partial X} = \sigma\frac{\partial}{\partial x}. \tag{3.II.17}$$

From eqns (3.II.15)–(3.II.17), we can show that Ψ satisfies

$$\frac{\partial \Psi}{\partial t} = \frac{\partial}{\partial x}\left(\sigma^2\frac{\partial \Psi}{\partial x} - V\Psi\right) = \frac{\partial}{\partial x}\frac{1}{\zeta}\left(k_B T\frac{\partial \Psi}{\partial x} + \Psi\frac{\partial U}{\partial x}\right) \tag{3.II.18}$$

which is the Smoluchowski equation.

Appendix 3.III The Oseen tensor

Defining the Fourier transform as

$$\boldsymbol{v}_{\boldsymbol{k}} = \frac{1}{V}\int \mathrm{d}\boldsymbol{r}\, \boldsymbol{v}(\boldsymbol{r})\mathrm{e}^{\mathrm{i}\boldsymbol{k}\cdot\boldsymbol{r}}\ldots \tag{3.III.1}$$

we can rewrite eqns (3.102) and (3.99) as

$$-\eta_s \boldsymbol{k}^2 \boldsymbol{v}_{\boldsymbol{k}} - \mathrm{i}\boldsymbol{k}P_{\boldsymbol{k}} = -\boldsymbol{g}_{\boldsymbol{k}}, \qquad \boldsymbol{k}\cdot\boldsymbol{v}_{\boldsymbol{k}} = 0, \tag{3.III.2}$$

which gives

$$\boldsymbol{v}_{\boldsymbol{k}} = \frac{1}{\eta_s \boldsymbol{k}^2}(\boldsymbol{I} - \hat{\boldsymbol{k}}\hat{\boldsymbol{k}})\cdot \boldsymbol{g}_{\boldsymbol{k}} \tag{3.III.3}$$

where $\hat{\boldsymbol{k}}$ indicates a unit vector in the direction of $\boldsymbol{k}$. Hence

$$\boldsymbol{v}(\boldsymbol{r}) = \int \mathrm{d}\boldsymbol{r}'\, \boldsymbol{H}(\boldsymbol{r}-\boldsymbol{r}')\cdot \boldsymbol{g}(\boldsymbol{r}') \tag{3.III.4}$$

where

$$\boldsymbol{H}(\boldsymbol{r}) = \frac{1}{(2\pi)^3}\int \mathrm{d}\boldsymbol{k}\,\frac{1}{\eta_s \boldsymbol{k}^2}(\boldsymbol{I} - \hat{\boldsymbol{k}}\hat{\boldsymbol{k}})\exp(-\mathrm{i}\boldsymbol{k}\cdot\boldsymbol{r}). \tag{3.III.5}$$

Since the tensor $\mathbf{H}(\boldsymbol{r})$ depends on the vector $\boldsymbol{r}$ only, it can be written in terms of the scalars A and B and the unit vector $\hat{\boldsymbol{r}}$ parallel to $\boldsymbol{r}$, as

$$H_{\alpha\beta}(\boldsymbol{r}) = A\delta_{\alpha\beta} + B\hat{r}_\alpha \hat{r}_\beta. \tag{3.III.6}$$

The scalars A and B are determined from the two equations

$$H_{\alpha\alpha} = 3A + B, \qquad H_{\alpha\beta}\hat{r}_\alpha\hat{r}_\beta = A + B, \tag{3.III.7}$$

i.e.,

$$3A + B = \frac{1}{(2\pi)^3}\int \mathrm{d}\boldsymbol{k}\,\frac{2}{\eta_s \boldsymbol{k}^2}\exp(-\mathrm{i}\boldsymbol{k}\cdot\boldsymbol{r}) \tag{3.III.8}$$

and

$$A + B = \frac{1}{(2\pi)^3}\int \mathrm{d}\boldsymbol{k}\,\frac{1-(\hat{\boldsymbol{k}}\cdot\hat{\boldsymbol{r}})^2}{\eta_s \boldsymbol{k}^2}\exp(-\mathrm{i}\boldsymbol{k}\cdot\boldsymbol{r}). \tag{3.III.9}$$

The integrals are easily evaluated by introducing the coordinates $t = \hat{\boldsymbol{k}}\cdot\hat{\boldsymbol{r}}$ and $\xi = |\boldsymbol{k}|\,|\boldsymbol{r}|$ to give

$$\begin{aligned} 3A + B &= \frac{2}{(2\pi)^3}\int_0^\infty \frac{\mathrm{d}\xi}{\eta_s r}2\pi\int_{-1}^{1}\mathrm{d}t\,\exp(-\mathrm{i}\xi t) = \frac{1}{\pi^2\eta_s r}\int_0^\infty \mathrm{d}\xi\,\frac{\sin\xi}{\xi} \\ &= \frac{1}{2\pi\eta_s r} \end{aligned} \tag{3.III.10}$$

and

$$A + B = \frac{1}{(2\pi)^3} \int_0^\infty \frac{d\xi}{\eta_s r} 2\pi \int_{-1}^{1} dt(1 - t^2)\exp(-i\xi t)$$

$$= \frac{1}{2\pi^2 \eta_s r} \int_0^\infty d\xi \left(1 + \frac{\partial^2}{\partial \xi^2}\right) \frac{\sin \xi}{\xi} = \frac{1}{4\pi \eta_s r}. \qquad (3.\text{III}.11)$$

From eqns (3.III.10) and (3.III.11), A and B are obtained as

$$A = B = \frac{1}{8\pi \eta_s r}. \qquad (3.\text{III}.12)$$

Hence

$$\boldsymbol{H}(\boldsymbol{r}) = \frac{1}{8\pi \eta_s r} (\boldsymbol{I} + \hat{\boldsymbol{r}}\hat{\boldsymbol{r}}). \qquad (3.\text{III}.13)$$

References

1. Wax, N., *Noise and Stochastic Processes.* Dover Publishing Co., New York (1954) represents a collection of classic papers.
2. Lax, M., *Rev. Mod. Phys.* **32,** 25 (1960); **38,** 541 (1966).
3. Kubo, R., *Rep. Prog. Phys.* **29,** 255 (1966).
4. Einstein, A., *Ann. Physik.* **17,** 549 (1905); **19,** 371 (1906). See also *Investigation on the Theory of the Brownian Movement.* E. P. Dutton and Copy Inc, New York (1926).
5. Batchelor, G. K., *An Introduction to Fluid Dynamics,* Chap. 4. Cambridge Univ. Press (1970).
6. Doi, M., *J. Chem. Phys.* **79,** 5080 (1983).
7. van Kampen, N. G. *Stochastic Processes in Physics and Chemistry.* North-Holland, Amsterdam (1981).
8. Risken, H., *The Fokker–Planck Equation.* Springer, Berlin (1984).
9. Zwanzig, R., *Adv. Chem. Phys.* **15,** 325 (1969).
10. Fixman, M., *J. Chem. Phys.* **69,** 1527 (1978).
11. Rainville, E. D., *Special Functions.* Macmillan, New York (1960).
12. Recent researches on the dynamics of interacting colloidal particles can be seen in *Faraday Discuss. Chem. Soc.* **76** (1983).
13. Happel, J., and Brenner, H., *Low Reynolds Number Hydrodynamics.* Prentice Hall, Englewood Cliffs, N. J. (1965).
14. Batchelor, G. K., *J. Fluid. Mech.* **74,** 1 (1976).
15. Felderhof, B. U., *Physica* **89A,** 373 (1977); *J. Phys.* **A11,** 929 (1978).
16. Mazur, P., and van Saarloos, W., *Physica* **110A,** 147 (1982).
17. For the rheology of suspensions see Maron, S. H., and Krieger, I. M., in *Rheology,* Vol 3. (ed. F. R. Eirich) Academic Press, New York (1960); and Bauer, W. H., and Collins, E. A., in *Rheology,* Vol 4. (ed. F. R. Eirich) Academic Press; New York (1967). A recent review on thixotropy is given by Mewis, J., *J. Non-Newtonian Fluid Mech.* **6** 1 (1979).

18. For the rheology of granular materials, see Shahinpoor, M., (ed.) *Advances in the Mechanics and the Flow of Granular Materials,* vol. I and II. Trans. Tech. Publications, Clausthal-Zellerfeld, Germany (1983).
19. For the rheology of polymeric liquids, see Bird, R. B., Armstrong, R. C., and Hassager, O., *Dynamics of Polymeric Liquids,* vol. 1. John Wiley, New York (1977).
20. Kirkwood, J. G., *Rec. Trav. Chim.* **68,** 649 (1949); see also Kirkwood, J. G., Macromolecules, in *Documents on Modern Physics.* Gordon & Breach, New York (1967).
21. Burgers, J. M., *Second Report on Viscosity and Plasticity,* Chap. 3. Amsterdam Academy of Sciences, North-Holland, Nordermann (1938),
22. Kuhn, W., and Kuhn, H., *Helv. Chim. Acta* **37,** 97 (1944); **38,** 1533 (1945); **39,** 71 (1946).
23. Kramers, H. A., *J. Chem. Phys.* **14,** 415 (1946).
24. A collection of classic papers is given in Hermans, J. J., (ed.) *Polymer Solutions Properties II, Hydrodynamics and Light Scattering.* Dowden Hutchinson & Ross Inc., Stroudsburg, Pa. (1978). Also a history of the development of the theory of suspensions is described in Frisch, H. L., and Simha, R., The viscosity of colloidal suspensions and macromolecular solutions, in *Rheology* Vol 1 (ed. F. R. Eirich). Academic Press, New York (1956).
25. Batchelor, G. K., *J. Fluid Mech.* **41,** 545 (1970).
26. Peterson, J. M., and Fixman, M., *J. Chem. Phys.* **39,** 2516 (1963).
27. See for example Meyer, R. E., (ed.) *Theory of Dispersed Multiphase Flow.* Academic Press, New York, (1983).
28. See for example, Goldstein, H., *Classical Mechanics.* Addison-Wesley, Reading, Mass (1959).
29. Erpenbeck, J. J., and Kirkwood, J. G., *J. Chem. Phys.* **29,** 909 (1958); **38,** 1023 (1963).
30. Fixman, M., and Kovac, J., *J. Chem. Phys.,* **61,** 4939 (1974); ibid. **61,** 4950 (1974); ibid. **63,** 935 (1975).
31. Hassager, O., *J. Chem. Phys.* **60,** 2111, 4001 (1974).
32. Nakajima, H., and Wada, Y., *Biopolymers* **16,** 875 (1977); **17,** 2291 (1978).
33. Doi, M., Nakajima, H., and Wada, Y., *Colloid Polym. Sci.* **253,** 905 (1975); **254,** 559 (1976).
34. Iwata, K., *J. Chem. Phys.* **71,** 931 (1979).
35. A recent review is given by Yamakawa, H., *Ann. Rev. Phys. Chem.* **35,** 23 (1984).
36. See for example, ref. 19 Appendix A.

4

DYNAMICS OF FLEXIBLE POLYMERS IN DILUTE SOLUTION

4.1 The Rouse model

4.1.1. Dynamics of a polymer with localized interaction

Having seen the general background of Brownian motion, we shall now discuss the dynamics of a polymer in solution. As we have seen in Chapter 2 the static properties of a polymer can be represented by a set of beads connected along a chain. It is natural to model the dynamics of the polymer by the Brownian motion of such beads. Such a model was first proposed by Rouse[1] and has been the basis of the dynamics of dilute polymer solutions.

Let $(\boldsymbol{R}_1, \boldsymbol{R}_2, \ldots, \boldsymbol{R}_N) \equiv \{\boldsymbol{R}_n\}$ be the positions of the beads (see Fig. 4.1*a*). The equation of motion of the beads is described by either the Smoluchowski equation (see eqn (3.21) or (3.110)).

$$\frac{\partial \Psi}{\partial t} = \sum_n \frac{\partial}{\partial \boldsymbol{R}_n} \cdot \boldsymbol{H}_{nm} \cdot \left[k_B T \frac{\partial \Psi}{\partial \boldsymbol{R}_m} + \frac{\partial U}{\partial \boldsymbol{R}_m} \Psi \right] \tag{4.1}$$

or the Langevin equation (see eqn (3.39))

$$\frac{\partial}{\partial t} \boldsymbol{R}_n(t) = \sum_m \boldsymbol{H}_{nm} \cdot \left(- \frac{\partial U}{\partial \boldsymbol{R}_m} + \boldsymbol{f}_m(t) \right) + \tfrac{1}{2} k_B T \sum_m \frac{\partial}{\partial \boldsymbol{R}_m} \cdot \boldsymbol{H}_{nm}. \tag{4.2}$$

In the Rouse model, the excluded volume interaction and the hydrodynamic interaction are disregarded and the mobility tensor and the interaction potential are written as

$$\boldsymbol{H}_{nm} = \frac{\boldsymbol{I}}{\zeta} \delta_{nm} \tag{4.3}$$

and

$$U = \frac{k}{2} \sum_{n=2}^{N} (\boldsymbol{R}_n - \boldsymbol{R}_{n-1})^2 \tag{4.4}$$

with

$$k = \frac{3 k_B T}{b^2}. \tag{4.5}$$

In this model the Langevin equation (4.2) becomes a linear equation for $\boldsymbol{R}_n$. For internal beads ($n = 2, 3, \ldots, N-1$),

$$\zeta \frac{\mathrm{d}\boldsymbol{R}_n}{\mathrm{d}t} = -k(2\boldsymbol{R}_n - \boldsymbol{R}_{n+1} - \boldsymbol{R}_{n-1}) + \boldsymbol{f}_n. \tag{4.6}$$

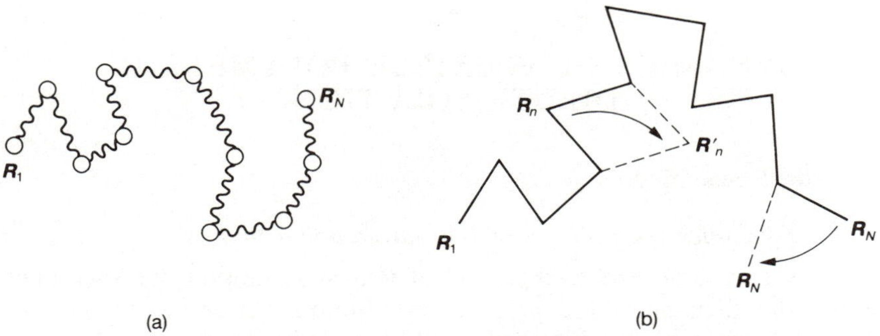

Fig. 4.1. (*a*) Rouse model and (*b*) local jump model.

For the end beads ($n = 1$ and N)

$$\zeta \frac{\mathrm{d}\boldsymbol{R}_1}{\mathrm{d}t} = -k(\boldsymbol{R}_1 - \boldsymbol{R}_2) + \boldsymbol{f}_1, \qquad \zeta \frac{\mathrm{d}\boldsymbol{R}_N}{\mathrm{d}t} = -k(\boldsymbol{R}_N - \boldsymbol{R}_{N-1}) + \boldsymbol{f}_N. \quad (4.7)$$

The distribution of the random force $\boldsymbol{f}_n$ is Gaussian, characterized by the moments given by eqns (3.40) and (3.41):

$$\langle \boldsymbol{f}_n(t) \rangle = 0,$$
$$\langle f_{n\alpha}(t) f_{m\beta}(t') \rangle = 2\zeta k_B T \delta_{nm} \delta_{\alpha\beta} \delta(t - t'). \quad (4.8)$$

As in the case of the Gaussian chain, the suffix n in the Rouse model can be regarded as a continuous variable. In the continuous limit, eqn (4.6) is rewritten as (see the transformation rule given in Table 2.1, Section 2.2)

$$\zeta \frac{\partial \boldsymbol{R}_n}{\partial t} = k \frac{\partial^2 \boldsymbol{R}_n}{\partial n^2} + \boldsymbol{f}_n. \quad (4.9)$$

To rewrite eqn (4.7) in the continuous limit, we note that eqn (4.7) is included in the general equation (4.6) if the hypothetical beads $\boldsymbol{R}_0$ and $\boldsymbol{R}_{N+1}$ are defined as

$$\boldsymbol{R}_0 = \boldsymbol{R}_1, \qquad \boldsymbol{R}_{N+1} = \boldsymbol{R}_N \quad (4.10)$$

which become, in the continuous limit,

$$\left.\frac{\partial \boldsymbol{R}_n}{\partial n}\right|_{n=0} = 0, \qquad \left.\frac{\partial \boldsymbol{R}_n}{\partial n}\right|_{n=N} = 0. \quad (4.11)$$

Also, the moments of the random forces are now given as

$$\langle \boldsymbol{f}_n(t) \rangle = 0,$$
$$\langle f_{n\alpha}(t) f_{m\beta}(t') \rangle = 2\zeta k_B T \delta(n - m) \delta_{\alpha\beta} \delta(t - t'). \quad (4.12)$$

Equations (4.9), (4.11), and (4.12) define the continuous Rouse model.

The results of the discrete model and the continuous model agree with each other for properties on a long time-scale, but do not for short times. The discrepancy, however, has no serious physical significance since the description of the polymer by discretized beads is an artefact, and the results which depend on the discrete nature of the beads have no validity.

It should be emphasized that the essence of the Rouse model is in the universal nature of the modelling of the dynamics of a connected object. The central assumption in the Rouse model is that the dynamics is governed by the interactions localized along the chain. In fact, if one assumes a linear Langevin equation for $\boldsymbol{R}_n$ with localized interaction, one ends up with the Rouse model in the long time-scale behaviour. To see this, consider the general form of the linearized Langevin equation

$$\frac{\mathrm{d}\boldsymbol{R}_n}{\mathrm{d}t} = \sum_m A_{nm}\boldsymbol{R}_m + \boldsymbol{g}_n \tag{4.13}$$

where A_{nm} is a constant matrix representing the interaction among the beads, and $\boldsymbol{g}_n$ is a random force. Since the system is homogeneous, A_{nm} depends only on $n-m$, so that eqn (4.13) can be written as

$$\frac{\mathrm{d}\boldsymbol{R}_n}{\mathrm{d}t} = \sum_m A_m\boldsymbol{R}_{n+m} + \boldsymbol{g}_n \tag{4.14}$$

with $A_m \equiv A_{n,n+m}$. In the long time-scale motion, $\boldsymbol{R}_n$ varies slowly with n, which allows $\boldsymbol{R}_{n+m}$ to be expanded with respect to m, giving

$$\begin{aligned}\sum_m A_m\boldsymbol{R}_{n+m} &= \sum_m A_m\left(\boldsymbol{R}_n + m\frac{\partial}{\partial n}\boldsymbol{R}_n + \tfrac{1}{2}m^2\frac{\partial^2}{\partial n^2}\boldsymbol{R}_n + \dots\right)\\ &= a_0\boldsymbol{R}_n + a_1\frac{\partial}{\partial n}\boldsymbol{R}_n + a_2\frac{\partial^2}{\partial n^2}\boldsymbol{R}_n + \dots\end{aligned} \tag{4.15}$$

where

$$a_0 = \sum_{m=-\infty}^{\infty} A_m, \qquad a_1 = \sum_{m=-\infty}^{\infty} mA_m, \qquad a_2 = \frac{1}{2}\sum_{m=-\infty}^{\infty} m^2A_m. \tag{4.16}$$

The assumption of local interaction guarantees that these sums converge to finite values. The coefficient a_0 must vanish since the equation must be invariant under a spatial translation ($\boldsymbol{R}_n \to \boldsymbol{R}_n + \boldsymbol{r}$), and a_1 vanishes since A_m is an even function of m (because the polymer cannot distinguish head and tail). Therefore, the asymptotic behaviour of eqn (4.14) is given by

$$\frac{\partial}{\partial t}\boldsymbol{R}_n = a_2\frac{\partial^2}{\partial n^2}\boldsymbol{R}_n + \boldsymbol{g}_n(t), \tag{4.17}$$

which is equivalent to eqn (4.9).

The Rouse model displays the general features of any model which assumes local interactions. One can conceive of other dynamical models.[2–5] For example, one can start from the freely jointed chain, and simulate its dynamics by allowing the local jump process depicted in Fig. 4.1*b*. This model can be shown to give the same results as the Rouse model for slow modes (see Appendix 4.I). It is generally believed[6] that the Rouse model represents the long time-scale behaviour of the 'local jump' model in the same way as the Gaussian chain represents the large length-scale properties of a polymer which has only short range interaction.†

4.1.2 Normal coordinates

Let us now study the consequence of the Rouse model. Equation (4.9) represents a Brownian motion of coupled oscillators. A standard way of treating such a system is to find the normal coordinates, each capable of independent motion. It is shown in Appendix 4.II, that in terms of the coordinates $\boldsymbol{X}_p$ defined by

$$\boldsymbol{X}_p \equiv \frac{1}{N}\int_0^N \mathrm{d}n \cos\left(\frac{p\pi n}{N}\right)\boldsymbol{R}_n(t) \quad \text{with} \quad p = 0, 1, 2, \ldots, \tag{4.18}$$

eqn (4.9) can be rewritten as

$$\zeta_p \frac{\partial}{\partial t}\boldsymbol{X}_p = -k_p\boldsymbol{X}_p + \boldsymbol{f}_p \tag{4.19}$$

where

$$\zeta_0 = N\zeta \quad \text{and} \quad \zeta_p = 2N\zeta \quad \text{for} \quad p = 1, 2, \ldots \tag{4.20}$$

$$k_p = 2\pi^2 k p^2/N = \frac{6\pi^2 k_B T}{Nb^2} p^2 \quad \text{for} \quad p = 0, 1, 2, \ldots \tag{4.21}$$

and the $\boldsymbol{f}_p$'s are the random forces which satisfy

$$\langle f_{p\alpha}\rangle = 0, \qquad \langle f_{p\alpha}(t) f_{q\beta}(t')\rangle = 2\delta_{pq}\delta_{\alpha\beta}\zeta_p k_B T \delta(t - t'). \tag{4.22}$$

Since the random forces are independent of each other, the motions of the $\boldsymbol{X}_p$'s are also independent of each other. Thus the motion of the polymer is decomposed into independent modes.

The time correlation functions of the normal coordinates can be calculated immediately from eqn (4.19) by using the results of Section

† However, this statement is a conjecture. Usually the dynamics of the local jump model becomes a nonlinear equation for $\boldsymbol{R}_n$ so that the proof given above does not apply to the general local jump model. In fact a counter example was given by Hilhorst and Deutch.[7]

3.5. For $p > 0$

$$\langle X_{p\alpha}(t) X_{q\beta}(0) \rangle = \delta_{pq}\delta_{\alpha\beta} \frac{k_B T}{k_p} \exp(-t/\tau_p) \tag{4.23}$$

where

$$\tau_p = \tau_1 / p^2 \tag{4.24}$$

and

$$\tau_1 = \frac{\zeta_1}{k_1} = \frac{\zeta N^2 b^2}{3\pi^2 k_B T}. \tag{4.25}$$

On the other hand for $p = 0$

$$\langle (X_{0\alpha}(t) - X_{0\alpha}(0))(X_{0\beta}(t) - X_{0\beta}(0)) \rangle = \delta_{\alpha\beta} \frac{2k_B T}{\zeta_0} t = \delta_{\alpha\beta} \frac{2k_B T}{N\zeta} t. \tag{4.26}$$

The inverse transform of eqn (4.18) is

$$\boldsymbol{R}_n = \boldsymbol{X}_0 + 2 \sum_{p=1}^{\infty} \boldsymbol{X}_p \cos\left(\frac{p\pi n}{N}\right). \tag{4.27}$$

Let us now consider the physical significance of the normal coordinates. The coordinate $\boldsymbol{X}_0$ represents the position of the centre of mass

$$\boldsymbol{R}_G \equiv \frac{1}{N} \int_0^N \mathrm{d}n \boldsymbol{R}_n = \boldsymbol{X}_0. \tag{4.28}$$

Thus the mean square displacement of $\boldsymbol{R}_G$ is calculated from eqn (4.26) as

$$\langle (\boldsymbol{R}_G(t) - \boldsymbol{R}_G(0))^2 \rangle = \sum_{\alpha = x,y,z} \langle (X_{0\alpha}(t) - X_{0\alpha}(0))^2 \rangle = 6 \frac{k_B T}{N\zeta} t. \tag{4.29}$$

The self diffusion constant of the centre of mass is defined by

$$D_G = \lim_{t \to \infty} \frac{1}{6t} \langle (\boldsymbol{R}_G(t) - \boldsymbol{R}_G(0))^2 \rangle. \tag{4.30}$$

From eqns (4.29) and (4.30)

$$D_G = \frac{k_B T}{N\zeta}. \tag{4.31}$$

The normal coordinate $\boldsymbol{X}_p$ with $p > 0$ represents the internal conformation of the polymer. Consider for example the end-to-end vector

$$\boldsymbol{P}(t) \equiv \boldsymbol{R}_N(t) - \boldsymbol{R}_0(t) \tag{4.32}$$

which is expressed by $\boldsymbol{X}_p$ as

$$\boldsymbol{P}(t) = -4 \sum_{p:\text{odd integer}} \boldsymbol{X}_p(t). \tag{4.33}$$

The time correlation function $\langle \boldsymbol{P}(t) \cdot \boldsymbol{P}(0) \rangle$ is calculated from eqns (4.23) and (4.33) as

$$\langle \boldsymbol{P}(t) \cdot \boldsymbol{P}(0) \rangle = 16 \sum_{p:\text{odd}} \langle \boldsymbol{X}_p(t) \cdot \boldsymbol{X}_p(0) \rangle \tag{4.34}$$

$$= 16 \sum_{p:\text{odd}} \frac{3k_BT}{k_p} \exp(-t/\tau_p)$$

$$= Nb^2 \sum_{p=1,3,\ldots} \frac{8}{p^2\pi^2} \exp(-tp^2/\tau_1). \tag{4.35}$$

Equation (4.35) indicates that the motion of the end-to-end vector is mainly governed by the first mode $\boldsymbol{X}_1$. In general, $\boldsymbol{X}_p$ represents the local motion of the chain which includes N/p segments and corresponds to the motion with the length-scale of the order $(Nb^2/p)^{1/2}$.

The rotational relaxation time τ_r of a polymer can be *defined* by the longest relaxation time of the correlation function $\langle \boldsymbol{P}(t) \cdot \boldsymbol{P}(0) \rangle$:†

$$\langle \boldsymbol{P}(t) \cdot \boldsymbol{P}(0) \rangle \propto \exp(-t/\tau_r) \quad \text{for} \quad t \gtrsim \tau_r. \tag{4.36}$$

From eqn (4.35) we see that

$$\tau_r = \tau_1 = \frac{\zeta N^2 b^2}{3\pi^2 k_B T}. \tag{4.37}$$

Since N is proportional to the molecular weight M, eqns (4.31) and (4.37) indicate that D_G and τ_r depend on the molecular weight M as

$$D_G \propto M^{-1}, \qquad \tau_r \propto M^2. \tag{4.38}$$

This prediction is not consistent with experimental results, which, in Θ conditions, are summarized as

$$D_G \propto M^{-1/2}, \qquad \tau_r \propto M^{3/2}. \tag{4.39}$$

This failure comes from the neglect of the hydrodynamic interaction, which will be discussed in the next section. Because of this failure the Rouse model is now regarded as inappropriate as a model in dilute solution. However, the model is conceptually quite important, and also it has turned out that it is a useful model in the dynamics of polymers in melts, which will be discussed in later chapters.

† It has been shown[8,9] that the most probable shape of the Gaussian chain is not spherical but an ellipsoid, the long axis being, on average, parallel to the end-to-end vector. Thus it is possible to talk about 'rotational motion' even for flexible polymers.

4.2 The Zimm model

4.2.1 Zimm model in Θ conditions

To describe the dynamics of polymers in dilute solution, we have to take into account the hydrodynamic interaction, which is expressed by the mobility matrix calculated in Chapter 3,

$$\boldsymbol{H}_{nn} = \boldsymbol{I}/\zeta$$

$$\boldsymbol{H}_{nm} = \frac{1}{8\pi\eta_s |\boldsymbol{r}_{nm}|}[\hat{\boldsymbol{r}}_{nm}\hat{\boldsymbol{r}}_{nm} + \boldsymbol{I}] \quad \text{for} \quad n \neq m \tag{4.40}$$

where $\boldsymbol{r}_{nm} \equiv \boldsymbol{R}_n - \boldsymbol{R}_m$ and $\hat{\boldsymbol{r}}_{nm}$ is the unit vector in the direction of $\boldsymbol{r}_{nm}$. For the tensor eqn (4.40), it can be shown that

$$\frac{\partial}{\partial \boldsymbol{R}_m} \cdot \boldsymbol{H}_{nm} = 0. \tag{4.41}$$

Thus the Langevin equation (4.2) becomes

$$\frac{\partial}{\partial t}\boldsymbol{R}_n = \sum_m \boldsymbol{H}_{nm} \cdot \left(-\frac{\partial U}{\partial \boldsymbol{R}_m} + \boldsymbol{f}_m(t) \right). \tag{4.42}$$

In particular, for the Θ condition, eqns (4.4) and (4.42) give (in the continuous limit)

$$\frac{\partial}{\partial t}\boldsymbol{R}_n = \sum_m \boldsymbol{H}_{nm} \cdot \left(k \frac{\partial^2}{\partial m^2}\boldsymbol{R}_m + \boldsymbol{f}_m(t) \right). \tag{4.43}$$

This model was first presented by Zimm.[10]

Since $\boldsymbol{H}_{nm}$ is a nonlinear function of $\boldsymbol{R}_n - \boldsymbol{R}_m$, eqn (4.43) is quite difficult to handle. To simplify the analysis, Zimm[10] introduced the preaveraging approximation, which replaces $\boldsymbol{H}_{nm}$ by its average,

$$\boldsymbol{H}_{nm} \to \langle \boldsymbol{H}_{nm} \rangle = \int \mathrm{d}\{\boldsymbol{R}_n\} \boldsymbol{H}_{nm} \Psi(\{\boldsymbol{R}_n\}, t). \tag{4.44}$$

If we are considering problems near equilibrium, which is the case in the subsequent part of this chapter, we may use the equilibrium distribution function $\Psi_{\mathrm{eq}}(\{\boldsymbol{R}_n\})$ in the average of eqn (4.44),

$$\boldsymbol{H}_{nm} \to \langle \boldsymbol{H}_{nm} \rangle_{\mathrm{eq}} \equiv \int \mathrm{d}\{\boldsymbol{R}_n\} \boldsymbol{H}_{nm} \Psi_{\mathrm{eq}}(\{\boldsymbol{R}_n\}). \tag{4.45}$$

Since the distribution of $\hat{\boldsymbol{r}}_{nm}$ is independent of $|\boldsymbol{r}_{nm}|$, $\langle \boldsymbol{H}_{nm} \rangle_{\mathrm{eq}}$ is written as

$$\langle \boldsymbol{H}_{nm} \rangle_{\mathrm{eq}} = \frac{1}{8\pi\eta_s} \left\langle \frac{1}{|\boldsymbol{r}_{nm}|} \right\rangle_{\mathrm{eq}} \langle \hat{\boldsymbol{r}}_{nm}\hat{\boldsymbol{r}}_{nm} + \boldsymbol{I} \rangle_{\mathrm{eq}}. \tag{4.46}$$

Using

$$\langle \hat{\boldsymbol{r}}_{nm}\hat{\boldsymbol{r}}_{nm}\rangle_{\text{eq}} = \frac{\boldsymbol{I}}{3} \tag{4.47}$$

we have

$$\langle \boldsymbol{H}_{nm}\rangle_{\text{eq}} = \frac{\boldsymbol{I}}{6\pi\eta_s}\left\langle\frac{1}{|\boldsymbol{R}_n - \boldsymbol{R}_m|}\right\rangle_{\text{eq}}. \tag{4.48}$$

In the Θ condition, the distribution of $\boldsymbol{R}_n - \boldsymbol{R}_m$ is Gaussian with the variance $|n-m|\,b^2$; hence

$$\begin{aligned}\langle \boldsymbol{H}_{nm}\rangle_{\text{eq}} &= \int_0^\infty \mathrm{d}r 4\pi r^2\left(\frac{3}{2\pi\,|n-m|\,b^2}\right)^{3/2}\exp\left(-\frac{3r^2}{2\,|n-m|\,b^2}\right)\frac{\boldsymbol{I}}{6\pi\eta_s r}\\ &= \frac{\boldsymbol{I}}{(6\pi^3\,|n-m|)^{1/2}\eta_s b} \equiv h(n-m)\boldsymbol{I}.\end{aligned} \tag{4.49}$$

Thus in the preaveraging approximation, eqn (4.43) becomes a linear equation for $\boldsymbol{R}_n$,

$$\frac{\partial}{\partial t}\boldsymbol{R}_n(t) = \sum_m h(n-m)\left(k\frac{\partial^2}{\partial m^2}\boldsymbol{R}_m(t) + \boldsymbol{f}_m(t)\right). \tag{4.50}$$

At first sight this approximation may appear quite crude. However, it has been shown that the results of this approximation are not very different from those of more sophisticated calculations which will be described later.

Note that since $h(n-m)$ decreases quite slowly ($h(n-m) \propto (|n-m|)^{-1/2}$), the moment a_2 of the interaction matrix A_{nm} in eqn (4.16) does not converge. Thus in the Zimm model the interaction among the segments is not localized. This gives the qualitative difference between the Rouse model and the Zimm model.

To analyse eqn (4.50) we rewrite it in terms of the Rouse normal coordinates $\boldsymbol{X}_p$ defined by eqn (4.18):

$$\frac{\partial}{\partial t}\boldsymbol{X}_p(t) = \sum_q h_{pq}(-k_q\boldsymbol{X}_q + \boldsymbol{f}_q) \tag{4.51}$$

where k_p is defined by eqn (4.21) and

$$h_{pq} = \frac{1}{N^2}\int_0^N \mathrm{d}n\int_0^N \mathrm{d}m\,\cos\left(\frac{p\pi n}{N}\right)\cos\left(\frac{q\pi m}{N}\right)h(n-m). \tag{4.52}$$

From eqns (4.49) and (4.52) h_{pq} is calculated as

$$h_{pq} = \frac{1}{N^2}\int_0^N dn \int_{-n}^{N-n} dm h(m)\cos\left(\frac{p\pi n}{N}\right)\cos\left(\frac{q\pi(n+m)}{N}\right) \tag{4.53}$$

$$= \frac{1}{N^2}\int_0^N dn\left[\cos\left(\frac{p\pi n}{N}\right)\cos\left(\frac{q\pi n}{N}\right)\underline{\int_{-n}^{N-n} dm h(m)\cos\left(\frac{q\pi m}{N}\right)}\right.$$

$$\left. - \cos\left(\frac{p\pi n}{N}\right)\sin\left(\frac{q\pi n}{N}\right)\underline{\int_{-n}^{N-n} dm h(m)\sin\left(\frac{q\pi m}{N}\right)}\right]. \tag{4.54}$$

The underlined integrals converge quickly to the following values if q is large:

$$\int_{-\infty}^{\infty} dm h(m)\cos\left(\frac{q\pi m}{N}\right) = \frac{\sqrt{N}}{(3\pi^3 q)^{1/2}\eta_s b} \tag{4.55}$$

and

$$\int_{-\infty}^{\infty} dm h(m)\sin\left(\frac{q\pi m}{N}\right) = 0.$$

If we replace the integrals by these asymptotic values, we obtain

$$h_{pq} \simeq \frac{\sqrt{N}}{(3\pi^3 q)^{1/2}\eta_s b}\frac{1}{N^2}\int_0^N dn \cos\left(\frac{p\pi n}{N}\right)\cos\left(\frac{q\pi n}{N}\right)$$

$$= \frac{\sqrt{N}}{(3\pi^3 p)^{1/2}\eta_s b}\frac{1}{2N}\delta_{pq}. \tag{4.56}$$

This equation shows that h_{pq} is nearly diagonal. Thus if we neglect the off diagonal component of h_{pq}, we have an equation which has the same structure as that of the Rouse model:[11]

$$\zeta_p \frac{\partial}{\partial t}\boldsymbol{X}_p(t) = -k_p \boldsymbol{X}_p + \boldsymbol{f}_p(t) \tag{4.57}$$

where

$$\zeta_p = (h_{pp})^{-1} = (12\pi^3)^{1/2}\eta_s(Nb^2 p)^{1/2} \quad (p = 1, 2, \ldots) \tag{4.58}$$

and

$$k_p = \frac{6\pi^2 k_B T}{Nb^2}p^2 \quad (p = 0, 1, 2, \ldots). \tag{4.59}$$

Equation (4.58) is not correct for $p = 0$, but ζ_0 is immediately calculated from eqn (4.52) as

$$\zeta_0 = (h_{00})^{-1} = \left[\frac{1}{N^2}\int_0^N \mathrm{d}n \int_0^N \mathrm{d}m h(|n-m|)\right]^{-1}$$
$$= \tfrac{3}{8}(6\pi^3)^{1/2}\eta_s b\sqrt{N}. \tag{4.60}$$

Given ζ_p and k_p, the diffusion constant and the relaxation times are obtained as

$$D_G = \frac{k_BT}{\zeta_0} = \frac{8k_BT}{3(6\pi^3)^{1/2}\eta_s b\sqrt{N}} = 0.196\frac{k_BT}{\eta_s \bar{R}} \tag{4.61}$$

and

$$\tau_p = \zeta_p/k_p = \tau_1 p^{-3/2} \tag{4.62}$$

with

$$\tau_1 = \tau_r = \frac{\eta_s(\sqrt{N}b)^3}{\sqrt{(3\pi)}k_BT} = 0.325\frac{\eta_s\bar{R}^3}{k_BT}. \tag{4.63}$$

where $\bar{R} = \sqrt{N}b$ has been used.

Equations (4.61) and (4.63) predict the molecular weight dependence of D_G and τ_r,

$$D_G \propto M^{-1/2}, \qquad \tau_r \propto M^{3/2} \tag{4.64}$$

which is in agreement with experimental results (eqn 4.39). A more accurate calculation for D_G and τ_r can be done by solving the eigenvalue problem associated with eqn (4.50).[10] Such a calculation gives slightly different numerical coefficients: 0.192 for D_G, and 0.398 for τ_r.

4.2.2 Zimm model in good solvent

So far we have been considering the Θ condition. The excluded volume interaction is taken into account if we add a potential

$$U_1 = \tfrac{1}{2}vk_BT\sum_{n,m}\delta(\boldsymbol{R}_n - \boldsymbol{R}_m). \tag{4.65}$$

If such a potential is introduced, the Langevin equation again becomes nonlinear. However, on the same level of the approximation as the preaveraging approximation, we can treat the problem simply by assuming the linear Langevin equation

$$\zeta_p\frac{\partial}{\partial t}\boldsymbol{X}_p(t) = -k_p\boldsymbol{X}_p + \boldsymbol{f}_p(t) \tag{4.66}$$

and include the excluded volume effect in the parameters ζ_p and k_p.[13,14] This approximation, which we shall refer to as a linearization approximation, works well at least qualitatively as will be shown later.

The parameter ζ_p is again determined by eqns (4.48) and (4.52),

$$\zeta_p = (h_{pp})^{-1} = \left[\frac{1}{N^2}\int_0^N dn \int_0^N dm \cos\left(\frac{p\pi n}{N}\right)\cos\left(\frac{p\pi m}{N}\right) \times \frac{1}{6\pi\eta_s}\left\langle\frac{1}{|\boldsymbol{R}_n - \boldsymbol{R}_m|}\right\rangle_{\text{eq}}\right]^{-1}. \quad (4.67)$$

To calculate ζ_p, we have to know the distribution of $\boldsymbol{R}_n - \boldsymbol{R}_m$ of the excluded volume chain. Since this is not yet known precisely, we assume, for the sake of simplicity, that the distribution of $\boldsymbol{R}_n - \boldsymbol{R}_m$ is the same as the distribution of the end-to-end vector of the excluded volume chain with $n - m$ segments, and has the following functional form:[15]†

$$\Psi_{n,m}(\boldsymbol{R}_n - \boldsymbol{R}_m) = F\left(\frac{|\boldsymbol{R}_n - \boldsymbol{R}_m|}{|n - m|^{\nu} b}\right). \quad (4.68)$$

This distribution function gives

$$\left\langle\frac{1}{|\boldsymbol{R}_n - \boldsymbol{R}_m|}\right\rangle_{\text{eq}} \simeq \frac{1}{|n - m|^{\nu} b}. \quad (4.69)$$

Substituting eqn (4.69) into eqn (4.67), we have

$$\zeta_0^{-1} \simeq \frac{1}{\eta_s N^{\nu} b} \quad (4.70)$$

and

$$\zeta_p^{-1} \simeq \frac{1}{2N}\int_{-\infty}^{+\infty} dm \cos\left(\frac{p\pi m}{N}\right)\frac{1}{\eta_s |m|^{\nu} b} \simeq \frac{N^{-\nu}}{\eta_s b} p^{\nu-1}. \quad (4.71)$$

On the other hand the parameter k_p can be determined from the equilibrium distribution of $\boldsymbol{X}_p$ as

$$k_p = \frac{3}{\langle \boldsymbol{X}_p^2 \rangle_{\text{eq}}} k_B T \quad \text{for} \quad p = 1, 2, \ldots \text{ and } k_0 = 0. \quad (4.72)$$

To calculate $\langle \boldsymbol{X}_p^2 \rangle_{\text{eq}}$, it is convenient to rewrite eqn (4.18) using integration by parts, so that

$$\boldsymbol{X}_p = -\frac{1}{p\pi}\int_0^N dn \sin\left(\frac{p\pi n}{N}\right)\frac{\partial \boldsymbol{R}_n}{\partial n}. \quad (4.73)$$

† This assumption has been shown to be incorrect by direct perturbation calculation[16] and neutron scattering experiment.[17] However, the conclusions (eqns (4.80)–(4.82)), will not be changed even if we use a more accurate form for $\Psi_{n,m}$.

Thus $\langle \boldsymbol{X}_p^2 \rangle_{\text{eq}}$ is written as

$$\langle \boldsymbol{X}_p^2 \rangle_{\text{eq}} = \frac{1}{p^2\pi^2} \int_0^N \mathrm{d}n \int_0^N \mathrm{d}m \sin\left(\frac{p\pi n}{N}\right) \sin\left(\frac{p\pi m}{N}\right) \left\langle \frac{\partial \boldsymbol{R}_n}{\partial n} \cdot \frac{\partial \boldsymbol{R}_m}{\partial m} \right\rangle_{\text{eq}}. \tag{4.74}$$

Using

$$\frac{\partial}{\partial n} \boldsymbol{R}_n \cdot \frac{\partial}{\partial m} \boldsymbol{R}_m = -\frac{1}{2} \frac{\partial^2}{\partial n\, \partial m} (\boldsymbol{R}_n - \boldsymbol{R}_m)^2 \tag{4.75}$$

we rewrite $\langle \boldsymbol{X}_p^2 \rangle_{\text{eq}}$ as

$$\langle \boldsymbol{X}_p^2 \rangle_{\text{eq}} = \frac{-1}{2p^2\pi^2} \int_0^N \mathrm{d}n \int_0^N \mathrm{d}m \sin\left(\frac{p\pi n}{N}\right) \sin\left(\frac{p\pi m}{N}\right) \frac{\partial^2}{\partial n\, \partial m} \langle (\boldsymbol{R}_n - \boldsymbol{R}_m)^2 \rangle_{\text{eq}}. \tag{4.76}$$

Thus

$$k_p^{-1} = \frac{-1}{6k_B T p^2 \pi^2} \int_0^N \mathrm{d}n \int_0^N \mathrm{d}m \sin\left(\frac{p\pi n}{N}\right) \sin\left(\frac{p\pi m}{N}\right) \times \frac{\partial^2}{\partial n\, \partial m} \langle (\boldsymbol{R}_n - \boldsymbol{R}_m)^2 \rangle_{\text{eq}}. \tag{4.77}$$

By using the same approximation as in eqn (4.54), we have

$$\begin{aligned} k_p^{-1} &= \frac{1}{6k_B T p^2 \pi^2} \int_0^N \mathrm{d}n \sin\left(\frac{p\pi n}{N}\right) \sin\left(\frac{p\pi n}{N}\right) \\ &\quad \times \int_{-\infty}^{\infty} \mathrm{d}(m-n) \cos\left(\frac{p\pi}{N}(m-n)\right) \frac{\partial^2}{\partial m\, \partial n} \langle (\boldsymbol{R}_n - \boldsymbol{R}_m)^2 \rangle_{\text{eq}} \\ &\simeq \frac{1}{6k_B T p^2 \pi^2} \frac{N}{2} \int_{-\infty}^{\infty} \mathrm{d}(m-n) \cos\left(\frac{p\pi}{N}(m-n)\right) \frac{\partial^2}{\partial m\, \partial n} \langle (\boldsymbol{R}_n - \boldsymbol{R}_m)^2 \rangle_{\text{eq}}. \end{aligned} \tag{4.78}$$

Since $\langle (\boldsymbol{R}_n - \boldsymbol{R}_m)^2 \rangle_{\text{eq}} \simeq |n-m|^{2\nu} b^2$ according to eqn (4.68), k_p is evaluated from eqn (4.78) as

$$k_p \simeq \frac{N^{-2\nu}}{b^2} k_B T p^{2\nu+1}. \tag{4.79}$$

The diffusion constant of the centre of mass is given by $k_B T/\zeta_0$,

$$D_G \simeq \frac{k_B T}{\eta_s N^\nu b} \tag{4.80}$$

and the relaxation times are given by $\tau_p = \zeta_p / k_p$, which are written as

$$\tau_p = \tau_1 / p^{3\nu} \tag{4.81}$$

with

$$\tau_1 = \tau_r = \zeta_1 / k_1 \simeq \eta_s N^{3\nu} b^3 / k_B T. \tag{4.82}$$

Using $R_g \simeq N^\nu b$, eqns (4.80) and (4.82) are again written in the same form as eqns (4.61) and (4.63),

$$D_G \simeq \frac{k_B T}{\eta_s R_g}, \qquad \tau_r \simeq \frac{\eta_s}{k_B T} R_g^3. \tag{4.83}$$

These results are the same as for rigid spheres:[18] the characteristic behaviour of the dilute solution is quite similar to the suspension of spheres of radius R_g.

Calculation of D_G based on the renormalization group theory is given by Oono *et al.*[19] In the good solvent limit this result is written as

$$D_G = 0.2030 \frac{k_B T}{\sqrt{6}\, \eta_s R_g} \tag{4.84}$$

The numerical factor is close to the 0.196 given in eqn (4.61).

4.3 Dynamical scaling

It has been proposed[20,21] that the static scaling law described in Section 2.6 can be generalized to dynamical phenomena. The hypothesis is that, for a polymer described by the Zimm model, when the parameters of the model are changed as

$$N \to N/\lambda, \qquad b \to b\lambda^\nu \tag{4.85}$$

any physical quantities, which may be static or dynamic, are changed as

$$A \to \lambda^x A. \tag{4.86}$$

The exponent ν is the same as that which appeared in the static scaling (ν is 1/2 in Θ solvent and about 3/5 in a good solvent).

Though experimental confirmation of dynamical scaling is not so satisfactory as static scaling,† and there is a delicate theoretical problem

† The situation is similar to that in critical phenomena, where although accepted, dynamical scaling[22] has often been found to be more delicate and less universal than static scaling. In the polymer problems, the scaling prediction is largely supported by various experiments, but minute discrepancies have been found, which, at this stage, it is difficult to tell whether they are due to the experiments being done outside the scaling regime, or they indicate the failure of the scaling (see the discussion in ref. 21, Chapter 6). The renormalization group calculation by Oono[23] indicates that the dynamical scaling is not rigorous, though the error is usually quite small.

related to the topological interaction which will be discussed later, the dynamical scaling is generally believed to hold for polymers of very large molecular weight.

To show the usefulness of the scaling argument, let us consider the diffusion constant D_G. As indicated by the Zimm model, the parameters appearing in the problem are N, b, k_BT and η_s. From the dimensional analysis, D_G is written as

$$D_G = \frac{k_BT}{\eta_s b} f(N). \tag{4.87}$$

Since D_G is invariant under the scaling transformation,

$$\frac{k_BT}{\eta_s b} f(N) = \frac{k_BT}{\eta_s b \lambda^\nu} f(N/\lambda). \tag{4.88}$$

For this to be true for arbitrary λ, $f(N)$ must have the following form

$$f(N) = \text{numerical constant} * N^{-\nu}. \tag{4.89}$$

Hence

$$D_G = \text{constant} * \frac{k_BT}{\eta_s b} N^{-\nu} \simeq \frac{k_BT}{\eta_s R_g}. \tag{4.90}$$

By similar argument, we can show that the rotational relaxation time of such a chain is

$$\tau_r \simeq \frac{\eta_s (N^\nu b)^3}{k_BT} \simeq \frac{\eta_s R_g^3}{k_BT}. \tag{4.91}$$

Equations (4.90) and (4.91) agree with eqn (4.83).

For the Rouse model, which has no hydrodynamic interaction, a similar scaling property exists. When the λ segments are grouped, the parameters in the Rouse model change as

$$N \to N\lambda, \qquad b \to b\lambda^\nu, \qquad \zeta \to \zeta\lambda. \tag{4.92}$$

Under this transformation, a physical quantity A changes as

$$A \to \lambda^x A. \tag{4.93}$$

From this it can be shown that D_G and τ_r depend on N as

$$D_G \propto N^{-1}, \qquad \tau_r \propto N^{1+2\nu}. \tag{4.94}$$

These results have been checked by computer simulation.[24]

4.4 Dynamic light scattering

The Brownian motion of polymers can be experimentally studied by dynamic light scattering.[25] By measuring the time correlation of the intensity of the scattered light, one can extract the dynamical structure factor

$$g(\boldsymbol{k}, t) \equiv \frac{1}{N} \sum_{n,m} \langle \exp[\mathrm{i}\boldsymbol{k} \cdot (\boldsymbol{R}_n(t) - \boldsymbol{R}_m(0))] \rangle. \tag{4.95}$$

The behaviour of $g(\boldsymbol{k}, t)$ has two limiting regimes.

The regime $kR_g \ll 1$. In this regime, only the overall translational motion of the polymer can be seen because $g(\boldsymbol{k}, t)$ is written as

$$g(\boldsymbol{k}, t) = \frac{1}{N} \sum_{n,m} \langle \exp[\mathrm{i}\boldsymbol{k} \cdot (\boldsymbol{R}_G(t) - \boldsymbol{R}_G(0)) + \underline{\mathrm{i}\boldsymbol{k} \cdot (\boldsymbol{R}_n(t) - \boldsymbol{R}_G(t))} - \underline{\mathrm{i}\boldsymbol{k} \cdot (\boldsymbol{R}_m(0) - \boldsymbol{R}_G(0))}] \rangle.$$

The underlined terms may be put at zero since they are of the order of kR_g, which is much less than unity. Thus

$$g(\boldsymbol{k}, t) \simeq N \langle \exp[\mathrm{i}\boldsymbol{k} \cdot (\boldsymbol{R}_G(t) - \boldsymbol{R}_G(0))] \rangle . \tag{4.96}$$

If t is large, the distribution of $\boldsymbol{R}_G(t) - \boldsymbol{R}_G(0)$ becomes Gaussian† with the variance $2D_G t$, hence

$$g(\boldsymbol{k}, t) = N \int \mathrm{d}\boldsymbol{r} \exp(\mathrm{i}\boldsymbol{k} \cdot \boldsymbol{r})(4\pi D_G t)^{-3/2} \exp\left(-\frac{\boldsymbol{r}^2}{4D_G t}\right)$$
$$= N \exp(-D_G \boldsymbol{k}^2 t). \tag{4.97}$$

Thus the decay of $g(\boldsymbol{k}, t)$ for a long time region is written as

$$g(\boldsymbol{k}, t) \propto \exp(-\Gamma_{\boldsymbol{k}} t) \tag{4.98}$$

with

$$\Gamma_{\boldsymbol{k}} = D_G \boldsymbol{k}^2. \tag{4.99}$$

On the other hand the initial decay rate of $g(\boldsymbol{k}, t)$ can be calculated rigorously by use of the formula (3.51).[26]

$$\Gamma_{\boldsymbol{k}}^{(0)} \equiv -\frac{\mathrm{d}}{\mathrm{d}t} \ln[g(\boldsymbol{k}, t)] \bigg|_{t=0}$$
$$= \frac{k_B T \sum_{m,n} \langle \boldsymbol{H}_{mn} \exp[\mathrm{i}\boldsymbol{k} \cdot (\boldsymbol{R}_m - \boldsymbol{R}_n)] \rangle : \boldsymbol{k}\boldsymbol{k}}{\sum_{m,n} \langle \exp[\mathrm{i}\boldsymbol{k} \cdot (\boldsymbol{R}_m - \boldsymbol{R}_n)] \rangle}. \tag{4.100}$$

For $kR_g \ll 1$, $\exp[\mathrm{i}\boldsymbol{k} \cdot (\boldsymbol{R}_m - \boldsymbol{R}_n)]$ can be replaced by unity, so that eqn (4.100) reduces to

$$\Gamma_{\boldsymbol{k}}^{(0)} = \frac{k_B T}{N^2} \sum_{m,n} \langle \boldsymbol{H}_{mn} \rangle_{\mathrm{eq}} : \boldsymbol{k}\boldsymbol{k} = D_G^{(\mathrm{K})} \boldsymbol{k}^2 \tag{4.101}$$

† This is a result of the central limit theorem mentioned in Section 2.1.3. Provided the correlation time τ_c of the internal motions is finite, the distribution of $\boldsymbol{R}_G(t) - \boldsymbol{R}_G(0)$ becomes Gaussian for $t \gg \tau_c$.

where

$$D_G^{(K)} = \frac{k_B T}{6\pi\eta_s}\int_0^N \frac{dn}{N}\int_0^N \frac{dm}{N}\left\langle\frac{1}{|\boldsymbol{R}_n - \boldsymbol{R}_m|}\right\rangle_{eq}. \tag{4.102}$$

Equation (4.102) was originally proposed by Kirkwood[27] as a rigorous expression for D_G. Though $D_G^{(K)}$ is generally not equal to D_G, it turned out that the difference between D_G and $D_G^{(K)}$ is usually quite small. In fact experimentally in the small $\boldsymbol{k}$ region no significant change is observed in the slope of $g(\boldsymbol{k}, t)$ beyond the uncertainties caused by the molecular weight distribution.

Calculation of $D_G^{(K)}$ has already been done in the previous section since in the linearization approximation D_G and $D_G^{(K)}$ are the same (compare eqn (4.67) with eqn (4.102)). For Θ condition the predicted molecular weight dependence $D_G \propto M^{-0.5}$ has been confirmed rather well,[28–31] but the experimental value of D_G is about 15% smaller than the theoretical value calculated by eqn (4.61) using R_g which is obtained from static light scattering.[32,33] Several explanations have been proposed, such as the non-Gaussian distribution,[34] the slowness to reach the asymptotic behaviours,[35,36] or the effect of hydrodynamic fluctuation[37,38] which has been neglected in the Zimm model. Though a clear conclusion has not yet been given (see ref. 33 for a discussion), this level of comparison displays the remarkable success of the Zimm model and the present level of accuracy in the theory of dilute solutions.

For good solvents where

$$D_G \propto M^{-\nu_D} \tag{4.103}$$

the measured exponent ν_D is about 0.55,[29,39] which is slightly smaller than the theoretical value $\nu \cong 0.6$. The discrepancy was first thought to indicate the failure of the dynamical scaling law, but the theoretical inequality proposed by des Cloizeaux[40] suggests that the exponent should eventually approach ν for very large molecular weight. The reason for the discrepancy is now thought to be due to the scaling regime in D_G starting at a much higher molecular weight than in R_g. Indeed this was indicated earlier by the perturbation expansion of $D_G^{(K)}$ with respect to the excluded volume parameter.[41,42] A semi-phenomenological theory which shows the slow approach to the asymptotic behaviour is given by Weill and des Cloizeaux.[43]

The regime $kR_g \gg 1$. In this regime, we can see the internal motion of the chain. Calculation of $g(\boldsymbol{k}, t)$ is involved even if the linearization approximation is used,[44,45] (see Appendix 4.III), but the characteristic features of $g(\boldsymbol{k}, t)$ can be obtained from the scaling argument.

By the same line of argument as in Section 4.3, we can show that

$g(\boldsymbol{k}, t)$ is written as

$$g(\boldsymbol{k}, t) = NF(kR_g, tD_G/R_g^2). \tag{4.104}$$

If $kR_g \gg 1$, $g(\boldsymbol{k}, t)$ should be independent of N since the local motion is independent of the total chain length. Since R_g and D_G depend on N as $R_g \propto N^{\nu}$ and $D_G \propto N^{-\nu_D}$($\nu_D = 1$ for the Rouse model and ν for the Zimm model), $g(\boldsymbol{k}, t)$ must have the following functional form,

$$g(\boldsymbol{k}, t) = N(kR_g)^{-1/\nu}F(tD_G R_g^{-2}(kR_g)^x) \tag{4.105}$$

where

$$x = 2 + \frac{\nu_D}{\nu}. \tag{4.106}$$

Thus if $g(\boldsymbol{k}, t)/g(\boldsymbol{k}, 0)$ is plotted against $t\Gamma_{\boldsymbol{k}}$ where

$$\Gamma_{\boldsymbol{k}} = D_G R_g^{-2}(kR_g)^x \tag{4.107}$$

one curve is obtained for all values of $\boldsymbol{k}$.

The dynamical structure factor (4.105) is characterized by the decay rate $\Gamma_{\boldsymbol{k}}$ which is given by

$$\begin{aligned}\Gamma_{\boldsymbol{k}} &= D_G k^4 R_g^2 \quad \text{for the Rouse model } (\nu_D = 1,\ \nu = 1/2)\\ &= D_G k^3 R_g \quad \text{for the Zimm model } (\nu_D = \nu).\end{aligned} \tag{4.108}$$

Note that in the Zimm model the relation $\Gamma_{\boldsymbol{k}} \propto k^3$ holds both in Θ solvents and in good solvents.

In the region $kR_g \gg 1$, the decay of $g(\boldsymbol{k}, t)$ is not a single exponential (see Appendix 4.III). The decay curve, however, is conveniently characterized by the initial decay rate $\Gamma_{\boldsymbol{k}}^{(0)}$. To evaluate eqn (4.100) for $kR_g \gg 1$, it is convenient to rewrite $\boldsymbol{H}_{nm}$ by use of the Fourier transform of the Oseen tensor (see eqn (3.III.5))

$$\boldsymbol{H}_{mn} = \frac{1}{\eta_s}\int \frac{\mathrm{d}\boldsymbol{q}}{(2\pi)^3}\left(\frac{\boldsymbol{I} - \hat{\boldsymbol{q}}\hat{\boldsymbol{q}}}{\boldsymbol{q}^2}\right)\exp[\mathrm{i}\boldsymbol{q}\cdot(\boldsymbol{R}_m - \boldsymbol{R}_n)]. \tag{4.109}$$

Then eqn (4.100) is rewritten as

$$\begin{aligned}\Gamma_{\boldsymbol{k}}^{(0)} &= \frac{k_B T}{g(\boldsymbol{k})\eta_s}\int \frac{\mathrm{d}\boldsymbol{q}}{(2\pi)^3}\left(\frac{\boldsymbol{k}^2 - (\boldsymbol{k}\cdot\hat{\boldsymbol{q}})^2}{\boldsymbol{q}^2}\right)g(\boldsymbol{k}+\boldsymbol{q})\\ &= \frac{k_B T}{g(\boldsymbol{k})\eta_s}\int \frac{\mathrm{d}\boldsymbol{q}}{(2\pi)^3}\left(\frac{\boldsymbol{k}^2}{(\boldsymbol{q}-\boldsymbol{k})^2} - \frac{(\boldsymbol{k}\cdot(\boldsymbol{q}-\boldsymbol{k}))^2}{(\boldsymbol{q}-\boldsymbol{k})^4}\right)g(\boldsymbol{q}).\end{aligned} \tag{4.110}$$

Integrating over the angle between $\boldsymbol{k}$ and $\boldsymbol{q}$, we have

$$\Gamma_{\boldsymbol{k}}^{(0)} = \frac{k_B T}{4\pi^2\eta_s}\int_0^\infty \mathrm{d}q q^2 \frac{g(\boldsymbol{q})}{g(\boldsymbol{k})}\left[\frac{q^2 + k^2}{2qk}\log\left|\frac{q+k}{q-k}\right| - 1\right]. \tag{4.111}$$

If $kR_g \gg 1$, the integral is dominated from the large q region. Hence using the asymptotic behaviour of $g(\boldsymbol{k}) \propto k^{-1/\nu}$ (see eqn (2.132)), we have

$$\Gamma_{\boldsymbol{k}}^{(0)} = \frac{k_B T}{4\pi^2 \eta_s} \int_0^\infty \mathrm{d}q q^2 \left(\frac{k}{q}\right)^{1/\nu} \left[\frac{q^2+k^2}{2qk} \log\left|\frac{q+k}{q-k}\right| - 1\right]$$

$$= -\frac{\nu(3\nu-1)}{4\pi(2\nu-1)(4\nu-1)} \tan\left(\frac{\pi}{2\nu}\right) \frac{k_B T}{\eta_s} k^3 \qquad (4.112)$$

which becomes

$$\Gamma_{\boldsymbol{k}}^{(0)} = 0.0625 \frac{k_B T}{\eta_s} k^3 \quad \text{for} \quad \nu = 0.5 \qquad (4.113)$$

$$= 0.0788 \frac{k_B T}{\eta_s} k^3 \quad \text{for} \quad \nu = 0.6. \qquad (4.114)$$

Note that eqns (4.113) and (4.114) are rigorous and include no adjustable parameters. Experimentally the k^3 dependence of $\Gamma_{\boldsymbol{k}}^{(0)}$ is well confirmed.[29,39,46] However, the experimental value is about 25% smaller than the theoretical value.[39] The reason for this discrepancy is thought to have the same origin as that for D_G.[33] The characteristic behaviour of the dynamical structure factor has also been confirmed by neutron scattering.[47–49]

4.5 Viscoelasticity

4.5.1 Introduction

The dynamics of polymers in solution can be studied by measuring their viscoelastic properties. Shear flows, for which the velocity components are given by

$$v_x(\boldsymbol{r}, t) = \kappa(t) r_y, \qquad v_y = v_z = 0, \qquad (4.115)$$

are commonly used for studying these properties. If the shear rate $\kappa(t)$ is small enough, the shear stress depends linearly on $\kappa(t)$ and can be written as[50]

$$\sigma_{xy}(t) = \int_{-\infty}^{t} \mathrm{d}t' G(t-t')\kappa(t') \qquad (4.116)$$

where $G(t)$ is called the shear relaxation modulus. For dilute solutions, in which the effect of the polymer is small, it is convenient to write eqn (4.116) as

$$\sigma_{xy}(t) = \eta_s \kappa(t) + \int_{-\infty}^{t} \mathrm{d}t' G^{(\mathrm{p})}(t-t')\kappa(t'). \qquad (4.117)$$

The first term represents the property of the pure solvent and the second term represents the effect of the polymers.

Two special cases are important.

(i) Steady shear flow:

$$\kappa(t) = \kappa = \text{constant}. \tag{4.118}$$

In this case, the shear stress is constant, which defines the steady state viscosity

$$\eta = \frac{1}{\kappa}\sigma_{xy}. \tag{4.119}$$

From eqns (4.117) and (4.118), it follows that

$$\eta = \eta_s + \int_0^\infty \mathrm{d}t G^{(\mathrm{p})}(t). \tag{4.120}$$

The increase in the viscosity due to the presence of polymers is usually expressed by the intrinsic viscosity, or viscosity number, defined by

$$[\eta] = \lim_{\rho \to 0} \frac{\eta - \eta_s}{\rho \eta_s} \tag{4.121}$$

where ρ is the weight of the polymer in the unit volume of solution. Using c (the number of segments per volume), N (the number of segments per polymer), M (the molecular weight) and N_A (the Avogadro number), ρ is written as

$$\rho = \frac{c}{N}\frac{M}{N_A}. \tag{4.122}$$

(ii) Oscillatory flow:

$$\kappa(t) = \kappa_0 \cos(\omega t) = \kappa_0 \,\mathrm{Re}(\mathrm{e}^{\mathrm{i}\omega t}) \tag{4.123}$$

(where Re stands for the real part.) The response for this flow defines the complex modulus $G^*(\omega)$:

$$\sigma_{xy}(t) = \kappa_0 \,\mathrm{Re}\left(\frac{G^*(\omega)}{\mathrm{i}\omega}\mathrm{e}^{\mathrm{i}\omega t}\right). \tag{4.124}$$

Since eqn (4.117) gives

$$\begin{aligned}\sigma_{xy}(t) &= \kappa_0 \,\mathrm{Re}\left(\mathrm{e}^{\mathrm{i}\omega t}\eta_s + \int_{-\infty}^{t} \mathrm{d}t' G^{(\mathrm{p})}(t-t')\mathrm{e}^{\mathrm{i}\omega t'}\right) \\ &= \kappa_0 \,\mathrm{Re}\left[\mathrm{e}^{\mathrm{i}\omega t}\left(\eta_s + \int_0^\infty \mathrm{d}t' G^{(\mathrm{p})}(t')\mathrm{e}^{-\mathrm{i}\omega t'}\right)\right].\end{aligned} \tag{4.125}$$

$G^*(\omega)$ is written as

$$G^*(\omega) = \mathrm{i}\omega\eta_s + \mathrm{i}\omega \int_0^\infty \mathrm{d}t e^{-\mathrm{i}\omega t} G^{(\mathrm{p})}(t)$$
$$\equiv G'(\omega) + \mathrm{i}G''(\omega). \tag{4.126}$$

The real part $G'(\omega)$ and the imaginary part $G''(\omega)$ are called the storage modulus and the loss modulus. Experimental results are often expressed by the dimensionless intrinsic moduli defined by:[50]

$$[G'(\omega)]_R = \lim_{\rho\to 0} \frac{M}{\rho RT} G'(\omega) = \lim_{\rho\to 0} \frac{M}{\rho RT} \int_0^\infty \mathrm{d}t\omega \sin(\omega t) G^{(\mathrm{p})}(t) \tag{4.127a}$$

$$[G''(\omega)]_R = \lim_{\rho\to 0} \frac{M}{\rho RT} (G''(\omega) - \omega\eta_s) = \lim_{\rho\to 0} \frac{M}{\rho RT} \int_0^\infty \mathrm{d}t\omega \cos(\omega t) G^{(\mathrm{p})}(t) \tag{4.127b}$$

where $R \equiv N_A k_B$ is the gas constant.

4.5.2 Microscopic expression for the stress tensor

Let us now study the viscoelastic properties using molecular models. As was discussed in Chapter 3, the macroscopic stress of the polymer solutions is written as (see eqn (3.133))

$$\sigma_{\alpha\beta} = \eta_s(\kappa_{\alpha\beta}(t) + \kappa_{\beta\alpha}(t)) + \sigma_{\alpha\beta}^{(\mathrm{p})} + P\delta_{\alpha\beta} \tag{4.128}$$

where

$$\sigma_{\alpha\beta}^{(\mathrm{p})} = -\frac{c}{N} \sum_{n=1}^{N} \langle F_{n\alpha} R_{n\beta} \rangle. \tag{4.129}$$

Here the factor c/N accounts for the number of polymers in the unit volume.

Since $\boldsymbol{F}_n$ is written as

$$\boldsymbol{F}_n = -\frac{\partial}{\partial \boldsymbol{R}_n}(k_B T \ln \Psi + U), \tag{4.130}$$

eqn (4.129) is rewritten as

$$\sigma_{\alpha\beta}^{(\mathrm{p})} = \frac{c}{N} \sum_n \int \mathrm{d}\{\boldsymbol{R}_n\} \Psi \frac{\partial}{\partial R_{n\alpha}} (k_B T \ln \Psi + U) R_{n\beta}$$
$$= \frac{c}{N} \sum_n \left[\underline{\int \mathrm{d}\{\boldsymbol{R}_n\} k_B T \frac{\partial \Psi}{\partial R_{n\alpha}} R_{n\beta}} + \left\langle \frac{\partial U}{\partial R_{n\alpha}} R_{n\beta} \right\rangle \right]. \tag{4.131}$$

The underlined term gives the isotropic stress $k_B T \delta_{\alpha\beta}$ by integration by parts, and can be dropped in the incompressible fluid (see Section 3.7.2).

Thus

$$\sigma_{\alpha\beta}^{(p)} = \frac{c}{N} \sum_n \left\langle \frac{\partial U}{\partial R_{n\alpha}} R_{n\beta} \right\rangle. \tag{4.132}$$

Under Θ conditions, U is given by eqn (4.4), hence

$$\begin{aligned}\sigma_{\alpha\beta}^{(p)} &= \frac{c}{N}\frac{3k_BT}{b^2} \sum_{n=1}^{N} \langle -(\boldsymbol{R}_{n+1} + \boldsymbol{R}_{n-1} - 2\boldsymbol{R}_n)_\alpha R_{n\beta} \rangle \\ &= \frac{c}{N}\frac{3k_BT}{b^2} \sum_{n=1}^{N} \langle (\boldsymbol{R}_{n+1} - \boldsymbol{R}_n)_\alpha (\boldsymbol{R}_{n+1} - \boldsymbol{R}_n)_\beta \rangle \end{aligned} \tag{4.133}$$

or in the continuous limit

$$\sigma_{\alpha\beta}^{(p)} = \frac{c}{N}\frac{3k_BT}{b^2} \int_0^N \mathrm{d}n \left\langle \frac{\partial R_{n\alpha}}{\partial n} \frac{\partial R_{n\beta}}{\partial n} \right\rangle. \tag{4.134}$$

In a good solvent, one has to add the excluded volume potential eqn (4.65) to (4.132). However, the stress arising from this potential can be neglected because

$$\begin{aligned}\left\langle \sum_n \frac{\partial U_1}{\partial R_{n\alpha}} R_{n\beta} \right\rangle &= \frac{v}{2} k_BT \sum_{n,m} \left\langle \left(\frac{\partial}{\partial R_{n\alpha}} \delta(\boldsymbol{R}_n - \boldsymbol{R}_m) \right) R_{n\beta} \right\rangle \\ &= \frac{v}{4} k_BT \sum_{n,m} \left\langle \left[\frac{\partial}{\partial R_{n\alpha}} \delta(\boldsymbol{R}_n - \boldsymbol{R}_m) \right] (R_{n\beta} - R_{m\beta}) \right\rangle \\ &= \frac{v}{4} k_BT \sum_{n,m} \left[\left\langle \frac{\partial}{\partial R_{n\alpha}} [\delta(\boldsymbol{R}_n - \boldsymbol{R}_m)(R_{n\beta} - R_{m\beta})] \right. \right. \\ &\qquad \left. \left. - \delta(\boldsymbol{R}_n - \boldsymbol{R}_m)\delta_{\alpha\beta} \right\rangle \right]. \end{aligned} \tag{4.135}$$

The first term is zero, and the second term can be omitted because it is isotropic. Therefore eqn (4.134) holds even for the chain with the excluded volume effect. (This of course does not mean that the excluded volume interaction plays no part in the viscoelastic properties. The excluded volume does affect the viscoelastic properties through the distribution function Ψ over which the average in eqn (4.134) is taken.)

Equation (4.134) can be rewritten using normal coordinates (4.18) as

$$\begin{aligned}\sigma_{\alpha\beta}^{(p)} &= \frac{c}{N}\frac{3k_BT}{b^2} \sum_{p,q} 4\left(\frac{p\pi}{N}\right)\left(\frac{q\pi}{N}\right) \langle X_{p\alpha}(t) X_{q\beta}(t) \rangle \int_0^N \mathrm{d}n \sin\left(\frac{p\pi n}{N}\right) \sin\left(\frac{q\pi n}{N}\right) \\ &= \frac{c}{N}\frac{3k_BT}{b^2} \sum_p \frac{2p^2\pi^2}{N} \langle X_{p\alpha}(t) X_{p\beta}(t) \rangle . \end{aligned} \tag{4.136}$$

In Θ conditions, this can be written using eqn (4.21)

$$\sigma_{\alpha\beta}^{(\mathrm{p})} = \frac{c}{N} \sum k_{\mathrm{p}} \langle X_{p\alpha} X_{p\beta} \rangle. \tag{4.137}$$

In a good solvent, eqn (4.137) is not rigorous since k_p is now given by eqn (4.72). However, use of eqn (4.137) may be justified in the linearization approximation, which is to assume that the potential U is given by

$$U = \tfrac{1}{2} k_B T \sum_p \frac{3\boldsymbol{X}_p^2}{\langle \boldsymbol{X}_p^2 \rangle_{\mathrm{eq}}} = \tfrac{1}{2} \sum_p k_p \boldsymbol{X}_p^2. \tag{4.138}$$

Equation (4.137) can be derived from this potential by the principle of virtual work described in Section 3.7.5. Thus we shall use eqn (4.137) for both Θ and good solvents.

4.5.3 Calculation of the intrinsic viscosity

We shall now calculate the stress using the linearization approximation. As was shown in Section 3.7.3, under the velocity field $\boldsymbol{v}(\boldsymbol{r}, t) = \boldsymbol{\kappa}(t) \cdot \boldsymbol{r}$, each bead has an additional velocity $\boldsymbol{\kappa}(t) \cdot \boldsymbol{R}_n$ (see eqn (3.120)). This gives the velocity $\boldsymbol{\kappa}(t) \cdot \boldsymbol{X}_p$ in normal coordinates. Thus the Langevin equation for $\boldsymbol{X}_p$ now becomes

$$\frac{\partial}{\partial t} \boldsymbol{X}_p(t) = -\frac{k_p}{\zeta_p} \boldsymbol{X}_p(t) + \frac{1}{\zeta_p} \boldsymbol{f}_p(t) + \boldsymbol{\kappa}(t) \cdot \boldsymbol{X}_p(t). \tag{4.139}$$

To calculate $\langle X_{p\alpha} X_{p\beta} \rangle$ it is convenient to rewrite this into an equivalent Smoluchowski equation.†

$$\frac{\partial \Psi}{\partial t} = \frac{1}{\zeta_p} \sum_p \frac{\partial}{\partial \boldsymbol{X}_p} \cdot \left(k_B T \frac{\partial \Psi}{\partial \boldsymbol{X}_p} + k_p \boldsymbol{X}_p \Psi \right) - \sum_p \frac{\partial}{\partial \boldsymbol{X}_p} \cdot \boldsymbol{\kappa}(t) \cdot \boldsymbol{X}_p \Psi. \tag{4.140}$$

If we multiply both sides of eqn (4.140) by $X_{p\alpha} X_{p\beta}$ and integrate over all the normal coordinates, we get, after integration by parts

$$\frac{\partial}{\partial t} \langle X_{p\alpha} X_{p\beta} \rangle = \Bigg\langle \sum_q \frac{1}{\zeta_q} \left[\frac{\partial}{\partial \boldsymbol{X}_q} \cdot k_B T \frac{\partial}{\partial \boldsymbol{X}_q} (X_{p\alpha} X_{p\beta}) - k_q \boldsymbol{X}_q \cdot \frac{\partial}{\partial \boldsymbol{X}_q} (X_{p\alpha} X_{p\beta}) \right]$$
$$+ \sum_q \boldsymbol{\kappa}(t) : \boldsymbol{X}_q \frac{\partial}{\partial \boldsymbol{X}_q} (X_{p\alpha} X_{p\beta}) \Bigg\rangle$$

† Equation (4.141) can also be derived from the Langevin equation if we use the relation,

$$\langle X_{p\alpha}(t) f_{p\beta}(t) \rangle = k_B T \delta_{\alpha\beta}$$

which follows from eqn (4.22), and the short time solution of eqn (4.139),

$$\boldsymbol{X}_p(t) = \boldsymbol{X}_p(t - \Delta t) + \Delta t \left(-\frac{k_p}{\zeta_p} \boldsymbol{X}_p(t - \Delta t) + \boldsymbol{\kappa}(t - \Delta t) \cdot \boldsymbol{X}_p(t - \Delta t) \right) + \int_{t - \Delta t}^{t} \mathrm{d}t' \frac{\boldsymbol{f}_p(t')}{\zeta_p}$$

$$= \frac{1}{\zeta_p}\left[2k_BT\delta_{\alpha\beta} - 2k_p\langle X_{p\alpha}X_{p\beta}\rangle\right] + \kappa_{\alpha\mu}\langle X_{p\mu}X_{p\beta}\rangle$$
$$+ \kappa_{\beta\mu}\langle X_{p\mu}X_{p\alpha}\rangle. \tag{4.141}$$

This equation can be solved for arbitrary $\kappa_{\alpha\beta}(t)$ (see Section 7.6.3), but here we consider a special case of the shear flow given by eqn (4.115). To calculate the shear stress $\sigma_{xy}(t)$, we have to solve

$$\frac{\partial}{\partial t}\langle X_{px}X_{py}\rangle = -2\frac{k_p}{\zeta_p}\langle X_{px}X_{py}\rangle + \kappa\langle X_{py}^2\rangle. \tag{4.142}$$

For small κ, $\langle X_{py}^2\rangle$ may be replaced by the equilibrium value k_BT/k_p. Hence eqn (4.142) becomes

$$\frac{\partial}{\partial t}\langle X_{px}X_{py}\rangle = -2\frac{k_p}{\zeta_p}\langle X_{px}X_{py}\rangle + \kappa\frac{k_BT}{k_p}. \tag{4.143}$$

In the steady state, eqn (4.143) gives

$$\langle X_{px}X_{py}\rangle = \frac{\zeta_p}{2k_p^2}k_BT\kappa. \tag{4.144}$$

The shear stress is calculated from eqns (4.137) and (4.144) as

$$\sigma_{xy}^{(\mathrm{p})} = \frac{c}{N}\sum_{p=1}^{\infty}k_p\langle X_{px}X_{py}\rangle = \frac{1}{2}\frac{c}{N}\sum_{p=1}^{\infty}\frac{\zeta_p}{k_p}k_BT\kappa. \tag{4.145}$$

Thus the intrinsic viscosity is given by

$$[\eta] = \frac{\sigma_{xy}^{(\mathrm{p})}}{\rho\eta_s\kappa} = \frac{k_BT}{2\rho\eta_s}\frac{c}{N}\sum_p\frac{\zeta_p}{k_p} \tag{4.146}$$

or, by eqn (4.122),

$$[\eta] = \frac{N_Ak_BT}{M\eta_s}\sum_{p=1}^{\infty}\frac{\zeta_p}{2k_p}. \tag{4.147}$$

The sum is evaluated for various models:

(i) The Rouse model: eqns (4.20) and (4.21) give

$$[\eta] = \frac{N_A}{M\eta_s}\frac{N^2b^2\zeta}{6\pi^2}\sum_{p=1}^{\infty}\frac{1}{p^2} = \frac{N_A}{M\eta_s}\frac{N^2b^2\zeta}{6\pi^2}\frac{\pi^2}{6} = \frac{N_A}{M\eta_s}\frac{N^2b^2\zeta}{36}. \tag{4.148}$$

(ii) The Zimm model for Θ solvent: eqns (4.58) and (4.59) give

$$[\eta] = \frac{N_A}{M}\frac{(\sqrt{N}b)^3}{\sqrt{(12\pi)}}\sum_{p=1}^{\infty}p^{-1.5} = \frac{N_A}{M}0.425(\sqrt{N}b)^3. \tag{4.149}$$

(iii) The Zimm model for good solvent: eqns (4.71) and (4.79) give

$$[\eta] \simeq \frac{N_A}{M} N^{3\nu} \sum_p p^{-3\nu} b^3 \simeq \frac{N_A}{M} N^{3\nu} b^3. \tag{4.150}$$

If we write the molecular weight dependence of $[\eta]$ as

$$[\eta] \propto M^{\nu_\eta} \tag{4.151}$$

we have

$$\nu_\eta = \begin{cases} 1 & \text{Rouse model } (\Theta \text{ solvent}), \\ 0.5 & \text{Zimm model } (\Theta \text{ solvent}), \\ 3\nu - 1 = 0.8 & \text{Zimm model (good solvent)}. \end{cases} \tag{4.152}$$

Experimental results show that in the Θ solvent ν_η is 0.5, in agreement with the Zimm model. The agreement is actually more quantitative. Eqn (4.149) is written in the form

$$[\eta] = \frac{\Phi_v}{M} (\sqrt{6}\, R_g)^3 \tag{4.153}$$

where the constant Φ_v, called the Flory–Fox parameter, is

$$\Phi_v = 0.425 N_A = 2.56 \times 10^{23}. \tag{4.154}$$

The experimental value of Φ_v is about 2.5×10^{23}. Perhaps this good agreement is fortuitous since a more accurate analysis[10] of the pre-averaged equation (4.50) gives $\Phi_v = 2.84 \times 10^{23}$. Various theoretical results[12] give Φ_v values ranging from 2.2×10^{23} to 2.87×10^{23}. In any case this level of agreement again indicates the success of the Zimm model in the dilute solution theory.

In good solvent, the experimental value of ν_η is slightly smaller than 0.8. This is perhaps because, as in the case of D_G, the molecular weight is not sufficiently high for the asymptotic behaviour to be observed. In such regions the Flory–Fox parameter decreases with increasing molecular weight and increasing excluded volume parameter. Detailed calculations for Φ_v are described in the book by Yamakawa.[12]

4.5.4 *Intrinsic moduli*

Next we consider the case that the shear rate is not constant. Equation (4.143) is solved for general $\kappa(t)$,

$$\langle X_{px} X_{py} \rangle = \frac{k_B T}{k_p} \int_{-\infty}^{t} dt' \exp(-(t-t')/\tau_p) \kappa(t') \tag{4.155}$$

where

$$\tau_p = \zeta_p / 2k_p = \tau_1 p^{-\mu} \tag{4.156}$$

with

$$\mu = \begin{cases} 2 & \text{Rouse model,} \\ 3/2 & \text{Zimm model } (\Theta \text{ condition}), \\ 3\nu & \text{Zimm model (good solvent).} \end{cases} \tag{4.157}$$

(Note that τ_1 is different from the rotational relaxation time τ_r by a factor 1/2.) From eqns (4.137) and (4.155), $G^{(p)}(t)$ is calculated as

$$G^{(p)}(t) = \frac{c}{N} k_B T \sum_p \exp(-t/\tau_p) = \frac{\rho RT}{M} \sum_p \exp(-t/\tau_p). \tag{4.158}$$

Hence

$$[G'(\omega)]_R = \int_0^\infty dt\omega \sin(\omega t) \sum_p \exp(-t/\tau_p) = \sum_{p=1}^\infty \frac{(\omega\tau_p)^2}{1+(\omega\tau_p)^2} \tag{4.159}$$

$$[G''(\omega)]_R = \int_0^\infty dt\omega \cos(\omega t) \sum_p \exp(-t/\tau_p) = \sum_{p=1}^\infty \frac{\omega\tau_p}{1+(\omega\tau_p)^2}. \tag{4.160}$$

The expressions are simplified in two cases.

(i) $\omega\tau_1 \ll 1$: In this case, $[G'(\omega)]_R$ and $[G''(\omega)]_R$ are approximated as

$$[G'(\omega)]_R = (\omega\tau_1)^2 \sum_{p=1}^\infty p^{-2\mu}, \tag{4.161}$$

$$[G''(\omega)]_R = \omega\tau_1 \sum_{p=1}^\infty p^{-\mu} \tag{4.162}$$

Hence $[G'(\omega)]_R$ and $[G''(\omega)]_R$ are proportional to ω^2 and ω respectively.

(ii) $\omega\tau_1 \gg 1$: In this case the sum over p can be replaced by an integral, so that

$$[G'(\omega)]_R = \int_0^\infty dp \frac{(\omega\tau_1)^2 p^{-2\mu}}{1+(\omega\tau_1)^2 p^{-2\mu}} = (\omega\tau_1)^{1/\mu} \frac{1}{\mu} \int_0^\infty dx \frac{x^{1-1/\mu}}{1+x^2}$$

$$= (\omega\tau_1)^{1/\mu} \frac{\pi}{2\mu \sin(\pi/2\mu)} \tag{4.163}$$

and

$$[G''(\omega)]_R = (\omega\tau_1)^{1/\mu} \frac{1}{\mu} \int_0^\infty dx \frac{x^{-1/\mu}}{1+x^2}$$

$$= (\omega\tau_1)^{1/\mu} \frac{\pi}{2\mu \cos(\pi/2\mu)}. \tag{4.164}$$

Thus

(a) $\mu = 2$ (Rouse model)

$$[G'(\omega)]_R = [G''(\omega)]_R = \frac{\pi}{2\sqrt{2}} (\omega\tau_1)^{1/2} = 1.11(\omega\tau_1)^{1/2}. \tag{4.165}$$

(b) $\mu = 3/2$ (Zimm model in Θ solvent)

$$[G'(\omega)]_R = 1.21(\omega\tau_1)^{2/3}: \qquad [G''(\omega)]_R = 2.09(\omega\tau_1)^{2/3}. \tag{4.166}$$

(c) $\mu = 9/5$ (Zimm model in good solvent)

$$[G'(\omega)]_R = 1.14(\omega\tau_1)^{5/9}: \qquad [G''(\omega)]_R = 1.38(\omega\tau_1)^{5/9}. \tag{4.167}$$

For Θ conditions, eqn (4.166) is well confirmed.[51–53] For good solvent, experimental data have been interpreted in terms of the so called 'draining parameter'[50], but the data might be interpreted by the above theory. Indeed eqns (4.166) and (4.167) indicate that the asymptotic slope of $[G'(\omega)]_R$ and $[G''(\omega)]_R$ decreases as the solubility of the polymer increases, which is consistent with the experimental results.

4.6 Variational bounds for the transport coefficients

4.6.1 *Introduction*

When the hydrodynamic interaction is introduced, rigorous analysis of the Smoluchowski equation or the Langevin equation becomes impossible. In the previous sections, we used the preaveraging approximation to avoid this difficulty. Another way of handling the hydrodynamic interaction is to use a variational principle.[54,55] This method gives rigorous bounds for the transport coefficients such as $[\eta]$ and D_G. The method has an advantage that the resulting formulae are easily applied to general polymers such as stiff polymers, branched polymers, or colloidal suspensions.

A cautionary remark has to be made. The variational principle in this section is based on the positive definiteness of the mobility matrix, i.e., for any vector $\{\boldsymbol{F}_n\}$:

$$\sum_{n,m} \boldsymbol{F}_n \cdot \boldsymbol{H}_{nm} \cdot \boldsymbol{F}_m \geqslant 0 \tag{4.168}$$

This condition is guaranteed for the correct mobility matrix. However, the mobility matrix given by eqn (4.40) is an approximate one, and does not satisfy the inequality (4.168) in a certain configuration in which the beads are too close to each other.[56] An improved formula which guarantees the inequality is proposed by Rotne and Prager.[57] However, this correction is irrelevant for the asymptotic behaviour of $N \gg 1$, which is determined by the hydrodynamic interaction between beads far apart from each other. Thus we shall use eqn (4.40) for $\boldsymbol{H}_{n,m}$.

4.6.2 *Bounds for the intrinsic viscosity*

Lower bound. According to the formalism described in Section 3.7, the intrinsic viscosity $[\eta]$ can be calculated in the following way (the shear flow (eqn 4.115) being considered):

(i) Firstly, the steady-state distribution function Ψ is obtained by solving the Smoluchowski equation in shear flow

$$\sum_{m,n} \frac{\partial}{\partial \boldsymbol{R}_m} \cdot \boldsymbol{H}_{mn} \cdot \left[k_B T \frac{\partial \Psi}{\partial \boldsymbol{R}_n} + \frac{\partial U}{\partial \boldsymbol{R}_n} \Psi \right] - \sum_n \frac{\partial}{\partial R_{nx}} \kappa R_{ny} \Psi = 0. \quad (4.169)$$

(ii) Secondly, $[\eta]$ is calculated from eqns (4.121) and (4.129)

$$[\eta] = \frac{1}{\rho \eta_s \kappa} \sigma_{xy}^{(\mathrm{p})} = -\frac{N_A}{\eta_s \kappa M} \sum_n \langle R_{ny} F_{nx} \rangle$$

$$= \frac{N_A}{\eta_s \kappa M} \int \mathrm{d}\{\boldsymbol{R}_n\} \Psi \sum_n R_{ny} \frac{\partial}{\partial R_{nx}} (k_B T \ln \Psi + U). \quad (4.170)$$

To put this into a variational formulation, we first rewrite eqn (4.169) using the equilibrium distribution function $\Psi_{\mathrm{eq}} \propto \exp(-U/k_B T)$ and the deviation Φ as

$$\Psi = \Psi_{\mathrm{eq}} \Phi. \quad (4.171)$$

Since the system is homogeneous, the function Φ satisfies translational invariance: i.e., for arbitrary $\boldsymbol{a}$,

$$\Phi(\{\boldsymbol{R}_n\}) = \Phi(\{\boldsymbol{R}_n + \boldsymbol{a}\}). \quad (4.172)$$

This condition is needed to ensure that the surface integrals appearing in the integral by parts in the following analysis vanish. Substituting eqn (4.171) into eqn (4.169) and retaining the terms linear in κ, we have

$$\sum_{m,n} \frac{\partial}{\partial \boldsymbol{R}_m} \cdot \left(\boldsymbol{H}_{mn} \cdot k_B T \Psi_{\mathrm{eq}} \frac{\partial \Phi}{\partial \boldsymbol{R}_n} \right) = \sum_n \frac{\partial}{\partial R_{nx}} \kappa R_{ny} \Psi_{\mathrm{eq}}. \quad (4.173)$$

To the first order in κ, eqn (4.170) is written as

$$[\eta] = \frac{N_A}{\eta_s \kappa M} \int \mathrm{d}\{\boldsymbol{R}_n\} \Psi_{\mathrm{eq}} k_B T \sum_n R_{ny} \frac{\partial \Phi}{\partial R_{nx}} \quad (4.174)$$

$$= \frac{N_A}{\eta_s \kappa M} k_B T \left\langle \sum_n R_{ny} \frac{\partial \Phi}{\partial R_{nx}} \right\rangle_{\mathrm{eq}} \quad (4.175)$$

where $\langle \ldots \rangle_{\mathrm{eq}}$ is the average over the equilibrium distribution function:

$$\langle \ldots \rangle_{\mathrm{eq}} = \int \mathrm{d}\{\boldsymbol{R}_n\} \Psi_{\mathrm{eq}}(\{\boldsymbol{R}_n\}) \ldots . \quad (4.176)$$

Now eqns (4.173) and (4.175) are converted to the following variational principle:[54,58] for arbitrary Φ satisfying eqn (4.172)

$$[\eta] \geqslant \frac{N_A}{\eta_s M \kappa^2} W[\Phi] \quad (4.177)$$

where

$$W[\Phi]=k_BT\left\langle -k_BT\sum_{m,n}\frac{\partial\Phi}{\partial\boldsymbol{R}_m}\cdot\boldsymbol{H}_{mn}\cdot\frac{\partial\Phi}{\partial\boldsymbol{R}_n}+2\sum_m\kappa\frac{\partial\Phi}{\partial R_{mx}}R_{my}\right\rangle_{\text{eq}}. \tag{4.178}$$

Equation (4.177) gives a lower bound for the intrinsic viscosity.

To prove eqn (4.177), we note that for the true solution Φ^* which satisfies eqn (4.173) the following equality holds for arbitrary Φ:

$$\left\langle k_BT\sum_{m,n}\frac{\partial\Phi^*}{\partial\boldsymbol{R}_m}\cdot\boldsymbol{H}_{mn}\cdot\frac{\partial\Phi}{\partial\boldsymbol{R}_n}\right\rangle_{\text{eq}}=\left\langle\sum_m\kappa\frac{\partial\Phi}{\partial R_{mx}}R_{my}\right\rangle_{\text{eq}}. \tag{4.179}$$

This can be proved by integration by parts. Using eqn (4.179) for the special case of $\Phi=\Phi^*$, we can show that

$$W[\Phi^*]=k_BT\left\langle\sum_m\kappa\frac{\partial\Phi^*}{\partial R_{mx}}R_{my}\right\rangle_{\text{eq}}=[\eta]\eta_s\kappa^2M/N_A. \tag{4.180}$$

Furthermore, from eqns (4.178), (4.180), and (4.168), it follows that

$$W[\Phi^*]-W[\Phi]=k_BT\left\langle k_BT\sum_{m,n}\left(\frac{\partial\Phi^*}{\partial\boldsymbol{R}_m}-\frac{\partial\Phi}{\partial\boldsymbol{R}_m}\right)\cdot\boldsymbol{H}_{mn}\cdot\left(\frac{\partial\Phi^*}{\partial\boldsymbol{R}_n}-\frac{\partial\Phi}{\partial\boldsymbol{R}_n}\right)\right\rangle_{\text{eq}}\geqslant 0. \tag{4.181}$$

Equation (4.177) follows from eqns (4.180) and (4.181).

A simple choice of the trial function is

$$\Phi=1+\xi\sum_n R'_{nx}R'_{ny} \tag{4.182}$$

where

$$\boldsymbol{R}'_n=\boldsymbol{R}_n-\boldsymbol{R}_G \tag{4.183}$$

and ξ is a variational parameter to be determined. Substituting eqn (4.182) into eqn (4.178), we have

$$W[\Phi]=-\frac{2}{\eta_s}(k_BT)^2\xi^2N^2R_H+\tfrac{2}{3}\kappa k_BT\xi NR_g^2 \tag{4.184}$$

where

$$R_H\equiv\frac{\eta_s}{N^2}\sum_{m,n}(\langle R'_{mx}R'_{ny}H_{mnyx}\rangle_{\text{eq}}+\langle R'_{mx}R'_{nx}H_{mnyy}\rangle_{\text{eq}}) \tag{4.185}$$

and we have used the relations

$$\begin{aligned}\langle R'_{mx}R'_{ny}H_{mnyx}\rangle_{\text{eq}}&=\langle R'_{my}R'_{nx}H_{mnxy}\rangle_{\text{eq}}\\ \langle R'_{mx}R'_{nx}H_{mnyy}\rangle_{\text{eq}}&=\langle R'_{my}R'_{ny}H_{mnxx}\rangle_{\text{eq}}.\end{aligned} \tag{4.186}$$

The best estimate for $[\eta]$ is obtained by maximizing eqn (4.184) with respect to ξ, which gives

$$[\eta] \geqslant \frac{N_A}{18M} \frac{R_g^4}{R_H}. \tag{4.187}$$

The right-hand side includes only the average over the equilibrium distribution function. It can be shown that eqn (4.187) gives a correct viscosity for two limiting cases, i.e., spherical particles and rigid rodlike particles. For flexible polymers in the Θ condition, however, the bound is rather weak: in terms of the Flory–Fox parameter, one obtains (after some tedious calculation)[58]

$$\Phi > \tfrac{5}{216}(6\pi^3)^{1/2} N_A = 1.90 \times 10^{23}. \tag{4.188}$$

Upper bound. A formula for the upper bounds for $[\eta]$ was given by Fixman[55]

$$[\eta] \leqslant \frac{N_A}{\eta_s M \kappa^2} \sum_{n,m} \langle (\boldsymbol{V}_n - \boldsymbol{\kappa} \cdot \boldsymbol{R}_n) \cdot (\boldsymbol{H}^{-1})_{nm} \cdot (\boldsymbol{V}_m - \boldsymbol{\kappa} \cdot \boldsymbol{R}_m) \rangle_{\text{eq}} \tag{4.189}$$

where $\boldsymbol{V}_n$ is a function of $\{\boldsymbol{R}_n\}$ satisfying

$$\sum_n \frac{\partial}{\partial \boldsymbol{R}_n} \cdot (\boldsymbol{V}_n \Psi_{\text{eq}}) = 0. \tag{4.190}$$

Use of this variational principle is not easy since it requires the evaluation of $(\boldsymbol{H}^{-1})_{nm}$. Nevertheless, the calculation can be done numerically,[59,60] and the results indicate that the error of the preaveraging approximation is less than 30% for $[\eta]$.

4.6.3 *Bounds for the diffusion constant*

Upper bound. Next we consider the translational diffusion constant. To calculate D_G, we consider that a weak constant field

$$U_{\text{ext}} = -FR_{Gz} = -\frac{1}{N} \sum_n FR_{nz} \tag{4.191}$$

is applied to the solution in uniform concentration. The field will cause a uniform motion of the centres of mass of each polymer with a constant velocity $\langle \boldsymbol{V}_G \rangle$ which is proportional to $\boldsymbol{F}$. According to the fluctuation dissipation theorem, D_G is obtained from $\langle \boldsymbol{V}_G \rangle$ by (see eqn (3.67))

$$D_G = \frac{\langle V_{Gz} \rangle k_B T}{F}. \tag{4.192}$$

To calculate $\langle V_{Gz} \rangle$ we have to solve the Smoluchowski equation for the steady state:

$$\sum_{n,m} \frac{\partial}{\partial \boldsymbol{R}_n} \cdot \boldsymbol{H}_{nm} \cdot \Psi_{\text{eq}}\left(k_B T \frac{\partial \Phi}{\partial \boldsymbol{R}_m} - \frac{\boldsymbol{F}}{N}\right) = 0 \tag{4.193}$$

where $\boldsymbol{F} = F\boldsymbol{e}_z$ ($\boldsymbol{e}_z$ being a unit vector in the direction of the z axis). Since we are considering the time-scale in which polymer concentration is homogeneous, only the translational invariant solution is needed; i.e.,

$$\Phi(\{\boldsymbol{R}_n\}) = \Phi(\{\boldsymbol{R}_n + \boldsymbol{a}\}). \tag{4.194}$$

(Mathematically this condition is needed again to ensure that the surface integral vanishes.) Now, since the velocity of the n-th bead is given as

$$\begin{aligned} \boldsymbol{V}_n &= -\sum_m \boldsymbol{H}_{nm} \cdot \frac{\partial}{\partial \boldsymbol{R}_m} (k_B T \ln \Psi + U + U_{\text{ext}}) \\ &= -\sum_m \boldsymbol{H}_{nm} \cdot \left(k_B T \frac{\partial \ln \Phi}{\partial \boldsymbol{R}_m} - \frac{\boldsymbol{F}}{N}\right), \end{aligned} \tag{4.195}$$

$\langle \boldsymbol{V}_G \rangle$ is calculated from

$$\langle \boldsymbol{V}_G \rangle = \frac{1}{N} \sum_n \langle \Phi \boldsymbol{V}_n \rangle_{\text{eq}} = \frac{1}{N} \sum_{n,m} \left\langle \boldsymbol{H}_{nm} \cdot \left(-k_B T \frac{\partial \Phi}{\partial \boldsymbol{R}_m} + \frac{\boldsymbol{F}}{N}\right)\right\rangle_{\text{eq}} \tag{4.196}$$

where terms of higher order in F have been neglected. The term $\partial \Phi / \partial \boldsymbol{R}_m$ denotes the driving force due to the deformation of the molecule. If we neglect this term, we have

$$\langle V_{Gz} \rangle = \frac{F}{N^2} \sum_{n,m} \langle H_{nmzz} \rangle_{\text{eq}} = \frac{F}{6\pi\eta_s N^2} \sum_{n,m} \left\langle \frac{1}{|\boldsymbol{R}_n - \boldsymbol{R}_m|} \right\rangle_{\text{eq}}. \tag{4.197}$$

This gives the Kirkwood formula

$$D_G^{(\text{K})} = \frac{k_B T}{6\pi\eta_s N^2} \sum_{n,m} \left\langle \frac{1}{|\boldsymbol{R}_n - \boldsymbol{R}_m|} \right\rangle_{\text{eq}}. \tag{4.198}$$

To see the effect of the deformation term $\partial \Phi / \partial \boldsymbol{R}_n$, we have to solve eqn (4.193). The effect can be calculated by a variational principle. It can be shown that for any Φ that satisfies eqn (4.194)

$$D_G \leqslant \frac{k_B T}{F^2} W_U[\Phi] \tag{4.199}$$

where

$$W_U[\Phi] = \sum_{n,m} \left\langle \left(k_B T \frac{\partial \Phi}{\partial \boldsymbol{R}_m} - \frac{\boldsymbol{F}}{N}\right) \cdot \boldsymbol{H}_{mn} \cdot \left(k_B T \frac{\partial \Phi}{\partial \boldsymbol{R}_n} - \frac{\boldsymbol{F}}{N}\right)\right\rangle_{\text{eq}} \tag{4.200}$$

Equation (4.199) indicates that the Kirkwood formula (4.198) corresponds to the choice of $\Phi = 1$ and is actually an upper bound for D_G.

Lower bound. A lower bound for the diffusion constant is given in a similar form to eqn (4.189).[55] Let $\boldsymbol{V}_n$ be the function which satisfies eqn (4.190), then

$$D_G \geqslant D_G^{(\mathrm{K})} - \frac{k_B T}{N^2}\left\langle \sum_{n,m} (\boldsymbol{V}_n - \boldsymbol{V}_n^{(0)}) \cdot (\boldsymbol{H}^{-1})_{nm} \cdot (\boldsymbol{V}_m - \boldsymbol{V}_m^{(0)}) \right\rangle_{\mathrm{eq}} \quad (4.201)$$

where

$$\boldsymbol{V}_n^{(0)} = \sum_m \boldsymbol{H}_{nm} \cdot \boldsymbol{e}_z. \quad (4.202)$$

Given the upper and lower bounds, one can estimate the error of the preaveraging approximation. Fixman[60] showed that the Kirkwood formula gives quite accurate estimation for D_G: for flexible polymers in Θ condition, the error is of the order of a few per cent.

4.7 Birefringence

4.7.1 *Birefringence of polymer solutions*

Polymer solutions are isotropic at equilibrium. If there is a velocity gradient, the statistical distribution of the polymer is deformed from the isotropic state, and the optical property of the solution becomes anisotropic. This phenomena is called flow birefringence (or the Maxwell effect).[61,62] Other external fields such as electric or magnetic fields also cause birefringence, which is called electric birefringence (or Kerr effect) and magnetic birefringence (Cotton–Mouton effect), respectively.

The birefringence of a material is expressed by the anisotropic dielectric tensor $\hat{\varepsilon}_{\alpha\beta}$ at optical frequency, or the refractive index tensor $\hat{n}_{\alpha\beta}$ which is the square root of $\hat{\varepsilon}_{\alpha\beta}$, i.e.,

$$\hat{n}_{\alpha\mu}\hat{n}_{\mu\beta} = \hat{\varepsilon}_{\alpha\beta}. \quad (4.203)$$

It is convenient to decompose $\hat{\varepsilon}_{\alpha\beta}$ and $\hat{n}_{\alpha\beta}$ into an isotropic tensor and a purely anisotropic tensor whose trace is zero:

$$\hat{\varepsilon}_{\alpha\beta} = \varepsilon\delta_{\alpha\beta} + \varepsilon_{\alpha\beta}, \qquad \hat{n}_{\alpha\beta} = n\delta_{\alpha\beta} + n_{\alpha\beta} \quad (4.204)$$

where

$$\varepsilon_{\alpha\alpha} = 0 \quad \text{and} \quad n_{\alpha\alpha} = 0. \quad (4.205)$$

In polymeric systems, the anisotropic part $\varepsilon_{\alpha\beta}$ is usually much smaller than the isotropic part $\varepsilon\delta_{\alpha\beta}$ ($|\varepsilon_{\alpha\beta}|/\varepsilon < 10^{-5}$). When this is so, eqn (4.203) gives

$$n_{\alpha\beta} = \frac{1}{2n}\varepsilon_{\alpha\beta}. \quad (4.206)$$

Let us now consider the flow birefringence. If the velocity gradient tensor $\kappa_{\alpha\beta}$ is small, the birefringence is linear to $\kappa_{\alpha\beta}$ and written as†

$$n_{\alpha\beta}(t) = \int_{-\infty}^{t} \mathrm{d}t' \mu(t-t')(\kappa_{\alpha\beta}(t') + \kappa_{\beta\alpha}(t')). \tag{4.207}$$

In a steady state

$$n_{\alpha\beta} = \lambda_M(\kappa_{\alpha\beta} + \kappa_{\beta\alpha}). \tag{4.208}$$

The constant λ_M is called the Maxwell constant. In dilute polymer solutions, experimental results are often expressed by the 'intrinsic Maxwell constant' defined by

$$[n] = \lim_{\rho \to 0} \frac{2\lambda_M}{\rho \eta_s}. \tag{4.209}$$

4.7.2 Molecular expression for birefringence

The birefringence of polymer solutions has two origins. Firstly, since the polymer segments have anisotropic polarizability, the orientation of the bond vector of the main chain causes birefringence, called the intrinsic birefringence. Secondly, since the dielectric constant of the polymer coil region is different from that of the solvent, the anisotropy of the shape of the polymer coil creates an anisotropic internal field, which also contributes to the birefringence. This is called the form birefringence.

Intrinsic birefringence. The intrinsic birefringence of polymeric materials was first calculated by Kuhn.[63] He considered a polymer segment whose end-to-end vector is fixed at $\boldsymbol{r}$, and calculated its average polarizability $\alpha_{\alpha\beta}(\boldsymbol{r})$ assuming that

(i) The segment is made up of n_s bonds connected by universal joints. Each bond has anisotropic polarizability, $\alpha_\parallel$ along itself and $\alpha_\perp$ perpendicular to itself (see Fig. 4.2*a*).

(ii) The total polarizability is the sum of the polarizability of individual bonds.

Kuhn's result is written as[61–63]

$$\hat{\alpha}_{\alpha\beta}(\boldsymbol{r}) = \frac{n_s}{3}(\alpha_\parallel + 2\alpha_\perp)\delta_{\alpha\beta} + \frac{3}{5}\frac{(\alpha_\parallel - \alpha_\perp)}{n_s b^2}(r_\alpha r_\beta - \tfrac{1}{3}\delta_{\alpha\beta}\boldsymbol{r}^2)\left[1 + \frac{12}{35}\left(\frac{|\boldsymbol{r}|}{n_s b}\right)^2 + \dots\right]. \tag{4.210}$$

If the polymer segment is not extremely extended, $|\boldsymbol{r}|$ is of the order of

† In electric (or magnetic) birefringence, the birefringence is not linear to the field, so that the response function is not written in the simple form (see Chapter 8).

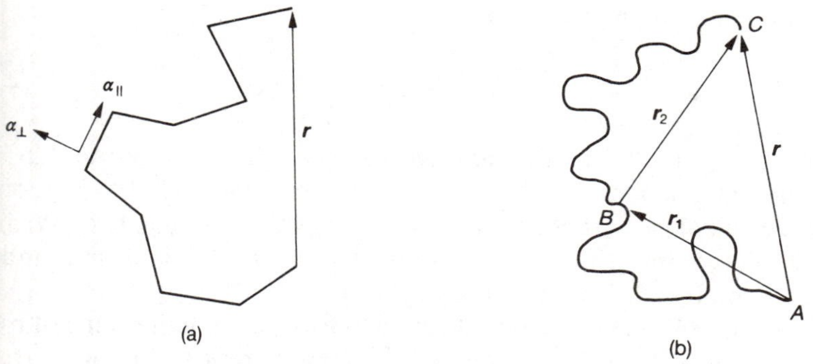

Fig. 4.2. (*a*) Kuhn's model, (*b*) Additivity assumption: if $\alpha_{\alpha\beta}(\boldsymbol{r}_1)$ and $\alpha_{\alpha\beta}(\boldsymbol{r}_2)$ are the polarizability of the part *AB* and *BC* respectively, the total polarizability is

$$\alpha_{\alpha\beta}(\boldsymbol{r}) = \langle \alpha_{\alpha\beta}(\boldsymbol{r}_1) + \alpha_{\alpha\beta}(\boldsymbol{r}_2)\rangle_{\boldsymbol{r}_1+\boldsymbol{r}_2=\boldsymbol{r}}$$

$\sqrt{n_s}\, b$, so that the underlined terms can be neglected. Thus the anisotropic part of the polarizability tensor

$$\alpha_{\alpha\beta}(\boldsymbol{r}) \equiv \hat{\alpha}_{\alpha\beta}(\boldsymbol{r}) - \tfrac{1}{3}\hat{\alpha}_{\mu\mu}(\boldsymbol{r})\delta_{\alpha\beta} \tag{4.211}$$

is given by

$$\alpha_{\alpha\beta}(\boldsymbol{r}) = A(r_\alpha r_\beta - \tfrac{1}{3}\delta_{\alpha\beta}\boldsymbol{r}^2) \tag{4.212}$$

where

$$A = \frac{3}{5}\frac{(\alpha_\parallel - \alpha_\perp)}{n_s b^2}. \tag{4.213}$$

An important result of this model is that in the Gaussian region A is independent of $\boldsymbol{r}$ and proportional to $1/n_s$. Actually this conclusion holds for more general models. In Appendix 4.IV, it is shown that the polarizability tensor is written in the form of eqn (4.212) with a constant A (proportional to n_s^{-1}) provided the following conditions are satisfied.

(i) The statistics of the chain is Gaussian.†

(ii) The polarizability $\alpha_{\alpha\beta}(\boldsymbol{r})$ is additive;‡ i.e., when the segment is divided into two parts, 1 and 2, the total polarizability is given by (see Fig. 4.2*b*)

$$\alpha_{\alpha\beta}(\boldsymbol{r}) = \langle \alpha_{\alpha\beta}(\boldsymbol{r}_1) + \alpha_{\alpha\beta}(\boldsymbol{r}_2)\rangle_{\boldsymbol{r}_1+\boldsymbol{r}_2=\boldsymbol{r}} \tag{4.214}$$

where $\alpha_{\alpha\beta}(\boldsymbol{r}_1)$ and $\alpha_{\alpha\beta}(\boldsymbol{r}_2)$ are the polarizability of each part and the average is taken for the equilibrium state with $\boldsymbol{r}_1 + \boldsymbol{r}_2$ being fixed at $\boldsymbol{r}$. In

† The effect of the excluded volume interaction on A is not known, but if the size of the segment is taken to be small, the effect will be only to change the constant A.

‡ The additivity is not satisfied if the dipole–dipole interaction within the segment is taken into account. It has been shown by Copic[66] that this effect is included in the form birefringence.

this general case, A is written as

$$A = \frac{\Delta\gamma}{\langle \boldsymbol{r}^2 \rangle_{\text{eq}}} \tag{4.215}$$

where $\langle \boldsymbol{r}^2 \rangle_{\text{eq}}$ is the equilibrium average of $\boldsymbol{r}^2$ in free state and $\Delta\gamma$ is a constant independent of the size of the segment: $\Delta\gamma$ depends on the detailed chemical structure and can be positive or negative. A microscopic derivation of $\Delta\gamma$ is described in refs 64 and 65, and the experimental values are summarized in ref. 61.

Now, eqn (4.212) can be used for each Rouse segment bounded by two beads, say n and $n-1$, then $\boldsymbol{r}$ and $\langle \boldsymbol{r}^2 \rangle_{\text{eq}}$ can be replaced by $\boldsymbol{R}_n - \boldsymbol{R}_{n-1} = \partial \boldsymbol{R}_n / \partial n$ and b^2, respectively. The polarizability of this Rouse segment is then written as

$$\alpha_{n\alpha\beta} = \frac{\Delta\gamma}{b^2} \left(\frac{\partial R_{n\alpha}}{\partial n} \frac{\partial R_{n\beta}}{\partial n} - \tfrac{1}{3}\delta_{\alpha b} \left(\frac{\partial \boldsymbol{R}_n}{\partial n} \right)^2 \right). \tag{4.216}$$

To relate the polarizability of individual segments to macroscopic birefringence, Kuhn[63] used the Clausius–Mossotti equation, according to which the (isotropic) refractive index n is related to the (isotropic) polarizability of individual molecules as:

$$\frac{n^2 - 1}{n^2 + 2} = \frac{4\pi}{3} \sum_i \alpha_i \tag{4.217}$$

where the summation is taken for all the molecules in unit volume. According to eqn (4.217), if α_i undergoes a slight change $\delta\alpha_i$, n changes to

$$\delta n = \frac{2\pi}{9n} (n^2 + 2)^2 \sum_i \delta\alpha_i. \tag{4.218}$$

Using this relation for each component of the *anisotropic* polarizability tensor $\alpha_{n\alpha\beta}$, which is much smaller than the isotropic part, one gets

$$n^{(\mathrm{i})}_{\alpha\beta} = \frac{2\pi}{9n} (n^2 + 2)^2 \sum_{\substack{\text{all segments} \\ \text{in unit volume}}} \alpha_{n\alpha\beta} \tag{4.219}$$

where the superscript (i) stands for the contribution of the intrinsic birefringence. Equations (4.216) and (4.219) give

$$n^{(\mathrm{i})}_{\alpha\beta} = K \left\langle \int_0^N \mathrm{d}n \frac{1}{b^2} \left[\frac{\partial R_{n\alpha}}{\partial n} \frac{\partial R_{n\beta}}{\partial n} - \tfrac{1}{3}\delta_{\alpha b} \left(\frac{\partial \boldsymbol{R}_n}{\partial n} \right)^2 \right] \right\rangle \tag{4.220}$$

where

$$K = \frac{2\pi}{9n} (n^2 + 2)^2 \frac{\rho N_A}{M} \Delta\gamma. \tag{4.221}$$

The factor $\rho N_A / M$ accounts for the number of polymers in a unit volume.

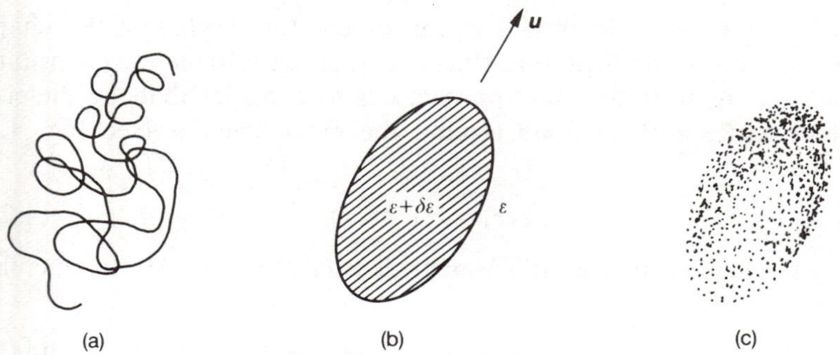

Fig. 4.3. (*a*) Polymer, (*b*) spheroidal model, and (*c*) fuzzy spheroidal model.

Form birefringence. The form birefringence arises from the difference in the isotropic part of the polarizability between polymer segments and solvent molecules. A simple way to understand this is to regard the polymer coil as an ellipsoid which has a different dielectric constant $\varepsilon + \delta\varepsilon$ from that of the outside (see Fig. 4.3).[67] For simplicity let us consider the coil as a prolate spheroid, and denote the direction of the long axis by a unit vector $\boldsymbol{u}$. The polarizability of the spheroid is larger when $\boldsymbol{u}$ is parallel to the electric field than when it is perpendicular to the field, so that if the distribution of $\boldsymbol{u}$ is not isotropic, the dielectric constant is not isotropic.

The average dielectric constant of a material which includes a small number of such spheroids is calculated by electrostatics (see ref. 68, for example) as

$$\hat{\varepsilon}_{\alpha\beta} = \varepsilon\delta_{\alpha\beta} + \varepsilon\phi\delta\varepsilon\left[\frac{\langle \delta_{\alpha\beta} - u_\alpha u_\beta\rangle}{\varepsilon + \delta\varepsilon(1 - N_d)/2} + \frac{\langle u_\alpha u_\beta\rangle}{\varepsilon + \delta\varepsilon N_d}\right] \tag{4.222}$$

where ϕ is the volume fraction of the spheroid and N_d is a numerical factor called the 'demagnetization factor' which takes values between 0 and 1/3 depending on the aspect ratio of the spheroid ($N_d = 1/3$ for a sphere and 0 for a rod). For small $\delta\varepsilon$, the anisotropic part $\varepsilon^{(\mathrm{f})}_{\alpha\beta} \equiv \hat{\varepsilon}_{\alpha\beta} - (\hat{\varepsilon}_{\mu\mu})\delta_{\alpha\beta}/3$ is calculated from eqn (4.222) to be

$$\varepsilon^{(\mathrm{f})}_{\alpha\beta} = \underline{\phi\frac{(\delta\varepsilon)^2}{\varepsilon}\tfrac{3}{2}(\tfrac{1}{3} - N_d)}(\langle u_\alpha u_\beta\rangle - \tfrac{1}{3}\delta_{\alpha\beta}). \tag{4.223}$$

Note that the underlined coefficient is always positive independent of the sign of $\delta\varepsilon$. Since the prolate orients toward the flow direction, the form birefringence always gives a positive contribution to the Maxwell constant.

Though the spheroid model is convenient for displaying the characteristic aspects of the form birefringence, it is not suitable for quantitative calculation. A more general approach† is to consider that the dielectric constant varies with position around the mean value $\bar{\varepsilon}$ (see Fig. 4.3*c*) such that

$$\varepsilon(\boldsymbol{r}) = \bar{\varepsilon} + \delta\varepsilon(\boldsymbol{r}). \tag{4.224}$$

If $\varepsilon(\boldsymbol{r})$ fluctuates, the dipole moment density $\boldsymbol{P} = (\varepsilon - 1)\boldsymbol{E}/4\pi$ fluctuates by

$$\delta P_\alpha(\boldsymbol{r}) = \delta\varepsilon(\boldsymbol{r})E_\alpha(\boldsymbol{r})/4\pi. \tag{4.225}$$

This causes a fluctuation of the electric field:

$$E_\alpha(\boldsymbol{r}) = \bar{E}_\alpha + \delta E_\alpha(\boldsymbol{r}), \tag{4.226}$$

$$\delta E_\alpha(\boldsymbol{r}) = \int \mathrm{d}\boldsymbol{r}'\, G_{\alpha\beta}(\boldsymbol{r} - \boldsymbol{r}')\delta P_\beta(\boldsymbol{r}') \tag{4.227}$$

where

$$G_{\alpha\beta}(\boldsymbol{r}) = \frac{3}{\bar{\varepsilon}r^3}(\hat{r}_\alpha\hat{r}_\beta - \tfrac{1}{3}\delta_{\alpha\beta}) \tag{4.228}$$

denotes the field created by a dipole moment. Thus the average electric displacement $\boldsymbol{D}$ is written as

$$\begin{aligned} D_\alpha &= \langle \varepsilon(\boldsymbol{r})E_\alpha(\boldsymbol{r})\rangle = \langle(\bar{\varepsilon} + \delta\varepsilon)(\bar{E}_\alpha + \delta E_\alpha(\boldsymbol{r}))\rangle \\ &= \bar{\varepsilon}\bar{E}_\alpha + \langle \delta\varepsilon(\boldsymbol{r})\delta E_\alpha(\boldsymbol{r})\rangle \\ &= \bar{\varepsilon}\bar{E}_\alpha + \frac{1}{4\pi}\int \mathrm{d}\boldsymbol{r}'\, G_{\alpha\beta}(\boldsymbol{r} - \boldsymbol{r}')\langle \delta\varepsilon(\boldsymbol{r})\delta\varepsilon(\boldsymbol{r}')E_\beta(\boldsymbol{r}')\rangle. \end{aligned} \tag{4.229}$$

Since the fluctuation of $\delta\varepsilon(\boldsymbol{r})$ is very small, $E_\beta(\boldsymbol{r}')$ on the right-hand side can be replaced by $\bar{E}_\beta$. Hence

$$D_\alpha = \left[\bar{\varepsilon}\delta_{\alpha\beta} + \frac{1}{4\pi}\int \mathrm{d}\boldsymbol{r}'\, G_{\alpha\beta}(\boldsymbol{r} - \boldsymbol{r}')\langle \delta\varepsilon(\boldsymbol{r})\delta\varepsilon(\boldsymbol{r}')\rangle\right]\bar{E}_\beta. \tag{4.230}$$

The second term becomes anisotropic if the correlation $\langle \delta\varepsilon(\boldsymbol{r})\delta\varepsilon(\boldsymbol{r}')\rangle$ is not isotropic. The form birefringence is thus given by (note $G_{\alpha\alpha} = 0$)

$$\varepsilon^{(\mathrm{f})}_{\alpha\beta} = \frac{1}{4\pi}\int \mathrm{d}\boldsymbol{r}\, G_{\alpha\beta}(\boldsymbol{r})\langle \delta\varepsilon(\boldsymbol{r})\delta\varepsilon(0)\rangle. \tag{4.231}$$

It can be shown that eqn (4.231) reduces to eqn (4.223) for dilute spheroid.

† The formulation presented here was adapted from the work of Onuki and Kawasaki[69] for the binary solution near the critical point. Here the static approximation is used for the electromagnetic field. This treatment is similar to that given in refs 66 and 70.

Equation (4.231) is rewritten using eqn (4.206) and $\varepsilon(\boldsymbol{r}) = n(\boldsymbol{r})^2$, in terms of the refractive index:

$$n_{\alpha\beta}^{(\mathrm{f})} = \frac{n}{2\pi}\int \mathrm{d}\boldsymbol{r} G_{\alpha\beta}(\boldsymbol{r})\langle \delta n(\boldsymbol{r})\delta n(0)\rangle. \tag{4.232}$$

In the polymer solutions, $\delta n(\boldsymbol{r})$ can be assumed, as in the theory of light scattering,[25] to be proportional to the local segment density $\delta c(\boldsymbol{r})$,

$$\delta n(\boldsymbol{r}) = \left(\frac{\partial n}{\partial c}\right)\delta c(\boldsymbol{r}) \tag{4.233}$$

Hence, eqn (4.232) is rewritten as

$$n_{\alpha\beta}^{(\mathrm{f})} = \frac{n}{2\pi}\left(\frac{\partial n}{\partial c}\right)^2 \int \mathrm{d}\boldsymbol{r} G_{\alpha\beta}(\boldsymbol{r})\langle \delta c(\boldsymbol{r})\delta c(0)\rangle. \tag{4.234}$$

If we use the Fourier transform of $G_{\alpha\beta}(\boldsymbol{r})$,

$$G_{\alpha\beta}(\boldsymbol{r}) = -\frac{4\pi}{n^2}\int \frac{\mathrm{d}\boldsymbol{k}}{(2\pi)^3}(\hat{k}_\alpha \hat{k}_\beta - \tfrac{1}{3}\delta_{\alpha\beta})\exp(\mathrm{i}\boldsymbol{k}\cdot\boldsymbol{r}), \tag{4.235}$$

and express $\langle \delta c(\boldsymbol{r})\delta c(0)\rangle$ by the structure factor $g(\boldsymbol{k})$,

$$\langle \delta c(\boldsymbol{r})\delta c(0)\rangle \equiv c\int \frac{\mathrm{d}\boldsymbol{k}}{(2\pi)^3}\exp(-\mathrm{i}\boldsymbol{k}\cdot\boldsymbol{r})g(\boldsymbol{k}), \tag{4.236}$$

we finally have

$$n_{\alpha\beta}^{(\mathrm{f})} = -2\frac{c}{n}\left(\frac{\partial n}{\partial c}\right)^2 \int \frac{\mathrm{d}\boldsymbol{k}}{(2\pi)^3}(\hat{k}_\alpha \hat{k}_\beta - \tfrac{1}{3}\delta_{\alpha\beta})g(\boldsymbol{k}). \tag{4.237}$$

The form birefringence is thus directly related to the anisotropy in the structure factor.

4.7.3 Flow birefringence

Having obtained the basic formula, we can now study the flow birefringence of dilute polymer solutions. The contribution from the intrinsic birefringence is easily calculated. Comparing eqn (4.220) with eqn (4.134), we note that the intrinsic birefringence is proportional to the stress:

$$n_{\alpha\beta}^{(\mathrm{i})} = C\sigma_{\alpha\beta}^{(\mathrm{p})} \tag{4.238}$$

where

$$C = \frac{2\pi}{27k_BT}\frac{(n^2+2)^2}{n}\Delta\gamma \tag{4.239}$$

is a constant called the stress optical coefficient, which depends only on the local structure of the polymer. Hence the contribution to $[n]$ from the intrinsic birefringence is easily calculated from eqn (4.146) as

$$[n^{(i)}] = 2C[\eta]. \tag{4.240}$$

However, calculation of the form birefringence is tedious (see refs 66 and 70, and also Section 5.5). Here we will give a simple approximate treatment. Under weak shear flow eqn (4.115), the deformation of the structure factor $g(\boldsymbol{k})$ is proportional to the dimensionless shear rate $\kappa/\Gamma_{\boldsymbol{k}}$, and will be written as

$$g(\boldsymbol{k}) = g_{\text{eq}}(\boldsymbol{k}) + \delta g(\boldsymbol{k}) \quad \text{with} \quad \delta g(\boldsymbol{k}) \simeq (\kappa/\Gamma_{\boldsymbol{k}})g_{\text{eq}}(\boldsymbol{k}) \tag{4.241}$$

where $g_{\text{eq}}(\boldsymbol{k})$ is the structure factor at equilibrium and $\Gamma_{\boldsymbol{k}}$ is the characteristic decay rate of the dynamical structure factor (see Section 4.4). From eqns (4.237) and (4.241), the form birefringence is estimated as

$$n_{xy}^{(\text{f})} \simeq \kappa \frac{c}{n}\left(\frac{\partial n}{\partial c}\right)^2 \int \frac{\mathrm{d}\boldsymbol{k}}{(2\pi)^3} \frac{g_{\text{eq}}(\boldsymbol{k})}{\Gamma_{\boldsymbol{k}}}. \tag{4.242}$$

Since $\Gamma_{\boldsymbol{k}}$ and $g_{\text{eq}}(\boldsymbol{k})$ are given by

$$\Gamma_{\boldsymbol{k}} = \frac{1}{\tau_r} F_1(kR_g), \qquad g_{\text{eq}}(\boldsymbol{k}) = NF_2(kR_g), \tag{4.243}$$

eqn (4.242) is evaluated as

$$\begin{aligned} n_{xy}^{(\text{f})} &\simeq \kappa \frac{c}{n}\left(\frac{\partial n}{\partial c}\right)^2 \frac{N\tau_r}{R_g^3} \simeq \frac{c}{n}\left(\frac{\partial n}{\partial c}\right)^2 N \frac{\kappa\eta_s}{k_B T} \\ &\simeq \frac{\rho}{n}\left(\frac{\partial n}{\partial \rho}\right)^2 M \frac{\kappa\eta_s}{RT}. \end{aligned} \tag{4.244}$$

This result is also derived from eqn (4.223) if we use

$$\phi(\delta\varepsilon)^2 \simeq \frac{c}{R_g^3}\left(\frac{\partial \varepsilon}{\partial c}\right)^2, \quad \langle u_x u_y \rangle \simeq \kappa\tau_r \simeq \kappa \frac{\eta_s R_g^3}{k_B T}, \quad \tfrac{1}{3} - N_d \simeq 1. \tag{4.245}$$

The contribution to $[n]$ from the form birefringence is

$$[n^{(f)}] = \frac{2}{\rho\eta_s\kappa} n_{xy}^{(f)} = k_f \frac{1}{n}\left(\frac{\partial n}{\partial \rho}\right)^2 \frac{M}{RT} \tag{4.246}$$

where k_f is a certain numerical constant. This formula was first derived by Janeschitz-Kriegl,[62,71] who estimated k_f from experimental results to be 0.34.

Equations (4.240) and (4.246) indicate that $[n^{(\text{i})}]$ and $[n^{(\text{f})}]$ increase with molecular weight M as

$$[n^{(i)}] \propto M^{3\nu-1}, \qquad [n^{(f)}] \propto M. \tag{4.247}$$

Hence for large molecular weight, the form birefringence dominates. From eqns (4.240) and (4.246), it follows that

$$\frac{[n]}{[\eta]} = \frac{[n^{(i)}] + [n^{(f)}]}{[\eta]} = 2C + k_f \frac{(\partial n/\partial \rho)^2}{nRT} \frac{M}{[\eta]}. \tag{4.248}$$

This relationship has been confirmed experimentally.[72]

The form birefringence $[n^{(f)}]$ is always positive while the intrinsic birefringence $[n^{(i)}]$ can be positive or negative depending on the sign of $\Delta\gamma$. Since the observed birefringence is the sum of the two, the analysis of the flow birefringence is not easy in dilute solutions. Indeed, in the case of $\Delta\gamma < 0$, the measured birefringence shows anomalous behaviour.[61,72] This complication can be avoided if the effect of the form birefringence is eliminated by choosing the solvent such that $\partial n/\partial \rho = 0$. The experimental results on flow birefringence of dilute polymer solutions are reviewed in refs 61 and 62.

Appendix 4.I The Verdier–Stockmayer model[2]

In this model the polymer is made up of N beads connected by $N-1$ bonds, each having constant length b (see Fig. 4.1*b*). In a small time interval Δt, each bead makes the following jump with probability $w\,\Delta t$.

(i) For the internal beads (i.e., beads $2, 3, \ldots, N-1$)

$$\boldsymbol{R}_n \to (\boldsymbol{R}_{n+1} + \boldsymbol{R}_{n-1} - \boldsymbol{R}_n). \tag{4.I.1}$$

(ii) For the end beads ($n = 1$ or N)

$$\boldsymbol{R}_1 \to \boldsymbol{R}_2 - \boldsymbol{v}_0 \tag{4.I.2}$$

$$\boldsymbol{R}_N \to \boldsymbol{R}_{N-1} + \boldsymbol{v}_N \tag{4.I.3}$$

where $\boldsymbol{v}_0$ and $\boldsymbol{v}_N$ are randomly chosen vectors of length b.

To analyse this model, it is convenient to look at the bond vector $\boldsymbol{v}_n = \boldsymbol{R}_{n+1} - \boldsymbol{R}_n$ rather than $\boldsymbol{R}_n$. The transition rule for $\boldsymbol{v}_n$ is

$$\boldsymbol{v}_n \to \begin{cases} \boldsymbol{v}_{n+1} & \text{This happens with probability } w\Delta t, \\ \boldsymbol{v}_{n-1} & \text{This happens with probability } w\Delta t, \\ \boldsymbol{v}_n & \text{This happens with probability } 1 - 2w\Delta t. \end{cases} \tag{4.I.4}$$

For $n = 1$ or $N-1$, $\boldsymbol{v}_0$ or $\boldsymbol{v}_{N+1}$ mean the random vector.

Now let us calculate the bond correlation function

$$C_{nm}(t) = \langle \boldsymbol{v}_n(t) \cdot \boldsymbol{v}_m(0) \rangle. \tag{4.I.5}$$

This is obtained as follows. If the jump $\boldsymbol{v}_n \to \boldsymbol{v}_{n+1}$ occurs in the time interval Δt, then $C_{nm}(t+\Delta t)$ becomes equal to $C_{n+1,m}(t)$. This happens with probability $w\Delta t$. The jump $\boldsymbol{v}_n \to \boldsymbol{v}_{n-1}$ occurs also with probability $w\Delta t$. The probability that no jump occurs is $1-2w\Delta t$; hence

$$C_{nm}(t+\Delta t) = C_{n+1,m}(t)w\Delta t + C_{n-1,m}(t)w\Delta t + C_{nm}(t)(1-2w\Delta t) \quad (4.I.6)$$

or for $\Delta t \to 0$

$$\frac{\partial}{\partial t} C_{nm}(t) = w(C_{n+1,m}(t) + C_{n-1,m}(t) - 2C_{nm}(t)) \quad (4.I.7)$$

$C_{0m}(t)$ and $C_{Nm}(t)$ are zero since $\boldsymbol{v}_0$ and $\boldsymbol{v}_N$ are chosen without any correlation to the other bond vectors, i.e.,

$$C_{0m}(t) = 0, \qquad C_{Nm}(t) = 0. \quad (4.I.8)$$

The initial condition is

$$C_{nm}(0) = b^2 \delta_{nm}. \quad (4.I.9)$$

By using the coordinate defined by

$$H_p(t) = \sum_n \left(\frac{2}{N}\right)^{1/2} \sin\left(\frac{p\pi n}{N}\right) C_{nm}(t) \quad (4.I.10)$$

the simultaneous difference equation (4.I.7) is solved giving

$$C_{nm}(t) = \frac{2}{N} b^2 \sum_p \sin\left(\frac{p\pi n}{N}\right) \sin\left(\frac{p\pi m}{N}\right) \exp(-\lambda_p t) \quad (4.I.11)$$

where

$$\lambda_p = 4w \sin^2\left(\frac{p\pi}{2N}\right). \quad (4.I.12)$$

On the other hand, for the Rouse model, the quantity corresponding to $C_{nm}(t)$ can be obtained by identifying $\boldsymbol{v}_n$ with $\partial \boldsymbol{R}_n / \partial n$:

$$C_{nm}^{(\mathrm{R})}(t) = \left\langle \frac{\partial \boldsymbol{R}_n(t)}{\partial n} \cdot \frac{\partial \boldsymbol{R}_m(0)}{\partial m} \right\rangle. \quad (4.I.13)$$

From eqns (4.23) and (4.27)

$$\begin{aligned} C_{nm}^{(\mathrm{R})}(t) &= 4 \sum_{p=1}^{\infty} \frac{3k_B T}{k_p} \left(\frac{p\pi}{N}\right)^2 \sin\left(\frac{p\pi n}{N}\right) \sin\left(\frac{p\pi m}{N}\right) \exp(-t/\tau_p) \\ &= \frac{2}{N} b^2 \sum_{p=1}^{\infty} \sin\left(\frac{p\pi n}{N}\right) \sin\left(\frac{p\pi m}{N}\right) \exp\left(-\frac{3\pi^2 k_B T}{\zeta N^2 b^2} p^2 t\right). \end{aligned} \quad (4.I.14)$$

This agrees with eqn (4.I.11) if one notes that for small p

$$\lambda_p = \frac{\pi^2 w}{N^2} p^2. \tag{4.I.15}$$

Thus w can be identified with $3k_BT/\zeta b^2$.

Appendix 4.II Derivation of the normal modes

To find the normal coordinates, we consider the linear transformation of $\boldsymbol{R}_n(t)$

$$\boldsymbol{X}_p(t) = \int_0^N \mathrm{d}n \phi_{pn} \boldsymbol{R}_n(t) \tag{4.II.1}$$

and choose ϕ_{pn} so that the equation of motion for $\boldsymbol{X}_p$ has the following form,

$$\zeta_p \frac{\partial}{\partial t} \boldsymbol{X}_p = -k_p \boldsymbol{X}_p + \boldsymbol{f}_p. \tag{4.II.2}$$

From eqns (4.9) and (4.II.1)

$$\zeta_p \frac{\partial \boldsymbol{X}_p}{\partial t} = \zeta_p \int_0^N \mathrm{d}n \phi_{pn} \frac{\partial}{\partial t} \boldsymbol{R}_n = \frac{\zeta_p}{\zeta} \int_0^N \mathrm{d}n \phi_{pn} \left(k \frac{\partial^2 \boldsymbol{R}_n}{\partial n^2} + \boldsymbol{f}_n \right). \tag{4.II.3}$$

Using integration by parts, we can rewrite this as

$$\text{rhs} = \frac{\zeta_p}{\zeta} \left[\phi_{pn} k \frac{\partial \boldsymbol{R}_n}{\partial n} \right]_0^N - \frac{\zeta_p}{\zeta} \left[k \frac{\partial \phi_{pn}}{\partial n} \boldsymbol{R}_n \right]_0^N + \frac{\zeta_p}{\zeta} \int_0^N \mathrm{d}n \left[k \frac{\partial^2 \phi_{pn}}{\partial n^2} \boldsymbol{R}_n + \phi_{pn} \boldsymbol{f}_n \right]. \tag{4.II.4}$$

The first term vanishes due to eqn (4.11). Hence eqn (4.II.2) is rewritten as

$$-\frac{\zeta_p}{\zeta} \left[k \frac{\partial \phi_{pn}}{\partial n} \boldsymbol{R}_n \right]_0^N + \frac{\zeta_p}{\zeta} \int_0^N \mathrm{d}n \left[k \frac{\partial^2 \phi_{pn}}{\partial n^2} \boldsymbol{R}_n + \phi_{pn} \boldsymbol{f}_n \right] = \int_0^N \mathrm{d}n (-k_p \phi_{pn} \boldsymbol{R}_n) + \boldsymbol{f}_p. \tag{4.II.5}$$

For eqn (4.II.5) to hold, we must have

$$\frac{\zeta_p}{\zeta} k \frac{\partial^2 \phi_{pn}}{\partial n^2} = -k_p \phi_{pn} \tag{4.II.6}$$

with

$$\frac{\partial \phi_{pn}}{\partial n} = 0 \quad \text{at} \quad n = 0 \text{ and } n = N \tag{4.II.7}$$

and

$$f_p = \frac{\zeta_p}{\zeta} \int_0^N \mathrm{d}n \phi_{pn} f_n. \tag{4.II.8}$$

Equations (4.II.6) and (4.II.7) are the eigenfunction equations for ϕ_{pn}, which are well known and have the solution

$$\phi_{pn} = \frac{1}{N} \cos\left(\frac{p\pi n}{N}\right) \qquad (p = 0, 1, 2, \ldots) \tag{4.II.9}$$

and

$$k_p = k \frac{\zeta_p}{\zeta} \left(\frac{p\pi}{N}\right)^2. \tag{4.II.10}$$

Now ζ_p can be chosen arbitrarily, so we choose ζ_p such that f_p satisfies the same formula as the oscillator (see eqn (3.82)); i.e.,

$$\langle f_{px}(t) f_{px}(0) \rangle = 2\zeta_p k_B T \delta(t). \tag{4.II.11}$$

The left-hand side can be calculated from eqns (4.II.8) and (4.II.9) as

$$\langle f_{p\alpha}(t) f_{q\beta}(0) \rangle = \frac{\zeta_p \zeta_q}{N^2 \zeta^2} \int_0^N \mathrm{d}n \int_0^N \mathrm{d}m \cos\left(\frac{p\pi n}{N}\right) \cos\left(\frac{q\pi m}{N}\right) \langle f_{n\alpha}(t) f_{m\beta}(0) \rangle \tag{4.II.12}$$

$$= \frac{\zeta_p \zeta_q}{N^2 \zeta^2} \int_0^N \mathrm{d}n \cos\left(\frac{p\pi n}{N}\right) \cos\left(\frac{q\pi n}{N}\right) 2\zeta k_B T \delta_{\alpha\beta} \delta(t)$$

$$= \frac{\zeta_p^2}{N^2 \zeta^2} \frac{1 + \delta_{p0}}{2} N \delta_{pq} 2\zeta k_B T \delta_{\alpha\beta} \delta(t) \tag{4.II.13}$$

where eqn (4.12) has been used. Comparing eqns (4.II.13) and (4.II.11) we have eqn (4.20) and thereby eqn (4.21) from eqn (4.II.10).

Appendix 4.III Dynamic structure factor

First we consider the Rouse model, for which $g(\boldsymbol{k}, t)$ can be calculated rigorously.[44] According to the theorem in Appendix 2.I, a linear combination of the Gaussian random variables obeys the Gaussian distribution. Since $\boldsymbol{R}_m(t) - \boldsymbol{R}_n(0)$ is a linear function of $f_n(t)$, which is Gaussian, the distribution of $\boldsymbol{R}_m(t) - \boldsymbol{R}_n(0)$ is also Gaussian. Hence, eqn

(2.I.20) gives

$$\langle \exp[i\boldsymbol{k}\cdot[\boldsymbol{R}_m(t)-\boldsymbol{R}_n(0)]]\rangle = \prod_{\alpha=x,y,z} \exp[-\tfrac{1}{2}k_\alpha^2\langle (R_{m\alpha}(t)-R_{n\alpha}(0))^2\rangle]$$

$$= \exp\left[-\frac{\boldsymbol{k}^2}{6}\phi_{mn}(t)\right] \qquad (4.\text{III}.1)$$

where

$$\phi_{mn}(t) = \langle (\boldsymbol{R}_m(t)-\boldsymbol{R}_n(0))^2\rangle. \qquad (4.\text{III}.2)$$

Using eqn (4.27), $\phi_{mn}(t)$ is written as

$$\phi_{mn}(t) = \left\langle \left[[\boldsymbol{X}_0(t)-\boldsymbol{X}_0(0)] + 2\sum_{p=1}^{\infty}\left(\cos\left(\frac{p\pi m}{N}\right)\boldsymbol{X}_p(t) - \cos\left(\frac{p\pi n}{N}\right)\boldsymbol{X}_p(0)\right)\right]^2\right\rangle. \qquad (4.\text{III}.3)$$

Since the correlation between different modes vanishes, eqn (4.III.3) can be rewritten as

$$\phi_{mn}(t) = \langle [\boldsymbol{X}_0(t)-\boldsymbol{X}_0(0)]^2\rangle + 4\sum_{p=1}^{\infty}\left[\cos^2\left(\frac{p\pi m}{N}\right)\langle \boldsymbol{X}_p(t)^2\rangle + \cos^2\left(\frac{p\pi n}{N}\right)\langle \boldsymbol{X}_p(0)^2\rangle - 2\cos\left(\frac{p\pi m}{N}\right)\cos\left(\frac{p\pi n}{N}\right)\langle \boldsymbol{X}_p(t)\cdot\boldsymbol{X}_p(0)\rangle\right]. \qquad (4.\text{III}.4)$$

From eqn (4.23) and

$$\sum_{p=1}^{\infty}\frac{1}{p^2}\left[\cos\left(\frac{p\pi m}{N}\right)-\cos\left(\frac{p\pi n}{N}\right)\right]^2 = \frac{\pi^2}{2N}|n-m|, \qquad (4.\text{III}.5)$$

we have

$$\phi_{mn}(t) = 6D_G t + |n-m|\,b^2 + \frac{4Nb^2}{\pi^2}\sum_{p=1}^{\infty}\frac{1}{p^2}\cos\left(\frac{p\pi m}{N}\right)\cos\left(\frac{p\pi n}{N}\right) \times (1-\exp(-p^2t/\tau_R)). \qquad (4.\text{III}.6)$$

where τ_R is the Rouse relaxation time given by eqn (4.25). Thus

$$g(\boldsymbol{k},t) = \frac{1}{N}\sum_{n,m}\exp\left[-\boldsymbol{k}^2D_Gt - \tfrac{1}{6}|n-m|\,\boldsymbol{k}^2b^2 - \frac{2Nb^2\boldsymbol{k}^2}{3\pi^2}\sum_{p=1}^{\infty}\frac{1}{p^2}\cos\left(\frac{p\pi n}{N}\right)\cos\left(\frac{p\pi m}{N}\right)[1-\exp(-tp^2/\tau_R)]\right]. \qquad (4.\text{III}.7)$$

The form of the $g(\boldsymbol{k},t)$ is simplified in the two limiting cases:

(i) Small angle regime: If $\boldsymbol{k}^2Nb^2 \ll 1$, the second and third terms in eqn (4.III.7) may be neglected since their magnitude is less than $\boldsymbol{k}^2Nb^2$.

Hence $g(\boldsymbol{k}, t)$ becomes

$$g(\boldsymbol{k}, t) = N \exp(-\boldsymbol{k}^2 D_G t) \tag{4.III.8}$$

which agrees with eqn (4.97).

(ii) Large angle regime: If $\boldsymbol{k}^2 N b^2 \gg 1$, we may limit consideration to the time region $t \ll \tau_R$ since $g(\boldsymbol{k}, t)$ becomes very small for $t \geq \tau_R$. Equation (4.III.6) can be rewritten as

$$\phi_{mn}(t) = 6D_G t + |n - m|\, b^2 + \frac{2Nb^2}{\pi^2} \sum_{p=1}^{\infty} \frac{1}{p^2} \times \left[\underline{\cos\left(\frac{p\pi(n+m)}{N}\right)} + \cos\left(\frac{p\pi(n-m)}{N}\right)\right](1 - \exp(-tp^2/\tau_R)). \tag{4.III.9}$$

For $t \ll \tau_R$, the sum is dominated by large p, for which the underlined term changes the sign rapidly and its contribution becomes very small. The remaining term is written by converting the sum over p to the integral:

$$g(\boldsymbol{k}, t) = \frac{1}{N} \int_0^N \mathrm{d}n \int_0^N \mathrm{d}m \exp\left[-\tfrac{1}{6}\boldsymbol{k}^2 |n - m|\, b^2 - \frac{\boldsymbol{k}^2 N b^2}{3\pi^2} \times \int_0^{\infty} \mathrm{d}p \frac{1}{p^2} \cos\left(\frac{p\pi(n-m)}{N}\right)(1 - \exp(-tp^2/\tau_R))\right]. \tag{4.III.10}$$

The integrand has a sharp peak at $n \simeq m$, so the double integral is evaluated as

$$\int_0^N \mathrm{d}n \int_0^N \mathrm{d}m = N \int_{-\infty}^{\infty} \mathrm{d}(n - m). \tag{4.III.11}$$

The final form is written as

$$g(\boldsymbol{k}, t) = \frac{12}{\boldsymbol{k}^2 b^2} \int_0^{\infty} \mathrm{d}u \exp[-u - (\Gamma_{\boldsymbol{k}} t)^{1/2} h(u(\Gamma_{\boldsymbol{k}} t)^{-1/2})] \tag{4.III.12}$$

where

$$\Gamma_{\boldsymbol{k}} = \frac{k_B T}{12\zeta} \boldsymbol{k}^4 b^2 \tag{4.III.13}$$

and

$$h(u) = \frac{2}{\pi} \int_0^{\infty} \mathrm{d}x \frac{\cos(xu)}{x^2}(1 - \exp(-x^2)). \tag{4.III.14}$$

For $\Gamma_k t \gg 1$, this expression is further simplified to

$$g(\boldsymbol{k}, t) = g(\boldsymbol{k}, 0)\int_0^\infty du \exp[-u - (\Gamma_k t)^{1/2} h(0)]$$
$$= g(\boldsymbol{k}, 0)\exp\left[-\frac{2}{\sqrt{\pi}}(\Gamma_k t)^{1/2}\right]. \quad (4.\text{III}.15)$$

For the Zimm model, the structure factor can be obtained in the same way if the linearization approximation[44,45] is used. The result for the Θ condition is

$$g(\boldsymbol{k}, t) = g(\boldsymbol{k}, 0)\int_0^\infty du \exp[-u - (\Gamma_k t)^{2/3} h(u(\Gamma_k t)^{-2/3})] \quad (4.\text{III}.16)$$

with

$$\Gamma_k = \frac{k_B T}{6\pi\eta_s} k^3 \quad (4.\text{III}.17)$$

and

$$h(u) = \frac{2}{\pi}\int_0^\infty dx \frac{\cos(xu)}{x^2}(1 - \exp(-x^{3/2}/\sqrt{2})). \quad (4.\text{III}.18)$$

For $\Gamma_k t \gg 1$,

$$g(\boldsymbol{k}, t) = g(\boldsymbol{k}, 0)\exp(-1.35(\Gamma_k t)^{2/3}). \quad (4.\text{III}.19)$$

Appendix 4.IV Polarizability tensor of a Gaussian chain

Let $\alpha_{\alpha\beta}(\boldsymbol{r}, N)$ be the average polarizability of a polymer chain consisting of N units with its end-to-end vector being fixed at $\boldsymbol{r}$. By symmetry, $\alpha_{\alpha\beta}(\boldsymbol{r}, N)$ is written as

$$\alpha_{\alpha\beta}(\boldsymbol{r}, N) = A(\boldsymbol{r}^2, N)(r_\alpha r_\beta - \tfrac{1}{3}\delta_{\alpha\beta}\boldsymbol{r}^2) \quad (4.\text{IV}.1)$$

where $A(\boldsymbol{r}^2, N)$ is a scalar. Equation (4.IV.1) can be expanded with respect to $\boldsymbol{r}^2$:

$$\alpha_{\alpha\beta}(\boldsymbol{r}, N) = (a_0(N) + a_1(N)\boldsymbol{r}^2 + \ldots)(r_\alpha r_\beta - \tfrac{1}{3}\delta_{\alpha\beta}\boldsymbol{r}^2). \quad (4.\text{IV}.2)$$

We shall show that $a_0(N)$ is proportional to $1/N$, and $a_1(N) = a_2(N) = \ldots = 0$ provided that (i) the statistics of the chain are Gaussian and that (ii) the polarizability is additive.

The second condition is written as follows. Suppose that the chain is divided into two parts each consisting of N_1 and N_2 units, then

$$\alpha_{\alpha\beta}(\boldsymbol{r}, N_1 + N_2) = \int d\boldsymbol{r}_1\, d\boldsymbol{r}_2 [\alpha_{\alpha\beta}(\boldsymbol{r}_1, N_1) + \alpha_{\alpha\beta}(\boldsymbol{r}_2, N_2)]\Phi(\boldsymbol{r}_1, \boldsymbol{r}_2) \quad (4.\text{IV}.3)$$

where $\Phi(\boldsymbol{r}_1, \boldsymbol{r}_2)$ is the equilibrium distribution of $\boldsymbol{r}_1$ and $\boldsymbol{r}_2$ under the constraint that $\boldsymbol{r}_1 + \boldsymbol{r}_2$ is fixed at $\boldsymbol{r}$. For the Gaussian chain, this is given by

$$\Phi(\boldsymbol{r}_1, \boldsymbol{r}_2) = \delta(\boldsymbol{r}_1 + \boldsymbol{r}_2 - \boldsymbol{r})\Psi(\boldsymbol{r}_1, N_1)\Psi(\boldsymbol{r}_2, N_2)/\Psi(\boldsymbol{r}, N_1 + N_2) \quad (4.\text{IV}.4)$$

where

$$\Psi(\boldsymbol{r}, N) = \left(\frac{3}{2\pi N b^2}\right)^{3/2} \exp\left(-\frac{3\boldsymbol{r}^2}{2Nb^2}\right). \quad (4.\text{IV}.5)$$

If we multiply eqn (4.IV.3) by $e^{i\boldsymbol{k}\cdot\boldsymbol{r}}\Psi(\boldsymbol{r}, N_1 + N_2)$ and integrate over $\boldsymbol{r}$, we have

$$\tilde{\alpha}_{\alpha\beta}(\boldsymbol{k}, N_1 + N_2)\tilde{\Psi}(\boldsymbol{k}, N_1 + N_2) = \tilde{\Psi}(\boldsymbol{k}, N_2)\tilde{\alpha}_{\alpha\beta}(\boldsymbol{k}, N_1)\tilde{\Psi}(\boldsymbol{k}, N_1) + \tilde{\Psi}(\boldsymbol{k}, N_1)\tilde{\alpha}_{\alpha\beta}(\boldsymbol{k}, N_2)\tilde{\Psi}(\boldsymbol{k}, N_2) \quad (4.\text{IV}.6)$$

where

$$\tilde{\Psi}(\boldsymbol{k}, N) = \exp(-Nb^2\boldsymbol{k}^2/6) \quad (4.\text{IV}.7)$$

and $\tilde{\alpha}_{\alpha\beta}(\boldsymbol{k}, N)$ is a differential operator:

$$\tilde{\alpha}_{\alpha\beta}(\boldsymbol{k}, N) = -\left(a_0 - a_1\frac{\partial^2}{\partial \boldsymbol{k}^2} + \dots\right)\left(\frac{\partial}{\partial k_\alpha}\frac{\partial}{\partial k_\beta} - \tfrac{1}{3}\delta_{\alpha\beta}\frac{\partial^2}{\partial \boldsymbol{k}^2}\right). \quad (4.\text{IV}.8)$$

Using eqn (4.IV.7), eqn (4.IV.6) is written as

$$\exp\left(\frac{b^2}{6}\boldsymbol{k}^2(N_1 + N_2)\right)\tilde{\alpha}_{\alpha\beta}(\boldsymbol{k}, N_1 + N_2)\exp\left(-\frac{b^2}{6}\boldsymbol{k}^2(N_1 + N_2)\right)$$
$$= \exp\left(\frac{b^2}{6}\boldsymbol{k}^2N_1\right)\tilde{\alpha}_{\alpha\beta}(\boldsymbol{k}, N_1)\exp\left(-\frac{b^2}{6}\boldsymbol{k}^2N_1\right) + \exp\left(\frac{b^2}{6}\boldsymbol{k}^2N_2\right)\tilde{\alpha}_{\alpha\beta}(\boldsymbol{k}, N_2)\exp\left(-\frac{b^2}{6}\boldsymbol{k}^2N_2\right). \quad (4.\text{IV}.9)$$

For this to be true for arbitrary N_1 and N_2, it must follow that

$$\exp\left(\frac{b^2}{6}\boldsymbol{k}^2N\right)\tilde{\alpha}_{\alpha\beta}(\boldsymbol{k}, N)\exp\left(-\frac{b^2}{6}\boldsymbol{k}^2N\right) = NB_{\alpha\beta}(\boldsymbol{k}) \quad (4.\text{IV}.10)$$

where $B_{\alpha\beta}(\boldsymbol{k})$ is a certain function of $\boldsymbol{k}$. Substitution of eqn (4.IV.8) and a straightforward calculation gives

$$\frac{b^4}{9}N^2(k_\alpha k_\beta - \tfrac{1}{3}\delta_{\alpha\beta}\boldsymbol{k}^2)\left[a_0(N) + \left(\frac{b^4N^2}{9}\boldsymbol{k}^2 - \tfrac{7}{3}b^2N\right)a_1(N) + \dots\right] = NB_{\alpha\beta}(\boldsymbol{k}). \quad (4.\text{IV}.11)$$

Comparison of the coefficients of the power of N indicates that this equality is satisfied only when

$$a_0(N) = c_0/N, \qquad a_1(N) = a_2(N) = \dots = 0. \quad (4.\text{IV}.12)$$

Choosing the constant c_0 as $\Delta\gamma/b^2$, we have

$$\alpha_{\alpha\beta}(\boldsymbol{r}, N) = \frac{\Delta\gamma}{Nb^2}(r_\alpha r_\beta - \tfrac{1}{3}\delta_{\alpha\beta}\boldsymbol{r}^2)$$

$$= \frac{\Delta\gamma}{\langle \boldsymbol{r}^2 \rangle_{\text{eq}}}(r_\alpha r_\beta - \tfrac{1}{3}\delta_{\alpha\beta}\boldsymbol{r}^2) \qquad (4.\text{IV}.13)$$

which is eqn (4.215).

References

1. Rouse, P. E., *J. Chem. Phys.* **21,** 1272 (1953).
2. Verdier, P. H., and Stockmayer, W. H., *J. Chem. Phys.* **36,** 227 (1962); and Verdier, P. H., *J. Chem. Phys.* **52,** 5512 (1970).
3. Orwoll, R. A., and Stockmayer, W. H., *Adv. Chem. Phys.* **15,** 305 (1969).
4. Iwata, K., *J. Chem. Phys.* **54,** 12 (1971).
5. Edwards, S. F., and Goodyear, A. G., *J. Phys.* **A5,** 965, 1188 (1972).
6. See general review, Stockmayer, W. H., Dynamics of chain molecules, in *Molecular Fluids* (eds R. Balian and G. Weill). Gordon & Breach, New York (1976).
7. Hilhorst, H. J., and Deutch, J. M., *J. Chem. Phys.* **63,** 5153 (1975).
8. Kuhn, W., *Kolloid Z.* **68,** 2 (1934).
9. Solc, K., and Stockmayer, W. H., *J. Chem. Phys.* **54,** 2756 (1971); see also Doi, M., and Nakajima, H., *Chem. Phys.* **6,** 124 (1974).
10. Zimm, B. H., *J. Chem. Phys.* **24,** 269 (1956).
11. Zimm, B. H., Roe, G. M., Epstein, L. F., *J. Chem. Phys.* **24,** 279 (1962). Hearst, H. E., *J. Chem. Phys.* **37,** 2547 (1962).
12. Yamakawa, H., *Modern Theory of Polymer Solutions,* Chapter 6. Harper & Row, New York (1971).
13. Edwards, S. F., and Freed, K. F., *J. Chem. Phys.* **61,** 1189 (1974).
14. Edwards, S. F., and Muthukumar, M., *Macromolecules* **17,** 586 (1984).
15. Peterlin, A., *J. Chem. Phys.* **23,** 2464 (1955).
16. Kurata, M., Yamakawa, H., and Teramoto, E., *J. Chem. Phys.* **28,** 785 (1958); Chikahisa, Y., and Tanake, T., *J. Polym. Sci.* **C 30,** 105 (1970).
17. Matsushita, Y., Noda, I., Nagasawa, M., Lodge, T. P., Amis, E. J., and Han, C. C., *Macromolecules* **17,** 1785 (1984).
18. Tanford, C., *Physical Chemistry of Macromolecules.* Wiley, New York 1961.
19. Oono, Y., and Kohmoto, M., *J. Chem. Phys.* **78,** 520 (1983).
20. de Gennes, P. G., *Macromolecules* **9,** 587, 594 (1976).
21. de Gennes, P. G., *Scaling Concepts in Polymer Physics,* Chapter 6. Cornell University Press, Ithaca (1979).
22. Hohenberg, P. C., and Halperin, B. I., *Rev. Mod. Phys.* **49,** 435 (1977).
23. Oono, Y., *J. Chem. Phys.* **79,** 4629 (1983).
24. Ceperley, D., Kalos, M. H., and Lebowitz, J. L., *Phys. Rev. Lett.* **41,** 313 (1978); *Macromolecules* **14,** 1472 (1981).
25. Berne, B. J., and Pecora, R., *Dynamic Light Scattering.* Wiley, New York (1976).

26. Akcasu, Z., and Gurol, H., *J. Polym. Sci. Phys. ed.* **14,** 1 (1976).
27. Kirkwood, J. G., *J. Polym. Sci.* **12,** 1 (1954).
28. King, T. A., Knox, A., Lee, W. I., and McAdam, J. D. G., *Polymer* **14,** 151 (1973).
29. Adam, M., and Delsanti, M., *J. Physique* **37,** 1045 (1976); *Macromolecules* **10,** 1229 (1977).
30. Nose, T., and Chu, B., *Macromolecules* **12,** 590 (1979).
31. Han, C. C., and Akcasu, A. Z., *Macromolecules* **14,** 1080 (1981).
32. Schmidt, M., and Burchard, W., *Macromolecules* **14,** 210 (1981).
33. Stockmayer, W. H., and Hammouda, B., *Pure & Appl. Chem.* **56,** 1373 (1984).
34. Guttman, C. M., McCrackin, F. L., and Han, C. C., *Macromolecules* **15,** 1205 (1982).
35. Yamakawa, H., *J. Chem. Phys.* **53,** 436 (1970).
36. Fixman, M., and Mansfield, M. L., *Macromolecules* **17,** 522 (1984).
37. Jasnow, D., and Moore, M. A., *J. Phys. (Paris)* **38,** L467 (1978); Al-Noaimi, G. F., Martinez-Mekler, G. C., and Wilson, C. A., *J. Phys. (Paris)* **39,** L373 (1978).
38. Lee, A., Baldwin, P. R., and Oono, Y., *Phys. Rev.* A **30,** 968 (1984).
39. Tsunashima, Y., Nemoto, N., and Kurata, M., *Macromolecules* **16,** 584, 1184 (1983); Nemoto, N., Makita, Y., Tsunashima, Y., and Kurata, M., *Macromolecules* **17,** 425 (1984).
40. des Cloizeaux, J., *J. Physique Lett.* **39,** L151 (1978).
41. Stockmayer, W. H., and Albrecht, A. C., *J. Polymer Sci.* **32,** 215 (1958).
42. Kurata, M., and Yamakawa, H., *J. Chem. Phys.* **29,** 311 (1958).
43. Weill, G., and des Cloizeaux, J., *J. Physique* **40,** 99 (1979).
44. de Gennes, P. G., *Physics* **3,** 37 (1967); Dubois-Violette, E., and de Gennes, P. G., *Physics* **3,** 181 (1967).
45. Akcasu, A. Z., Benmouna, M., and Han, C. C., *Polymer* **21,** 866 (1980).
46. Schaefer, D. W., and Han, C. C., in *Dynamic Light Scattering* (ed. R. Pecora). Plenum Press, New York (1985).
47. Allen, G., Ghosh, R., Higgins, J. S., Cotton, J. P., Farnoux, B., Jannink, G., and Weill, G., *Chem. Phys. Lett.* **38,** 577 (1976).
48. Richter, D., Hayter, J. B., Mezei, F., and Ewen, B., *Phys. Rev. Lett.* **41,** 1484 (1978).
49. Nicholson, L. K., Higgins, J. S., and Hayter, J. B., *Macromolecules* **14,** 836 (1981).
50. Ferry, J. D., *Viscoelastic Properties of Polymers.* (3rd edn). Wiley, New York (1980).
51. Johnson, R. M., Schrag, J. L., and Ferry, J. D., *Polymer J.* 742 (1970).
52. Tanaka, H., Sakanishi, A., Kaneko, M., and Furuichi, J., *J. Polymer Sci.* **C15,** 317 (1966); Sakanishi, A., *J. Chem. Phys.* **48,** 3850 (1968).
53. Osaki, K., *Adv. Polym. Sci.* **12,** 1 (1973).
54. Prager, S., *J. Chem. Phys.* **75,** 72 (1971).
55. Fixman, M., *J. Chem. Phys.* **78,** 1588 (1983). Wilemski, G., and Tanaka, G., *Macromolecules* **14,** 1531 (1981).
56. De Wames, R. E., Holland, W. F., and Shen, M. C., *J. Chem. Phys.* **40,** 2782 (1967); see also Zwanzig, R., Kiefer, J., and Weiss, G. H., *Proc. Natl. Acad. Sci. US* **60,** 381 (1968).

57. Rotne, J., and Prager, S., *J. Chem. Phys.* **50,** 4831 (1969).
58. Doi, M., unpublished work.
59. Zimm, B. H., *Macromolecules* **13,** 592 (1980).
60. Fixman, M., *J. Chem. Phys.* **78,** 1594 (1983).
61. Tsvetkov, V. N., in *Newer Methods of Polymer Characterization* (ed. B. Ke), *Polymer Rev.* 6. Interscience, New York (1964).
62. Janeschitz-Kriegl, H., Flow birefringence of elastico viscous polymer solutions. *Adv. Polym. Sci.* **6,** 170 (1969).
63. Kuhn, W., *Kolloid Z* **68,** 2 (1934); Kuhn, W., and Grün, F., *Kolloid Z.* **101,** 248 (1942); Kuhn, W., *J. Polym. Sci.* **1,** 360 (1946).
64. Volkenstein, M. V., *Configurational Statistics of Polymeric Chains.* Interscience, New York (1963).
65. Flory, P., *Statistical Mechanics of Chain Molecules.* Interscience, New York (1969).
66. Copic, M., *J. Chem. Phys.* **26,** 1382 (1957).
67. Peterlin, A., and Stuart, H. A., *Z. Physik* **112,** 1 (1939); see also Peterlin, A., in *Rheology,* (ed. F. R. Eirich) Vol 1, p. 615. Academic Press, New York (1956).
68. Landau, L. D., and Lifshitz, E. M., *Electrodynamics of Continuous Media.* Pergamon Press, Oxford (1960).
69. Onuki, A., and Kawasaki, K., *Physica* **111A,** 607 (1982); Onuki, A., and Doi, M., *J. Chem. Phys.* (1986) to be published.
70. Koyama, R., *J. Phys. Soc. Jpn* **16,** 1366 (1961); **19,** 1709 (1964). Daudi, S., *J. Physique* **38** 1301 (1977).
71. Janeschitz-Kriegl, H., *Macromol. Chem.* **40,** 140, (1959).
72. Frisman, E. V., and Tsvetkov, V. N., *J. Polym. Sci.* **30,** 297 (1958).

5

MANY CHAIN SYSTEMS

5.1 Semidilute and concentrated solutions

The physical properties of a polymer solution depend on solvent, temperature, and concentration. The solvents for polymers are broadly classified into two categories, good and poor solvents. Good solvents have strong attractive energy with polymers and dissolve polymers over a wide range of temperature. In such a solvent, the net interaction between polymer segments is repulsive (since they tend to contact with solvent molecules rather than themselves), and the excluded volume parameter v is positive and large. Poor solvents, on the other hand, are less keen to accommodate polymers, and precipitate polymers when the temperature is changed or the polymer concentration is increased. The excluded volume v in such a solvent can be positive, zero, or negative, depending on the temperature.

For the purpose of the discussion, polymer solutions in good solvents can be divided into three regions: dilute, semidilute, and concentrated (see Fig. 5.1).

A dilute solution is defined as one of sufficiently low concentration that the polymers are separated from each other; each polymer on average occupying a spherical region of radius R_g. (Fig. 5.1*a*) In this solution, the polymer–polymer interaction has only a small effect, and any physical property is expressed as a power series with respect to the polymer concentration ρ (weight of polymer in unit volume). Consider for example the osmotic pressure Π and the viscosity η. In the dilute limit, these are written as

$$\Pi = \frac{\rho}{M} RT, \tag{5.1a}$$

$$\eta = \eta_s(1 + \rho[\eta]) \tag{5.1b}$$

where R is the gas constant, M the molecular weight, T the temperature and η_s, the solvent viscosity. Equation (5.1*a*) is van't Hoff's law and eqn (5.1*b*) is the definition of the intrinsic viscosity $[\eta]$. The interaction among the polymers gives terms of order ρ^2, so that eqns (5.1*a*) and (5.1*b*) become

$$\Pi = RT\left(\frac{\rho}{M} + A_2\rho^2 + \ldots\right), \tag{5.2a}$$

$$\eta = \eta_s(1 + \rho[\eta] + k_H\rho^2[\eta]^2 + \ldots). \tag{5.2b}$$

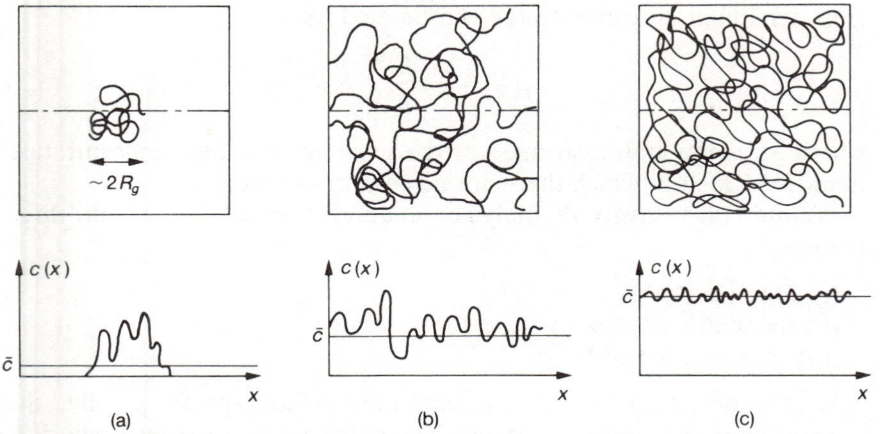

Fig. 5.1. Three concentration regimes in good solvent: (*a*) dilute, (*b*) semidilute, and (*c*) concentrated. $c(x)$ denotes the concentration profile along the dot-dashed lines.

The parameters A_2 and k_H are called the second virial coefficient and the Huggins coefficient, respectively.

As the concentration increases, the polymer coils come closer and start to overlap each other. Since the number of polymers in a unit volume is $\rho N_A/M$ the concentration ρ^* at which the overlap starts is estimated as

$$\frac{\rho^* N_A}{M}\tfrac{4}{3}\pi R_g^3 \simeq 1. \tag{5.3}$$

Note that the concentration ρ^* can be quite low. As R_g is proportional to M^ν, ρ^* depends on the molecular weight as

$$\rho^* \propto M^{1-3\nu} \simeq M^{-4/5} \quad (\text{for } \nu = 3/5). \tag{5.4}$$

Thus for large molecular weight, ρ^* becomes quite small; e.g., for polystyrene of $M \simeq 10^6$, ρ^* is about 0.005g/ml (about 0.5% in weight). Hence we can easily get a solution in which the molecules are strongly overlapped, but still occupy a small volume fraction. Such a solution is called semidilute.

A semidilute solution is characterized by the large and strongly correlated fluctuations in the segment density such as we have in dilute solutions. Although the fluctuations decrease with increased polymer concentration, the semidilute solution still retains the same character as critical phenomena in statistical mechanics, and presents the same kind of problem as we discussed in Sections 2.5 and 2.6.[1]

If the concentration becomes sufficiently large, the fluctuations become small and can be treated by a simple mean field theory. Such a solution is called concentrated. As we shall show later, the cross-over concentration from semidilute to concentrated is estimated by

$$\rho^{**} \simeq \frac{v(M_s/N_A)}{b^6} \tag{5.5}$$

where M_s is the molecular mass of the segment. The high concentration limit is the melt in which there are no solvent molecules.

We may thus classify the polymer solutions in good solvent into three regimes:

(i) dilute; $\rho < \rho^*$.
(ii) semidilute $\rho^* < \rho < \rho^{**}$.
(iii) concentrated $\rho^{**} < \rho$.

It may be noted that this classification is conceptual: usually, the crossover between various regimes is not sharp, and experimentally it is often difficult to identify the cross-over concentration.

In a poor solvent, the situation is more complicated. When the temperature is lowered, the excluded volume parameter v becomes negative, i.e., the net force between the segments becomes attractive. If the attractive force is strong, the polymers aggregate and phase separation takes place. The characteristic feature of the phase separation is described by the theory of Flory[2] and Huggins[3] and is shown schematically in Fig. 5.2.

The theory predicts that for large molecules, the critical temperature T_c

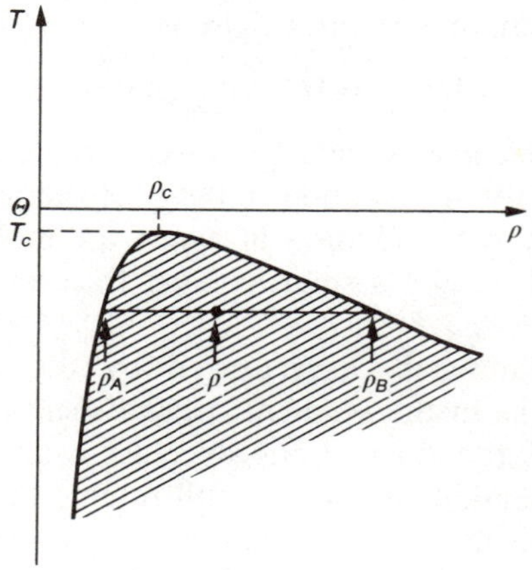

Fig. 5.2. Phase diagram of polymer solutions in poor solvent. The shaded area denotes the biphasic region; the system at the concentration ρ in the figure will separate into two phases of the concentration, ρ_A and ρ_B.

becomes very close to Θ temperature:

$$\Theta - T_c \propto M^{-1/2} \tag{5.6}$$

and the critical concentration ρ_c is small:

$$\rho_c \propto M^{-1/2}. \tag{5.7}$$

In the temperature region near to or less than the Θ temperature, it is no longer valid to use a pseudo-potential, and the detailed structure of the segmental interaction matters. For example, polymers in dilute solutions collapse well below the Θ temperature due to the intramolecular attractive force.[4] The radius of gyration of a fully collapsed polymer depends on the hard core radius b_h of the segment, and will be written as

$$R_g^3 \simeq N b_h^3, \tag{5.8}$$

and the attractive forces play no part.

In the temperature region $|T - \Theta| \lesssim \Theta - T_c$, classification of the solutions becomes involved since the properties depend on a subtle combination of the chain length, temperature, concentration, and additional molecular parameters such as the third virial coefficient w for segments (see eqn (2.90)), and the chain stiffness. Since the pioneering work of Daoud and Jannink,[5] a considerable amount of work has been done[6–8] in this region. However, as both the theoretical results and experimental results have not yet been settled and as the region is limited to a small area in the concentration–temperature diagram, we shall focus our attention only on good solvents in this book.

5.2 Gaussian approximation for concentration fluctuations

5.2.1 Collective coordinates

First we consider static properties of concentrated solutions in equilibrium. Let $\boldsymbol{R}_{an}$ be the position of the n-th segment of the a-th chain. The equilibrium distribution function for $\boldsymbol{R}_{an}$ is written as a natural generalization of eqn (2.95)

$$\Psi[\boldsymbol{R}_{an}] \propto \exp[-(U_0[\boldsymbol{R}_{an}] + U_1[\boldsymbol{R}_{an}])/k_B T] \tag{5.9}$$

where

$$U_0[\boldsymbol{R}_{an}]/k_B T = \sum_{a,n} \frac{3}{2b^2} (\boldsymbol{R}_{an} - \boldsymbol{R}_{an-1})^2 \tag{5.10}$$

is the energy of the chain connectivity and

$$U_1[\boldsymbol{R}_{an}]/k_B T = \tfrac{1}{2} \sum_{\substack{a,b \\ n,m}} v\delta(\boldsymbol{R}_{an} - \boldsymbol{R}_{bm}) \tag{5.11}$$

is the excluded volume interaction, which includes both the intramolecular interaction $a = b$ and the intermolecular interaction $a \neq b$. Since eqn (5.11) uses the pseudo potential, this model needs some alterations at high concentration where the detailed chemical structure matters, but here we will proceed using the simplest form.

Although the starting point is clear, theoretical development is not easy since calculation of any average from eqn (5.9) involves a highly nonseparable integration which is common to many body problems. However, under certain conditions we can progress by using collective coordinates.[9,10]

Instead of describing the problem in terms of $\boldsymbol{R}_{an}$, we focus our attention on the local segment density $c(\boldsymbol{r})$ defined by

$$c(\boldsymbol{r}) = \sum_{a,n} \delta(\boldsymbol{r} - \boldsymbol{R}_{an}) \tag{5.12}$$

and consider the distribution function $\Psi[c(\boldsymbol{r})]$ for $c(\boldsymbol{r})$. This method is effective if the physical quantiy under consideration is expressed by $c(\boldsymbol{r})$

For the mathematical development, it is convenient to use the Fourier transform of $c(\boldsymbol{r})$:

$$c_{\boldsymbol{k}} = \frac{1}{V}\int \mathrm{d}\boldsymbol{r} \exp(\mathrm{i}\boldsymbol{k}\cdot\boldsymbol{r})c(\boldsymbol{r}) = \frac{1}{V}\sum_{a,n} \exp[\mathrm{i}\boldsymbol{k}\cdot\boldsymbol{R}_{an}] \tag{5.13}$$

$$c(\boldsymbol{r}) = \sum_{\boldsymbol{k}} c_{\boldsymbol{k}} \exp(-\mathrm{i}\boldsymbol{k}\cdot\boldsymbol{r}) = \frac{V}{(2\pi)^3}\int \mathrm{d}\boldsymbol{k} c_{\boldsymbol{k}} \mathrm{e}^{-\mathrm{i}\boldsymbol{k}\cdot\boldsymbol{r}} \tag{5.14}$$

where V is the volume of the system.

The component c_0 is equal to the average concentration c, and does not fluctuate at all, while $c_{\boldsymbol{k}}$ with $\boldsymbol{k} \neq 0$ denotes the fluctuation of the local concentration.

A technical remark has to be made here. In the representation of $\{c_{\boldsymbol{k}}\}$, not all $c_{\boldsymbol{k}}$ are independent of each other since $c_{\boldsymbol{k}}$ and $c_{-\boldsymbol{k}}$ are related as

$$c_{-\boldsymbol{k}} = c_{\boldsymbol{k}}^* \tag{5.15}$$

where $c_{\boldsymbol{k}}^*$ is the complex conjugate of $c_{\boldsymbol{k}}$.

We shall choose those $c_{\boldsymbol{k}}$ with positive k_x values as the independent components, and use the abbreviated symbol $\boldsymbol{k} > 0$ to denote the set of the independent components. When the component $c_{\boldsymbol{k}}$ with negative k_x appears in the subsequent calculation, it should be remembered that it is $c_{\boldsymbol{k}}^*$. Thus for example

$$I = \sum_{\boldsymbol{k}>0} c_{\boldsymbol{k}} c_{-\boldsymbol{k}} \tag{5.16}$$

is the same as the sum

$$I = \sum_{k_x>0} \sum_{k_y=-\infty}^{\infty} \sum_{k_z=-\infty}^{\infty} c_{\boldsymbol{k}} c_{\boldsymbol{k}}^* \tag{5.17}$$

which is expressed by the sum over the entire $\boldsymbol{k}$ space as

$$\sum_{\boldsymbol{k}>0} c_{\boldsymbol{k}} c_{-\boldsymbol{k}} = \tfrac{1}{2}\sum_{\boldsymbol{k}} c_{\boldsymbol{k}} c_{-\boldsymbol{k}} - \tfrac{1}{2}c_0^2 = \tfrac{1}{2}\sum_{\boldsymbol{k}} c_{\boldsymbol{k}} c_{\boldsymbol{k}}^* - \tfrac{1}{2}c_0^2. \tag{5.18}$$

Now the distribution function of $c_{\boldsymbol{k}}$ is given by

$$\Psi(\{c_{\boldsymbol{k}}\}) \propto \int \prod_a \delta \boldsymbol{R}_{an} \exp(-(U_0[\boldsymbol{R}_{an}] + U_1[\boldsymbol{R}_{an}])/k_B T) \times \prod_{\boldsymbol{k}>0} \delta\Big(c_{\boldsymbol{k}} - \frac{1}{V}\sum_{a,n} \exp(\mathrm{i}\boldsymbol{k}\cdot\boldsymbol{R}_{an})\Big). \tag{5.19}$$

By use of the formula

$$\delta(\boldsymbol{r}-\boldsymbol{r}') = \frac{1}{V}\sum_{\boldsymbol{k}} \exp(\mathrm{i}\boldsymbol{k}\cdot(\boldsymbol{r}-\boldsymbol{r}')), \tag{5.20}$$

$U_1[\boldsymbol{R}_{an}]$ is rewritten as

$$\begin{aligned} U_1[\boldsymbol{R}_{an}]/k_B T &= \frac{v}{2V}\sum_{\boldsymbol{k}} \sum_{\substack{a,b\\n,m}} \exp[\mathrm{i}\boldsymbol{k}\cdot(\boldsymbol{R}_{an}-\boldsymbol{R}_{bm})] \\ &= \frac{vV}{2}\sum_{\boldsymbol{k}} \underline{\Big(\frac{1}{V}\sum_{a,n}\exp(\mathrm{i}\boldsymbol{k}\cdot\boldsymbol{R}_{an})\Big)} \underline{\Big(\frac{1}{V}\sum_{b,m}\exp(-\mathrm{i}\boldsymbol{k}\cdot\boldsymbol{R}_{bm})\Big)}. \end{aligned} \tag{5.21}$$

If this is substituted into eqn (5.19), the underlined terms can be replaced by $c_{\boldsymbol{k}}$ and $c_{-\boldsymbol{k}}$ because of the delta functions. Equation (5.19) is thus rewritten as

$$\Psi(\{c_{\boldsymbol{k}}\}) \propto \exp\Big[-\frac{V}{2}v\sum_{\boldsymbol{k}} c_{\boldsymbol{k}} c_{-\boldsymbol{k}}\Big] \times \int \prod_a \delta\boldsymbol{R}_{an} \exp(-U_0[\boldsymbol{R}_{an}]/k_B T) \prod_{\boldsymbol{k}>0} \delta\Big(c_{\boldsymbol{k}} - \frac{1}{V}\sum_{a,n}\exp(\mathrm{i}\boldsymbol{k}\cdot\boldsymbol{R}_{an})\Big). \tag{5.22}$$

Let $U_0(\{c_{\boldsymbol{k}}\})$ and $U_1(\{c_{\boldsymbol{k}}\})$ be defined by

$$U_0(\{c_{\boldsymbol{k}}\})/k_B T = -\ln \int \prod_a \delta\boldsymbol{R}_{an} \exp(-U_0[\boldsymbol{R}_{an}]/k_B T) \times \prod_{\boldsymbol{k}>0} \delta\Big(c_{\boldsymbol{k}} - \frac{1}{V}\sum_{a,n}\exp(\mathrm{i}\boldsymbol{k}\cdot\boldsymbol{R}_{an})\Big) \tag{5.23}$$

$$U_1(\{c_{\boldsymbol{k}}\})/k_B T = \frac{V}{2}v\sum_{\boldsymbol{k}} c_{\boldsymbol{k}} c_{-\boldsymbol{k}} \tag{5.24}$$

then

$$\Psi(\{c_{\boldsymbol{k}}\}) \propto \exp[-(U_0(\{c_{\boldsymbol{k}}\}) + U_1(\{c_{\boldsymbol{k}}\}))/k_B T]. \tag{5.25}$$

Calculation of $U_0(\{c_{\boldsymbol{k}}\})$ is difficult and it is only possible to give the explicit form as a series. A systematic method[9] is described in Appendix 5.I. (Another method is described in ref. 11). Here we use the Gaussian approximation (or random phase approximation), which is to say that the distribution of $c_{\boldsymbol{k}}$ is Gaussian, i.e., the free energy is written as

$$U_0(\{c_{\boldsymbol{k}}\})/k_B T = \sum_{\boldsymbol{k}} A_{0\boldsymbol{k}} c_{\boldsymbol{k}} c_{-\boldsymbol{k}} + \text{higher order terms in } c_{\boldsymbol{k}} \tag{5.26}$$

where $A_{0\boldsymbol{k}}$ is a constant to be determined. This approximation is justified in the concentrated solutions where the density fluctuations are small, while it is not adequate in the semidilute solutions.

To determine $A_{0\boldsymbol{k}}$, we use the fact that for noninteracting polymers ($v = 0$), $\langle c_{\boldsymbol{k}} c_{-\boldsymbol{k}} \rangle$ can be calculated exactly as a sum of the correlation functions of independent Gaussian chains,

$$\langle c_{\boldsymbol{k}} c_{-\boldsymbol{k}} \rangle_0 = \frac{c}{V} g_0(\boldsymbol{k}) \tag{5.27}$$

where $\langle \ldots \rangle_0$ denotes the average for the state of $v = 0$, and $g_0(\boldsymbol{k})$ is the scattering function of ideal polymers (see eqn (2.80)),

$$g_0(\boldsymbol{k}) = \frac{72N}{(Nb^2\boldsymbol{k}^2)^2}\left[\exp\left(-\frac{Nb^2\boldsymbol{k}^2}{6}\right) + \frac{Nb^2\boldsymbol{k}^2}{6} - 1\right]. \tag{5.28}$$

On the other hand, for $v = 0$, the distribution of $c_{\boldsymbol{k}}$ becomes

$$\Psi_0(\{c_{\boldsymbol{k}}\}) \propto \exp\left[-\sum_{\boldsymbol{k}} A_{0\boldsymbol{k}} c_{\boldsymbol{k}} c_{-\boldsymbol{k}}\right] = \exp\left[-2\sum_{\boldsymbol{k}>0} A_{0\boldsymbol{k}} c_{\boldsymbol{k}} c_{\boldsymbol{k}}^*\right] \tag{5.29}$$

which gives (see eqn (2.I.31))

$$\langle c_{\boldsymbol{k}} c_{-\boldsymbol{k}} \rangle_0 = \langle c_{\boldsymbol{k}} c_{\boldsymbol{k}}^* \rangle_0 = \frac{1}{2A_{0\boldsymbol{k}}}. \tag{5.30}$$

Comparing eqn (5.30) with eqn (5.27) we obtain $A_{0\boldsymbol{k}}$ as

$$A_{0\boldsymbol{k}} = \frac{V}{2cg_0(\boldsymbol{k})}. \tag{5.31}$$

From eqns (5.24), (5.26), and (5.31) the total energy

$$U(\{c_{\boldsymbol{k}}\}) = U_0(\{c_{\boldsymbol{k}}\}) + U_1(\{c_{\boldsymbol{k}}\}) \tag{5.32}$$

is obtained as

$$U(\{c_{\boldsymbol{k}}\})/k_B T = \frac{V}{2}\sum_{\boldsymbol{k}}\left(\frac{1}{cg_0(\boldsymbol{k})} + v\right) c_{\boldsymbol{k}} c_{-\boldsymbol{k}}. \tag{5.33}$$

For large $\boldsymbol{k}$, the asymptotic form of $g_0(\boldsymbol{k})$ gives

$$\frac{1}{cg_0(\boldsymbol{k})} + v = \frac{\boldsymbol{k}^2 b^2}{12c} + v. \tag{5.34}$$

This expression is also valid for small $\boldsymbol{k}$ because $1/cg_0(0) = 1/cN$ is negligibly small compared to v in the concentrated solution. Thus for the entire $\boldsymbol{k}$ values we have

$$U(\{c_{\boldsymbol{k}}\})/k_BT = \frac{V}{2}\sum_{\boldsymbol{k}}\left(\frac{\boldsymbol{k}^2b^2}{12c} + v\right)c_{\boldsymbol{k}}c_{-\boldsymbol{k}} = \frac{V}{2}\sum_{\boldsymbol{k}}\frac{b^2}{12c}(\boldsymbol{k}^2 + \xi^{-2})c_{\boldsymbol{k}}c_{-\boldsymbol{k}} \quad (5.35)$$

where

$$\xi = (b^2/12cv)^{1/2} \quad (5.36)$$

is called the correlation length.

The distribution function for $c_{\boldsymbol{k}}$ is

$$\Psi(\{c_{\boldsymbol{k}}\}) \propto \exp\left[-\frac{V}{2}\sum_{\boldsymbol{k}}\frac{b^2}{12c}(\boldsymbol{k}^2 + \xi^{-2})c_{\boldsymbol{k}}c_{-\boldsymbol{k}}\right]. \quad (5.37)$$

We shall now examine some consequences of this expression.

5.2.2 *Pair correlation function*

From eqn (5.37), $\langle c_{\boldsymbol{k}}c_{-\boldsymbol{k}}\rangle$ is given by

$$\langle c_{\boldsymbol{k}}c_{-\boldsymbol{k}}\rangle = \frac{1}{V}\frac{12c}{b^2(\boldsymbol{k}^2 + \xi^{-2})} \quad (5.38)$$

whence the scattering function per segment is

$$g(\boldsymbol{k}) \equiv \frac{V}{c}\langle c_{\boldsymbol{k}}c_{-\boldsymbol{k}}\rangle = \frac{12}{b^2(\boldsymbol{k}^2 + \xi^{-2})}. \quad (5.39)$$

The Fourier transform of $g(\boldsymbol{k})$ gives the pair correlation function:[9]

$$\begin{aligned}\langle c(\boldsymbol{r})c(0)\rangle - c^2 &= c\int\frac{d\boldsymbol{k}}{(2\pi)^3}g(\boldsymbol{k})e^{-i\boldsymbol{k}\cdot\boldsymbol{r}} \\ &= c\int_0^\infty\frac{4\pi k^2\,dk}{(2\pi)^3}\frac{12}{b^2(k^2+\xi^{-2})}\frac{\sin(kr)}{kr} \\ &= \frac{3c}{\pi b^2 r}\exp(-r/\xi). \quad (5.40)\end{aligned}$$

Equation (5.40) indicates the physical significance of ξ: ξ represents the correlation length of the concentration fluctuation.

Experimentally, the correlation length can be determined from the behaviour of $g(\boldsymbol{k})$ in the small $\boldsymbol{k}$ region:

$$\frac{g(0)}{g(\boldsymbol{k})} = 1 + \boldsymbol{k}^2\xi_{\text{app}}^2 \quad \text{for} \quad \boldsymbol{k}\to 0. \quad (5.41)$$

The definition of this apparent correlation length can be extended to dilute regimes, in which case ξ_{app} gives $R_g/\sqrt{3}$ (see eqn (2.72)). The behaviour of ξ_{app} is entirely different in dilute solutions and concentrated solutions. In dilute solutions, $\xi_{app} \simeq R_g$ increases with the molecular weight and the excluded volume, while in concentrated solutions $\xi_{app} = \xi$ is independent of the molecular weight and decreases as a function of concentration and excluded volume (see eqn (5.36)). The reason can be easily understood from Fig. 5.1: once polymers overlap each other, the excluded volume interaction tends to make the concentration homogeneous.

The simple theory given above is valid only at rather high concentration or at small excluded volume, i.e., near the Θ condition. At both these limits there are additional difficulties. At very high concentration the precise form of the potential matters. Also, near the Θ conditions the precise details of the interaction, in the sense of a cluster expansion passed to the two-body term, can also matter. In both of these limits it is possible to make an appropriate improvement,[6,7] and the results have been found to be in good agreement with experiments.[12,13]

5.2.3 Osmotic pressure

The free energy A of the system is obtained from the integration of eqn (5.9). It is convenient to take the system of non-interacting polymers ($v = 0$) as a reference state. The difference in the free energy $A - A_0$ between the two states is

$$\exp(-(A - A_0)/k_BT) = \frac{\int \prod_a \delta \boldsymbol{R}_{an} \exp(-(U_0[\boldsymbol{R}_{an}] + U_1[\boldsymbol{R}_{an}])/k_BT)}{\int \prod_a \delta \boldsymbol{R}_{an} \exp(-U_0[\boldsymbol{R}_{an}]/k_BT)} \tag{5.42}$$

which can be rewritten as

$$\exp(-(A - A_0)/k_BT)$$
$$= \frac{\int \prod_{k>0} \mathrm{d}c_k \delta\left(c_k - \frac{1}{V}\sum_{a,n} \exp(\mathrm{i}\boldsymbol{k}\cdot\boldsymbol{R}_{an})\right) \int \prod_a \delta\boldsymbol{R}_{an} \times \exp(-(U_0[\boldsymbol{R}_{an}] + U_1[\boldsymbol{R}_{an}])/k_BT)}{\int \prod_{k>0} \mathrm{d}c_k \delta\left(c_k - \frac{1}{V}\sum_{a,n} \exp(\mathrm{i}\boldsymbol{k}\cdot\boldsymbol{R}_{an})\right) \int \prod_a \delta\boldsymbol{R}_{an} \exp(-(U_0[\boldsymbol{R}_{an}]/k_BT)}$$
$$= \frac{\int \prod_{k>0} \mathrm{d}c_k \exp(-(U_0(\{c_k\}) + U_1(\{c_k\}))/k_BT)}{\int \prod_{k>0} \mathrm{d}c_k \exp(-U_0(\{c_k\})/k_BT)}. \tag{5.43}$$

In the Gaussian approximation, the integrals on the right-hand side can be performed rigorously (see Appendix 5.II). Given the free energy, the osmotic pressure is calculated by

$$\Pi = -\frac{\partial A}{\partial V} = k_B T\left(\frac{c}{N}\right) - \frac{\partial}{\partial V}(A - A_0). \tag{5.44}$$

The result is[9]

$$\Pi = k_B T\left(\frac{c}{N} + \tfrac{1}{2}vc^2 - \frac{\sqrt{3}}{\pi}\frac{(cv)^{3/2}}{b^3}\right). \tag{5.45}$$

The first two terms are easily understood: the first term represents van't Hoff's law $(c/N)k_BT$ (c/N being the number of polymers in unit volume), and the second term is the excluded volume interaction between the segments. The last term represents the correction to the second term due to the chain connectivity: i.e., the effect that the intramolecular excluded volume interaction does not contribute to the osmotic pressure.

From eqn (5.45) it will be seen that as $c \to 0$, the osmotic compressibility $(\partial\Pi/\partial c)$ becomes negative, which is of course incorrect. It follows that the theory presented above is only valid if

$$vc^2 \gtrsim \frac{(cv)^{3/2}}{b^3} \quad \text{i.e.,} \quad c \gtrsim c^{**} \simeq \frac{v}{b^6}. \tag{5.46}$$

This gives eqn (5.5). The situation $c < c^{**}$ is described by the theory of semidilute solutions.

Equation (5.45) indicates that as a result of the chain connectivity, the slope in the log Π – log c graph is larger than 2, the value predicted by the simple theory which disregards the correlation in the segmental distribution. This effect of correlation exists also in semidilute solutions, and as we shall see in the next section, the slope there is indeed larger than two. Thus the present theory smoothly crosses over to the semidilute regime. A theory which interpolates the two regimes in a more explicit way has been given by Muthukumar and Edwards,[14] and in the forthcoming book by des Cloizeaux and Jannink.[15]

5.2.4 *Size of a single chain*

In dilute solutions of good solvent, the size of a polymer is much larger than that at the Θ condition due to the excluded volume interaction. How does the polymer size behave in the concentrated solution? To answer this question let us suppose that added to the concentrated polymer solution is a test polymer of the same chemical structure and the same molecular weight. If the conformation of the test polymer is $\boldsymbol{R}_n$, the energy of the whole system is written as

$$U_{\text{tot}}[\boldsymbol{R}_n, c(\boldsymbol{r})]/k_BT = U[c(\boldsymbol{r})]/k_BT + \frac{3}{2b^2}\sum_n (\boldsymbol{R}_n - \boldsymbol{R}_{n-1})^2 + \frac{v}{2}\sum_{n,m}\delta(\boldsymbol{R}_n - \boldsymbol{R}_m) + v\sum_n c(\boldsymbol{R}_n). \tag{5.47}$$

The first term is the energy of the host polymer, which is given in the previous section, the second and third terms are the energy of the test polymer, and the last term denotes the interaction between the test polymer and the host polymer. In terms of $c_{\boldsymbol{k}}$, eqn (5.47) is written as

$$U_{\text{tot}}[\boldsymbol{R}_n, \{c_{\boldsymbol{k}}\}]/k_BT = U(\{c_{\boldsymbol{k}}\})/k_BT + \frac{3}{2b^2}\sum_n (\boldsymbol{R}_n - \boldsymbol{R}_{n-1})^2 + \frac{v}{2}\sum_{n,m}\delta(\boldsymbol{R}_n - \boldsymbol{R}_m) + v\sum_n\sum_{\boldsymbol{k}} c_{\boldsymbol{k}}\exp(-\mathrm{i}\boldsymbol{k}\cdot\boldsymbol{R}_n). \tag{5.48}$$

The distribution function for $\boldsymbol{R}_n$ is obtained by integrating over $c_{\boldsymbol{k}}$:

$$\Psi[\boldsymbol{R}_n] = \int \prod_{\boldsymbol{k}>0} \mathrm{d}c_{\boldsymbol{k}} \exp(-U_{\text{tot}}[\boldsymbol{R}_n, \{c_{\boldsymbol{k}}\}]/k_BT). \tag{5.49}$$

The functional integral over $c_{\boldsymbol{k}}$ is carried out using eqn (2.I.30)

$$\begin{aligned}
&\int \prod_{\boldsymbol{k}>0} \mathrm{d}c_{\boldsymbol{k}} \exp\Big[-U(\{c_{\boldsymbol{k}}\})/k_BT - v\sum_n\sum_{\boldsymbol{k}} c_{\boldsymbol{k}}\exp(-\mathrm{i}\boldsymbol{k}\cdot\boldsymbol{R}_n)\Big] \\
&= \int \prod_{\boldsymbol{k}>0} \mathrm{d}c_{\boldsymbol{k}} \exp\Big[-\sum_{\boldsymbol{k}>0}\Big(\frac{c_{\boldsymbol{k}}c_{-\boldsymbol{k}}}{\langle c_{\boldsymbol{k}}c_{-\boldsymbol{k}}\rangle} + c_{\boldsymbol{k}}v\sum_n \exp(-\mathrm{i}\boldsymbol{k}\cdot\boldsymbol{R}_n) \\
&\qquad + c_{\boldsymbol{k}}^* v\sum_m \exp(\mathrm{i}\boldsymbol{k}\cdot\boldsymbol{R}_m)\Big)\Big] \\
&\propto \exp\Big[\sum_{n,m}\sum_{\boldsymbol{k}>0}\langle c_{\boldsymbol{k}}c_{-\boldsymbol{k}}\rangle v^2 \exp[-\mathrm{i}\boldsymbol{k}\cdot(\boldsymbol{R}_n - \boldsymbol{R}_m)]\Big] \\
&= \exp\Big[\frac{1}{2}\sum_{n,m}\sum_{\boldsymbol{k}}\langle c_{\boldsymbol{k}}c_{-\boldsymbol{k}}\rangle v^2 \exp[-\mathrm{i}\boldsymbol{k}\cdot(\boldsymbol{R}_n - \boldsymbol{R}_m)]\Big].
\end{aligned} \tag{5.50}$$

Hence

$$\Psi[\boldsymbol{R}_n] \propto \exp\Big[-\frac{3}{2b^2}\sum_n (\boldsymbol{R}_n - \boldsymbol{R}_{n-1})^2 - \tfrac{1}{2}\sum_{n,m}\tilde{v}(\boldsymbol{R}_n - \boldsymbol{R}_m)\Big] \tag{5.51}$$

where

$$\tilde{v}(\boldsymbol{r}) = v\delta(\boldsymbol{r}) - \sum_{\boldsymbol{k}}\langle c_{\boldsymbol{k}}c_{-\boldsymbol{k}}\rangle v^2 \mathrm{e}^{-\mathrm{i}\boldsymbol{k}\cdot\boldsymbol{r}}. \tag{5.52}$$

Using eqns (5.36) and (5.38), we get

$$\begin{aligned}\tilde{v}(\boldsymbol{r}) &= v\left[\delta(\boldsymbol{r}) - \frac{1}{V}\sum_{\boldsymbol{k}}\frac{\xi^{-2}}{\boldsymbol{k}^2+\xi^{-2}}\exp(-\mathrm{i}\boldsymbol{k}\cdot\boldsymbol{r})\right] \\ &= v\int\frac{\mathrm{d}\boldsymbol{k}}{(2\pi)^3}\exp(-\mathrm{i}\boldsymbol{k}\cdot\boldsymbol{r})\left[1-\frac{\xi^{-2}}{\boldsymbol{k}^2+\xi^{-2}}\right] \\ &= v\left(\delta(\boldsymbol{r}) - \frac{\exp(-r/\xi)}{4\pi r\xi^2}\right). \end{aligned} \tag{5.53}$$

The potential $\tilde{v}(\boldsymbol{r})$ denotes an effective potential between the segments of the test polymer. The effective potential consists of a strong repulsive part ($v\delta(\boldsymbol{r})$) of very short interaction range, and a weak attractive part ($-v\exp(-r/\xi)/4\pi\xi^2 r$) of interaction range ξ. In the length-scale larger than ξ, these two parts cancel precisely: indeed from eqn (5.53) it can be shown that

$$\int \mathrm{d}\boldsymbol{r}\,\tilde{v}(\boldsymbol{r}) = 0. \tag{5.54}$$

Thus there is no excluded volume interaction among the segments whose mean separation is larger than ξ. This effect is called the screening of the excluded volume interaction.

As a consequence of the screening, the distribution of the conformation of the test polymer becomes Gaussian. Indeed given the effective excluded volume potential $\tilde{v}(\boldsymbol{r})$, it is easy to calculate the mean size of the polymer by use of the method described in Section 2.5.[16] For example, the straightforward perturbation calculation gives (see Appendix 5.III)

$$\langle \boldsymbol{R}^2\rangle = Nb^2\left[1+\frac{12}{\pi}\frac{v\xi}{b^4}\right]. \tag{5.55}$$

Note that the expansion coefficient $\langle \boldsymbol{R}^2\rangle/Nb^2$ does not depend on N. This is essentially different from the situation in dilute solutions, and indicates the screening of the excluded volume effect. Flory[17] first conjectured that the conformation of a polymer in concentrated solutions and melts becomes Gaussian, and the present theoretical calculation is that given by Edwards.[16]

If eqn (5.36) and the temperature dependence of v (eqn (2.94)) are substituted in eqn (5.55), the change in the size of the polymer in the concentrated solution is written as

$$\frac{\langle \boldsymbol{R}^2\rangle}{\langle \boldsymbol{R}^2\rangle_\Theta} - 1 \propto v\xi \propto v^{1/2}c^{-1/2} \propto \left(1-\frac{\Theta}{T}\right)^{1/2} c^{-1/2}. \tag{5.56}$$

The general aspect of this theoretical prediction was confirmed by neutron scattering of labelled polymers,[18] though a quantitative comparison needs more careful analysis of the other terms discussed in the previous sections. Approximate formulae which interpolate the dilute solution behaviour and the concentrated solution behaviour are given in refs 14, 15, 16.

5.3 Scaling theory—statics

As already mentioned, the theory in the previous section becomes inadequate in the semidilute region where the concentration fluctuations are large. This problem falls into the class of critical phenomena of statistical mechanics, and therefore there have been many calculations of it based on the renormalization group (see the reviews[19,20] for general references). However, it is possible to get the main results using the scaling arguments given in Chapter 2, and this is how it was first approached by de Gennes *et al.*[1,21]

Scaling theory considers how a physical quantity changes when we group λ segments into one. The transformation changes the parameters N and b as already discussed in Chapter 2 (eqn (2.124)). In the semidilute solution we have the additional parameter c, which changes from c to c/λ. Thus the rule of changes of the basic parameters is

$$N \to N/\lambda, \qquad c \to c/\lambda, \qquad b \to b\lambda^{\nu}. \tag{5.57}$$

If we know how a certain physical quantity changes under this transformation, we can discuss the dependence of this quantity on those parameters.

As an example, let us consider the apparent correlation length ξ_{app} defined by eqn (5.41). We can take the same line of argument as given in Section 2.6.

(i) From dimensional analysis ξ_{app} is written as

$$\xi_{app} = bF(N, cb^3). \tag{5.58}$$

(ii) Since ξ_{app} is invariant under the transformation (5.57) we have

$$b\lambda^{\nu}F(N/\lambda, cb^3\lambda^{3\nu-1}) = F(N, cb^3). \tag{5.59}$$

For this to be satisfied ξ_{app} must be written as†

$$\xi_{app} = N^{\nu}bF(cb^3N^{3\nu-1}). \tag{5.60}$$

This is rewritten by using the radius of gyration in infinite dilution

† Here a single symbol F is used to denote various functions for the sake of notational simplicity.

$R_g^{(0)} \simeq N^\nu b$ and the segment density c^* at the overlap concentration

$$c^* \simeq N/(R_g^{(0)})^3 \simeq N^{1-3\nu}/b^3, \tag{5.61}$$

as

$$\xi_{\text{app}} = R_g^{(0)} F(c/c^*) \text{ (dilute and semidilute).} \tag{5.62}$$

This equation is valid in both the dilute and semidilute regimes.

In the semidilute regime, the functional form can be further specified.

(iii) In the overlapped state of polymers, the correlation length ξ ($=\xi_{\text{app}}$ in this region) should be a function of the segment density c only, and be independent of N. (To see this, imagine that the polymers in the highly entangled state shown in Fig. 5.1*b* are cut into halves. The behaviour of the concentration fluctuation would not be changed by this operation.) For eqn (5.60) to be independent of N, it must be written as

$$\xi \simeq N^\nu b(cb^3 N^{3\nu-1})^x \tag{5.63}$$

with

$$\nu + (3\nu - 1)x = 0, \quad \text{i.e.,} \quad x = -\frac{\nu}{3\nu - 1}. \tag{5.64}$$

Thus

$$\xi \simeq R_g^{(0)}\left(\frac{c}{c^*}\right)^x \propto c^{-\nu/(3\nu-1)} \propto c^{-3/4} \quad \text{(semidilute).} \tag{5.65}$$

Next we consider the structure factor $g(\boldsymbol{k})$. Since $g(\boldsymbol{k})$ is dimensionless, and changes as $g(\boldsymbol{k}) \to g(\boldsymbol{k})/\lambda$ under the scaling transformation, it is written as

$$g(\boldsymbol{k}) = F(kb, N, cb^3) \tag{5.66}$$

with

$$F(kb\lambda^\nu, N\lambda^{-1}, cb^3\lambda^{3\nu-1}) = \lambda^{-1}F(kb, N, cb^3). \tag{5.67}$$

Thus $g(\boldsymbol{k})$ can be written as

$$g(\boldsymbol{k}) = cb^3N^{3\nu}F(kN^\nu b, cb^3N^{3\nu-1}), \tag{5.68}$$

or using eqn (5.60)

$$g(\boldsymbol{k}) = c\xi_{\text{app}}^3 F(k\xi_{\text{app}}, c/c^*) \quad \text{(dilute and semidilute).} \tag{5.69}$$

In the semidilute regime, $g(\boldsymbol{k})$ must be independent of N, which gives

$$g(\boldsymbol{k}) = c\xi^3 F(k\xi) \quad \text{(semidilute).} \tag{5.70}$$

Similarly it can be shown that the osmotic pressure is given by

$$\Pi = \frac{ck_BT}{N}F(c/c^*) \quad \text{(dilute and semidilute).} \tag{5.71}$$

In the semidilute regime, Π will again be independent of N, so that eqn (5.71) predicts

$$\Pi \simeq \frac{ck_BT}{N}(c/c^*)^{1/(3\nu-1)} \propto c^{9/4} \text{ (semidilute).} \tag{5.72}$$

To compare these results with experiments, it is convenient to express the concentration by ρ (the weight of polymer in unit volume). Equation (5.71) is then rewritten as

$$\Pi = \frac{\rho RT}{M} F(\rho/\rho^*). \tag{5.73}$$

In the dilute region, this must be expanded with respect to ρ

$$\Pi = \frac{\rho RT}{M}\left[1 + a_2\frac{\rho}{\rho^*} + a_3\left(\frac{\rho}{\rho^*}\right)^2\right] \tag{5.74}$$

where a_2, a_3 are some numerical factors. Comparing eqn (5.74) with eqn (5.1), we have

$$A_2 = \frac{a_2}{M\rho^*} \propto M^{3\nu-2} \propto M^{-1/5}. \tag{5.75}$$

In the semidilute region, eqn (5.72) gives

$$\Pi \simeq \frac{\rho RT}{M}\left(\frac{\rho}{\rho^*}\right)^{1/(3\nu-1)} \propto \rho^{9/4}. \tag{5.76}$$

These results are confirmed by the experiments of Noda *et al.*[22] (see Fig. 5.3).

The radius of gyration of a single chain is also derived by the scaling argument. By dimensional analysis and the scaling argument, one can show that the radius of gyration must be written as

$$R_g = R_g^{(0)} F(c/c^*). \tag{5.77}$$

In the semidilute region, the statistical distribution of the chain becomes Gaussian and R_g must be proportional to $N^{1/2}$. This requirement gives

$$R_g = R_g^{(0)}(c/c^*)^{(1-2\nu)/2(1-3\nu)} \propto N^{1/2}c^{-1/8}. \tag{5.78}$$

This prediction on the shrinkage of the polymer chains in semidilute solution was confirmed by neutron scattering.[21]

Though the scaling argument gives only a qualitative feature of the parameter dependence, explicit functional form can be calculated by the renormalization group method.[19,23,24] For example Ohta *et al.*[25] gives by ε

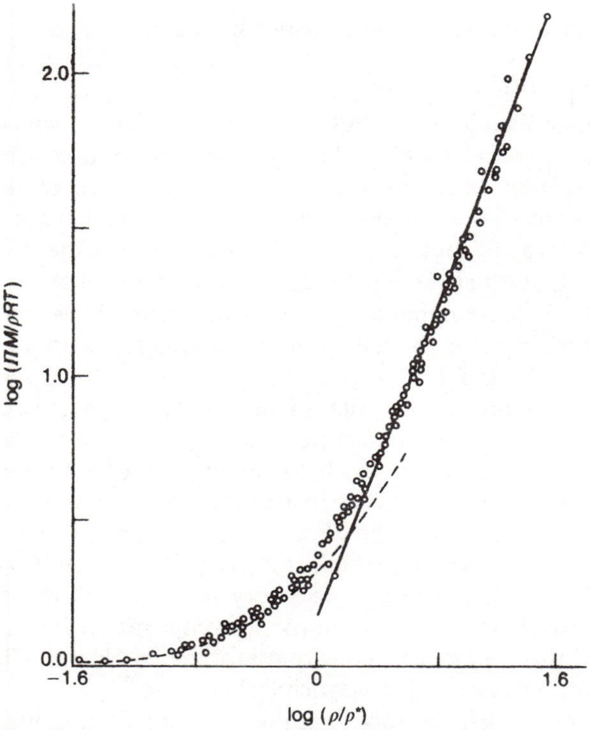

Fig. 5.3. Reduced osmotic pressure of poly(α methylstyrenes) in toluene at 25°C is plotted against reduced concentration ρ/ρ^*. Polymers of four molecular weights ranging from 7.1×10^4 to 1.2×10^6 are shown. The slope of the full line is 1.32. Reproduced from ref. 22.

expansion

$$\frac{M}{RT}\left(\frac{\partial \Pi}{\partial \rho}\right) = 1 + \frac{1}{8}\left[9X - 2 + \frac{2\ln(1+X)}{X}\right] \times \exp\left[\frac{1}{4}\left(\frac{1}{X} + (1 - X^{-2})\ln(1+X)\right)\right] \quad (5.79)$$

where X is a parameter proportional to $c/c^* = \rho/\rho^*$, or is more explicitly expressed by the second virial coefficient A_2,

$$X = \frac{16}{9} A_2 \rho M. \quad (5.80)$$

This prediction is in good agreement with the experimental result of Wiltzius *et al.*[26] Explicit calculation for $\xi_{\rm app}$ was done by Nakanishi and Ohta.[27]

5.4 Topological interaction in polymer dynamics

5.4.1 Entanglement effect

Having discussed static properties, we now consider dynamical problems in many chain systems. Here we have to consider another very important type of interaction which arises from the very nature of the polymer: polymers are one-dimensionally connected objects and cannot cross each other. A good way of looking at this is to imagine that the polymers have no thickness, and no attractive force, like a mathematical curve in space. Clearly the excluded volume of such a chain is zero. However, even such chains can interact strongly due to the topological constraints that chains cannot cross each other.

Consider two loops initially placed separately in space as in Fig. 5.4*a*. If one wants to calculate the partition functions of such systems, one has to exclude the configurations such as shown in Fig. 5.4*b* and *c* because such configurations are not accessible due to the topological constraints. The net effect of this is that the effective interaction between the loops is repulsive; the second virial coefficient A_2 is positive even if $v = 0$.[28,29]

The topological interaction is also very important in the problems of rubbers, in which the configurations of composite chains are severely restricted by the topological constraints of other chains. This gives an additional contribution to the elasticity of rubbers.[30,31]

For linear polymers, the topological constraints do not affect the static problems at all since all configurations are accessible. However, the topological interaction seriously affects the dynamical properties since it imposes constraints on the motion of polymers. Indeed it is a crucial factor in the dynamics of polymer solutions above the overlap concentration.†

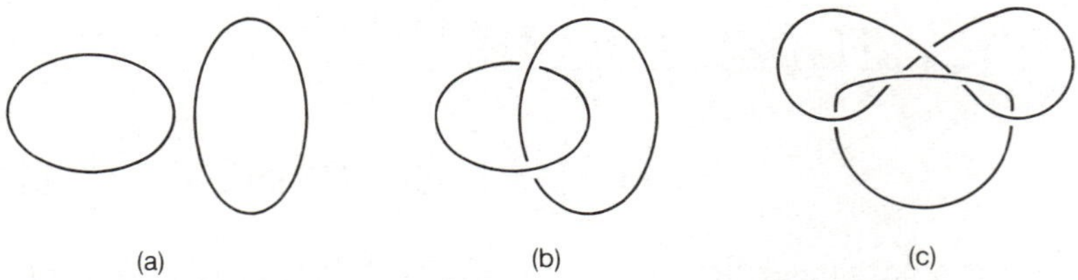

Fig. 5.4. Two loops in various topological states.

† The topological constraints exist in the single chain problem, but the effect is generally considered not to be serious because the properties in dilute solutions are usually dominated by the external modes (translation and rotation of the chain as a whole) for which the topological constraints are not important. This conjecture is supported by the scaling theory and results of a computer simulation;[32] both indicate that the topological constraints affect the numerical factor only, although the conclusion is not yet definitive.

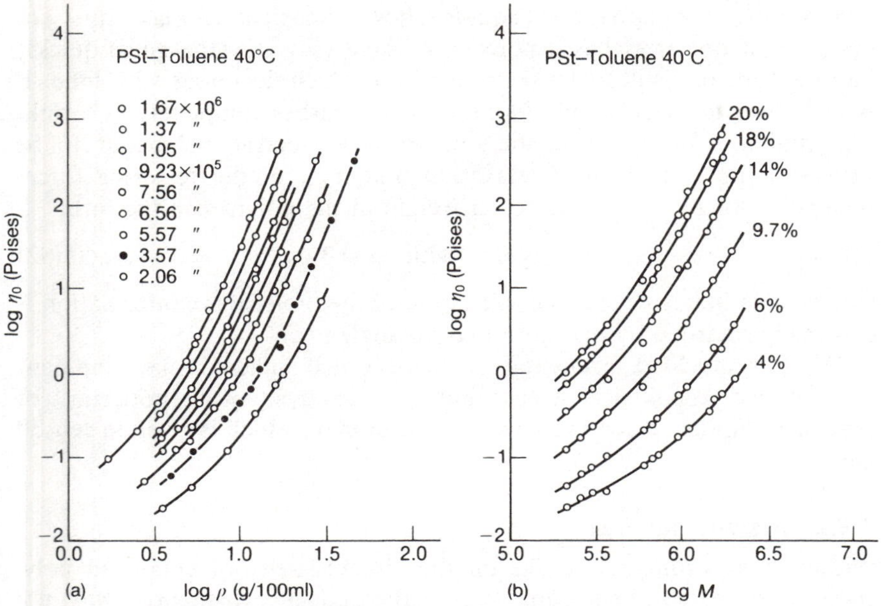

Fig. 5.5. Steady state viscosity at zero shear rate of polystyrene in toluene is plotted against (*a*) concentration and (*b*) molecular weight. Reproduced from ref. 66.

This has been well realized in the study of mechanical properties, which become quite conspicuous above the overlap concentration.[30,31]

Firstly, the viscosity of polymer solutions increases steeply (roughly in proportion to $\rho^{4\sim5}$) above the overlap concentration, and becomes much larger than the solvent viscosity. An example is shown in Fig. 5.5*a*. Notice that for high molecular weight polymers, even at the concentration of only 10%, the viscosity is several orders of magnitude larger than the solvent viscosity, which is about 0.01 poise. Such high viscosity is a result of molecular entanglement in the state shown in Fig. 5.1*b*. It can be easily imagined that the viscosity in such a state will depend strongly on the chain length. Indeed experiments indicate that at constant concentration ρ, the viscosity depends on the molecular weight as (see Fig. 5.5*b*)

$$\eta \propto M^x \tag{5.81}$$

where the exponent x is about 3.4. This behaviour is rather universal, independent of temperature, solvent and the molecular species (as long as the polymer is linear and flexible), which indicates that the phenomena are governed by the general nature of polymers.

Secondly, the polymeric liquids show conspicuous elasticity. For example, if one stretches a polymeric liquid (say chewing gum) quickly and holds it, one will feel a restoring force which decreases with time. If one releases the specimen before the force vanishes completely, it shrinks like rubber. The characteristic time for such elastic behaviour to be observed (i.e., the longest relaxation time τ_{max} of the restoring force) depends strongly on the molecular weight of the polymer in the form

$$\tau_{max} \propto M^x \quad \text{with} \quad x \simeq 3.4 \tag{5.82}$$

which again indicates that the relaxation of the molecular conformation is extremely retarded by the molecular entanglement.

There is much experimental evidence which indicates the dominant role of the topological constraints in the dynamical properties of polymeric liquids, a comprehensive discussion on which is given in refs 30 and 31.

5.4.2 Rigorous approach

Although experimental results on the viscoelasticity of entangled polymers were established quite early, theoretical explanation was not successful for a long time. (Some of the early theories are reviewed in ref. 31). This is due to the difficulty of handling the topological constraints.

From the theoretical point of view, the topological constraints provide a unique class of problems. This interaction is quite singular: for example, in linear polymers its effect is null for static properties but quite serious for dynamical properties. The interaction has no parameter which characterizes the strength. Rigorous theoretical treatment of such interaction is quite difficult. This may be seen in a simple example.

Consider the problem of calculating the second virial coefficients A_2 of the ring polymers shown in Fig. 5.4. The first step in such a calculation is to find a mathematical expression which distinguishes the non-entangled state (such as (*a*)) from the entangled state (such as (*b*)). How can we do that? A way of doing it is to use topological invariants which remain the same as long as the polymers do not cross each other. A classical quantity found by Gauss is the integral for two loops:[33]

$$I = \frac{1}{4\pi}\oint \mathrm{d}n \oint \mathrm{d}m \left[\left(\frac{\mathrm{d}\boldsymbol{R}_{1n}}{\mathrm{d}n}\right) \times \left(\frac{\mathrm{d}\boldsymbol{R}_{2m}}{\mathrm{d}m}\right)\right] \cdot \frac{(\boldsymbol{R}_{1n} - \boldsymbol{R}_{2m})}{|\boldsymbol{R}_{1n} - \boldsymbol{R}_{2m}|^3} \tag{5.83}$$

where the integral is done over the closed contour of the loops 1 and 2. The integral is 0 for the configuration (*a*) in Fig. 5.4, and 1 for (*b*) and remains the same if the loops do not cross each other. (This can be proved by using Ampère's law in magnetism: the integral is equal to the number of times that the (directed) loop 1 passes through loop 2.) However, the Gaussian invariant (5.83) is not enough to specify the topological class of loops because, for example, the configuration (*c*) gives the same value of I as the configuration (*a*). A better specification is achieved by the use of Alexander polynomials,[34] which amounts to an analysis of the order of crossing of the projections of the curves on the surfaces,[35] but the expression becomes too complicated to be used for analytical calculation.

Despite the difficulty, progress has been made in the static problems (e.g., the calculation of A_2 of ring polymers[36–39] and the rubber elasticity[40,41]), by using the Gauss integral as a principal index for the topological class. Rigorous results can be obtained for the entanglement between a polymer on the plane and a line standing on it,[42–44] and the two-dimensional problem can generally be solved using Riemann surfaces.[45]

In the case of linear polymers, the entanglement effect appears only in the dynamical properties, and it appears at first sight one does not know where to start. Edwards[46] suggested that the entanglement effect can be described by the Smoluchowski equation,

$$\frac{\partial \Psi}{\partial t} = \sum_{\substack{an \\ bm}} \frac{\partial}{\partial \boldsymbol{R}_{an}} \cdot \boldsymbol{H}_{anbm} \cdot \left(k_B T \frac{\partial \Psi}{\partial \boldsymbol{R}_{bm}} + \frac{\partial U}{\partial \boldsymbol{R}_{bm}} \Psi \right), \tag{5.84}$$

because the mobility tensor $\boldsymbol{H}_{anbm}$ is zero for the pair of segments which are going to cross each other. To see how this works, consider the one-dimensional motion of two particles, 1 and 2 (see Fig. 5.6). If we

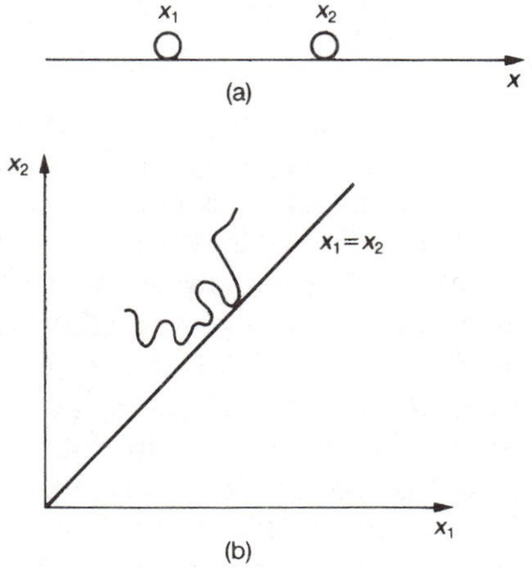

Fig. 5.6. (*a*) One-dimensional Brownian motion of two particles which cannot cross each other, (*b*) phase space of the system.

assume that the diffusion constant $D_{ij} = k_B T H_{ij}$ is a function of $|x_i - x_j|$ only, the Smoluchowski equation is written as

$$\frac{\partial \Psi}{\partial t} = \frac{\partial}{\partial x_1} D(0) \frac{\partial \Psi}{\partial x_1} + \frac{\partial}{\partial x_2} D(0) \frac{\partial \Psi}{\partial x_2} + \frac{\partial}{\partial x_1} D(x_1 - x_2) \frac{\partial \Psi}{\partial x_2} + \frac{\partial}{\partial x_2} D(x_2 - x_1) \frac{\partial \Psi}{\partial x_1}. \quad (5.85)$$

Using

$$X_1 = \frac{1}{\sqrt{2}}(x_1 + x_2), \qquad X_2 = \frac{1}{\sqrt{2}}(x_1 - x_2) \quad (5.86)$$

eqn (5.85) is rewritten as

$$\frac{\partial \Psi}{\partial t} = (D(0) + D(X_2)) \frac{\partial^2 \Psi}{\partial X_1^2} + \frac{\partial}{\partial X_2} \underline{(D(0) - D(X_2))} \frac{\partial \Psi}{\partial X_2}. \quad (5.87)$$

The underlined diffusion constant, which corresponds to the relative motion of the particles, vanishes when $X_2 = 0$, i.e., when $x_1 = x_2$. This guarantees that the flux of the representative point across the line $x_1 = x_2$ is zero, i.e., the particles cannot cross each other. Notice that this conclusion holds for any functional form of D (provided it is finite at $X_2 = 0$). An attractive feature of this approach is that it guarantees that the equilibrium properties with the topological constraints are the same as the ideal chains. However, rigorous solution of eqn (5.84) is not easy, and approximate treatment such as the preaveraging approximation does not give the correct molecular weight dependence for η and τ_{max}. At this moment, this idea still remains to be explored.

5.4.3 The tube model

Though rigorous theory for the topological interaction is extremely difficult, it has been shown that a highly entangled state can be treated by an effective model, the tube model. This model assumes that due to the topological constraints, the motion of the chain is essentially confined in a tube-like region made of the surrounding polymers (see Fig. 1.6). This model, originally proposed for the problem of rubbers,[40] offers a basis for the dynamics of chains in a network,[47] which has been quite successful in explaining many dynamical properties of entangled polymers.[48] We shall give a detailed account of it in the subsequent chapters.

Though the tube model is successful, our present understanding of the dynamics in entangled systems is still incomplete. Agreement between theory and experiments is not yet complete as we shall discuss later. More seriously, the tube model does not describe all aspects of the dynamics: it describes properties which depend on a single chain dynamics, but cannot so far handle problems which involves the collective motion of many chains. Progress can be made on aspects of this problem, particularly in concentrated and semidilute solutions, as against melts, and this we shall study in the rest of this chapter.

5.5 Dynamics of concentration fluctuations

5.5.1 Kinetic equation

In Section 5.2, it was shown that the Gaussian approximation for the collective coordinates c_k is quite useful for describing the static properties of polymers in concentrated solutions. It is a natural temptation to generalize this approach to dynamical problems. The central assumption to be made in this approach is that the set of coordinates $\{c_k\}$ are good variables for characterizing the state of the system, and that therefore a closed equation can be constructed for their time evolution. This approach is quite analogous to the critical dynamics for binary solutions of low molecular weight,[49] where the dynamics of the system is described by the phenomenological Langevin equation

$$\frac{\partial}{\partial t} c_k = \sum_{k'} - L_{kk'} \frac{\partial U(\{c_k\})}{\partial c_{k'}} + r_k \tag{5.88}$$

where $U(\{c_k\})$ is the free energy, $L_{kk'}$ are phenomenological kinetic coefficients and r_k are Gaussian random variables satisfying

$$\langle r_k \rangle = 0, \qquad \langle r_k(t) r_{k'}(t') \rangle = 2k_B T L_{kk'} \delta(t - t'). \tag{5.89}$$

In the polymer problem, the validity of this equation is not obvious since the description of the polymeric system by $\{c_k\}$ disregards the chain connectivity and, therefore, neglects the entanglement effect. However, as far as the dynamics in the short time-scale is concerned, this will not be a serious problem† since, as we shall discuss later, the topological constraints are not important in the short time-scale dynamics. Indeed it will be shown that the initial slope in the dynamical structure factor is correctly described in this approach. In the long time-scale, on the other hand, the validity of eqn (5.88) is not clear, and it may well be that the theory has to be modified in future. Fortunately, many experiments related to concentration fluctuations are concerned with the short time-scale motion, so that it is worthwhile to pursue the idea in detail.

† Here short time means that it is shorter than the reptation time τ_d which will be discussed in Chapter 6. For the phenomena which involve long time-scale, or large length-scale motion of polymers (as happens in the problem of phase separation[50] and Θ regimes[51]), the present theory is not valid.

Now for the polymer problems, $U(\{c_k\})$ has been calculated in Section 5.2.1. To determine $L_{kk'}$, we use the Langevin equation for $\boldsymbol{R}_{an}$ (see eqn (4.42)).

$$\frac{\partial}{\partial t}\boldsymbol{R}_{an} = \sum_{b,m} \boldsymbol{H}_{anbm} \cdot \left(-\frac{\partial U[\boldsymbol{R}_{an}]}{\partial R_{bm}} + \boldsymbol{f}_{bm}\right)$$
$$= -\sum_{b,m} \boldsymbol{H}_{anbm} \cdot \frac{\partial U[\boldsymbol{R}_{an}]}{\partial \boldsymbol{R}_{bm}} + \boldsymbol{r}_{an} \tag{5.90}$$

where $\boldsymbol{H}_{anbm} = \boldsymbol{H}(\boldsymbol{R}_{an} - \boldsymbol{R}_{bm})$ and

$$\boldsymbol{r}_{an} = \sum_{bm} \boldsymbol{H}_{anbm} \cdot \boldsymbol{f}_{bm} \tag{5.91}$$

whose time correlation function is

$$\langle \boldsymbol{r}_{an}(t)\rangle = 0,\ \langle \boldsymbol{r}_{an}(t)\boldsymbol{r}_{bm}(t')\rangle = 2k_BT\boldsymbol{H}_{anbm}\delta(t-t'). \tag{5.92}$$

Equations (5.13) and (5.90) give

$$\frac{\partial}{\partial t}c_k(t) = \frac{1}{V}\sum_{a,n} \mathrm{i}\boldsymbol{k} \cdot \frac{\partial \boldsymbol{R}_{an}}{\partial t}\exp(\mathrm{i}\boldsymbol{k}\cdot\boldsymbol{R}_{an})$$
$$= \frac{1}{V}\sum_{an,bm} \mathrm{i}\boldsymbol{k} \cdot \left(-\boldsymbol{H}_{anbm} \cdot \frac{\partial U[\boldsymbol{R}_{an}]}{\partial \boldsymbol{R}_{bm}} + \boldsymbol{r}_{an}\right)\exp(\mathrm{i}\boldsymbol{k}\cdot\boldsymbol{R}_{an}). \tag{5.93}$$

Comparing eqn (5.93) with eqn (5.88), we have

$$r_k(t) = \frac{1}{V}\sum_{an} \mathrm{i}\boldsymbol{k} \cdot \boldsymbol{r}_{an}(t)\exp(\mathrm{i}\boldsymbol{k}\cdot\boldsymbol{R}_{an}). \tag{5.94}$$

From eqn (5.92) and (5.94), it follows that

$$\langle r_k(t) r_{k'}(t')\rangle = \frac{-1}{V^2}\sum_{an,bm} \boldsymbol{k} \cdot \langle \boldsymbol{r}_{an}(t)\boldsymbol{r}_{bm}(t')\rangle \cdot \boldsymbol{k}'$$
$$\times \exp(\mathrm{i}\boldsymbol{k}\cdot\boldsymbol{R}_{an} + \mathrm{i}\boldsymbol{k}'\cdot\boldsymbol{R}_{bm})\rangle \tag{5.95}$$
$$= \frac{-1}{V^2}\sum_{an,bm} 2k_BT\delta(t-t')\boldsymbol{k}\boldsymbol{k}' : \boldsymbol{H}_{anbm}$$
$$\times \exp(\mathrm{i}\boldsymbol{k}\cdot\boldsymbol{R}_{an} + \mathrm{i}\boldsymbol{k}'\cdot\boldsymbol{R}_{bm}).$$

Using

$$\boldsymbol{H}_{anbm} = \int \frac{\mathrm{d}\boldsymbol{q}}{(2\pi)^3}\frac{\boldsymbol{I} - \hat{\boldsymbol{q}}\hat{\boldsymbol{q}}}{\eta_s \boldsymbol{q}^2}\exp(\mathrm{i}\boldsymbol{q}\cdot(\boldsymbol{R}_{an} - \boldsymbol{R}_{bm})) \tag{5.96}$$

we have

$$\langle r_{\boldsymbol{k}}(t) r_{\boldsymbol{k}'}(t')\rangle = \frac{-2\delta(t-t')}{V^2} k_B T \int \frac{\mathrm{d}\boldsymbol{q}}{(2\pi)^3} \frac{\boldsymbol{k}\cdot\boldsymbol{k}' - (\boldsymbol{k}\cdot\hat{\boldsymbol{q}})(\boldsymbol{k}'\cdot\hat{\boldsymbol{q}})}{\eta_s q^2}$$
$$\times \sum_{an,bm} \exp[\mathrm{i}(\boldsymbol{k}+\boldsymbol{q})\cdot\boldsymbol{R}_{an} + i(\boldsymbol{k}'-\boldsymbol{q})\cdot\boldsymbol{R}_{bm}]$$
$$= -2\delta(t-t') k_B T \int \frac{\mathrm{d}\boldsymbol{q}}{(2\pi)^3} \frac{\boldsymbol{k}\cdot\boldsymbol{k}' - (\boldsymbol{k}\cdot\hat{\boldsymbol{q}})(\boldsymbol{k}'\cdot\hat{\boldsymbol{q}})}{\eta_s q^2} c_{\boldsymbol{k}+\boldsymbol{q}} c_{\boldsymbol{k}'-\boldsymbol{q}}. \tag{5.97}$$

From eqns (5.89) and (5.97) we have

$$L_{\boldsymbol{k}\boldsymbol{k}'} = -\int \frac{\mathrm{d}\boldsymbol{q}}{(2\pi)^3} \frac{\boldsymbol{k}\cdot\boldsymbol{k}' - (\boldsymbol{k}\cdot\hat{\boldsymbol{q}})(\boldsymbol{k}'\cdot\hat{\boldsymbol{q}})}{\eta_s q^2} c_{\boldsymbol{k}+\boldsymbol{q}} c_{\boldsymbol{k}'-\boldsymbol{q}}. \tag{5.98}$$

Equations (5.88) and (5.98) give a nonlinear equation for $c_{\boldsymbol{k}}$. To proceed further, we use the preaveraging approximation, i.e., replace $L_{\boldsymbol{k}\boldsymbol{k}'}$ by its average in equilibrium

$$L_{\boldsymbol{k}\boldsymbol{k}'} \to \langle L_{\boldsymbol{k}\boldsymbol{k}'}\rangle_{\mathrm{eq}} = -\int \frac{\mathrm{d}\boldsymbol{q}}{(2\pi)^3} \frac{\boldsymbol{k}\cdot\boldsymbol{k}' - (\boldsymbol{k}\cdot\hat{\boldsymbol{q}})(\boldsymbol{k}'\cdot\hat{\boldsymbol{q}})}{\eta_s q^2} \langle c_{\boldsymbol{k}+\boldsymbol{q}} c_{\boldsymbol{k}'-\boldsymbol{q}}\rangle_{\mathrm{eq}}. \tag{5.99}$$

Since $\langle c_{\boldsymbol{k}+\boldsymbol{q}} c_{\boldsymbol{k}'-\boldsymbol{q}}\rangle_{\mathrm{eq}}$ is given by $(c/V)\delta_{\boldsymbol{k}-\boldsymbol{k}'} g(\boldsymbol{k}+\boldsymbol{q})$, $\langle L_{\boldsymbol{k}\boldsymbol{k}'}\rangle_{\mathrm{eq}}$ is written as

$$\langle L_{\boldsymbol{k}\boldsymbol{k}'}\rangle_{\mathrm{eq}} = \delta_{\boldsymbol{k}-\boldsymbol{k}'} L_{\boldsymbol{k}} \tag{5.100}$$

where

$$L_{\boldsymbol{k}} = \frac{c}{V}\int \frac{\mathrm{d}\boldsymbol{q}}{(2\pi)^3} \frac{k^2 - (\boldsymbol{k}\cdot\hat{\boldsymbol{q}})^2}{\eta_s q^2} g(\boldsymbol{k}+\boldsymbol{q}). \tag{5.101}$$

Therefore the Langevin equation becomes

$$\frac{\partial}{\partial t} c_{\boldsymbol{k}}(t) = -L_{\boldsymbol{k}} \frac{\partial U(\{c_{\boldsymbol{k}}\})}{\partial c_{-\boldsymbol{k}}} + r_{\boldsymbol{k}}(t). \tag{5.102}$$

The equivalent form of the Smoluchowski equation is

$$\frac{\partial}{\partial t}\Psi(\{c_{\boldsymbol{k}}\}) = \sum_{\boldsymbol{k}} \frac{\partial}{\partial c_{\boldsymbol{k}}} L_{\boldsymbol{k}} \left[k_B T \frac{\partial \Psi}{\partial c_{-\boldsymbol{k}}} + \Psi \frac{\partial U(\{c_{\boldsymbol{k}}\})}{\partial c_{-\boldsymbol{k}}} \right]. \tag{5.103}$$

In the Gaussian approximation, $U(\{c_{\boldsymbol{k}}\})$ is given by

$$U(\{c_{\boldsymbol{k}}\}) = \frac{V}{2c} k_B T \sum_{\boldsymbol{k}} \frac{1}{g(\boldsymbol{k})} c_{\boldsymbol{k}} c_{-\boldsymbol{k}}. \tag{5.104}$$

Thus the Langevin equation (5.102) becomes

$$\frac{\partial}{\partial t} c_{\boldsymbol{k}} = -\Gamma_{\boldsymbol{k}} c_{\boldsymbol{k}} + r_{\boldsymbol{k}}(t) \tag{5.105}$$

where

$$\Gamma_{\boldsymbol{k}} = \frac{V}{c} k_B T \frac{L_{\boldsymbol{k}}}{g(\boldsymbol{k})}, \tag{5.106}$$

or using eqn (5.101)

$$\Gamma_{\boldsymbol{k}} = k_B T \int \frac{\mathrm{d}\boldsymbol{q}}{(2\pi)^3} \frac{g(\boldsymbol{k}+\boldsymbol{q})}{g(\boldsymbol{k})} \frac{\boldsymbol{k}^2 - (\boldsymbol{k}\cdot\hat{\boldsymbol{q}})^2}{\eta_s \boldsymbol{q}^2}. \tag{5.107}$$

Equation (5.105) is the basic equation for the dynamics of concentration fluctuation. We now study the consequence of this equation.

5.5.2 Dynamic light scattering

Equation (5.105) is a linear Langevin equation which has been studied in Section 3.5. The time correlation function $\langle c_{\boldsymbol{k}}(t) c_{-\boldsymbol{k}}(0)\rangle$ is thus easily calculated as

$$\langle c_{\boldsymbol{k}}(t) c_{-\boldsymbol{k}}(0)\rangle = \langle c_{\boldsymbol{k}} c_{-\boldsymbol{k}}\rangle \exp(-\Gamma_{\boldsymbol{k}} t). \tag{5.108}$$

Hence the dynamical structure factor is

$$g(\boldsymbol{k}, t) \equiv \frac{V}{c} \langle c_{\boldsymbol{k}}(t) c_{-\boldsymbol{k}}(0)\rangle = g(\boldsymbol{k}) \exp(-\Gamma_{\boldsymbol{k}} t). \tag{5.109}$$

In this approximation, the decay of $g(\boldsymbol{k}, t)$ is single exponential.

Note that $\Gamma_{\boldsymbol{k}}$ given by eqn (5.107) is precisely the same as the exact initial decay rate given in Chapter 4 (eqn (4.110)). Thus eqn (5.109) is exact for $t \to 0$. The explicit form of $\Gamma_{\boldsymbol{k}}$ is evaluated from eqns (5.39) and (5.107) (see eqn (4.111)[49]) as

$$\begin{aligned}\Gamma_{\boldsymbol{k}} &= \frac{k_B T}{4\pi^2 \eta_s} \int_0^\infty \mathrm{d}q \frac{1 + k^2\xi^2}{1 + q^2\xi^2} q^2 \left[\frac{k^2 + q^2}{2kq} \log \left| \frac{k+q}{k-q} \right| - 1 \right] \\ &= \frac{k_B T}{6\pi\eta_s \xi} k^2 F(k\xi) \end{aligned} \tag{5.110}$$

where

$$F(x) = \frac{3}{4} \frac{1 + x^2}{x^3} (x + (x^2 - 1)\tan^{-1}(x)). \tag{5.111}$$

The function is plotted in Fig. 5.7.

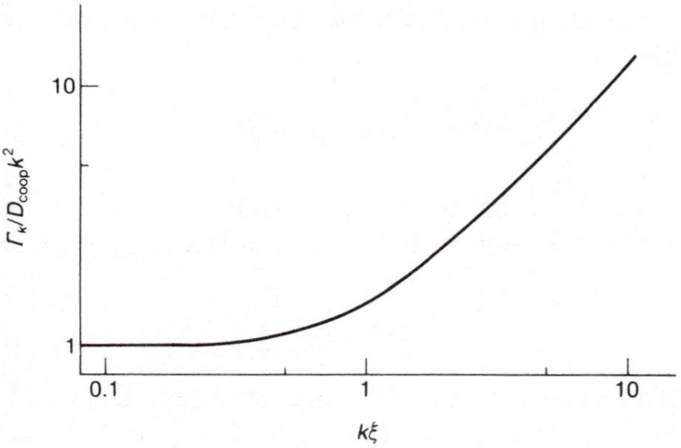

Fig. 5.7. $F(x) = \Gamma_k/D_{coop}k^2$ is plotted against $x = k\xi$.

The expression is simplified in the two limiting cases

(i) For $x \gg 1$, $F(x) = 3\pi x/8$, whence,

$$\Gamma_{\boldsymbol{k}} = \frac{k_B T}{16\eta_s} k^3, \quad k\xi \gg 1. \tag{5.112}$$

This is precisely the same as eqn (4.113). This must be so since in the region of $k\xi \gg 1$, the dynamics is dominated by the single chain behaviour. The condition of $k\xi \gg 1$ is not easily attainable by light scattering, but the result has been confirmed by neutron scattering.[52]

(ii) For $x \ll 1$: $F(0) = 1$, so that

$$\Gamma_{\boldsymbol{k}} = \frac{k_B T}{6\pi\eta_s\xi} \boldsymbol{k}^2 \qquad k\xi \ll 1. \tag{5.113}$$

The $\boldsymbol{k}$ dependence of $\Gamma_{\boldsymbol{k}}$ is again the same as in dilute solutions (eqn (4.99)). In fact this $\boldsymbol{k}$ dependence holds quite generally in the small $\boldsymbol{k}$ region, for consider the Taylor series expansion of $\Gamma_{\boldsymbol{k}}$ with respect to $\boldsymbol{k}$. In the isotropic state, the most general form is

$$\Gamma_{\boldsymbol{k}} = a_0 + a_1\boldsymbol{k}^2 + a_2(\boldsymbol{k}^2)^2 + \dots.$$

Since $\Gamma_{\boldsymbol{k}}$ must vanish at $\boldsymbol{k} = 0$, $a_0 = 0$, whence the above equation is written as

$$\Gamma_{\boldsymbol{k}} = D_{app}\boldsymbol{k}^2 \quad (k \to 0). \tag{5.114}$$

This defines an apparent diffusion constant D_{app} for the entire concentration region.

In dilute solutions, it was shown that D_{app} agrees with the self-diffusion constant defined by

$$D_G = \lim_{t\to\infty} \frac{1}{6t} \langle (\boldsymbol{R}_G(t) - \boldsymbol{R}_G(0))^2 \rangle \tag{5.115}$$

where $\boldsymbol{R}_G$ is the position of a test polymer. On the other hand, in concentrated solutions, eqn (5.113) indicates that D_{app} is given by[1,53]

$$D_{coop} \equiv \frac{k_B T}{6\pi\eta_s \xi} \tag{5.116}$$

which is called the cooperative diffusion constant. It should be noticed that D_{coop} has no relation to D_G. Indeed D_G should decrease with concentration (because of the molecular entanglement), while D_{coop} increases with the concentration. The increase of D_{coop} with concentration results from the fact that the restoring force for the concentration fluctuations is larger at higher concentration.

The concentration dependence of D_{coop} is obtained from eqns (5.36) and (5.116):

$$D_{coop} \propto c^{1/2} \text{ (Gaussian approximation).} \tag{5.117a}$$

On the other hand, in the semidilute regime, use of eqn (5.65) gives

$$D_{coop} \propto c^{3/4}. \tag{5.117b}$$

From experiments, the exponent in $D_{coop} \propto c^x$ has been found to be between 0.5 and 0.75[54–57] and this is discussed in detail by Schaefer and Han.[58]

It must be mentioned that although the behaviour of the dynamic light scattering in the short time-scale is well understood, theory is still lacking for the behaviour in the whole time-scale. Experimentally it has been observed[59–61] that in some systems the structure factor does not decrease in a single exponential manner and has a long tail. The long time-scale behaviour is considered to be related to the topological interaction, but quantitative theory is not yet given.

5.5.3 Form birefringence

As mentioned in Chapter 4, the birefringence of polymer solutions has two origins: the intrinsic birefringence which arises from the orientation of polymer segments and the form birefringence which comes from the anisotropy in the correlation of the segment density. In dilute solutions, the relative contribution of each of these terms depends both on polymer molecules and solvent molecules. In concentrated solutions, the form birefringence decreases with increasing concentration,[62] while the intrinsic birefringence increases steeply like the viscosity. (Note that the intrinsic birefringence is proportional to the stress (see eqn (4.238).) This leads to an important relation called the stress optical law[63] which states that the stress and birefringence are proportional to each other. Detailed discussion of this relationship is given in Chapter 7. Here, as an application of the Langevin equation (5.105), we shall calculate the form birefringence and show that it actually decreases with the concentration.

According to eqn (4.237), the form birefringence is given by

$$n_{\alpha\beta}^{(f)} = -2\frac{c}{n}\left(\frac{\partial n}{\partial c}\right)^2 \frac{1}{(2\pi)^3}\int \mathrm{d}\boldsymbol{k}(\hat{k}_\alpha \hat{k}_\beta - \tfrac{1}{3}\delta_{\alpha\beta})g_t(\boldsymbol{k}) \tag{5.118}$$

where $g_t(\boldsymbol{k})$ is the (static) structure factor at time t,

$$g_t(\boldsymbol{k}) \equiv \frac{V}{c}\langle c_{\boldsymbol{k}}(t)c_{-\boldsymbol{k}}(t)\rangle. \tag{5.119}$$

To calculate $g_t(\boldsymbol{k})$, we first consider the Langevin equation under the macroscopic velocity field

$$v_\alpha(\boldsymbol{r}, t) = \kappa_{\alpha\beta}(t)r_\beta. \tag{5.120}$$

The macroscopic velocity causes a flow of $c(\boldsymbol{r}, t)$,

$$\left.\frac{\partial c(\boldsymbol{r}, t)}{\partial t}\right|_{\text{drift}} = -\frac{\partial}{\partial \boldsymbol{r}}\cdot \boldsymbol{\kappa}(t)\cdot \boldsymbol{r}c(\boldsymbol{r}, t). \tag{5.121}$$

The Fourier transform of which gives the drift for $c_{\boldsymbol{k}}(t)$ as

$$\left.\frac{\partial}{\partial t}c_{\boldsymbol{k}}(t)\right|_{\text{drift}} = \boldsymbol{k}\cdot\boldsymbol{\kappa}(t)\cdot\frac{\partial}{\partial \boldsymbol{k}}c_{\boldsymbol{k}}(t). \tag{5.122}$$

Adding this to eqn (5.105), we have the Langevin equation in the presence of the velocity gradient,

$$\frac{\partial}{\partial t}c_{\boldsymbol{k}}(t) = -\Gamma_{\boldsymbol{k}}c_{\boldsymbol{k}}(t) + r_{\boldsymbol{k}}(t) + \boldsymbol{k}\cdot\boldsymbol{\kappa}(t)\cdot\frac{\partial}{\partial \boldsymbol{k}}c_{\boldsymbol{k}}(t). \tag{5.123}$$

The equation for $g_t(\boldsymbol{k})$ is obtained from eqn (5.123) as

$$\begin{aligned}\frac{\partial}{\partial t}g_t(\boldsymbol{k}) &= \frac{V}{c}\left[\left\langle\frac{\partial c_{\boldsymbol{k}}(t)}{\partial t}c_{-\boldsymbol{k}}(t)\right\rangle + \left\langle c_{\boldsymbol{k}}(t)\frac{\partial c_{-\boldsymbol{k}}(t)}{\partial t}\right\rangle\right] \\ &= -2\Gamma_{\boldsymbol{k}}g_t(\boldsymbol{k}) + \underline{\frac{V}{c}[\langle r_{\boldsymbol{k}}(t)c_{-\boldsymbol{k}}(t)\rangle + \langle c_{\boldsymbol{k}}(t)r_{-\boldsymbol{k}}(t)\rangle]} \\ &\quad + \boldsymbol{k}\cdot\boldsymbol{\kappa}(t)\cdot\frac{\partial}{\partial \boldsymbol{k}}g_t(\boldsymbol{k}).\end{aligned} \tag{5.124}$$

The underlined terms can be calculated in the same way as in eqn (4.141), and gives $2k_BTL_{\boldsymbol{k}}$. Thus, using eqn (5.106) we have

$$\frac{\partial}{\partial t}g_t(\boldsymbol{k}) = -2\Gamma_{\boldsymbol{k}}(g_t(\boldsymbol{k}) - g_{\text{eq}}(\boldsymbol{k})) + \boldsymbol{k}\cdot\boldsymbol{\kappa}(t)\cdot\frac{\partial}{\partial \boldsymbol{k}}g_t(\boldsymbol{k}) \tag{5.125}$$

where $g_{\text{eq}}(\boldsymbol{k})$ is the equilibrium structure factor. Since eqn (5.125) is a first-order differential equation, it can be solved by a standard method,† but here we consider the simple case of $\boldsymbol{\kappa}(t)$ small and independent of time.

The steady-state solution of eqn (5.125) is written as

$$g_t(\boldsymbol{k}) = g_{\text{eq}}(\boldsymbol{k}) + g_1(\boldsymbol{k}) \tag{5.126}$$

with

$$g_1(\boldsymbol{k}) = \frac{1}{2\Gamma_{\boldsymbol{k}}}\boldsymbol{k}\cdot\boldsymbol{\kappa}\cdot\frac{\partial}{\partial \boldsymbol{k}}g_{\text{eq}}(\boldsymbol{k}) = \frac{\kappa_{\alpha\beta}\hat{k}_\alpha\hat{k}_\beta}{2\Gamma_{\boldsymbol{k}}}k\frac{\partial}{\partial k}g_{\text{eq}}(\boldsymbol{k}). \tag{5.127}$$

Substituting this into eqn (5.118), we have

$$n_{\alpha\beta}^{(\text{f})} = -2\frac{c}{n}\left(\frac{\partial n}{\partial c}\right)^2\frac{1}{(2\pi)^3}\int \mathrm{d}\boldsymbol{k}(\hat{k}_\alpha\hat{k}_\beta - \tfrac{1}{3}\delta_{\alpha\beta})\frac{\hat{k}_\mu\hat{k}_\nu\kappa_{\mu\nu}}{2\Gamma_{\boldsymbol{k}}}k\frac{\partial}{\partial k}g_{\text{eq}}(\boldsymbol{k}). \tag{5.128}$$

The integral over the direction $\hat{\boldsymbol{k}}$ can be calculated by using

$$\frac{1}{4\pi}\int \mathrm{d}\hat{\boldsymbol{k}}\hat{k}_\alpha\hat{k}_\beta = \tfrac{1}{3}\delta_{\alpha\beta} \tag{5.129}$$

$$\frac{1}{4\pi}\int \mathrm{d}\hat{\boldsymbol{k}}\hat{k}_\alpha\hat{k}_\beta\hat{k}_\mu\hat{k}_\nu = \tfrac{1}{15}(\delta_{\alpha\beta}\delta_{\mu\nu} + \delta_{\alpha\mu}\delta_{\beta\nu} + \delta_{\alpha\nu}\delta_{\beta\mu}). \tag{5.130}$$

The result is written as

$$n_{\alpha\beta}^{(\text{f})} = \lambda_M^{(\text{f})}(\kappa_{\alpha\beta} + \kappa_{\beta\alpha}) \tag{5.131}$$

with

$$\lambda_M^{(\text{f})} = -\frac{c}{n}\left(\frac{\partial n}{\partial c}\right)^2\frac{1}{30\pi^2}\int_0^\infty \mathrm{d}k k^3\frac{1}{\Gamma_k}\frac{\partial}{\partial k}g_{\text{eq}}(k). \tag{5.132}$$

† Let $F_{\alpha\beta}(t, t')$ be the solution of the equation

$$\frac{\partial}{\partial t}F_{\alpha\beta}(t, t') = F_{\alpha\mu}(t, t')\kappa_{\beta\mu}(t)$$

with the boundary condition $F_{\alpha\beta}(t, t) = \delta_{\alpha\beta}$, then the solution of eqn (5.125) is

$$g_t(\boldsymbol{k}) = 2\frac{V}{c}k_BT\int_{-\infty}^{t}\mathrm{d}t'\exp\left[-2\int_{t'}^{t}\mathrm{d}t''\Gamma_{\boldsymbol{F}(t,t'')\cdot\boldsymbol{k}}\right]L_{\boldsymbol{F}(t,t')\cdot\boldsymbol{k}}.$$

In semidilute solutions, Γ_k and $g_{eq}(\boldsymbol{k})$ are given by the following scaling form (see 5.3 and 5.6):

$$\Gamma_k = \frac{k_B T}{\eta_s \xi} k^2 F_1(k\xi), \qquad g_{eq}(\boldsymbol{k}) = c\xi^3 F_2(k\xi). \tag{5.133}$$

Equations (5.132) and (5.133) give

$$\lambda_M^{(f)} \simeq \eta_s \frac{c\xi^3}{k_B T} \frac{c}{n} \left(\frac{\partial n}{\partial c}\right)^2. \tag{5.134}$$

If $\partial n/\partial c$ is regarded as independent of concentration, $\lambda_M^{(f)}$ depends on the concentration as

$$\lambda_M^{(f)} \propto c^2 \xi^3 \propto c^{-1/4} \tag{5.135}$$

which decreases with increasing concentration.

On the other hand, for concentrated solutions, substitution of eqns (5.39) and (5.110) gives

$$\lambda_M^{(f)} \simeq \eta_s \frac{1}{k_B T} \left(\frac{\xi}{b}\right)^2 \frac{c}{n} \left(\frac{\partial n}{\partial c}\right) \simeq \frac{\eta_s}{k_B T} \frac{1}{v} \frac{1}{n} \left(\frac{\partial n}{\partial c}\right)^2 \tag{5.136}$$

which also decreases with concentration since $\partial n/\partial c$ decreases with concentration.

Therefore in both the semidilute and concentrated solutions, the contribution from the form birefringence decreases with the concentration. Experimental results are summarized in refs 62 and 63.

5.6 Scaling theory—dynamics

The dynamical scaling law discussed in Section 4.3 can also be used for semidilute solutions.[1,53] The hypothesis is that any physical quantities characterizing the dynamics satisfy the scaling transformation

$$A \to \lambda^x A \tag{5.137}$$

when the basic parameters are changed as

$$N \to N/\lambda, \qquad b \to b\lambda^\nu, \qquad c \to c/\lambda. \tag{5.138}$$

It is worthwhile to ask whether dynamical scaling is valid in a system in which the topological constraints are important since it is well known that topology is extremely dimension dependent; for example, vorticity is conserved in two dimensions, not in three dimensions; knots exist in three dimensions but not in four dimensions, etc. Such effects do not appear in the renormalization group calculations which are quite continuous in variation of the dimensionality of the system. On the other hand, one can argue that since the topological constraints introduce no new length-scale, it would not upset the scaling arguments; rather the scaling argument would give a common restriction on the theoretical results for systems both with and without topological constraints (the latter of course being a hypothetical system in which chains can pass through each other like phantoms.) Here we shall take this viewpoint. Various experimental results are quite consistent with the scaling assumption in good solvents, though the validity is not yet established in Θ solvents.[51]

As an example let us consider the apparent diffusion constant D_{app} measured by dynamic light scattering (see eqn (5.114)). First we use a dimensional analysis. The relevant parameters in the problem are b, c, N, k_BT, and η_s, so D_{app} is written as

$$D_{app} = \frac{k_BT}{\eta_s b} F(cb^3, N). \tag{5.139}$$

This must be invariant under the transformation (5.138), whence

$$\frac{1}{b} F(cb^3, N) = \frac{1}{\lambda^\nu b} F(cb^3\lambda^{3\nu-1}, N/\lambda) \tag{5.140}$$

which leads to

$$D_{app} = \frac{k_BT}{\eta_s N^\nu b} F(cb^3 N^{-1+3\nu}). \tag{5.141}$$

Using the diffusion coefficient in the dilute limit $D_G^{(0)} \simeq k_BT/N^\nu b$ and $c^* \simeq N/(N^\nu b)^3$, eqn (5.141) is written as

$$D_{app} = D_G^{(0)} F(c/c^*) \quad \text{(dilute and semidilute).} \tag{5.142}$$

This equation is valid both in dilute and semidilute solutions. Unlike with dilute solutions, the scaling argument does not predict molecular weight dependence explicitly. (This must be so since one can derive the same formula for the self-diffusion constant D_G. Further information is needed to distinguish the functional form of D_{app} from that of D_G.) If one assumes that D_{app} is independent of the molecular weight, then one has

$$D_{app} \simeq \frac{k_BT}{\eta_s b}(cb^3)^{\nu/(3\nu-1)} \simeq \frac{k_BT}{\eta_s \xi} \propto c^{3/4} \tag{5.143}$$

which agrees with eqn (5.117*b*). On the other hand, D_G depends on the molecular weight. In the next section we shall show that $D_G \propto N^{-2}$ by the tube model. For this to be written as $D_G^{(0)}F(c/c^*)$, D_G must be written as

$$D_G \simeq \frac{k_BT}{\eta_s N^\nu b (c/c^*)^{(2-\nu)/(3\nu-1)}} \propto N^{-2} c^{-7/4} \tag{5.144}$$

which decreases with concentration. The self-diffusion coefficient in dilute and semidilute solutions has been measured by forced Rayleigh scattering.[64] The result is shown in Fig. 5.8, which includes the data of D_{coop} and displays the different concentration dependence of D_G and D_{coop}.

Likewise the viscosity η of the solution is shown to be

$$\eta = \eta_s F(c/c^*). \tag{5.145}$$

Again, the molecular weight dependence is not determined by the scaling argument alone, but eqn (5.145) predicts that if η/η_s is plotted against

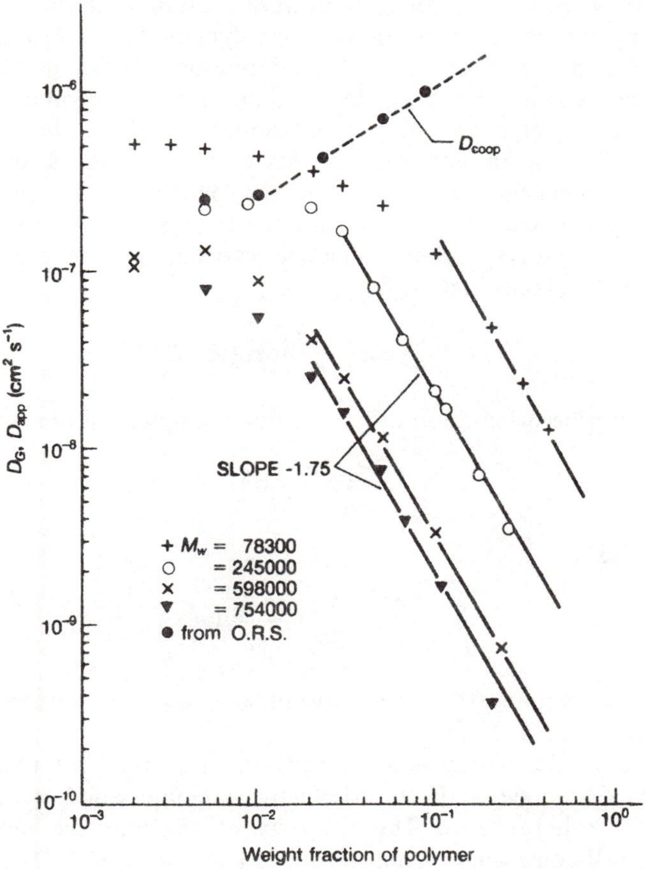

Fig. 5.8. The apparent diffusion constant D_{app} obtained by the dynamic light scattering[56] and the self-diffusion constant D_G obtained by the forced Rayleigh scattering[64] of the chains for four different molecular weights. Reproduced from ref. 64.

$c/c^* = \rho/\rho^*$, then the viscosity curves of various molecular weight can be superimposed. Such superposition had been known prior to the scaling theory.[65–67] If one uses $\rho^*[\eta] = \text{constant}$, eqn (5.145) can be rewritten as

$$\eta = \eta_s F(\rho[\eta]). \tag{5.146}$$

This form also has been established experimentally for a long time.[68]

5.7 Effective medium theory

5.7.1 Failure of the scalar field description

The collective coordinate formalism given in Section 5.5.1 is not capable of describing the whole aspect of polymer dynamics. For example, the quantities which are related to the orientation of the bond vector $\partial \boldsymbol{R}_n/\partial n$, such as stress, birefringence, and electric displacement, are not expressed by $\{c_{\boldsymbol{k}}(t)\}$, and therefore cannot be dealt with in this framework. This is in contrast to the critical dynamics of binary solutions[49] in which the dynamics is entirely specified by an equation of motion for the concentration fluctuation. To discuss the orientation dynamics of polymers, other collective coordinates are needed. A possible collective coordinate is[69]

$$c_{\boldsymbol{k},p}(t) = \frac{1}{V}\sum_{a,n} \cos\left(\frac{p\pi n}{N}\right)\exp(i\boldsymbol{k}\cdot\boldsymbol{R}_{an}) \tag{5.147}$$

which can describe orientation of bonds. For example, the tensor

$$\sum_{a,n}\left\langle \frac{\partial R_{an\alpha}}{\partial n}\frac{\partial R_{an\beta}}{\partial n}\right\rangle \tag{5.148}$$

is expressed as

$$\sum_{p,\boldsymbol{k}}\left(\frac{p\pi}{N}\right)^2\left\langle \frac{\partial c_{\boldsymbol{k}p}}{\partial k_\alpha}\frac{\partial c_{-\boldsymbol{k}p}}{\partial k_\beta}\right\rangle. \tag{5.149}$$

However, the usefulness of such coordinates has not yet been fully worked out.

An alternative idea proposed by Edwards and Freed[70] is to consider the motion of a single chain in an effective medium which includes the effect of the other chains. The property of the effective medium is determined self-consistently from the single chain dynamics. Though this method fails to describe the entanglement effect appropriately, it indicates an important aspect of the hydrodynamic interaction in the concentrated system, which is the screening of the hydrodynamic interaction.

5.7.2 Hydrodynamic screening

The concept of hydrodynamic screening is seen if one considers the velocity field created by a point force in a fluid (see Fig. 5.9). If the fluid is a pure solvent of viscosity η_s, the velocity field $\boldsymbol{v}(\boldsymbol{r})$ is long range, decreasing like $1/\eta_s r$ as predicted by the Oseen formula. On the other hand if there are polymers in the fluid, the situation is different. On a small length-scale, the velocity profile will still be like $1/\eta_s r$ since it is not disturbed by the polymers. However, on a large length-scale, the velocity profile will obey the macroscopic hydrodynamics, so that the velocity must decrease like $1/\eta r$, where η is the *macroscopic* viscosity of the solution. If $\eta \gg \eta_s$, the velocity field falls quickly and the effect caused by the point force becomes very weak beyond a certain characteristic length ξ_H called the hydrodynamic screening length (see Fig. 5.9). In such a situation, the hydrodynamic interaction becomes negligible between

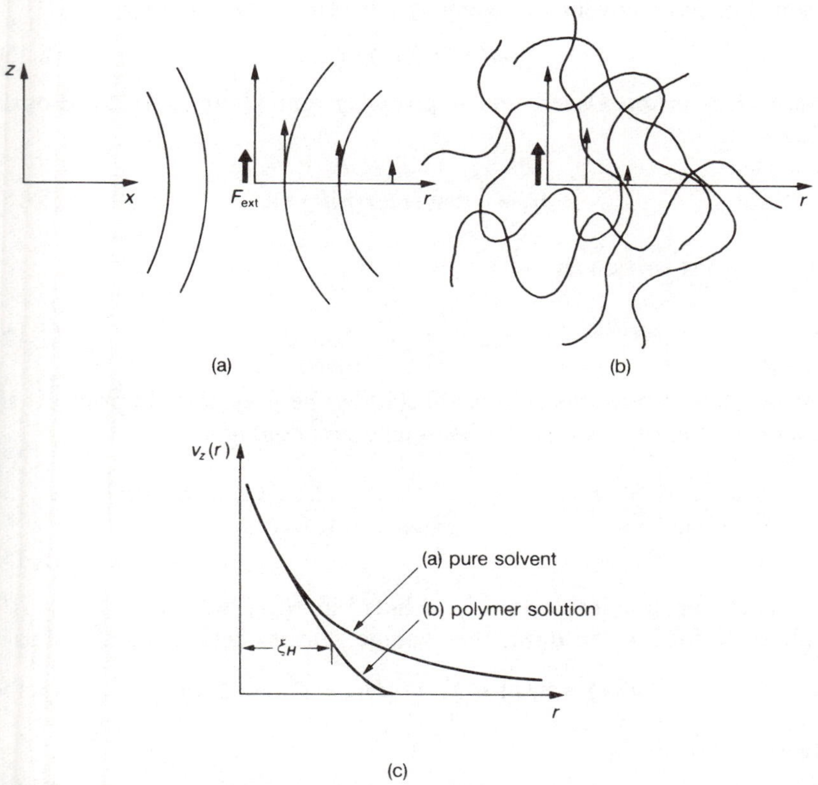

Fig. 5.9. (*a*), (*b*) Velocity profiles caused by a point force in a pure solvent and a polymer solution. (*c*) The velocity $v_z(r)$ on the axis shown in (*a*), (*b*) is plotted against r.

segments whose distance apart is larger than ξ_H. This is called the hydrodynamic screening. As a consequence of the hydrodynamic screening the dynamical behaviour of a single chain becomes Rouse-like as against the Zimm-like behaviour in dilute solution.

Though the idea of hydrodynamic screening is easy to understand, the actual formulation needs somewhat elaborate methods. In the following, we shall explain the method using the crudest approximation. The reader who is interested in a complete form of the formulation is advised to see the original papers.[10,70–72]

5.7.3 Effective medium theory

First let us consider how the velocity profile caused by a point force is affected when a small number of polymers are present in solution. For simplicity we consider the steady state in the velocity field, though this assumption is not essential.[71,73] Let $\boldsymbol{F}_{\text{ext}}$ be the point force acting at the origin. For pure solvent, the velocity perturbance is given by

$$\boldsymbol{v}_0(\boldsymbol{r}) = \boldsymbol{H}(\boldsymbol{r}) \cdot \boldsymbol{F}_{\text{ext}} \tag{5.150}$$

where $\boldsymbol{H}(\boldsymbol{r})$ is the Oseen tensor given by eqn (3.106). By the Fourier transform,

$$\boldsymbol{v}_{0\boldsymbol{k}} \equiv \int \mathrm{d}\boldsymbol{r} \boldsymbol{v}_0(\boldsymbol{r}) \exp(\mathrm{i}\boldsymbol{k} \cdot \boldsymbol{r}). \tag{5.151}$$

eqn (5.150) is written as

$$\boldsymbol{v}_{0\boldsymbol{k}} = \frac{\boldsymbol{I} - \hat{\boldsymbol{k}}\hat{\boldsymbol{k}}}{\eta_s \boldsymbol{k}^2} \cdot \boldsymbol{F}_{\text{ext}}. \tag{5.152}$$

Our problem is how this field is affected by the polymers. In principle the answer is given by solving the Smoluchowski equation

$$\sum_{a,n} \frac{\partial}{\partial \boldsymbol{R}_{an}} \cdot \left[\sum_{b,m} \boldsymbol{H}_{anbm} \cdot \left(k_B T \frac{\partial \Psi}{\partial \boldsymbol{R}_{bm}} + \Psi \frac{\partial U}{\partial \boldsymbol{R}_{bm}} \right) - \boldsymbol{v}_0(\boldsymbol{R}_{an}) \Psi \right] = 0. \tag{5.153}$$

The hydrodynamic interaction is included in $\boldsymbol{H}_{anbm} \equiv \boldsymbol{H}(\boldsymbol{R}_{an} - \boldsymbol{R}_{bm})$. If Ψ is obtained for a given $\boldsymbol{v}_0(\boldsymbol{r})$, the average velocity field is calculated as

$$\bar{\boldsymbol{v}}(\boldsymbol{r}) = \boldsymbol{v}_0(\boldsymbol{r}) + \sum_a \sum_m \langle \boldsymbol{H}(\boldsymbol{r} - \boldsymbol{R}_{am}) \cdot \boldsymbol{F}_{am} \rangle \tag{5.154}$$

where

$$\boldsymbol{F}_{am} = -\frac{\partial}{\partial \boldsymbol{R}_{am}} (k_B T \ln \Psi + U) \tag{5.155}$$

and the average $\langle \ldots \rangle$ is taken for the distribution function Ψ.

Now it is difficult to solve the diffusion equation (5.153) which includes polymer–polymer interactions. So we first focus our attention on a particular polymer, which we shall call the test polymer, assuming that the velocity field created by the other polymers is known. Let $\bar{\boldsymbol{v}}_a(\boldsymbol{r})$ be the velocity field created by the external force and the polymers excluding the test polymer, i.e.,

$$\bar{\boldsymbol{v}}_a(\boldsymbol{r}) = \boldsymbol{v}_0(\boldsymbol{r}) + \sum_{b \neq a} \sum_n \langle \boldsymbol{H}(\boldsymbol{r} - \boldsymbol{R}_{bn}) \cdot \boldsymbol{F}_{bn} \rangle. \tag{5.156}$$

The effect of the test polymer is solved by the Smoluchowski equation for the single chain:

$$\sum_{n,m} \frac{\partial}{\partial \boldsymbol{R}_{an}} \cdot \left[\boldsymbol{H}_{anam} \cdot \left(k_B T \frac{\partial \Psi}{\partial \boldsymbol{R}_{am}} + \Psi \frac{\partial U}{\partial \boldsymbol{R}_{am}} \right) \right] - \sum_n \frac{\partial}{\partial \boldsymbol{R}_{an}} \cdot [\bar{\boldsymbol{v}}_a(\boldsymbol{R}_{an})\Psi] = 0. \tag{5.157}$$

The velocity perturbation created by the test polymer is then calculated by

$$\delta \bar{\boldsymbol{v}}_a(\boldsymbol{r}) = \sum_m \langle \boldsymbol{H}(\boldsymbol{r} - \boldsymbol{R}_{am}) \cdot \boldsymbol{F}_{am} \rangle. \tag{5.158}$$

Since $\delta \bar{\boldsymbol{v}}_a(\boldsymbol{r})$ depends on $\bar{\boldsymbol{v}}_a(\boldsymbol{r})$ linearly, it must be written using a spatial response function $\Xi(\boldsymbol{r})$,

$$\delta \bar{\boldsymbol{v}}_a(\boldsymbol{r}) = -\frac{1}{V} \int \mathrm{d}\boldsymbol{r}' \Xi(\boldsymbol{r} - \boldsymbol{r}') \bar{\boldsymbol{v}}_a(\boldsymbol{r}'). \tag{5.159}$$

Equation (5.156) is then written as

$$\bar{\boldsymbol{v}}_a(\boldsymbol{r}) = \boldsymbol{v}_0(\boldsymbol{r}) + \sum_b \delta \bar{\boldsymbol{v}}_b(\boldsymbol{r}) = \boldsymbol{v}_0(\boldsymbol{r}) - \frac{1}{V} \sum_b \int \mathrm{d}\boldsymbol{r}' \Xi(\boldsymbol{r} - \boldsymbol{r}') \bar{\boldsymbol{v}}_b(\boldsymbol{r}'). \tag{5.160}$$

Now we assume that the perturbation $\delta \bar{\boldsymbol{v}}_a(\boldsymbol{r})$ is not large, and replace $\bar{\boldsymbol{v}}_a(\boldsymbol{r})$ and $\bar{\boldsymbol{v}}_b(\boldsymbol{r}')$ by $\bar{\boldsymbol{v}}(\boldsymbol{r})$ and $\bar{\boldsymbol{v}}(\boldsymbol{r}')$ respectively:

$$\begin{aligned} \bar{\boldsymbol{v}}(\boldsymbol{r}) &= \boldsymbol{v}_0(\boldsymbol{r}) - \frac{1}{V} \sum_b \int \mathrm{d}\boldsymbol{r}' \Xi(\boldsymbol{r} - \boldsymbol{r}') \bar{\boldsymbol{v}}(\boldsymbol{r}') \\ &= \boldsymbol{v}_0(\boldsymbol{r}) - \frac{c}{N} \int \mathrm{d}\boldsymbol{r}' \Xi(\boldsymbol{r} - \boldsymbol{r}') \bar{\boldsymbol{v}}(\boldsymbol{r}') \end{aligned} \tag{5.161}$$

where c/N is the number of polymers in unit volume.

Equation (5.161) is easily solved by the Fourier transform as

$$\bar{\boldsymbol{v}}(\boldsymbol{k}) = \frac{\boldsymbol{v}_{0k}}{1 + \dfrac{c}{N} \Xi(\boldsymbol{k})} = \bar{\boldsymbol{H}}(\boldsymbol{k}) \cdot \boldsymbol{F}_{\text{ext}}$$

where

$$\bar{\boldsymbol{H}}(\boldsymbol{k}) = \frac{\boldsymbol{I} - \hat{\boldsymbol{k}}\hat{\boldsymbol{k}}}{\eta_s \boldsymbol{k}^2\left[1 + \frac{c}{N}\Xi(\boldsymbol{k})\right]} \tag{5.162}$$

and

$$\Xi(\boldsymbol{k}) = \int \mathrm{d}\boldsymbol{r}\Xi(\boldsymbol{r})\mathrm{e}^{\mathrm{i}\boldsymbol{k}\cdot\boldsymbol{r}}. \tag{5.163}$$

The function $\bar{\boldsymbol{H}}(\boldsymbol{k})$ represents the effect of the point force in the presence of the polymers. Comparing eqn (5.162) with eqn (5.152) we may interpret

$$\eta_{\text{eff}}(\boldsymbol{k}) = \eta_s\left[1 + \frac{c}{N}\Xi(\boldsymbol{k})\right] \tag{5.164}$$

as an effective viscosity of the polymer solution on the length-scale of $1/|\boldsymbol{k}|$. The macroscopic viscosity is given in the limit of $\boldsymbol{k} \to 0$ as

$$\eta = \eta_s\left[1 + \frac{c}{N}\Xi(\boldsymbol{k})\right]\Bigg|_{\boldsymbol{k}=0}. \tag{5.165}$$

So far we have been considering the dilute solution. This allowed us to use the Oseen tensor $\boldsymbol{H}(\boldsymbol{r})$ in solving the single chain problem, eqn (5.157). In the concentrated solution, however, the hydrodynamic interactions among the segments are affected by the surrounding polymers. Thus we have to replace $\boldsymbol{H}_{anam}$ in eqn (5.157) by the effective interaction $\bar{\boldsymbol{H}}_{anam}$ which is derived from $\bar{\boldsymbol{H}}(\boldsymbol{k})$ given by eqn (5.162).

$$\boldsymbol{H}_{anam} \to \bar{\boldsymbol{H}}_{anam} = \bar{\boldsymbol{H}}(\boldsymbol{R}_{an} - \boldsymbol{R}_{am}) = \int \frac{\mathrm{d}\boldsymbol{k}}{(2\pi)^3}\exp[\mathrm{i}\boldsymbol{k}\cdot(\boldsymbol{R}_{an} - \boldsymbol{R}_{am})]\bar{\boldsymbol{H}}(\boldsymbol{k}). \tag{5.166}$$

This gives a closed equation for the effective hydrodynamic interaction.

5.7.4 Example

As an example, we consider a solution in Θ condition. For simplicity of notation we omit the suffix a to denote the quantity of the test polymer. Since the general solution of eqn (5.157) is involved, we use a crude approximation: we assume that in the velocity field $\bar{\boldsymbol{v}}(\boldsymbol{r})$, the test polymer behaves as a rigid chain, and moves with constant velocity $\boldsymbol{V}$ (the rotational motion being neglected for simplicity.). The equations determining $\boldsymbol{V}$ are

$$\boldsymbol{V} = \sum_m \bar{\boldsymbol{H}}_{nm}\cdot\boldsymbol{F}_m + \bar{\boldsymbol{v}}(\boldsymbol{R}_m) \tag{5.167}$$

and

$$\sum_m \boldsymbol{F}_m = 0. \tag{5.168}$$

Equation (5.167) indicates that the fluid velocity at $\boldsymbol{R}_m$ must be $\boldsymbol{V}$ and eqn (5.168) represents the condition that the total force acting on the polymer is zero.

To solve the equations we use the preaveraging approximation

$$\bar{\boldsymbol{H}}_{nm} \simeq \langle \bar{\boldsymbol{H}}_{nm} \rangle_{\text{eq}} \equiv \bar{h}_{nm}\boldsymbol{I} = \int \frac{\mathrm{d}\boldsymbol{k}}{(2\pi)^3} \langle \exp(\mathrm{i}\boldsymbol{k}\cdot(\boldsymbol{R}_n - \boldsymbol{R}_m)) \rangle_{\text{eq}} \bar{\boldsymbol{H}}(\boldsymbol{k}). \tag{5.169}$$

From eqns (5.162) and (5.169) it follows that

$$\bar{h}_{nm} = \int \frac{\mathrm{d}\boldsymbol{k}}{(2\pi)^3} \frac{2}{3\eta_s \boldsymbol{k}^2} \frac{1}{1 + \dfrac{c}{N}\Xi(\boldsymbol{k})} \langle \exp(\mathrm{i}\boldsymbol{k}\cdot(\boldsymbol{R}_n - \boldsymbol{R}_m)) \rangle_{\text{eq}}. \tag{5.170}$$

Equations (5.167) and (5.168) then give

$$\boldsymbol{F}_m = -\sum_n \zeta_{mn} \boldsymbol{v}(\boldsymbol{R}_n) \tag{5.171}$$

where

$$\zeta_{mn} = (\bar{h}^{-1})_{mn} - \Big(\sum_{i,j} (\bar{h}^{-1})_{ni} (\bar{h}^{-1})_{mj}\Big) \Big/ \Big(\sum_{i,j} (\bar{h}^{-1})_{ij}\Big). \tag{5.172}$$

Thus the perturbed flow is given by eqns (5.158) and (5.171) as

$$\begin{aligned}
\delta\bar{\boldsymbol{v}}(\boldsymbol{r}) &= -\sum_{n,m} \langle \boldsymbol{H}(\boldsymbol{r} - \boldsymbol{R}_m) \cdot \zeta_{mn} \bar{\boldsymbol{v}}(\boldsymbol{R}_n) \rangle_{\text{eq}} \\
&= -\sum_{m,n} \int \mathrm{d}\boldsymbol{r}' \int \mathrm{d}\boldsymbol{r}'' \langle \boldsymbol{H}(\boldsymbol{r} - \boldsymbol{r}'') \delta(\boldsymbol{r}'' - \boldsymbol{R}_m) \delta(\boldsymbol{r}' - \boldsymbol{R}_n) \rangle_{\text{eq}} \zeta_{mn} \cdot \bar{\boldsymbol{v}}(\boldsymbol{r}') \\
&= -\sum_{m,n} \int \mathrm{d}\boldsymbol{r}' \int \mathrm{d}\boldsymbol{r}'' \frac{1}{6\pi\eta_s |\boldsymbol{r} - \boldsymbol{r}''|} \langle \delta(\boldsymbol{r}'' - \boldsymbol{R}_m) \delta(\boldsymbol{r}' - \boldsymbol{R}_n) \rangle_{\text{eq}} \zeta_{mn} \bar{\boldsymbol{v}}(\boldsymbol{r}').
\end{aligned} \tag{5.173}$$

Comparing eqn (5.173) with eqn (5.159), we get

$$\Xi(\boldsymbol{r}) = V \sum_{n,m} \int \mathrm{d}\boldsymbol{r}'' \frac{1}{6\pi\eta_s |\boldsymbol{r} - \boldsymbol{r}''|} \langle \delta(\boldsymbol{r}'' - \boldsymbol{R}_m) \delta(\boldsymbol{R}_n) \rangle_{\text{eq}} \zeta_{nm} \tag{5.174}$$

or in Fourier transform,

$$\Xi(\boldsymbol{k}) = \int \mathrm{d}\boldsymbol{r} e^{\mathrm{i}\boldsymbol{k}\cdot\boldsymbol{r}} \Xi(\boldsymbol{r}) = \frac{2}{3\eta_s \boldsymbol{k}^2} \sum_{n,m} \langle \exp[\mathrm{i}\boldsymbol{k}\cdot(\boldsymbol{R}_m - \boldsymbol{R}_n)] \rangle_{\text{eq}} \zeta_{nm}. \tag{5.175}$$

Equations (5.170), (5.172), and (5.175) give a nonlinear equation for $\Xi(\boldsymbol{k})$.

To proceed further, we use the diagonal approximation for the normal coordinate representation of h_{mn}:

$$(\bar{h}^{-1})_{mn} = \frac{1}{N^2} \sum_{p=0}^{\infty} \frac{1}{\bar{h}_p} \cos\left(\frac{p\pi n}{N}\right) \cos\left(\frac{p\pi m}{N}\right) \tag{5.176}$$

where

$$\bar{h}_p \equiv \frac{1}{N^2} \sum_{n,m} \cos\left(\frac{p\pi n}{N}\right) \cos\left(\frac{p\pi m}{N}\right) \bar{h}_{nm}. \tag{5.177}$$

In this approximation, eqn (5.172) gives

$$\zeta_{mn} = \frac{1}{N^2} \sum_{p=1}^{\infty} \frac{1}{\bar{h}_p} \cos\left(\frac{p\pi n}{N}\right) \cos\left(\frac{p\pi m}{N}\right). \tag{5.178}$$

(Note that the term of $p = 0$ is omitted from the summation.) Thus in the Θ condition, eqn (5.175) becomes

$$\Xi(\boldsymbol{k}) = \frac{2}{3\eta_s} \boldsymbol{k}^{-2} \sum_{p,m,n} \exp(-\tfrac{1}{6} b^2 \boldsymbol{k}^2 |n-m|) \frac{1}{N^2} \frac{1}{\bar{h}_p} \cos\left(\frac{p\pi n}{N}\right) \cos\left(\frac{p\pi m}{N}\right). \tag{5.179}$$

The sum over n and m are calculated as in eqn (4.52)

$$\begin{aligned}
&\sum_{n,m} \cos\left(\frac{p\pi n}{N}\right) \cos\left(\frac{p\pi m}{N}\right) \exp(-\tfrac{1}{6} b^2 \boldsymbol{k}^2 |n-m|) \\
&\simeq \int_0^N \mathrm{d}n \int_{-\infty}^{\infty} \mathrm{d}m' \cos\left(\frac{p\pi n}{N}\right) \cos\left(\frac{p\pi}{N}(n+m')\right) \exp(-\tfrac{1}{6} b^2 \boldsymbol{k}^2 |m'|) \\
&= \frac{N}{2} \int_{-\infty}^{\infty} \mathrm{d}m' \cos\left(\frac{p\pi m'}{N}\right) \exp(-\tfrac{1}{6} b^2 \boldsymbol{k}^2 |m'|) = \frac{N}{2} \frac{2(b^2\boldsymbol{k}^2/6)}{(p\pi/N)^2 + (b^2\boldsymbol{k}^2/6)^2}.
\end{aligned} \tag{5.180}$$

Then

$$\Xi(\boldsymbol{k}) = \frac{b^2}{9\eta_s N} \sum_{p=1}^{\infty} \frac{1}{(p\pi/N)^2 + (b^2\boldsymbol{k}^2/6)^2} \frac{1}{\bar{h}_p}. \tag{5.181}$$

Similarly, $\bar{h}_p$ is calculated from eqns (5.170) and (5.177),

$$\begin{aligned}
\bar{h}_p &= \frac{1}{N^2} \sum_{n,m} \int \frac{\mathrm{d}\boldsymbol{k}}{(2\pi)^3} \frac{2}{3\eta_s \boldsymbol{k}^2 \left(1 + \frac{c}{N} \Xi(\boldsymbol{k})\right)} \\
&\quad \times \exp(-\tfrac{1}{6} b^2 \boldsymbol{k}^2 |n-m|) \cos\left(\frac{p\pi n}{N}\right) \cos\left(\frac{p\pi m}{N}\right) \\
&= \frac{1}{2N} \int \frac{\mathrm{d}\boldsymbol{k}}{(2\pi)^3} \frac{2}{3\eta_s \boldsymbol{k}^2 \left(1 + \frac{c}{N} \Xi(\boldsymbol{k})\right)} \frac{2(b^2\boldsymbol{k}^2/6)}{(p\pi/N)^2 + (b^2\boldsymbol{k}^2/6)^2}
\end{aligned}$$

i.e.,

$$\bar{h}_p = \frac{b^2}{9\eta_s N}\int \frac{d\boldsymbol{k}}{(2\pi)^3}\frac{1}{1+(c/N)\Xi(\boldsymbol{k})}\frac{1}{(p\pi/N)^2+(b^2\boldsymbol{k}^2/6)^2}. \tag{5.182}$$

Equations (5.181) and (5.182) determine $\Xi(\boldsymbol{k})$. Since this is a nonlinear integral equation the general solution is quite hard to obtain,[74] but in special cases, the characteristic results can be discussed. Without going into the detail, we quote the main results.

(i) In dilute solution, eqns (5.181) and (5.182) can be solved by expanding $\Xi(\boldsymbol{k})$ with respect to c:

$$\Xi(\boldsymbol{k}) = \Xi_0(\boldsymbol{k}) + c\Xi_1(\boldsymbol{k}). \tag{5.183}$$

The first term gives the intrinsic viscosity $[\eta]$ and the second term gives the Huggins coefficient (see eqn (5.2)). The result is

$$[\eta] = \frac{N_A}{M}\frac{(\sqrt{N}\,b)^3}{\sqrt{(6\pi)}}\left(\sum_{p=1}^{\infty} p^{-1.5}\right) \tag{5.184}$$

and

$$k_H = \frac{\pi^3}{6}\left(\sum_{p=1}^{\infty} p^{-1.5}\right)^{-2} = 0.757\ldots \tag{5.185}$$

Equation (5.184) is larger than eqn (4.149) by a factor $\sqrt{2}$. Equation (5.185) is slightly larger than the experimental value which is between 0.4 and 0.6.[75] Thus the numerical factors differ by about 50%, but this can be improved by removing the rigid chain approximation used in the calculation.

(ii) Given the effective hydrodynamic interaction, one can calculate the relaxation time τ_p for the normal mode as

$$\tau_p = \tau_1^{(0)} p^{-3/2}\left(1 + Ac\left(\frac{p}{N}\right)^{-\mu} + \ldots\right) \tag{5.186}$$

where $\tau_1^{(0)}$ is the longest relaxation time in the dilute limit $(c \to 0)$ and μ is about 0.5 in Θ condition. Equation (5.186) explicitly shows that the p dependence of τ_p changes from the Zimm-like $(\tau_p \propto p^{-3/2})$ to Rouse-like $(\tau_p \propto p^{-2})$ as the concentration increases. This result has been compared with experiments on dynamic flow birefringence.[76]

(iii) At higher concentration, over a wide region of $\boldsymbol{k}$, $\bar{\boldsymbol{H}}(\boldsymbol{k})$ is expressed as[70]

$$\bar{\boldsymbol{H}}(\boldsymbol{k}) = \frac{\boldsymbol{I} - \hat{\boldsymbol{k}}\hat{\boldsymbol{k}}}{\eta_s(\boldsymbol{k}^2 + \xi_H^{-2})} \tag{5.187}$$

or in the r space†

$$\bar{H}(r) \simeq \exp(-r/\xi_H)/\eta_s r \tag{5.188}$$

which explicitly shows the screening of the hydrodynamic interaction. The length ξ_H is called the hydrodynamic screening length, and given by[10]

$$\xi_H = 2/\pi c b^2. \tag{5.189}$$

In this regime the steady-state viscosity is obtained[10] as

$$\eta \simeq \eta_s\left(1 + \frac{c^2 b^6}{24} N\right). \tag{5.190}$$

This again indicates that the molecular weight dependence of the viscosity becomes Rouse-like ($\eta \propto M$) for high concentration.

Appendix 5.I Transformation to collective coordinates

The free energy U_0 is formally calculated from

$$\exp(-U_0(\{c_{\boldsymbol{k}}\})/k_B T) = \int \prod_a \delta \boldsymbol{R}_{an} \prod_{k>0} \delta\left[c_{\boldsymbol{k}} - \frac{1}{V}\sum \exp(\mathrm{i}\boldsymbol{k}\cdot\boldsymbol{R}_{an})\right] \times \exp(-U_0[\boldsymbol{R}_{an}]/k_B T). \tag{5.I.1}$$

From the definition of the complex integral given in eqn (2.I.28), it is shown that

$$\prod_{k>0} \delta(c_{\boldsymbol{k}} - A_{\boldsymbol{k}}) = \int \left(\prod_{k>0} \frac{\mathrm{d}\phi_{\boldsymbol{k}}}{\pi^2}\right) \exp\left(\mathrm{i}\sum_{k\neq 0} (c_{\boldsymbol{k}} - A_{\boldsymbol{k}})\phi_{-\boldsymbol{k}}\right) \tag{5.I.2}$$

where $\phi_{-\boldsymbol{k}}$ means $\phi_{\boldsymbol{k}}{}^*$. Using this, eqn (5.I.1) can be rewritten as

$$\exp(-U_0(\{c_{\boldsymbol{k}}\})/k_B T) \propto \int \prod_{k>0} \mathrm{d}\phi_{\boldsymbol{k}} \int \prod_a \delta \boldsymbol{R}_{an} \times \exp\left(\mathrm{i}\sum_{k\neq 0} \phi_{-\boldsymbol{k}} c_{\boldsymbol{k}} - \frac{\mathrm{i}}{V}\sum_{\substack{k\neq 0\\ a,n}} \phi_{-\boldsymbol{k}} \exp(\mathrm{i}\boldsymbol{k}\cdot\boldsymbol{R}_{an})\right) \times \exp(-U_0[\boldsymbol{R}_{an}]/k_B T). \tag{5.I.3}$$

To evaluate the integral on $\boldsymbol{R}_{an}$, let

$$\phi(\boldsymbol{r}) = -\frac{\mathrm{i}}{V}\sum_{k\neq 0} \phi_{-\boldsymbol{k}} e^{\mathrm{i}\boldsymbol{k}\cdot\boldsymbol{r}}. \tag{5.I.4}$$

† Note that eqn (5.188) is not correct for $r \gg \xi_H$ since the asymptotic behaviour of $\bar{H}(r)$ must be $1/\eta r$, but the important region of integration in (5.181) and (5.182) is not this region, so (5.187) is a good self-consistent solution.

Then eqn (2.46) gives

$$\int \delta \boldsymbol{R}_{an} \exp\left[\sum_n \phi(\boldsymbol{R}_{an}) - \frac{3}{2b^2}\int_0^N \mathrm{d}n\left(\frac{\mathrm{d}\boldsymbol{R}_{an}}{\mathrm{d}n}\right)^2\right] = \int \mathrm{d}\boldsymbol{r}\,\mathrm{d}\boldsymbol{r}'G(\boldsymbol{r}, \boldsymbol{r}', N; [\phi])$$

where $G(\boldsymbol{r}, \boldsymbol{r}', N; [\phi])$ satisfies

$$\left[\frac{\partial}{\partial N} - \frac{b^2}{6}\nabla^2 - \phi(\boldsymbol{r})\right]G(\boldsymbol{r}, \boldsymbol{r}', N; [\phi]) = \delta(\boldsymbol{r} - \boldsymbol{r}')\delta(N). \qquad (5.I.5)$$

Thus

$$\exp(-U_0(\{c_{\boldsymbol{k}}\})/k_BT) = \int \prod_{\boldsymbol{k}>0} \mathrm{d}\phi_{\boldsymbol{k}} \exp\left[\mathrm{i}\sum_{\boldsymbol{k}\neq 0} \phi_{-\boldsymbol{k}}c_{\boldsymbol{k}}\right] \times \left[\int \mathrm{d}\boldsymbol{r}\int \mathrm{d}\boldsymbol{r}'G(\boldsymbol{r}, \boldsymbol{r}', N; [\phi])\right]^{N_p} \qquad (5.I.6)$$

where $N_p = cV/N$ is the total number of polymers in the system. Equation (5.I.5) can be solved in a power series of ϕ as

$$G(\boldsymbol{r}, \boldsymbol{r}', N; [\phi]) = G_0(\boldsymbol{r} - \boldsymbol{r}'; N) + \int_0^N \mathrm{d}n_1 \times \int \mathrm{d}\boldsymbol{r}_1 G_0(\boldsymbol{r} - \boldsymbol{r}_1; N - n_1)\phi(\boldsymbol{r}_1)G_0(\boldsymbol{r}_1 - \boldsymbol{r}'; n_1) + \int_0^N \mathrm{d}n_1\int_0^{n_1} \mathrm{d}n_2$$
$$\times \int \mathrm{d}\boldsymbol{r}_1\int \mathrm{d}\boldsymbol{r}_2 G_0(\boldsymbol{r} - \boldsymbol{r}_1; N - n_1)\phi(\boldsymbol{r}_1)G_0(\boldsymbol{r}_1 - \boldsymbol{r}_2; n_1 - n_2)\phi(\boldsymbol{r}_2)G_0(\boldsymbol{r}_2 - \boldsymbol{r}'; n_2) + \dots \qquad (5.I.7)$$

Using the relation

$$\int \mathrm{d}\boldsymbol{r}G_0(\boldsymbol{r} - \boldsymbol{r}'; N) = 1$$

we have

$$\int \mathrm{d}\boldsymbol{r}\int \mathrm{d}\boldsymbol{r}'G(\boldsymbol{r}, \boldsymbol{r}', N; [\phi]) = V + N\int \mathrm{d}\boldsymbol{r}\phi(\boldsymbol{r}) + \int_0^N \mathrm{d}n_1\int_0^{n_1} \mathrm{d}n_2\int \mathrm{d}\boldsymbol{r}_1 \times \int \mathrm{d}\boldsymbol{r}_2\phi(\boldsymbol{r}_1)\phi(\boldsymbol{r}_2)G_0(\boldsymbol{r}_1 - \boldsymbol{r}_2; n_1 - n_2). \qquad (5.I.8)$$

Using eqn (5.I.4) and the definition of $g_0(\boldsymbol{k})$ (see eqn (2.70)), this is written as

$$\int \mathrm{d}\boldsymbol{r}\int \mathrm{d}\boldsymbol{r}'G(\boldsymbol{r}, \boldsymbol{r}', N; [\phi]) = V\left[1 - \frac{N}{V^2}\sum_{\boldsymbol{k}>0} \phi_{\boldsymbol{k}}\phi_{-\boldsymbol{k}}g_0(\boldsymbol{k})\right] \qquad (5.I.9)$$

Thus

$$\exp(-U_0(\{c_k\})/k_BT) = \int \prod_{k>0} \mathrm{d}\phi_k \exp\Big[\mathrm{i}\sum_{k\neq 0} \phi_{-k}c_k + N_p \log\Big(1 - \frac{N}{V^2}\sum_{k>0} \phi_k\phi_{-k}g_0(k)\Big)\Big]. \qquad (5.I.10)$$

Assuming that the fluctuation in ϕ_k is small, we expand the logarithm and evaluate the integral over ϕ_k:

$$\begin{aligned}\text{rhs} &= \int \prod_{k>0} \mathrm{d}\phi_k \exp\Big[\mathrm{i}\sum_{k\neq 0} \phi_{-k}c_k - \frac{N_pN}{V^2}\sum_{k>0} \phi_k\phi_{-k}g_0(k)\Big)\Big] \\ &= \int \prod_{k>0} \mathrm{d}\phi_k \exp\Big[\sum_{k>0}\Big(\mathrm{i}\phi_k^*c_k + \mathrm{i}\phi_kc_k^* - \frac{c}{V}\phi_k\phi_{-k}g_0(k)\Big)\Big] \\ &= \exp\Big[-\sum_{k>0} \frac{Vc_kc_k^*}{cg_0(k)}\Big] \qquad (5.I.11)\end{aligned}$$

which gives eqn (5.31).

Appendix 5.II Osmotic pressure in concentrated solution

Using the formula in Appendix 2.I, we have

$$\begin{aligned}\exp(-(A - A_0)/k_BT) &= \exp\Big[-\frac{V}{2}vc^2\Big] \frac{\int \prod_{k>0} \mathrm{d}c_k \exp\Big(-V\sum_{k>0}\frac{1}{cg(k)}c_kc_k^*\Big)}{\int \prod_{k>0} \mathrm{d}c_k \exp\Big(-V\sum_{k>0}\frac{1}{cg_0(k)}c_kc_k^*\Big)} \\ &= \exp\Big[-\frac{V}{2}vc^2\Big]\prod_{k>0}(g(k)/g_0(k)) \\ &= \exp\Big[-\frac{V}{2}vc^2 + \sum_k \tfrac{1}{2}\log\Big(\frac{g(k)}{g_0(k)}\Big)\Big]. \qquad (5.II.1)\end{aligned}$$

Hence

$$(A - A_0)/k_BT = \frac{V}{2}vc^2 + \frac{1}{2}\frac{V}{(2\pi)^3}\int \mathrm{d}k \log\Big(\frac{k^2 + \xi^{-2}}{k^2}\Big). \qquad (5.II.2)$$

The osmotic pressure is thus given by

$$\begin{aligned}\Pi/k_BT &= \frac{c}{N} - \frac{\partial(A - A_0)}{\partial V}\Big|_{T,cV\ \text{constant}} \\ &= \frac{c}{N} + \frac{v}{2}c^2 - \frac{1}{2}\frac{1}{(2\pi)^3}\int \mathrm{d}k\Big[\log\Big(\frac{k^2+\xi^{-2}}{k^2}\Big) + \frac{V}{k^2+\xi^{-2}}\frac{\partial\xi^{-2}}{\partial V}\Big] \\ &= \frac{c}{N} + \frac{v}{2}c^2 - \frac{1}{2}\frac{1}{(2\pi)^3}\int_0^\infty \mathrm{d}k 4\pi k^2\Big[\log\Big(1 + \frac{1}{k^2\xi^2}\Big) - \frac{1}{1 + k^2\xi^2}\Big]. \qquad (5.II.3)\end{aligned}$$

The underlined integral is performed by the integration by parts:

$$\int_0^\infty \mathrm{d}x x^2\left[\log\left(1+\frac{1}{x^2}\right)-\frac{1}{1+x^2}\right] = -\int_0^\infty \mathrm{d}x \tfrac{1}{3}x^3 \frac{\partial}{\partial x} \times \left[\log\left(1+\frac{1}{x^2}\right)-\frac{1}{1+x^2}\right] = \frac{\pi}{6}. \quad (5.\text{II}.4)$$

Hence

$$\Pi = k_B T\left[\frac{c}{N}+\frac{v}{2}c^2-\frac{1}{24\pi\xi^3}\right]. \quad (5.\text{II}.5)$$

Appendix 5.III Perturbation calculation of $\langle \boldsymbol{R}^2 \rangle$

Let $\langle \ldots \rangle_0$ be the average for the ideal chain and $\tilde{U}_1$ be defined by

$$\tilde{U}_1 = \tfrac{1}{2}\sum_{n,m} \tilde{v}(\boldsymbol{R}_n - \boldsymbol{R}_m) \quad (5.\text{III}.1)$$

then the perturbation method described in Appendix 2.III gives

$$\langle \boldsymbol{R}^2 \rangle = \langle (\boldsymbol{R}_N - \boldsymbol{R}_0)^2 \rangle_0 [1 + \langle \tilde{U}_1 \rangle_0] - \langle (\boldsymbol{R}_N - \boldsymbol{R}_0)^2 \tilde{U}_1 \rangle_0. \quad (5.\text{III}.2)$$

The first term gives

$$\langle (\boldsymbol{R}_N - \boldsymbol{R}_0)^2 \rangle_0 [1 + \langle \tilde{U}_1 \rangle_0] = Nb^2\left(1 + \int_0^N \mathrm{d}m \int_0^m \mathrm{d}n \langle \tilde{v}(\boldsymbol{R}_n - \boldsymbol{R}_m) \rangle_0\right) \quad (5.\text{III}.3)$$

and the second term becomes

$$\langle (\boldsymbol{R}_N - \boldsymbol{R}_0)^2 \tilde{U}_1 \rangle_0 = \int_0^N \mathrm{d}m \int_0^m \mathrm{d}n \langle \tilde{v}(\boldsymbol{R}_n - \boldsymbol{R}_m)[(N-m)b^2 + (\boldsymbol{R}_m - \boldsymbol{R}_n)^2 + nb^2] \rangle_0. \quad (5.\text{III}.4)$$

Hence

$$\begin{aligned}\langle \boldsymbol{R}^2 \rangle &= Nb^2 + \int_0^N \mathrm{d}m \int_0^m \mathrm{d}n \langle \tilde{v}(\boldsymbol{R}_n - \boldsymbol{R}_m)[(m-n)b^2 - (\boldsymbol{R}_m - \boldsymbol{R}_n)^2] \rangle_0 \\ &= Nb^2 + \int_0^N \mathrm{d}m \int_0^m \mathrm{d}n \int \mathrm{d}\boldsymbol{r} \tilde{v}(\boldsymbol{r})((m-n)b^2 - \boldsymbol{r}^2) G_0(\boldsymbol{r}, m-n) \quad (5.\text{III}.5)\end{aligned}$$

where

$$G_0(\boldsymbol{r}, n) = (3/2\pi nb^2)^{3/2} \exp\left(-\frac{3\boldsymbol{r}^2}{2nb^2}\right). \quad (5.\text{III}.6)$$

Since the integrand of eqn (5.III.5) depends only $m - n$, this is rewritten as

$$\langle \boldsymbol{R}^2 \rangle = Nb^2 + \int_0^N \mathrm{d}n(N-n) \int \mathrm{d}\boldsymbol{r}\tilde{v}(\boldsymbol{r})(nb^2 - \boldsymbol{r}^2)G_0(\boldsymbol{r}, n). \quad (5.\text{III}.7)$$

By use of the Fourier transform

$$G_0(\boldsymbol{r}, n)(nb^2 - \boldsymbol{r}^2) = \int \frac{\mathrm{d}\boldsymbol{k}}{(2\pi)^3} \frac{b^4\boldsymbol{k}^2}{9} n^2 \exp\left(-\frac{b^2\boldsymbol{k}^2}{6} n\right) \mathrm{e}^{\mathrm{i}\boldsymbol{k}\cdot\boldsymbol{r}} \quad (5.\text{III}.8)$$

and eqn (5.53), eqn (5.III.7) is written as

$$\langle \boldsymbol{R}^2 \rangle = Nb^2 + v \int \frac{\mathrm{d}\boldsymbol{k}}{(2\pi)^3} \int_0^N \mathrm{d}n(N-n) \underline{\frac{b^4\boldsymbol{k}^2}{9} n^2 \exp\left(-\frac{b^2\boldsymbol{k}^2}{6} n\right) \frac{\boldsymbol{k}^2}{\boldsymbol{k}^2 + \xi^{-2}}}. \quad (5.\text{III}.9)$$

Since the underlined part decreases quickly with n, the integral over n is evaluated for large N as

$$\int_0^N \mathrm{d}n(N-n) \ldots = N \int_0^\infty \mathrm{d}n \ldots. \quad (5.\text{III}.10)$$

Hence, we finally get

$$\begin{aligned} \langle \boldsymbol{R}^2 \rangle &= Nb^2 \left[1 + v \int \frac{\mathrm{d}\boldsymbol{k}}{(2\pi)^3} \frac{2b^2\boldsymbol{k}^2}{9} \left(\frac{6}{\boldsymbol{k}^2 b^2} \right)^3 \frac{\boldsymbol{k}^2}{\boldsymbol{k}^2 + \xi^{-2}} \right] \\ &= Nb^2 \left[1 + \frac{12}{\pi} \frac{v\xi}{b^4} \right]. \end{aligned} \quad (5.\text{III}.11)$$

References

1. de Gennes, P. G., *Scaling Concepts in Polymer Physics,* Chapter 3. Cornell Univ. Press, Ithaca, New York (1979).
2. Flory, P., *J. Chem. Phys.* **10,** 51 (1942); see also *Principles of Polymer Chemistry,* Chapter 12. Cornell Univ. Press, Ithaca, New York (1953).
3. Huggins, M. L., *J. Phys. Chem.* **46,** 151 (1942); *J. Am. Chem. Soc.* **64,** 1712 (1942).
4. A review on polymer collapse is given by Williams, C., Brochard, F., and Frisch, H. L., *Ann. Rev. Phys. Chem.* **32,** 433 (1981).
5. Daoud, M., and Jannink G., *J. Phys.* (*Paris*) **37,** 973 (1976).
6. Schaefer, D. W., Joanny J. F., and Pincus, P., *Macromolecules* **13,** 1280 (1980).
7. Moore, M. A., *J. Phys.* (*Paris*) **38,** 265 (1977).

8. Schäfer, L., *Macromolecules* **17,** 1357 (1984).
9. Edwards, S. F., *Proc. Phys. Soc. London* **88,** 265 (1966).
10. Edwards, S. F., and Muthukumar, M., *Macromolecules* **17,** 586 (1984).
11. Leibler, L., *Macromolecules* **13,** 1602 (1980).
12. Okano, K., Wada, E., Kurita, K., Hiramatsu, H., and Fukuro, H., *J. Physique Lett.* **40,** L171 (1979).
13. Okano, K., Wada, E., Kurita, K., and Fukuro, H., *J. Appl. Cryst.* **11,** 507 (1978).
14. Muthukumar, M., and Edwards, S. F., *J. Chem. Phys.* **76,** 2720 (1982).
15. Des Cloizeaux J., and Jannink G., *Les Polymères en Solution: Modelisation et leur structures,* Chapter 13, § 2.5.5. Edition de Physique, Paris (1986 or 7).
16. Edwards, S. F., *J. Phys.* **A8,** 1670 (1975); Edwards, S. F., and Jeffers, E. F., *J. C. S. Faraday Trans. II* **75,** 1020 (1979).
17. Flory, P., *J. Chem. Phys.* **17,** 303 (1949); see also *Principles of Polymer Chemistry,* Chapter 12. Cornell Univ. Press, Ithaca, New York (1953).
18. Richards, R. W., Maconnachie, A., and Allen, G., *Polymer* **19,** 266 (1978).
19. Oono, Y., *Adv. Chem. Phys.* **61,** 301 (1985).
20. Freed, K. F., *Accounts Chem. Res.,* to be published.
21. Daoud, M., Cotton, J. P., Farnoux, B., Jannink, G., Sarma, G., Benoit, H., Duplessix, R., Picot, C., and de Gennes, P. G., *Macromolecules* **8,** 804 (1975).
22. Noda, I., Kato, N., Kitano, T., and Ngasawa, M., *Macromolecules* **14,** 668 (1981).
23. des Cloizeaux J., and Noda, I., *Macromolecules* **15,** 1505 (1982).
24. Schäfer, L., *Macromolecules* **15,** 652 (1982).
25. Ohta, T., and Oono, Y., *Phys. Lett.* **89A,** 460 (1982).
26. Wiltzius, P., Haller, H. R., Cannell, D. S., and Schaefer, D. W., *Phys. Rev. Lett.* **51,** 1183 (1983).
27. Nakanishi, A., and Ohta., T., *J. Phys.* **A18,** 127 (1985).
28. Vologodskii, A., Lukashin, A., Frank-Kamenetskii, M. F., and Anshelevich, V., *Sov. Phys. JETP* **39,** 1059 (1975); **40,** 932 (1975).
29. Roovers, J., and Toporowski, P. M., *Macromolecules* **16,** 843 (1983).
30. Ferry, J. D., *Viscoelastic Properties of Polymers* (3rd edn). Wiley, New York (1980).
31. Graessley, W. W., *Adv. Polym. Sci.* **16,** 1 (1974).
32. Baumgärtner, A., *Polymer* **22,** 1308 (1981).
33. Edwards, S. F., *J. Phys.* **A1,** 15 (1968).
34. Alexander, J. W., *Trans. Am. Math. Soc.* **30,** 275 (1928).
35. Ball, R., and Mehta, M. L., *J. Physique.* **42,** 1193 (1981).
36. Brereton, M. G., and Shah, S., *J. Phys.* **A14,** L51 (1981).
37. des Cloizeaux, J., *J. Phys. (Paris)* **42,** L433 (1981).
38. Iwata, K., and Kimura, T., *J. Chem. Phys.* **74,** 2039 (1981); Iwata, K., *J. Chem. Phys.* **78,** 2778 (1983); *Macromolecules* **18,** 115 (1985).
39. Tanaka, F., *Prog. Theor. Phys.* **68,** 148, 164 (1982).
40. Edwards, S. F., *Proc. Phys. Soc.* **92,** 9 (1967).
41. Iwata, K., *J. Chem. Phys.* **76,** 6363, 6375 (1982).
42. Edwards, S. F., *Proc. Phys. Soc.* **91,** 513 (1967).
43. Prager, S., and Frisch, H. L., *J. Chem. Phys.* **46,** 1475 (1967).
44. Saito, N., and Chen, Y., *J. Chem. Phys.* **59,** 3701 (1973).

45. Ito, K., and McKean, H. P., *Diffusion Processes and their Simple Paths*. Springer, Berlin (1965).
46. Edwards, S. F., *Proc. R. Soc. Lond.* **A385,** 267 (1982).
47. de Gennes, P. G., *J. Chem. Phys.* **55,** 572 (1971).
48. Doi, M., and Edwards, S. F., *J. C. S. Faraday Trans. II* **74,** pp. 1789, 1802, and 1818 (1978).
49. Kawasaki, K., *Annals Phys.* **61,** 1 (1970); Kawasaki, K., and Gunton, J. D., Critical dynamics, in *Progress in Liquid Physics* (ed. C. Croxton). Wiley, New York (1978).
50. Pincus, P., *J. Chem. Phys.* **75,** 1996 (1981).
51. Brochard, F., and de Gennes, P. G., *Macromolecules* **10,** 1157 (1977); Brochard, F., *J. Physique* **44,** 39 (1983).
52. Richter, D., Hayter, J. B., Mezei, F., and Ewen, B., *Phys. Rev. Lett.* **41,** 1484 (1978).
53. de Gennes, P. G., *Macromolecules* **9,** 587, 594 (1976).
54. King, T. A., Knox, A., and McAdam, J. D. G., *Polymer* **14,** 293 (1973).
55. Ford, N. C., Karasz, F. E., and Owen, J. E. M., *Discuss. Faraday Soc.* **49,** 228 (1970).
56. Adam, M., and Delsanti, M., *Macromolecules* **10,** 1229 (1977).
57. Nose, T., and Chu, B., *Macromolecules* **12,** pp. 590, 599, and 1122 (1979); **13,** 122 (1980); Nemoto, N., Makita, Y., Tsunashima, Y., and Kurata, M., *Macromolecules* **17,** 2629 (1984).
58. Schaefer, D. W., and Han, C. C., in *Dynamic Light Scattering* (ed. R. Pecora). Plenum Press, New York (1985).
59. Amis, E. J., and Han, C. C., *Polymer* **23,** 1403 (1982); see also Amis, E. J., Janmey, P. A., Ferry, J. D., and Yu, H., *Macromolecules* **16,** 441 (1983).
60. Eisele, M., and Burchard, W., *Macromolecules,* **17,** 1636 (1984).
61. Brown, W., *Macromolecules* **17,** 66 (1984).
62. Tsvetkov, V. N., in *Newer Methods of Polymer Characterization* (ed. B. Ke). Interscience, New York (1964).
63. Janeschitz-Kriegl, H., *Adv. Polym. Sci.* **6,** 170 (1969).
64. Léger, L., Hervet, H., and Rondelez, F., *Macromolecules* **14,** 1732 (1981).
65. Debye, P., *J. Chem. Phys.* **14,** 636 (1946); see also ref. 75.
66. Onogi, S., Kimura, S., Kato, T., Masuda, T., and Miyanaga, N., *J. Polym. Sci.* C 381 (1966).
67. Onogi, S., Masuda, T., Miyanaga, N., and Kimura, S., *J. Polym. Sci. A2* **5,** 899 (1967).
68. See for example, Dreval, V. E., Malkin, A. Y., and Botvinnik, G. O., *J. Polym. Sci. Phys. ed.* **11,** 1055 (1973).
69. Edwards, S. F., *Faraday Symp. Chem. Soc.* **18,** 145 (1983).
70. Edwards, S. F., and Freed, K. F., *J. Chem. Phys.* **61,** 1189 (1974); Freed, K. F., and Edwards, S. F., *J. Chem. Phys.* **61,** 3626 (1974); ibid **62,** 4032 (1975).
71. Freed, K. F., Polymer dynamics and the hydrodynamics of polymer solutions, in *Progress in Liquid Physics* (ed C. Croxton). Wiley, New York (1978).
72. Muthukumar, M., and Freed, K. F., *Macromolecules* **10,** 899 (1977); **11,** 843 (1978).
73. Perico, A., and Freed, K. F., *J. Chem. Phys.* **81,** 1466, 1475 (1984).

74. A discussion for the solution is given by Yoshimura, T., *J. Math. Phys.* **24,** 2056 (1983).
75. Frisch, H. L., and Simha, R., The viscosity of colloidal suspensions and macromolecular solutions, in *Rheology* (ed. F. R. Eirich) Vol. 1, Chapter 14. Academic Press, New York (1956); see also Bohdanecky, M., and Kovar, J., *Viscosity of Polymer Solutions,* Chapter 3. Elsevier, Amsterdam (1982).
76. Martel, C. J. T., Lodge, T. P., Dibbs, M. G., Stokich, T. M., Sammler, R. L., Carriere C. J., and Schrag, J. L., *Faraday Symp. Chem. Soc.* **18,** 173 (1983).

6

DYNAMICS OF A POLYMER IN A FIXED NETWORK

6.1 Tube model

6.1.1 Tube model in crosslinked systems

As mentioned in Section 5.4.3, the highly entangled state of polymers can be effectively described by the tube model. The idea of the tube model originated in studying the problem of rubber elasticity.[1,2] A rubber is a huge molecular network which is formed when a polymeric liquid is crosslinked by chemical bonds.[3] An important problem in the theory of rubber elasticity is to calculate the entropy, which is essentially the number of allowed conformations of the chains constituting the rubber. The topological constraints play an important role in such a problem.

Consider a lightly crosslinked rubber which consists of long strands of polymers between crosslinks. A strand in such a rubber is schematically shown in Fig. 6.1. In Fig. 6.1*b* the strand is placed on a plane and the cross-sections of other strands are shown by dots. Due to the topological constraints, the strand cannot cross the dots, so that the number of conformations allowed for the strand is much less than that in free space. How can we estimate it?

Suppose for a moment that the other chains are frozen, then the dots can be regarded as fixed obstacles. One can see that the allowed conformation of the strand is almost confined in a tube-like region shown by the dotted lines: the conformations which go outside the tube are likely to violate the topological constraints. The axis of the tube can be defined as the shortest path connecting the two ends of the strand with the same topology as the strand itself relative to the obstacles. Such a path represents a group of conformations which are accessible to each other without violating the topological constraints imposed by the other chains, and is called the primitive path.[2] If the topological constraints are replaced by the tube, the number ω of the allowed conformations can be calculated easily by the method described in Chapter 2 (see Section 6.4).

In real rubbers, the situation is more complicated since the other strands are mobile. However, even in such a case, a self-consistent picture will be that the range in which each part of the strand can move around will remain finite. The range is perhaps larger than the mean separation between the frozen strands discussed above. What diameter one should assign to the tube is a question which has not been answered with absolute certainty. However, as long as the strand is long enough, the diameter is determined by local conditions, and will be independent

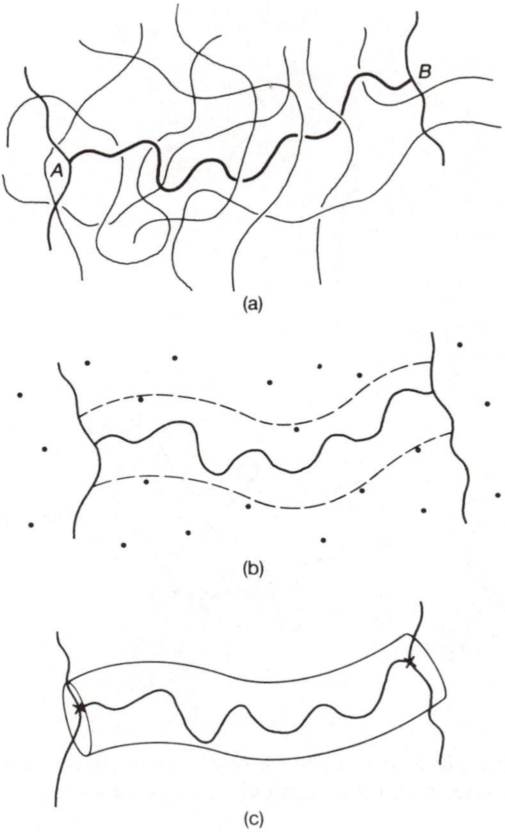

Fig. 6.1. (*a*) A strand in a rubber. *A* and *B* denote the crosslinks. (*b*) Schematic picture of (*a*): the strand under consideration is placed on a plane and the other strands intersecting the plane are shown by dots. (*c*) The tube model.

of the length of the strand. Detailed discussions[1,2,4] on rubber elasticity based on this idea and its appraisals[5,6] are given in the literature. An important point here is the proposition that the tube concept will be a self-consistent picture in the system of topologically interacting system.

6.1.2 Tube model in uncrosslinked systems

The idea of the tube is intuitively appealing, and one can imagine that the same picture will be useful for uncrosslinked system such as polymer melts. However, one has to face a new problem that, in melts, the tube itself changes with time because all conformations in a melt are accessible. A key concept to solve this problem was introduced by de Gennes[7] who discussed the Brownian motion of an unattached chain

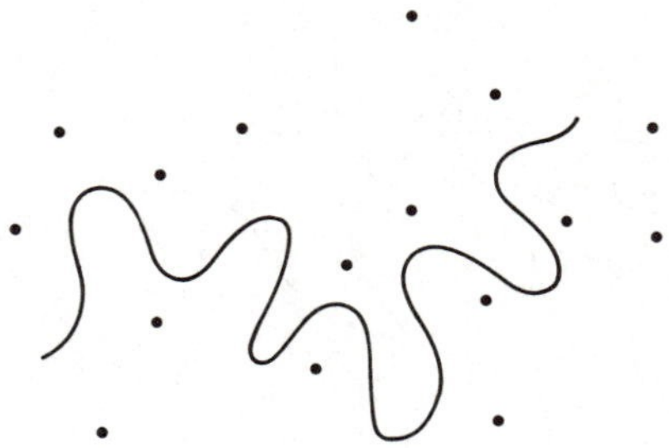

(a)

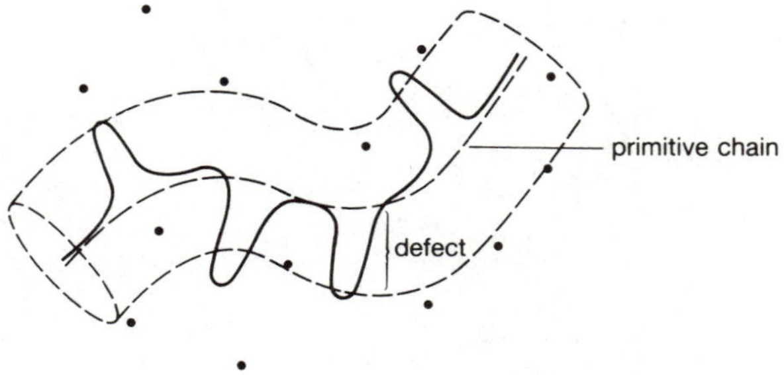

(b)

Fig. 6.2. (*a*) A chain in a fixed network of obstacles (denoted by dots). (*b*) The defects and the primitive chain (dashed line).

moving through a fixed network (see Fig. 6.2). His idea was that in the situation shown in Fig. 6.2*a,* the motion of the chain is almost confined in a tube-like region denoted by dotted lines in Fig. 6.2*b*. Since the chain is rather longer than the tube, the slack will constitute a series of 'defects' which can flow up and down the tube (Fig. 6.2*b*). De Gennes visualized this as a gas of non-interacting defects running along like the arch in a caterpillar (see Fig. 6.3*a*). As a result of such motion, the tube itself changes with time (Fig. 6.3*b*): for example if the chain moves right, the part B_0B can choose a random direction, and create a new part of the tube which will be a constraint for the rest of the chain, while the part of the previous tube A_0A becomes empty and disappears. This type of motion was called reptation by de Gennes after the Latin *reptare,* to creep.

Our current understanding of the dynamics of the highly entangled state is based on the concept of reptation. This picture is rigorously correct for the system that has been presented, i.e., a single chain in a

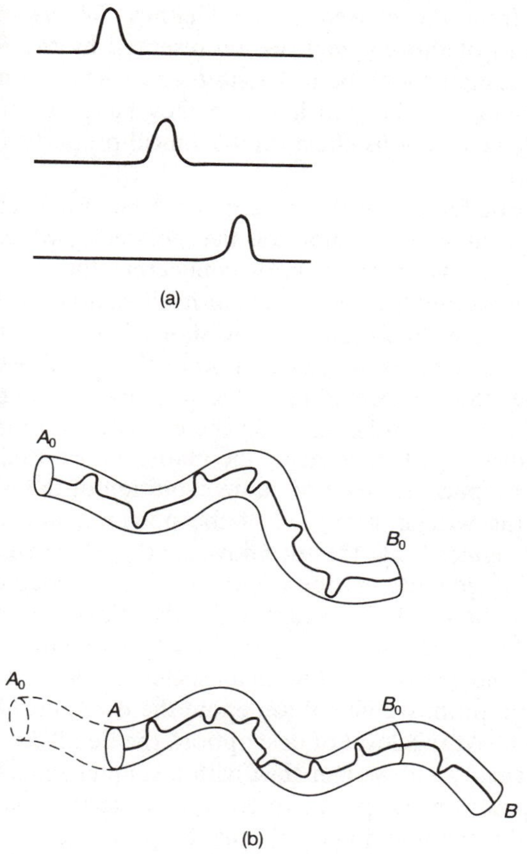

Fig. 6.3. (*a*) Motion of a single defect (*b*) Motion of the tube.

fixed network. Whether the picture does indeed hold for concentrated polymer solutions or melts still remains a matter for debate, but many experimental results suggest that reptation is the dominant mechanism for the dynamics of a chain in the highly entangled state. Leaving the detailed discussion of this problem to the next chapter, we shall first consider the simple situation of a chain moving in a fixed network.

6.2 Reptation

6.2.1 Primitive chain

Let us consider a polymer moving in a fixed network of obstacles. For the convenience of later discussion, we shall specify the problem in slightly different terms from those used by de Gennes. We assume that the intrinsic properties of the polymer are represented by the Rouse model consisting of N segments with bond length b and friction constant ζ. The obstacles are assumed to be thin lines, so they have no effect on static properties, but have a serious effect on dynamical properties by imposing topological constraints.

The characteristic feature of the dynamics can be visualized by the tube model. For a given conformation of the polymer, we can draw the primitive path, i.e., the shortest path connecting the two ends of the chain with the same topology as the chain itself relative to the obstacles (see the dashed line in Fig. 6.2*b*). In the short time-scale the motion of the polymer is regarded as wriggling around the primitive path. On a longer time-scale, the conformation of the primitive path changes as the polymer moves, creating and destroying the ends of the primitive path.

Even though such a picture is clear, the mathematical treatment of the problem is still complicated since the time evolution of the primitive path is governed by the wriggling motion of the polymer, and the wriggling motion itself is limited by the primitive path. However, if we are interested in the large-scale motion of long chains, we may disregard the small-scale fluctuations, and discuss only the time evolution of the primitive path. Since the primitive path at any moment represents the conformation of the chain with the small-scale fluctuations omitted, we shall use the term 'primitive chain' to denote the dynamical equivalent of the primitive path. At this level of description, the details of the wriggling motion are irrelevant, and we can start with a simpler model.

To denote a point on the primitive chain, we use the contour length s measured from the chain end and call this the primitive chain segment s. If $\boldsymbol{R}(s, t)$ is its position at time t, the vector

$$\boldsymbol{u}(s, t) = \frac{\partial}{\partial s}\boldsymbol{R}(s, t) \tag{6.1}$$

is the unit vector tangent to the primitive chain.

The dynamics of the primitive chain is characterized by the following assumptions.

(i) The primitive chain has constant contour length L.

(ii) The primitive chain can move back and forth only along itself with a certain diffusion constant D_c.

(iii) The correlation of the tangent vectors $\boldsymbol{u}(s, t)$ and $\boldsymbol{u}(s', t)$ decreases quickly with $|s - s'|$.

The first assumption corresponds to neglecting the fluctuations of the contour length. The second states that the motion of the primitive chain is reptation. The third guarantees that the conformation of the primitive chain becomes Gaussian† on a large length-scale. This assumption introduces a new parameter into the problem. Since the mean square distance between two points on the Gaussian chain is proportional to $|s - s'|$, it is written as

$$\langle (\boldsymbol{R}(s, t) - \boldsymbol{R}(s', t))^2 \rangle = a\,|s - s'| \quad \text{for} \quad |s - s'| \gg a. \tag{6.2}$$

The length a is called the step length of the primitive chain.

The primitive chain is thus characterized by three parameters L, D_c and a, which must be expressed by the Rouse model parameters N, b, ζ and the parameters characterizing the network. The parameter D_c can be identified as the diffusion coefficient of the Rouse model

$$D_c = \frac{k_B T}{N\zeta} \tag{6.3}$$

because the motion of the primitive chain corresponds to the overall translation of the Rouse chain along the tube. The length L is expressed by a since the mean square end-to-end vector of the primitive chain, which is La according to eqn (6.2), must be the same as that of the Rouse chain Nb^2. Thus

$$L = \frac{Nb^2}{a}. \tag{6.4}$$

We are left with a single parameter a, which depends on the statistical nature of the network. Though precise calculation of this parameter is difficult, it is obvious that a is of the order of the mesh size of the network and much less than L. This knowledge is enough for the purpose of the present discussion.

6.2.2 Simple application

We now study the dynamics of the primitive chain and show that certain time correlation functions can be calculated by a straightforward method.[7] For example, consider the time correlation function of the end-to-end vector $\boldsymbol{P}(t) \equiv \boldsymbol{R}(L, t) - \boldsymbol{R}(0, t)$. Figure 6.4 explains the principle of calculating this correlation function. At $t = 0$, the chain is trapped in a certain tube. As time passes, the primitive chain reptates and at a certain later time (Fig. 6.4*d*), the part of the chain CD remains in the original tube while the parts AC and DB are in a new tube. To calculate

† It must be remembered that despite the Gaussian behaviour with a large length-scale, the primitive chain cannot be modelled by a continuous Gaussian chain since the contour length of the continuous Gaussian chain is infinite and has no physical significance, while the contour length of the primitive chain has a definite physical significance and appears in various dynamical results.

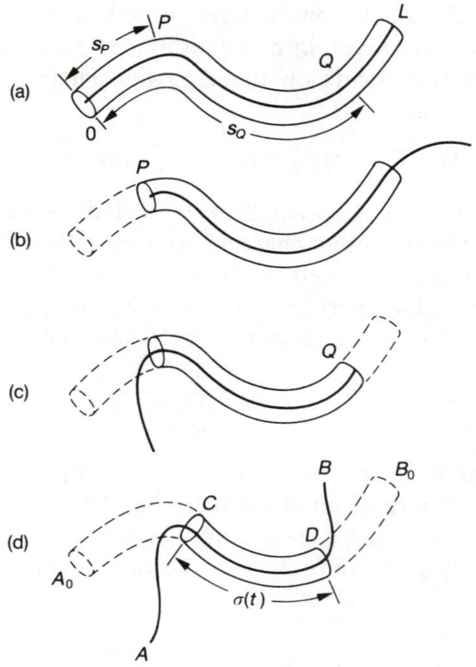

Fig. 6.4. Four successive situations of a reptating chain. (*a*) The initial conformation of the primitive chain and the tube which we call the original tube. (*b*) and (*c*) As the chain moves right or left, some parts of the chain leave the original tube. The parts of the original tube which have become empty of the chain disappear (dotted line). (*d*) The conformation at a later time *t*. The tube segment vanishes when it is reached by either of the chain ends: e.g., the tube segment *P* and *Q* vanish at the instance (*b*) when $\xi(t)=s_P$ and at (*c*) when $\xi(t)=s_Q-L$, respectively.

$\langle \boldsymbol{P}(t)\cdot\boldsymbol{P}(0)\rangle$, we express $\boldsymbol{P}(t)$ and $\boldsymbol{P}(0)$ as

$$\boldsymbol{P}(0)=\overrightarrow{A_0C}+\overrightarrow{CD}+\overrightarrow{DB_0} \tag{6.5}$$

$$\boldsymbol{P}(t)=\overrightarrow{AC}+\overrightarrow{CD}+\overrightarrow{DB} \tag{6.6}$$

Since the vectors $\overrightarrow{AC}$ and $\overrightarrow{DB}$ are uncorrelated with $\boldsymbol{P}(0)$, $\langle \boldsymbol{P}(t)\cdot\boldsymbol{P}(0)\rangle$ will have the form

$$\langle \boldsymbol{P}(t)\cdot\boldsymbol{P}(0)\rangle=\langle\overrightarrow{CD}^2\rangle=a\langle\sigma(t)\rangle \tag{6.7}$$

where $\sigma(t)$ is the contour length of *CD*, i.e., the part in the original tube.

To calculate $\langle\sigma(t)\rangle$ we focus attention on a certain segment *s* of the original tube. This tube segment disappears when it is reached by either end of the primitive chain. Let $\psi(s, t)$ be the probability that this tube segment remains at time t. The average $\langle\sigma(t)\rangle$ is calculated as

$$\langle\sigma(t)\rangle = \int_0^L \mathrm{d}s\,\psi(s,t). \tag{6.8}$$

Let $\Psi(\xi, t; s)$ be the probability that the primitive chain moves the distance ξ while its ends have not reached the segment s of the original tube. The probability satisfies the one-dimensional diffusion equation

$$\frac{\partial\Psi}{\partial t} = D_c\frac{\partial^2\Psi}{\partial\xi^2} \tag{6.9}$$

with the initial condition

$$\Psi(\xi, 0; s) = \delta(\xi). \tag{6.10}$$

When $\xi = s$, the tube segment s is reached by the end of the primitive chain and $\Psi(\xi, t; s)$ vanishes (see Fig. 6.4). Similarly when $\xi = s - L$, the tube segment is reached by the other end and $\Psi(\xi, t; s)$ vanishes.† Thus

$$\Psi(\xi, t; s) = 0 \quad \text{at} \quad \xi = s \quad \text{and} \quad \xi = s - L. \tag{6.11}$$

The solution of eqn (6.9) with these boundary conditions is

$$\Psi(\xi, t; s) = \sum_{p=1}^{\infty}\frac{2}{L}\sin\left(\frac{p\pi s}{L}\right)\sin\left(\frac{p\pi(s-\xi)}{L}\right)\exp(-p^2t/\tau_d) \tag{6.12}$$

where

$$\tau_d = L^2/D_c\pi^2. \tag{6.13}$$

For the tube segment s to remain, ξ can be anywhere between $s - L$ and s, so that

$$\begin{aligned}\psi(s,t) &= \int_{s-L}^{s}\mathrm{d}\xi\,\Psi(\xi, t; s)\\ &= \sum_{p;\,\mathrm{odd}}\frac{4}{p\pi}\sin\left(\frac{p\pi s}{L}\right)\exp(-p^2t/\tau_d).\end{aligned} \tag{6.14}$$

Thus from eqns (6.7), (6.8), and (6.14)

$$\langle \boldsymbol{P}(t)\cdot\boldsymbol{P}(0)\rangle = La\psi(t) = Nb^2\psi(t) \tag{6.15}$$

where

$$\psi(t) = \frac{1}{L}\int_0^L \mathrm{d}s\,\psi(s,t) = \sum_{p;\,\mathrm{odd}}\frac{8}{p^2\pi^2}\exp(-p^2t/\tau_d). \tag{6.16}$$

† Strictly speaking this argument is valid in the limit of $a \to 0$. If a is finite, the boundary condition is not written in a simple form, but the correction is of the order of a/L.

The longest relaxation time of $\langle \boldsymbol{P}(t) \cdot \boldsymbol{P}(0) \rangle$ is given by τ_d. This is called the reptation or disengagement time, since it is the time needed for the primitive chain to disengage from the tube it was confined to at $t = 0$.

Equation (6.15) can be compared with eqn (4.35) for the Rouse chain without constraints:

$$\langle \boldsymbol{P}(t) \cdot \boldsymbol{P}(0) \rangle = Nb^2 \sum_{p;\,\mathrm{odd}} \frac{8}{p^2\pi^2} \exp(-p^2 t/\tau_R) \tag{6.17}$$

where τ_R is the Rouse relaxation time,

$$\tau_R = \frac{\zeta N^2 b^2}{3\pi^2 k_B T}. \tag{6.18}$$

On the other hand, eqn (6.13) is rewritten by eqns (6.3) and (6.4) as

$$\tau_d = \frac{1}{\pi^2} \frac{\zeta N^3 b^4}{k_B T a^2}. \tag{6.19}$$

Note that τ_d is proportional to N^3 and becomes much larger than τ_R for large N. This demonstrates the crucial effect of topological constraints on the conformational change of polymers.

Let us define the number of steps in a primitive chain by

$$Z = \frac{L}{a} = \frac{Nb^2}{a^2}. \tag{6.20}$$

Then the ratio between τ_d and τ_R is written as

$$\tau_d/\tau_R = 3Z. \tag{6.21}$$

Equation (6.19) has been confirmed by computer simulation.[8–10]

The function $\psi(s, t)$ will appear frequently in the subsequent discussions. This function has been defined as the probability that the original tube segment s remains at time t. As will be shown in Section 6.3.3, $\psi(s, t)$ also represents the probability that the primitive chain segment s is in the original tube at time t. (Note the distinction between the *tube segment* and the *primitive chain segment*; the former is fixed in space, while the latter moves with the primitive chain.)

The behaviour of $\psi(s, t)$ is shown in Fig. 6.5. The tube segments in the middle ($s \simeq L/2$) have long lifetimes of order τ_d, while the tube segments near the chain ends have very short lifetimes: the end segment is almost instantaneously replaced. This fact will be used in the subsequent discussions. The function $\psi(t)$ represents the average fraction of the original tube that remains at time t. This function is also equal to the average fraction of the primitive chain contour that remains in the original tube.

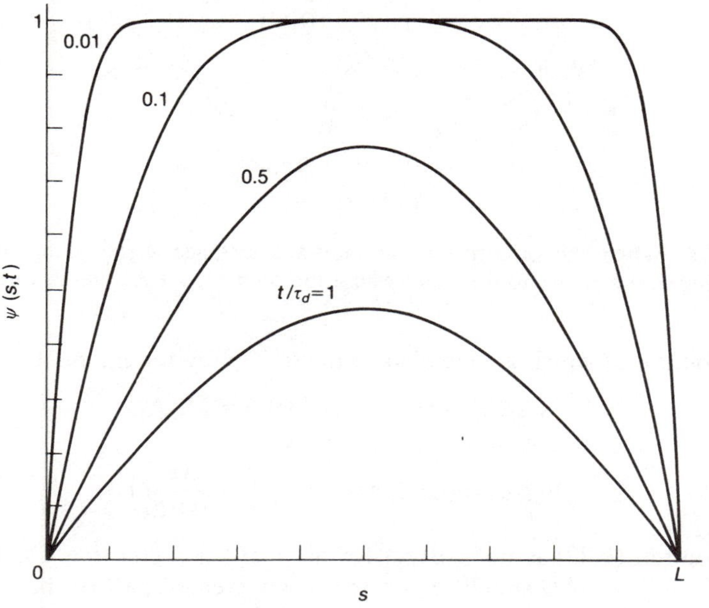

Fig. 6.5. The probability $\psi(s, t)$ that the tube segment s is remaining at time t. This is also equal to the probability that the primitive chain segment s remains in the original tube at time t.

6.3 Reptation dynamics

6.3.1 Stochastic equation for reptation dynamics

Although the above probabilistic description is quite useful in understanding the essence of reptation dynamics, it becomes progressively more difficult to proceed with the calculation for other types of time correlation function. For example, it is not easy to calculate the mean square displacement of a primitive chain segment $\langle(\boldsymbol{R}(s, t) - \boldsymbol{R}(s, 0))^2\rangle$ by this method. In this section we shall describe a convenient method[11] for calculating general time correlation functions.

First we derive a simple mathematical equation for reptation dynamics. Let $\Delta\xi(t)$ be the distance that the primitive chain moves in a time interval between t and $t + \Delta t$, then

$$\boldsymbol{R}(s, t + \Delta t) = \boldsymbol{R}(s + \Delta\xi(t), t), \tag{6.22}$$

which states simply that if the primitive chain moves the distance $\Delta\xi(t)$ along itself, the segment s comes to the point where the segment $s + \Delta\xi$ was at time t (see Fig. 6.6). The variable $\Delta\xi(t)$ takes random values, the

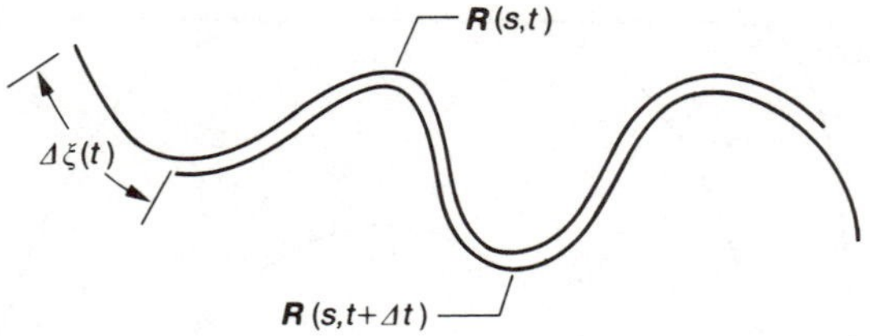

Fig. 6.6. When the primitive chain moves a distance $\Delta\xi(t)$ along itself, the segment s comes to the point where the segment $s + \Delta\xi$ was at time t.

distribution of which is Gaussian characterized by the moments

$$\langle \Delta\xi(t) \rangle = 0, \qquad \langle \Delta\xi(t)^2 \rangle = 2D_c\Delta t, \tag{6.23}$$

i.e.,

$$\Psi(\Delta\xi) = (4\pi D_c \Delta t)^{-1/2} \exp\left(-\frac{\Delta\xi^2}{4D_c\Delta t}\right). \tag{6.24}$$

Equation (6.22) is not correct for all s: if $s + \Delta\xi(t)$ is not between 0 and L, $\boldsymbol{R}(s, t + \Delta t)$ should be on the newly created part of the tube (see Fig. 6.6). Writing down this condition in a formal mathematical equation is cumbersome. However, such an expression is not needed in practice since the condition can usually be accounted for by the fact that the distribution of the tangent vectors at the chain ends $\boldsymbol{u}(0, t)$ and $\boldsymbol{u}(L, t)$ are independent of the previous conformation of the primitive chain since their correlation time is infinitesimal.†

6.3.2 Segmental motion

To see how eqn (6.22) works, we calculate the mean square displacement of a primitive chain segment $s \langle (\boldsymbol{R}(s, t) - \boldsymbol{R}(s, 0))^2 \rangle$.

In this case it is more convenient to calculate the following time correlation function,

$$\phi(s, s'; t) = \langle (\boldsymbol{R}(s, t) - \boldsymbol{R}(s', 0))^2 \rangle. \tag{6.25}$$

The time evolution equation for $\phi(s, s'; t)$ is written as

$$\begin{aligned}\phi(s, s'; t + \Delta t) &= \langle [\boldsymbol{R}(s + \Delta\xi(t), t) - \boldsymbol{R}(s', 0)]^2 \rangle \\ &= \langle \phi(s + \Delta\xi(t), s'; t) \rangle.\end{aligned} \tag{6.26}$$

The bracket in the last expression represents the average over $\Delta\xi(t)$. To

† It must be remembered, however, that the dynamics of the chain end is extremely important in reptation dynamics since eqn (6.22) only describes the one-dimensional motion, and the three-dimensional property of the primitive chain is entirely governed by the dynamics of the chain ends.

calculate this average, we expand the right-hand side of eqn (6.26) and use eqn (6.23):

$$\begin{aligned}\langle \phi(s+\Delta\xi, s'; t)\rangle &= \left\langle \left(1 + \Delta\xi \frac{\partial}{\partial s} + \frac{\Delta\xi^2}{2}\frac{\partial^2}{\partial s^2}\right)\phi(s, s'; t)\right\rangle \\ &= \left(1 + \langle \Delta\xi\rangle \frac{\partial}{\partial s} + \frac{\langle \Delta\xi^2\rangle}{2}\frac{\partial^2}{\partial s^2}\right)\phi(s, s'; t) \\ &= \left(1 + D_c \Delta t \frac{\partial^2}{\partial s^2}\right)\phi(s, s'; t). \end{aligned} \tag{6.27}$$

From eqns (6.26) and (6.27) we have

$$\frac{\partial}{\partial t}\phi(s, s'; t) = D_c \frac{\partial^2}{\partial s^2}\phi(s, s'; t). \tag{6.28}$$

To solve this, we need the initial condition and the boundary condition:

(i) The initial condition is given by eqn (6.2),

$$\phi(s, s'; t)|_{t=0} = |s - s'|\, a. \tag{6.29}$$

(ii) To obtain the boundary condition, we write $\partial\phi/\partial s$ at $s = L$ as

$$\begin{aligned}\frac{\partial}{\partial s}\phi(s, s'; t)\bigg|_{s=L} &= 2\langle \boldsymbol{u}(L, t)\cdot(\boldsymbol{R}(L, t) - \boldsymbol{R}(s', 0))\rangle \\ &= 2\langle \boldsymbol{u}(L, t)\cdot(\boldsymbol{R}(L, t) - \boldsymbol{R}(s'', t))\rangle \\ &\quad \underline{+\,2\langle \boldsymbol{u}(L, t)\cdot(\boldsymbol{R}(s'', t) - \boldsymbol{R}(s', 0))\rangle} \end{aligned} \tag{6.30}$$

where s'' is an arbitrary value betwen 0 and L. Since the correlation time of $\boldsymbol{u}(L, t)$ is infinitesimal, the underlined part gives

$$\langle \boldsymbol{u}(L, t)\cdot(\boldsymbol{R}(s'', t) - \boldsymbol{R}(s', 0))\rangle = \langle \boldsymbol{u}(L, t)\rangle \cdot \langle(\boldsymbol{R}(s'', t) - \boldsymbol{R}(s', 0))\rangle = 0. \tag{6.31}$$

The first term in eqn (6.30) gives

$$\begin{aligned}2\langle \boldsymbol{u}(L, t)\cdot(\boldsymbol{R}(L, t) - \boldsymbol{R}(s'', t))\rangle &= \frac{\partial}{\partial s}\langle(\boldsymbol{R}(s, t) - \boldsymbol{R}(s'', t))^2\rangle\bigg|_{s=L} \\ &= \frac{\partial}{\partial s} a(s - s'')\bigg|_{s=L} = a. \end{aligned} \tag{6.32}$$

Hence

$$\frac{\partial}{\partial s}\phi(s, s'; t) = a \quad \text{at} \quad s = L. \tag{6.33}$$

Similarly

$$\frac{\partial}{\partial s}\phi(s, s'; t) = -a \quad \text{at} \quad s = 0. \tag{6.34}$$

The solution of eqn (6.28) with the boundary conditions (6.33) and (6.34), and the initial condition (6.29) is

$$\phi(s, s'; t) = |s - s'|\, a + 2\frac{a}{L} D_c t + \sum_{p=1}^{\infty} \frac{4La}{p^2\pi^2} \times \cos\left(\frac{p\pi s}{L}\right)\cos\left(\frac{p\pi s'}{L}\right)[1 - \exp(-tp^2/\tau_d)]. \tag{6.35}$$

Hence

$$\phi(s, s; t) = \frac{2}{Z} D_c t + \sum_{p=1}^{\infty} \frac{4La}{p^2\pi^2} \cos\left(\frac{p\pi s}{L}\right)^2 [1 - \exp(-tp^2/\tau_d)]. \tag{6.36}$$

The expression becomes simple in two extreme cases:

(i) For $t \ll \tau_d$, $\phi(s, s:t)$ is dominated by the terms with large p. Replacing $\cos^2(p\pi s/L)$ by the average 1/2, and converting the sum into the integral, we have

$$\begin{aligned}\phi(s, s; t) &= \int_0^{\infty} \mathrm{d}p \frac{4La}{p^2\pi^2} \cdot \tfrac{1}{2}(1 - \exp(-tp^2/\tau_d)) \\ &= 2a(D_c t/\pi)^{1/2}.\end{aligned} \tag{6.37}$$

This result is easily derived, for if $t \ll \tau_d$ the polymer segment remains in the initial tube, so that if the chain moves $\xi(t)$ along the tube, the mean square displacement in the three-dimensional space is given by $a\langle|\xi(t)|\rangle$ (see eqn (6.2)). Hence $\phi(s, s; t)$ is given by

$$\begin{aligned}\phi(s, s; t) &= a\langle|\xi(t)|\rangle \\ &= a \int_{-\infty}^{\infty} \mathrm{d}\xi\, |\xi|\, (4\pi D_c t)^{-1/2} \exp\left(-\frac{\xi^2}{4D_c t}\right) \\ &= 2a(D_c t/\pi)^{1/2}.\end{aligned} \tag{6.38}$$

(ii) For $t > \tau_d$, the first term in eqn (6.36) dominates, so that

$$\phi(s, s; t) = 2D_c t/Z. \tag{6.39}$$

Hence the diffusion constant D_G of the centre of mass is given as

$$D_G = \lim_{t\to\infty} \frac{\phi(s, s; t)}{6t} = D_c/3Z = \frac{k_B T a^2}{3N^2\zeta b^2}. \tag{6.40}$$

Thus the self-diffusion constant is proportional to N^{-2}, which has also been confirmed by computer experiment.[9,10]

6.3.3 Correlation function of the tangent vectors

Next let us consider the time correlation function of the tangent vector $\boldsymbol{u}(s, t)$:

$$G(s, s'; t) \equiv \langle \boldsymbol{u}(s, t) \cdot \boldsymbol{u}(s', 0) \rangle \tag{6.41}$$

which can be related to $\phi(s, s'; t)$ by eqns (6.1) and (6.25):

$$\begin{aligned} G(s, s'; t) &= \left\langle \frac{\partial \boldsymbol{R}(s, t)}{\partial s} \cdot \frac{\partial \boldsymbol{R}(s', 0)}{\partial s'} \right\rangle \\ &= -\frac{1}{2} \frac{\partial^2}{\partial s \, \partial s'} \langle (\boldsymbol{R}(s, t) - \boldsymbol{R}(s', 0))^2 \rangle \\ &= -\frac{1}{2} \frac{\partial^2}{\partial s \, \partial s'} \phi(s, s'; t). \end{aligned} \tag{6.42}$$

Substituting eqn (6.35), one gets

$$\begin{aligned} G(s, s'; t) &= a\delta(s - s') - \sum_{p=1}^{\infty} \frac{2a}{L} \sin\left(\frac{p\pi s}{L}\right) \sin\left(\frac{p\pi s'}{L}\right) [1 - \exp(-tp^2/\tau_d)] \\ &= \sum_{p=1}^{\infty} \frac{2a}{L} \sin\left(\frac{p\pi s}{L}\right) \sin\left(\frac{p\pi s'}{L}\right) \exp(-tp^2/\tau_d). \end{aligned} \tag{6.43}$$

An important conclusion is derived from eqn (6.43) by a geometrical interpretation of $\langle \boldsymbol{u}(s, t) \cdot \boldsymbol{u}(s', 0) \rangle$. First note that if $\boldsymbol{u}(s)$ and $\boldsymbol{u}(s')$ are the tangent vectors of a Gaussian chain of step length a, the integral

$$I = \frac{1}{a} \int_{s_2}^{s_1} \mathrm{d}s \langle \boldsymbol{u}(s) \cdot \boldsymbol{u}(s') \rangle \qquad (s_1 > s_2) \tag{6.44}$$

is equal to 1 when s' is between s_1 and s_2. Now consider

$$\psi(s', t) = \frac{1}{a} \int_0^L \mathrm{d}s \langle \boldsymbol{u}(s, t) \cdot \boldsymbol{u}(s', 0) \rangle. \tag{6.45}$$

This is equal to unity at time $t = 0$. As time passes, the original tube segment s' becomes empty of the primitive chain. If this happens $\boldsymbol{u}(s', 0)$ becomes totally uncorrelated to $\boldsymbol{u}(s, t)$ and the contribution from such case to $\psi(s', t)$ is zero. Therefore $\psi(s', t)$ represents the probability that the original tube segment s' is remaining at time t. Indeed from eqns (6.43) and (6.45), it follows that

$$\psi(s', t) = \sum_{p;\,\mathrm{odd}} \frac{4}{p\pi} \sin\left(\frac{p\pi s'}{L}\right) \exp(-p^2 t/\tau_d) \tag{6.46}$$

which agrees with eqn (6.14).

Now, one can show by a similar argument that the integral

$$\tilde{\psi}(s, t) = \frac{1}{a}\int_0^L ds' \langle \boldsymbol{u}(s, t) \cdot \boldsymbol{u}(s', 0) \rangle \tag{6.47}$$

represents the probability that the primitive chain segment s is in the original tube. From eqns (6.43), (6.45), and (6.47) it follows that

$$\tilde{\psi}(s, t) = \psi(s, t) \tag{6.48}$$

i.e., the probability that the original tube segment s remains is equal to the probability that the primitive chain segment s is in the original tube that remains.

6.3.4 *Dynamic structure factor*

As the final example, we shall consider the dynamic structure factor of a single chain:

$$g(\boldsymbol{k}, t) = \frac{N}{L^2}\int_0^L ds \int_0^L ds' \langle \exp[i\boldsymbol{k} \cdot (\boldsymbol{R}(s, t) - \boldsymbol{R}(s', 0))] \rangle . \tag{6.49}$$

To calculate this quantity, we again consider $\phi(s, s'; t)$ defined by

$$\phi(s, s'; t) = \langle \exp[i\boldsymbol{k} \cdot (\boldsymbol{R}(s, t) - \boldsymbol{R}(s', 0))] \rangle. \tag{6.50}$$

By the same trick as that used in Section 6.3.2, we can show that

$$\frac{\partial}{\partial t}\phi(s, s'; t) = D_c \frac{\partial^2}{\partial s^2}\phi(s, s'; t). \tag{6.51}$$

Since the distribution of $\boldsymbol{R}(s, 0) - \boldsymbol{R}(s', 0)$ is Gaussian with the variance $a\,|s - s'|$, the initial condition for $\phi(s, s'; t)$ is obtained as (see eqn (2.79))

$$\phi(s, s'; t)|_{t=0} = \exp\left(-\frac{\boldsymbol{k}^2}{6} a\,|s - s'|\right). \tag{6.52}$$

The boundary condition for $\phi(s, s'; t)$ is again obtained by the same method as in eqn (6.30):

$$\begin{aligned}\left.\frac{\partial}{\partial s}\phi(s, s'; t)\right|_{s=L} &= i\boldsymbol{k} \cdot \langle \boldsymbol{u}(L, t)\exp[i\boldsymbol{k} \cdot (\boldsymbol{R}(L, t) - \boldsymbol{R}(s', 0))] \rangle \\ &= i\boldsymbol{k} \cdot \langle \underline{\boldsymbol{u}(L, t)\exp[i\boldsymbol{k} \cdot (\boldsymbol{R}(L, t) - \boldsymbol{R}(s'', t))]} \\ &\quad \times \exp[i\boldsymbol{k} \cdot (\boldsymbol{R}(s'', t) - \boldsymbol{R}(s', 0))] \rangle\end{aligned} \tag{6.53}$$

where s'' is again an arbitrary value between 0 and L. If s'' is taken to be close to L, the average of the underlined part can be taken separately from the rest since its correlation time is very short. Hence

$$\begin{aligned}\frac{\partial}{\partial s}\phi(s,s';t)\bigg|_{s=L} &= \mathrm{i}\boldsymbol{k}\cdot\langle \boldsymbol{u}(L,t)\exp[\mathrm{i}\boldsymbol{k}\cdot(\boldsymbol{R}(L,t)-\boldsymbol{R}(s'',t))]\rangle \\ &\quad\times\langle\exp[\mathrm{i}\boldsymbol{k}\cdot(\boldsymbol{R}(s'',t)-\boldsymbol{R}(s',0))]\rangle \\ &= \frac{\partial}{\partial s}\langle\exp[\mathrm{i}\boldsymbol{k}\cdot(\boldsymbol{R}(s,t)-\boldsymbol{R}(s'',t))]\rangle\bigg|_{s=L}\phi(s'',s';t) \\ &= -\tfrac{1}{6}\boldsymbol{k}^2a\exp[-\tfrac{1}{6}\boldsymbol{k}^2a(L-s'')]\phi(s'',s';t).\end{aligned} \tag{6.54}$$

In the limit of $s''\to L$, this gives

$$\frac{\partial}{\partial s}\phi(s,s';t)\bigg|_{s=L} = -\tfrac{1}{6}\boldsymbol{k}^2a\phi(L,s';t). \tag{6.55}$$

Similarly

$$\frac{\partial}{\partial s}\phi(s,s';t)\bigg|_{s=0} = \tfrac{1}{6}\boldsymbol{k}^2a\phi(0,s';t). \tag{6.56}$$

The solution of eqn (6.51) under these conditions is

$$\begin{aligned}\phi(s,s';t) = \sum_{p=1}^{\infty}\bigg[&\frac{2\mu}{\mu^2+\alpha_p^2+\mu}\cos\left[\frac{2\alpha_p}{L}\left(s-\frac{L}{2}\right)\right] \\ &\times\cos\left[\frac{2\alpha_p}{L}\left(s'-\frac{L}{2}\right)\right]\exp\left(\frac{-4D_ct\alpha_p^2}{L^2}\right)+\frac{2\mu}{\mu^2+\beta_p^2+\mu} \\ &\times\sin\left[\frac{2\beta_p}{L}\left(s-\frac{L}{2}\right)\right]\sin\left[\frac{2\beta_p}{L}\left(s'-\frac{L}{2}\right)\right]\exp\left(\frac{-4D_ct\beta_p^2}{L^2}\right)\bigg]\end{aligned} \tag{6.57}$$

where

$$\mu = \frac{\boldsymbol{k}^2}{12}La = \frac{\boldsymbol{k}^2}{12}Nb^2 = \tfrac{1}{2}(kR_g)^2 \tag{6.58}$$

and α_p and β_p are the positive solutions of the equations

$$\alpha_p\tan\alpha_p = \mu, \qquad \beta_p\cot\beta_p = -\mu. \tag{6.59}$$

From eqn (6.57) we get[11]

$$g(\boldsymbol{k},t) = \sum_{p=1}^{\infty}\frac{2\mu N}{\alpha_p^2(\mu^2+\alpha_p^2+\mu)}\sin^2\alpha_p\exp\left(-\frac{4D_ct\alpha_p^2}{L^2}\right). \tag{6.60}$$

The expressions are simplified for the two limiting cases.

(i) $\mu \ll 1$, i.e., $kR_g \ll 1$. Here eqn (6.60) is dominated by the term including $\alpha_1 \simeq \sqrt{\mu} \ll 1$, thus

$$g(\boldsymbol{k}, t) = N \exp(-4D_c t\mu/L^2) = N \exp\left(-\frac{D_c a t}{3L}\boldsymbol{k}^2\right) = N \exp(-D_G \boldsymbol{k}^2 t) \tag{6.61}$$

which is natural since, for $kR_g \ll 1$, the dynamical structure factor is governed by the diffusion of the centre of mass.

(ii) $\mu \gg 1$, $kR_g \gg 1$. In this case, the α_p are approximated as $(p - 1/2)\pi$, and eqn (6.60) reduces to

$$\begin{aligned} g(\boldsymbol{k}, t) &= \frac{96N}{\boldsymbol{k}^2 a L \pi^2} \sum_{p:\,\mathrm{odd}} \frac{1}{p^2} \exp(-tp^2/\tau_d) \\ &= \frac{12}{\boldsymbol{k}^2 b^2} \psi(t). \end{aligned} \tag{6.62}$$

This result is also derived by a simple argument. In the limit $kR_g \gg 1$, the average of

$$\frac{N}{L}\int_0^L \mathrm{d}s' \exp[\mathrm{i}\boldsymbol{k}\cdot(\boldsymbol{R}(s', t) - \boldsymbol{R}(s, 0))]$$

is equal to $12/\boldsymbol{k}^2 b^2$ if $\boldsymbol{R}(s', t)$ remains in the original tube, and zero if it has already disengaged from it. Thus using the probability $\psi(s, t)$ that it remains, we have

$$g(\boldsymbol{k}, t) = \int_0^L \frac{\mathrm{d}s}{L} \frac{12}{\boldsymbol{k}^2 b^2} \psi(s, t) = \frac{12}{\boldsymbol{k}^2 b^2} \psi(t). \tag{6.63}$$

Notice that the decay rate of $g(\boldsymbol{k}, t)$ is $1/\tau_d$ and depends strongly on the molecular weight. This behaviour is entirely different from the result of the Rouse dynamics, according to which $g(\boldsymbol{k}, t)$ becomes independent of the chain length for $kR_g \gg 1$ (see eqn (4.III.12)).

6.3.5 General time correlation functions

Time correlation functions of more complicated form can be calculated in a similar way. For example, the time correlation function

$$C(s, s', s''; t) = \langle (\boldsymbol{R}(s, t) - \boldsymbol{R}(s'', 0)) \cdot (\boldsymbol{R}(s', t) - \boldsymbol{R}(s'', 0)) \rangle \tag{6.64}$$

can be calculated from

$$\frac{\partial}{\partial t} C = D_c \left(\frac{\partial}{\partial s} + \frac{\partial}{\partial s'}\right)^2 C \tag{6.65}$$

under appropriate boundary conditions. In general the time correlation function $C(s_1, s_2, \ldots, s_m, t)$ which depends on the m-points on the primitive chain is obtained from

$$\frac{\partial}{\partial t} C = D_c \left(\sum_{p=1}^{m} \frac{\partial}{\partial s_p} \right)^2 C \tag{6.66}$$

with appropriate boundary conditions.

6.4 Contour length fluctuation

6.4.1 Statistical distribution of the contour length

In the previous sections we regarded the primitive chain as an inextensible string of contour length L. In reality, the contour length of the primitive chain fluctuates with time, and the fluctuation sometimes plays an important role in various dynamical processes.

First we consider the statistical distribution of the contour length. Since the primitive chain represents a set of conformations of the Rouse chain, the probability that a certain conformation of the primitive chain is realized is proportional to ω, the number of the conformations of the Rouse chain which are represented by that primitive chain. The simplest hypothesis is to take the polymer as a random walk confined within a tube. Then ω is calculated by the method described in Section 2.3.2 (see Appendix 6.I). The result is

$$\omega(L) = \omega_0 \exp\left(-\frac{3L^2}{2Nb^2} - \alpha_0 \frac{Nb^2}{a_0^2} \right) \tag{6.67}$$

where ω_0 is the number of the configuration in the free space, a_0, the diameter of the tube, and α_0, a certain numerical factor which depends on the shape of the cross-section of the tube. The logarithm of ω gives the entropy of the primitive chain

$$S(L) = S_0 - k_B \left(\frac{3L^2}{2Nb^2} + \alpha_0 \frac{Nb^2}{a_0^2} \right). \tag{6.68}$$

At first sight eqn (6.68) may seem to imply a paradoxical result: since the entropy increases with decreasing L, the chain will contract to the state of $L = 0$ if its ends are not fixed. This argument is wrong. The collapse happens if the chain is confined in an infinitely long tube of given conformation, but does not happen in the case of a network, where the Rouse chain explores many tubes. To calculate the statistical distribution of L in the network, one has to take into account the multiplicity of the state specified by L. Let $\Omega(L)$ be the number of primitive paths which have length L, then the probability that a primitive chain has a contour length L is

$$\Psi(L) \propto \omega(L)\Omega(L). \tag{6.69}$$

$\Omega(L)$ can be estimated by the number of random walks consisting of L/a_0 steps:

$$\Omega(L) = \exp\left(\alpha_1 \frac{L}{a_0}\right) \tag{6.70}$$

where α_1 is a certain numerical constant which depends on the structure of the network. From eqns (6.67), (6.69), and (6.70)

$$\Psi(L) \propto \exp\left(-\frac{3L^2}{2Nb^2} + \alpha_1 \frac{L}{a_0}\right) \propto \exp\left[-\frac{3}{2Nb^2}(L - \bar{L})^2\right] \tag{6.71}$$

which has a maximum at

$$\bar{L} = \frac{\alpha_1}{3} \frac{Nb^2}{a_0}. \tag{6.72}$$

From eqns (6.4) and (6.72)

$$a = \frac{3}{\alpha_1} a_0. \tag{6.73}$$

This shows explicitly that the step length a of the primitive chain is of the same order as the tube diameter a_0. The average of the fluctuation is calculated from eqn (6.71) as

$$\Delta\bar{L} \equiv \langle \Delta L^2 \rangle^{1/2} = \left[\int_0^\infty \mathrm{d}L\,\Psi(L)(L - \bar{L})^2\right]^{1/2} = (Nb^2/3)^{1/2} \quad \text{for} \quad \bar{L} \gg \sqrt{N}\,b. \tag{6.74}$$

This statistical property of the primitive chain has been studied by computer simulation.[10] Under certain conditions, the statistical distribution of the primitive chain can be calculated analytically.[12–14] Both studies show the relations

$$\langle L \rangle \propto N \quad \text{and} \quad \langle \Delta L^2 \rangle \propto N \tag{6.75}$$

in agreement with eqns (6.72) and (6.74).

6.4.2 Dynamics of the contour length fluctuation

Having studied the static distribution of the contour length, we now examine the dynamics of the contour length fluctuation. As before, we use the Rouse model for the dynamics (see Fig. 6.7). Let s_n be the curvilinear coordinate of the n-th Rouse segment measured from a

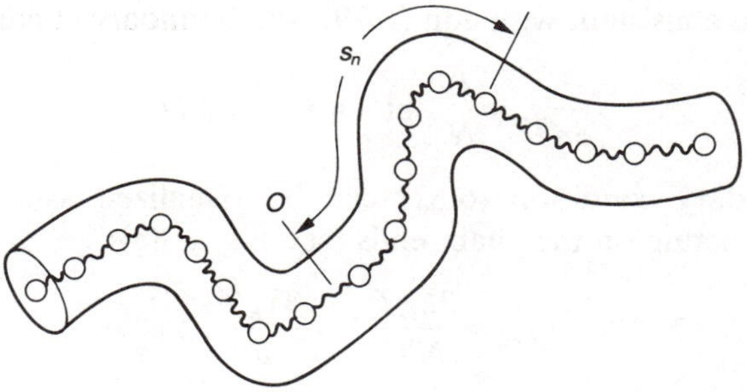

Fig. 6.7. A Rouse model describing the contour length fluctuation. The point O denotes the origin of the curvilinear coordinate s_n.

certain fixed point on the tube. The contour length of the primitive chain is defined by

$$L(t) = s_N(t) - s_0(t). \tag{6.76}$$

The dynamics of s_n are described by the Langevin equation for the Rouse model,

$$\zeta \frac{\partial}{\partial t} s_n = \frac{3k_B T}{b^2} \frac{\partial^2}{\partial n^2} s_n(t) + f_n(t) \tag{6.77}$$

where

$$\langle f_n(t) \rangle = 0 \quad \text{and} \quad \langle f_n(t) f_m(t') \rangle = 2\zeta k_B T \delta(n-m)\delta(t-t'). \tag{6.78}$$

Since eqn (6.77) is the same linear equation as appeared in the Rouse dynamics, analysis of this equation is straightforward. An important difference, however, must be mentioned. In the present case, the equilibrium average of the contour length is $\bar{L}$:

$$\langle s_N - s_0 \rangle = \bar{L}, \tag{6.79}$$

while in the Rouse model the corresponding quantity $\langle \boldsymbol{R}_N - \boldsymbol{R}_0 \rangle$ is zero. To get eqn (6.79) we have to modify the boundary condition which corresponds to eqn (4.11). This is obtained as follows.

If we take the average of both sides of eqn (6.77) for the equilibrium state, we have

$$\zeta \frac{\partial}{\partial t} \langle s_n \rangle = \frac{3k_B T}{b^2} \frac{\partial^2}{\partial n^2} \langle s_n(t) \rangle + \langle f_n(t) \rangle . \tag{6.80}$$

Both $\partial \langle s_n \rangle / \partial t$ and $\langle f_n(t) \rangle$ vanish at equilibrium, so that

$$\frac{\partial^2}{\partial n^2} \langle s_n \rangle = 0. \tag{6.81}$$

For this to be consistent with eqn (6.79), the boundary condition must be

$$\frac{\partial}{\partial n} s_n = \frac{\bar{L}}{N} \quad \text{at} \quad n = 0 \text{ and } N. \tag{6.82}$$

The boundary condition (6.82) can be visualized as a hypothetical tensile force acting on the chain ends (see Fig. 6.8*a*):

$$F_{\text{eq}} = \frac{3k_BT}{Nb^2} \bar{L} = \frac{3k_BT}{a}. \tag{6.83}$$

The origin of this force is again the multiplicity of the tube. One can intuitively understand it by considering the dynamical process at the chain end. Suppose in a time Δt the chain end moves one step in a random direction (see Fig. 6.8*b*). If it moves to the $(z-1)$ positions A_1, $A_2, \ldots, A_{z-1}$, the contour length increases, whilst if it moves to A_0, the contour length decreases. Thus there is an imbalance in the change, which tends to increase the contour length and causes the force (6.83).

For eqn (6.82) to be satisfied, the normal mode is now defined as

$$Y_0 = \frac{1}{N} \int_0^N \mathrm{d}n s_n, \tag{6.84}$$

$$Y_p = \frac{1}{N} \int_0^N \mathrm{d}n \cos\left(\frac{p\pi n}{N}\right)\left(s_n - \frac{n\bar{L}}{N}\right) \quad \text{for} \quad p = 1, 2, \ldots, \tag{6.85}$$

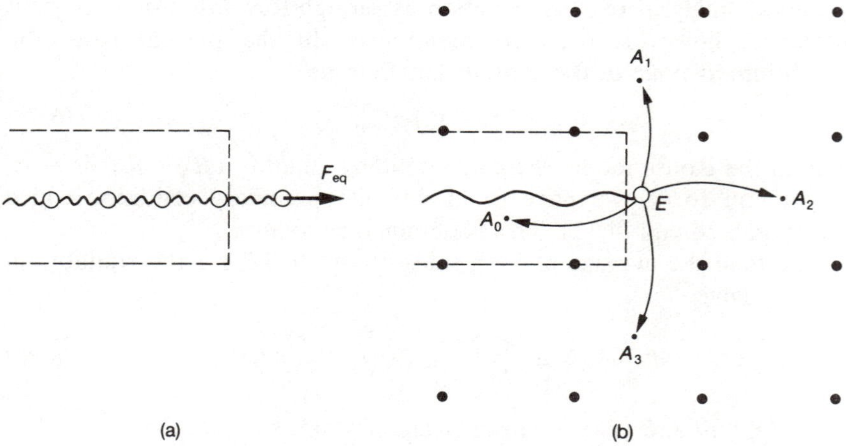

Fig. 6.8. (*a*) Equilibrium tension acting on the primitive chain. (*b*) Physical origin of the force F_{eq}. The case of $z = 4$ is shown.

or

$$s_n = Y_0 + 2\sum_{p=1}^{\infty} Y_p \cos\left(\frac{p\pi n}{N}\right) + \frac{n\bar{L}}{N}. \tag{6.86}$$

The coordinate Y_0 represents the position of the 'curvilinear center of mass',

$$s_G(t) = \frac{1}{N}\int_0^N \mathrm{d}n s_n(t) = Y_0(t), \tag{6.87}$$

while the other coordinates Y_p $(p > 0)$ represents fluctuations along the tube.

The equation of motion for Y_p is precisely the same as eqn (4.19):

$$\zeta_p \frac{\partial Y_p}{\partial t} = -k_p Y_p + f_p \tag{6.88}$$

with

$$\zeta_0 = N\zeta, \qquad \zeta_p = 2N\zeta \quad \text{for} \quad p = 1, 2, \ldots; \tag{6.89}$$

$$k_p = \frac{6\pi^2 k_B T}{Nb^2} p^2; \tag{6.90}$$

and

$$\langle f_p(t)\rangle = 0, \quad \langle f_p(t) f_q(t')\rangle = 2\zeta_p k_B T \delta_{pq}\delta(t - t'). \tag{6.91}$$

Thus the time correlation functions for Y_p are obtained as

$$\langle (Y_0(t) - Y_0(0))^2\rangle = 2\frac{k_B T}{N\zeta} t, \tag{6.92}$$

$$\langle Y_p(t) Y_p(0)\rangle = \frac{Nb^2}{6\pi^2 p^2}\exp(-tp^2/\tau_R), \tag{6.93}$$

where τ_R is the Rouse relaxation time given by eqn (6.18).

Equation (6.92) justifies the previous assumption that the diffusion coefficient D_c is equal to $k_B T/N\zeta$, the diffusion constant of the Rouse chain.

From eqn (6.86) the contour length of the primitive chain is written as

$$L(t) = s_N(t) - s_0(t) \tag{6.94}$$

$$= \bar{L} - 4\sum_{p:\,\mathrm{odd}} Y_p(t). \tag{6.95}$$

The time correlation function of $L(t)$ is calculated from eqns (6.93) and (6.95). The result is

$$\langle L(t)L(0)\rangle = \bar{L}^2 + \frac{8Nb^2}{3\pi^2}\sum_{p:\,\mathrm{odd}} \frac{1}{p^2}\exp(-tp^2/\tau_R). \tag{6.96}$$

In particular,

$$\langle \Delta L^2 \rangle = \langle L^2 \rangle - \bar{L}^2 = \frac{8Nb^2}{3\pi^2} \sum_{p:\,\mathrm{odd}} \frac{1}{p^2} = \frac{Nb^2}{3} \tag{6.97}$$

which agrees with eqn (6.74). The characteristic time of the contour length fluctuation is τ_R, the Rouse relaxation time. Since

$$\frac{\tau_R}{\tau_d} \simeq \frac{1}{Z}, \qquad \frac{\Delta \bar{L}}{\bar{L}} \simeq \left(\frac{a}{\bar{L}}\right)^{1/2} \simeq \frac{1}{\sqrt{Z}}, \tag{6.98}$$

the effect of the contour length fluctuation can be neglected for $Z \gg 1$. It is at this limit that the dynamics described in Section 6.3 is valid.

6.4.3 Effect of the contour length fluctuation on reptation

Though the contour length fluctuation becomes negligible for very large Z, its effect is not negligible for usual values of Z, which are typically less than 100. An important effect is found in the disengagement time τ_d.[15] Consider the two situations shown in Fig. 6.9, one represents the motion of the chain with fluctuations, and the other without. Obviously the life time of the tube becomes shorter for the chain with fluctuation than for the chain without, i.e., the contour length fluctuation reduces the disengagement time. This effect is estimated as follows.

If the contour length fluctuation is neglected, the disengagement time is estimated as the time necessary for the chain to move the distance $\bar{L}$,

$$\tau_d^{(\mathrm{NF})} \simeq \bar{L}^2 / D_c \tag{6.99}$$

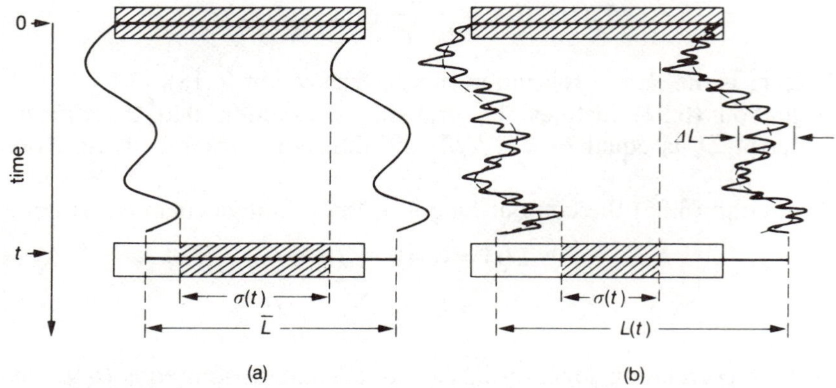

Fig. 6.9. The Brownian motion of a primitive chain with (*a*) fixed contour length, and (*b*) fluctuating contour length. The oblique lines denote the region that has not been reached by either end of the primitive chain. Obviously the length of this region $\sigma(t)$ decreases faster in (*b*) than in (*a*). Reproduced from ref. 15.

(The superscript (NF) stands for 'no fluctuation'.) On the other hand if there are fluctuations, the chain can disengage from the tube when it moves the distance $\bar{L} - \Delta\bar{L}$ because the chain ends are fluctuating rapidly over the distance $\Delta\bar{L}$. Hence the disengagement time is estimated as

$$\tau_d^{(\mathrm{F})} \simeq (\bar{L} - \Delta\bar{L})^2/D_c. \tag{6.100}$$

(The superscript (F) stands for 'fluctuation'.) From eqns (6.98) and (6.100), it follows that

$$\tau_d^{(\mathrm{F})} \simeq \tau_d^{(\mathrm{NF})}\left(1 - \frac{X}{\sqrt{Z}}\right)^2 \tag{6.101}$$

where X is a certain numerical constant.

Precise calculation of $\tau_d^{(\mathrm{F})}$ requires the first passage problem in multidimensional phase space. A variational calculation[16] for the Rouse model shows that X is larger that 1.47. Hence the effect of the contour length fluctuation is significant even if Z is as large as 100.

The effect of the contour length fluctuation has been studied for slightly different models[17,18] and it has been reported that the effect is less significant than in the case of the Rouse model. The discrepancy is perhaps due to the difference in the dynamics of the fluctuation, especially the short-time dynamics, which is quite important in the first passage time problem.[19,20]

6.4.4 *Other small-scale fluctuations and their effects on the segmental motion*

Due to the small-scale fluctuations around the primitive path, the actual dynamics of the Rouse segment in a network is much more complicated than that described in Section 6.3.2. As an example, let us consider the mean square displacement of a Rouse segment.

$$\phi_n(t) = \langle(\boldsymbol{R}_n(t) - \boldsymbol{R}_n(0))^2\rangle. \tag{6.102}$$

The precise calculation of $\phi_n(t)$ is difficult, but its characteristic features can be inferred easily.

(i) For a very short time the segment does not feel the constraints of the network, so that $\phi_n(t)$ is the same as that calculated for the Rouse model in free space. Hence $\phi_n(t)$ is given by eqn (4.III.6) as

$$\phi_n(t) = 6D_c t + \frac{4Nb^2}{\pi^2}\sum_{p=1}^{\infty}\frac{1}{p^2}\cos^2\left(\frac{p\pi n}{N}\right)(1 - \exp(-tp^2/\tau_R)). \tag{6.103}$$

Since $t \ll \tau_R$, eqn (6.103) is approximated in the same way as in eqn

(6.37),

$$\phi_n(t) = \frac{4Nb^2}{\pi^2}\int_0^\infty \mathrm{d}p\,\frac{1}{p^2}\tfrac{1}{2}(1-\exp(-tp^2/\tau_R))$$

$$= 2Nb^2(t/\pi^3\tau_R)^{1/2} \simeq (k_BTb^2t/\zeta)^{1/2}. \quad (6.104)$$

This formula is correct when the average displacement is much less than the tube diameter. Let τ_e be the time at which the segmental displacement becomes comparable to a:

$$a^2 \simeq \phi_n(\tau_e) \simeq (k_BTb^2/\zeta)^{1/2}\sqrt{\tau_e} \quad (6.105)$$

i.e.,

$$\tau_e \simeq a^4\zeta/k_BTb^2. \quad (6.106)$$

The time τ_e denotes the onset of the effect of tube constraints: for $t < \tau_e$, the chain behaves as a Rouse chain in free space, while for $t > \tau_e$ the chain feels the constraints imposed by the tube,

(ii) For $t > \tau_e$ the motion of the Rouse segment perpendicular to the primitive path is restricted, but the motion along the primitive path is free. The mean square displacement along the tube is calculated from eqns (6.86)–(6.93) as

$$\langle (s_n(t)-s_n(0))^2\rangle = \langle (Y_0(t)-Y_0(0))^2\rangle + 4\sum_{p=1}^{\infty}\cos^2\left(\frac{p\pi n}{N}\right)\langle (Y_p(t)-Y_p(0))^2\rangle$$

$$= 2\frac{k_BT}{N\zeta}t + \frac{4Nb^2}{3\pi^2}\sum_{p=1}^{\infty}\frac{1}{p^2}\cos^2\left(\frac{p\pi n}{N}\right)(1-\exp(-tp^2/\tau_R))$$

$$\simeq \begin{cases} (k_BTb^2t/\zeta)^{1/2} & t \lesssim \tau_R, \quad (6.107) \\ k_BTt/N\zeta & t \gtrsim \tau_R. \quad (6.108) \end{cases}$$

From this, $\phi_n(t)$ is estimated as in Section 6.3.2. If the segment moves $s_n(t) - s_n(0)$ along the tube, the mean square displacement in the three-dimensional space is $a\,|s_n(t)-s_n(0)|$. Hence

$$\phi_n(t) = a\langle |s_n(t)-s_n(0)|\rangle \simeq a\langle (s_n(t)-s_n(0))^2\rangle^{1/2}. \quad (6.109)$$

From eqns (6.107)–(6.109)

$$\phi_n(t) \simeq \begin{cases} a(k_BTb^2t/\zeta)^{1/4} & \tau_e \lesssim t \lesssim \tau_R, \\ a(k_BTt/N\zeta)^{1/2} & \tau_R \lesssim t \lesssim \tau_d. \end{cases} \quad (6.110)$$

Note that $\phi_n(t)$ is proportional to $t^{1/4}$ in the time region $\tau_e \lesssim t \lesssim \tau_R$. This specific diffusion behaviour, first predicted by de Gennes,[7] is a consequence of the two effects, the Rouse-like diffusion equation (6.107), and the tube constraints equation (6.109). On the other hand, the behaviour

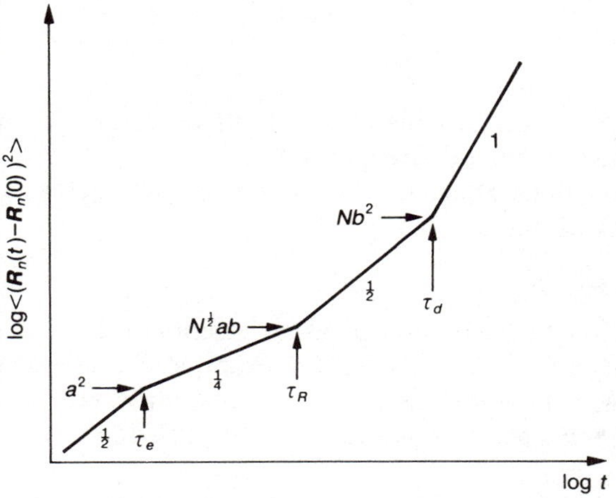

Fig. 6.10. Mean square displacement of a chain segment is plotted logarithmically against time.

for $\tau \gtrsim \tau_R$ agrees with that predicted by the primitive chain dynamics equation (6.37).

(iii) For $t \gtrsim \tau_d$, the dynamics is governed by the reptation process. As discussed in the previous section

$$\phi_n(t) = \frac{2}{Z} D_c t \simeq \frac{k_B T a^2}{N^2 \zeta b^2} t \qquad t \gtrsim \tau_d \tag{6.111}$$

Equations (6.104), (6.110), and (6.111) are summarized as follows[21]

$$\phi_n(t) \simeq \begin{cases} Nb^2(t/\tau_R)^{1/2} & t \lesssim \tau_e, \\ Nb^2(t/Z^2\tau_R)^{1/4} & \tau_e \lesssim t \lesssim \tau_R, \\ Nb^2(t/\tau_d)^{1/2} & \tau_R \lesssim t \lesssim \tau_d, \\ Nb^2(t/\tau_d) & \tau_d \lesssim t, \end{cases} \tag{6.112}$$

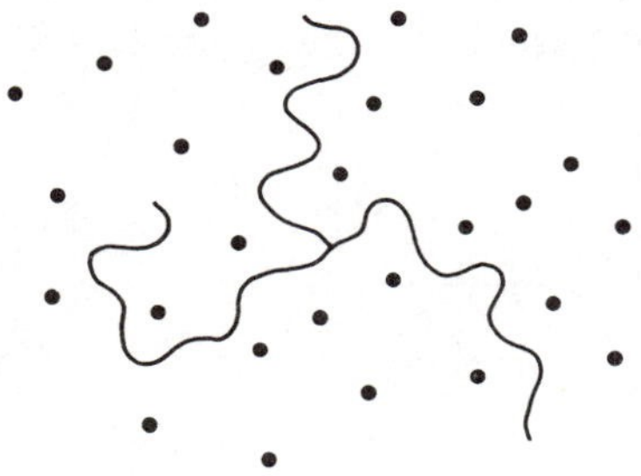

Fig. 6.11. Motion of a star polymer. Reproduced from ref. 28.

with

$$\tau_e \simeq Z^{-2}\tau_R, \qquad \tau_d \simeq Z\tau_R.$$

Figure 6.11 shows the schematic behaviour of $\phi_n(t)$. The behaviour has been confirmed by computer experiment.[22]

These fluctuations affect the dynamic structure factor. This has been studied in refs 21 and 22.

6.4.5 Branched polymers

The contour length fluctuation plays an essential role in the dynamics of branched polymers.[23,24] Consider for example the star-shaped polymer shown in Fig. 6.11. Obviously simple reptation is not possible, but the polymer can change its conformation by utilizing the contour length fluctuation.

For example, if the polymer withdraws an arm down the tube to its branching point, the arm can go into a new tube (see Fig. 6.12*a–c*). This process can be described by the model given in Section 6.4.2. Though it is possible to develop a rather elaborate calculation,[25] here we will consider only a crude estimation.[26]

Let us consider that the polymer has f arms of equal length each consisting of $N_a = N/f$ Rouse segments. For simplicity we assume that the central segment is fixed, and consider the fluctuation of the contour length L_a of the arm. Since the probability distribution of L_a at equilibrium is given by

$$\Psi_{\text{eq}}(L_a) \propto \exp\left[-\frac{3}{2N_a b^2}(L_a - \bar{L}_a)^2\right], \tag{6.113}$$

the motion of L_a can be visualized as the Brownian motion of a particle in a harmonic potential,

$$U(L_a) = \frac{3k_B T}{2N_a b^2}(L_a - \bar{L}_a)^2. \tag{6.114}$$

The disengagement time τ_d of the arm can be estimated by the mean first passage time of the particle to the point $L_a = 0$. If we use the simple

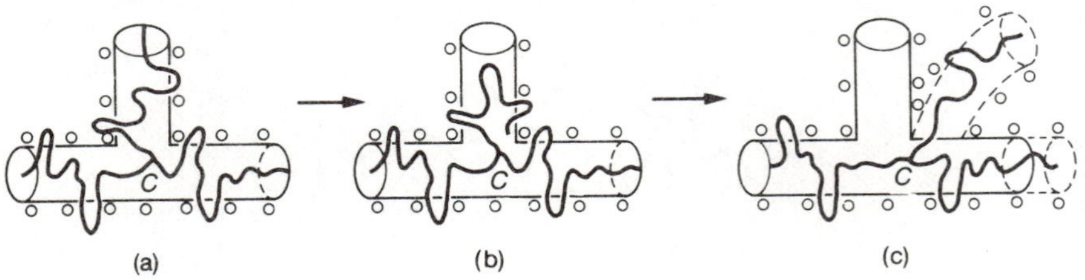

Fig. 6.12. A star polymer in the network of obstacles.

theory of the activation process, we have

$$\tau_d \propto \tau_0 \exp([U(L_a = 0) - U(\bar{L}_a)]/k_B T) \tag{6.115}$$

where the front factor τ_0 depends on L weakly. Equations (6.114) and (6.115) give

$$\tau_d \propto \tau_0 \exp(3\bar{L}_a^2/2N_a b^2) \propto \tau_0 \exp\left[\tfrac{3}{2}N_a\left(\frac{b}{a}\right)^2\right]. \tag{6.116}$$

Hence τ_d increases exponentially with the arm length N_a. This result, first predicted by de Gennes,[23] has been confirmed by computer simulation,[14] and by experiments.[28]

The diffusion constant D_G of the star polymers is estimated by the following argument.[27] For the central segment to move to the next stable point, the polymer has to withdraw $(f-2)$ arms to the branching point. The 'activation' energy required for such a process is

$$\Delta U \simeq (f-2)(U(L_a = 0) - U(L_a = \bar{L}_a)) = \tfrac{3}{2}N_a(f-2)k_B T\left(\frac{b}{a}\right)^2. \tag{6.117}$$

Hence

$$D_G \simeq a^2/[\tau_0 \exp(\Delta U/k_B T)] \simeq \frac{a^2}{\tau_0}\exp\left[-\tfrac{3}{2}(f-2)N_a\left(\frac{b}{a}\right)^2\right]. \tag{6.118}$$

Thus D_G decreases exponentially with the arm length N_a. This result has been confirmed by computer simulation[14] and by the diffusion experiment of a star polymer in a large molecular weight matrix.[28]

In the above discussion, we assumed that the position of the central segment is essentially fixed. This assumption can be removed,[28] but the essential result remains unchanged.

Appendix 6.I Entropy of a polymer in a tube

Consider a polymer confined in a straight tube of square cross-section, with both its ends fixed at $\boldsymbol{R}$ and $\boldsymbol{R}'$, respectively (see Fig. 6.13). The number ω of the allowed conformation of the polymer is proportional to

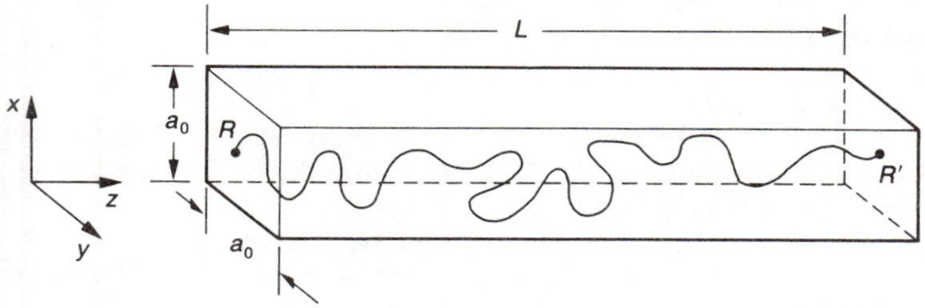

Fig. 6.13. A polymer confined in a straight tube of square cross-section with its ends fixed at $\boldsymbol{R}$ and $\boldsymbol{R}'$, respectively.

the Green function $G(\boldsymbol{R}, \boldsymbol{R}'; N)$:

$$\omega = \omega_0 G(\boldsymbol{R}, \boldsymbol{R}'; N) \tag{6.I.1}$$

where ω_0 is the number of the allowed conformation in free space. The Green function of a polymer in a box has already been calculated in Section 2.3.2 as

$$G(\boldsymbol{R}, \boldsymbol{R}'; N) = \prod_{\alpha=x,y,z} g_\alpha(R_\alpha, R'_\alpha; N) \tag{6.I.2}$$

where g_x and g_y are given as (for $L_x = L_y = a_0 \ll \sqrt{N}\, b$)

$$g_\alpha(R_\alpha, R'_\alpha; N) = \frac{2}{a_0} \sin\left(\frac{\pi R_\alpha}{a_0}\right) \sin\left(\frac{\pi R'_\alpha}{a_0}\right) \times \exp\left(-\frac{\pi^2 N b^2}{6a_0^2}\right) \quad \text{for} \quad \alpha = x, y, \tag{6.I.3}$$

while g_z is given by that in the free space

$$g_z(R_z, R'_z; N) = \left(\frac{3}{2\pi N b^2}\right)^{1/2} \exp\left(-\frac{3(R_z - R'_z)^2}{2Nb^2}\right). \tag{6.I.4}$$

Consider that the end points are fixed at

$$\boldsymbol{R} = \left(\frac{a_0}{2}, \frac{a_0}{2}, L\right) \quad \text{and} \quad \boldsymbol{R}' = \left(\frac{a_0}{2}, \frac{a_0}{2}, 0\right). \tag{6.I.5}$$

Substituting this into eqns (6.I.3) and (6.I.5), we have

$$\omega = \omega_0 \left(\frac{2}{a_0}\right)^2 \exp\left(-\frac{\pi^2 N b^2}{3a_0^2}\right) \left(\frac{3}{2\pi N b^2}\right)^{1/2} \exp\left(-\frac{3L^2}{2Nb^2}\right) \tag{6.I.6}$$

which is eqn (6.67).

References

1. Edwards, S. F., *Proc. Phys. Soc.* **92,** 9 (1967).
2. Edwards, S. F., *Polymer* **9,** 140 (1977).
3. Treloar, L. R. G., *The Physics of Rubber Elasticity* (3rd edn). Clarendon Press, Oxford (1975).
4. Deam, R. T., and Edwards, S. F., *Phil. Trans. R. Soc. London* **280,** 317 (1976); see also Ball, R. C., Doi, M., Edwards, S. F., and Warner, M., *Polymer* **22,** 1010 (1981).
5. Eichinger, B. E., *Ann. Rev. Phys. Chem.* **34,** 359 (1983).
6. Gottlieb, M., and Gaylord, R. J., *Polymer* **24,** 1644 (1983); *Macromolecules* **17,** 2024 (1984).
7. de Gennes, P. G., *J. Chem. Phys.* **55,** 572 (1971).
8. Agren, G., *J. Chem. Phys.* **69,** 329 (1972).

9. Doi, M., *Polymer J.* **5,** 288 (1973).
10. Evans, K. E., and Edwards, S. F., *J. Chem. Soc. Faraday Trans. 2,* **77,** 1891 (1981).
11. Doi, M., and Edwards, S. F., *J. Chem. Soc. Faraday Trans. 2,* **74,** 1789 (1978).
12. Evans, K. E., and Edwards, S. F., *J. Chem. Soc. Faraday Trans. 2,* **77,** 1929 (1981).
13. Helfand, E., and Pearson, D. S., *J. Chem. Phys.* **79,** 2054 (1983); Rubinstein, M., and Helfand, E., *J. Chem. Phys.* **82,** 2477 (1985).
14. Needs, R. J., and Edwards, S. F., *Macromolecules* **16,** 1492 (1983).
15. Doi, M., *J. Polym. Sci. Lett.* **19,** 265 (1981).
16. Doi, M., *J. Polym. Sci.* **21,** 667 (1983).
17. des Cloizeaux, J., *J. Physique Lett.* **45,** 17 (1984).
18. Needs, R. J., *Macromolecules* **17,** 437 (1984).
19. Doi, M., *Chem. Phys.* **11,** 107, 115 (1975).
20. de Gennes, P. G., *J. Chem. Phys.* **76,** 3316, 3322 (1982).
21. de Gennes, P. G., *J. Phys.* (*Paris*) **42,** 735 (1981).
22. Kremer, K., *Macromolecules* **16,** 1632 (1983).
23. de Gennes, P. G., *J. Phys.* (*Paris*) **36,** 1199 (1975).
24. Graessley, W. W., Masuda, T., Roovers, J. E. L., and Hadjichristidis, N., *Macromolecules* **9,** 127 (1976).
25. Pearson, D. S., and Helfand, E., *Macromolecules* **17,** 888 (1984).
26. Doi, M., and Kuzuu, N. Y., *J. Polym. Sci. Lett.* **18,** 775 (1980).
27. Graessley, W. W., *Adv. Polym. Sci.* **16,** 1 (1974).
28. Klein, J., Fletcher, D., and Fetters, L. J., *Faraday Symp. Chem. Soc.* **18,** 159 (1983); *Nature* (*London*) **304,** 526 (1983).

7

MOLECULAR THEORY FOR THE VISCOELASTICITY OF POLYMERIC LIQUIDS

7.1 Tube model in concentrated solutions and melts

In the previous chapter, we discussed the dynamics of a polymer in a fixed network. We shall now discuss the polymer dynamics in concentrated solutions and melts. In these systems, though all polymers are moving simultaneously it can be argued that the reptation picture will also hold. Consider the motion of a certain test polymer arbitrarily chosen in melts. If the test polymer moves perpendicularly to its own contour, it drags many other chains surrounding it and will feel a large resistance. On the other hand the movement of the test polymer along its contour will be much easier. It will be thus plausible to assume that the polymer is confined in a tube-like region, and the major mode of the dynamics is reptation.

Physically the tube represents the following situation: as long as the displacement of the segment of the test polymer is small, the environment behaves as a liquid characterized by a certain local viscosity, which determines the friction constant ζ. On the other hand the segment cannot make a large displacement in arbitrary direction which would distort many surrounding chains. Thus there must be a certain characteristic length which distinguishes the two cases. This length is represented by a, the Kuhn step length of the tube, which approximately represents the diameter of the tube.

Clearly it is a crude simplification to characterize the effect of the environment by a single parameter a. A more appropriate description would be to assign a viscoelastic character to the environment. However, we proceed here using the simplest possible model. As we shall show later various experimental results indicate that this model is adequate for linear polymers with narrow distribution of molecular weight. Limitations of the model will be discussed later.

Within this framework, one can draw a rather simplified picture about the dynamics of the polymers in the entangled state. If the characteristic length scale of a motion is smaller than a, the entanglement effect is not important, and the dynamics is well described by the Rouse model (or the Zimm model if the hydrodynamic interaction is not screened). On the other hand, if the length-scale of the motion becomes larger than a, the dynamics is governed by reptation.

That the short time-scale (or small length-scale) motion of the polymer is not affected by entanglement is supported by various experimental observations:

(i) The viscoelasticity of polymer melts can be explained by the Rouse model if the molecular weight is less than a certain value, or if the time-scale is shorter than a certain characteristic time.[1–3]

(ii) The dielectric relaxation of the melts of polymers which have dipole moments along the main chain is described by the Rouse model if the molecular weight is less than a certain value.[4]

(iii) The dynamical structure factor of neutron scattering, whose time-scale usually corresponds to the motion inside the tube, is fitted by the Rouse model in melts and the Zimm model in solutions.[5,6,7]

On the other hand there is much experimental evidence which indicates that reptation is the dominant motion governing the dynamics in the highly entangled state.

Clear evidence comes from the study of diffusion. Klein *et al.*[8,9] measured the diffusion constant of a deutrated polyethylene in a polyethylene matrix, and found

$$D_G \propto M^{-2} \tag{7.1}$$

in agreement with the result of reptation dynamics (eqn 6.40). The self-diffusion constant of a labelled polymer in solutions and melts has been measured by other techniques (forced Rayleigh scattering[10] and field gradient Nuclear Magnetic Resonance,[11,12]) and eqn (7.1) is fully confirmed in these systems. (A comprehensive review is given by Tirrell.[13])

More evidence comes from the study of viscoelasticity, which has been done extensively in the past and established the characteristic aspects common to all flexible polymers.[2,3] The reptation model has succeeded in explaining many of these features and also predicting some of the behaviour in nonlinear viscoelasticity. In this chapter we shall describe the reptation theory[14,15] for viscoelasticity in detail, and discuss the validity of the reptation model in solutions and melts.

The viscoelastic properties of polymers are quite important in polymer technology. A great deal of experimental work has been done, and at the same time phenomenological theories have been developed to a highly sophisticated level. These are summarized in various monographs.[2,16–18] An overview of this field can be obtained in the excellent textbook by Bird *et al.*[19] Here we shall limit ourselves to the molecular aspects of the problem, i.e., how the viscoelastic properties are related to the molecular dynamics and how they depend on molecular parameters such as molecular weight, concentration, and molecular structure.

7.2 Microscopic expression for the stress tensor

7.2.1 Stress in polymeric liquids

To understand the mechanical properties of materials, it is important to consider first the microscopic origin of stress. In the usual gas or liquid of small molecules, the stress comes from the momentum transfer due to the intermolecular collision. In polymeric liquids, the stress is mainly due to the intramolecular force, and is directly related to the orientation of the bond vectors of the polymer. This idea, originated from the theory of rubber elasticity,[20] is fundamental in the physics of polymeric materials.†

To demonstrate the point, let us first consider a dilute polymer solution. As shown in Section 4.5.2, the stress tensor is written as

$$\sigma_{\alpha\beta}(t) = (\kappa_{\alpha\beta}(t) + \kappa_{\beta\alpha}(t))\eta_s + \sigma^{(p)}_{\alpha\beta}(t) + P\delta_{\alpha\beta} \tag{7.2}$$

where η_s is the viscosity of the solvent, $\kappa_{\alpha\beta}$ is the velocity gradient tensor and $\sigma^{(p)}_{\alpha\beta}$ represents the contribution from the polymer

$$\sigma^{(p)}_{\alpha\beta}(t) = \frac{c}{N}\sum_n \frac{3k_BT}{b^2}\left\langle \frac{\partial R_{n\alpha}(t)}{\partial n}\frac{\partial R_{n\beta}(t)}{\partial n}\right\rangle. \tag{7.3}$$

This is directly related to the orientation of the vector $\partial \mathbf{R}_n/\partial n$, and it is this term that gives the viscoelasticity of dilute solutions.

In the dilute solution, the viscoelasticity has only a small effect; the major contribution to the stress is the purely viscous stress given by the first term in eqn (7.2). The situation changes with increasing polymer concentration. The contribution of $\sigma^{(p)}_{\alpha\beta}$ increases steeply (because the polymers become more easily oriented due to the entanglement), while the contribution of the first term remains essentially unchanged. When $\sigma^{(p)}_{\alpha\beta}$ dominates the first term, the total stress is simply related to the orientation of the bond vectors:

$$\sigma_{\alpha\beta}(t) = \frac{c}{N}\sum_n \frac{3k_BT}{b^2}\left\langle \frac{\partial R_{n\alpha}(t)}{\partial n}\frac{\partial R_{n\beta}(t)}{\partial n}\right\rangle. \tag{7.4}$$

Note that eqn (7.4) is valid even if the excluded volume effect is accounted for since, as shown in eqn (4.135), the pseudo-potential described by the delta function does not change the expression for the stress tensor (apart from the isotropic stress, i.e., the pressure).

† Historically, the molecular theory for the mechanical properties of polymeric materials was first constructed for rubbers, and it was realized that the stress is related to the orientation of the polymer segments. That the stress in polymer melts has the same molecular origin as in rubbers was noted by Green and Tobolsky,[21] who regarded the polymer melt as a kind of network with temporary junctions. This idea has been the base of successful phenomenological theories[18,22] and it is indeed used in the present theory.

The above argument is essentially for semidilute solutions, and may not apply to concentrated solutions and melts in which the intermolecular interaction cannot be described by the pseudo-potential. However, since the intermolecular interaction does not have any specific polymeric effect (for example the associated relaxation times will be as short as in a liquid monomer), one can argue that their effect will be only to give a constant viscous stress. Indeed eqn (7.4) is derived from the fact that even in the highly entangled state the short-time dynamics of polymers is described by the Rouse model. For, as was discussed in Sections 3.7.4 and 3.8.4, the expression of the stress tensor is entirely determined by the short-time dynamics: for example, the elastic stress is given by the change in the free energy for an *instantaneous* deformation. Therefore if the short-time dynamics is given by the Rouse model, the expression for the elastic stress must be the same as for the Rouse model, which is eqn (7.4). Thus eqn (7.4) holds quite generally in polymeric liquids in which the viscous stress is negligibly small.†

Since the short-time dynamics of polymers is unaffected by the global molecular structure, such as branching or crosslinks, the same argument will also hold for branched polymers and gels. In these systems, the stress is written as

$$\sigma_{\alpha\beta}(t) = \sum_{\substack{\text{all segments}\\ \text{in unit volume}}} \frac{3k_BT}{b^2}\left\langle \frac{\partial R_{n\alpha}(t)}{\partial n}\frac{\partial R_{n\beta}(t)}{\partial n}\right\rangle \tag{7.5}$$

where $\boldsymbol{R}_n$ is the position of the segments which are chosen small compared to the distance between the branching points (or crosslink points).

7.2.2 *Stress optical law*

The above argument is crucially supported by the stress optical law,[24,25] which is obtained by comparing eqn (7.4) with the formula for the intrinsic birefringence (eqn (4.238))

$$n^{(i)}_{\alpha\beta} = C\sigma_{\alpha\beta}(t) \tag{7.6}$$

where

$$C = \frac{2\pi(n^2+2)^2}{27k_BTn}\Delta\gamma. \tag{7.7}$$

† That the viscous stress of polymeric liquids is negligibly small compared to the elastic stress is well established experimentally. Indeed this fact has been taken as a basic postulate in Coleman's phenomenological theory[23] (see also Chapter 4, ref. 17). It must be noted that this is the result of the elastic stress contribution becoming so large relative to the viscous stress: the absolute magnitude of the viscous stress will not differ considerably between simple liquids and polymeric liquids.

In concentrated solutions and melts, the intrinsic birefringence $n^{(i)}_{\alpha\beta}$ can be regarded as the total birefringence since the form birefringence is negligibly small as was shown in Chapter 5. Thus eqn (7.6) is written as

$$n_{\alpha\beta} = C\sigma_{\alpha\beta} + \text{isotropic tensor}. \tag{7.8}$$

This is the stress optical law, which has been found experimentally to hold for rubbers and the more general problem of polymeric liquids.[26]

Two important aspects of eqn (7.8) must be mentioned.

(i) The simple linear relation between the stress and the birefringence holds even when the relation between the stress and the velocity gradient becomes quite complex (generally a nonlinear functional). Thus eqn (7.8) has much deeper physical significance than the similar relation for amorphous solids, which is derived by the symmetry argument and is limited to the Hookian range of elasticity.

(ii) The proportional constant C is a function of the local condition such as the temperature, solvent, and polymer concentration but is quite independent of the features of molecular structure on a large length-scale such as the molecular weight, molecular weight distribution, branching, and degree of crosslinking.†

These results are fully confirmed by experiment. The only explanation for such a general relation is that both the stress and the birefringence have the same physical origin, i.e., the orientational ordering in the polymer segments.

In deriving eqn (7.8) we assumed that (a) the relation between the orientation of the bond vectors and that of the end-to-end vector of the Rouse segments is linear (eqn (4.IV.13)) and that (b) the form birefringence is neglected. The stress optical law breaks down when either of these conditions is not met.[24] For example, the first condition is not satisfied when the stress is very large, as often happens in experiments near the glass transition temperature. The second condition is not satisfied when the sample includes a spatial inhomogeneity as in the case of block copolymers or polymer blends which include microphase separation. Except for these situations, the stress optical law holds quite generally in polymeric liquids.

7.3 Linear viscoelasticity

7.3.1 Background of phenomenological theory

The viscoelastic properties of a material are entirely characterized by the constitutive equation which relates the stress tensor $\sigma_{\alpha\beta}(t)$ to the velocity

† Though the excluded volume interaction (and other Van der Waals' type of interaction among the segments) do not violate the validity of eqns (7.4) and (7.8),[27] they do affect the stress optical coefficient C. For example C is sensitive to the nematic-like interaction which tends to orient the neighbouring segments in the same direction.[28,29]

gradient tensor $\kappa_{\alpha\beta}(t)$. In usual fluids the relationship is simple,

$$\sigma_{\alpha\beta}(t) = (\kappa_{\alpha\beta}(t) + \kappa_{\beta\alpha}(t))\eta. \tag{7.9}$$

In polymeric liquids the constitutive equation becomes quite complicated: the stress depends on the previous values of the velocity gradient tensor and the dependence is generally not linear; hence the relation is written only by a nonlinear functional relation:

$$\sigma_{\alpha\beta}(t) = F_{\alpha\beta}[\kappa_{\alpha\beta}] \quad \text{for} \quad \kappa_{\alpha\beta}(t') \text{ with } t' < t. \tag{7.10}$$

However, in the special case that the perturbation by the velocity gradient is small, the relation between the stress and the velocity gradient becomes linear and is written as[2,19]

$$\sigma_{\alpha\beta}(t) = \int_{-\infty}^{t} \mathrm{d}t' G(t-t')(\kappa_{\alpha\beta}(t') + \kappa_{\beta\alpha}(t')). \tag{7.11}$$

This constitutive equation includes only one material function $G(t)$, called the shear relaxation modulus.

Though the applicability of eqn (7.11) is limited, linear viscoelasticity has been studied in great detail as it represents a property of the material in a well-defined limit.

For shear flow

$$v_x = \kappa(t)y, \qquad v_y = 0, \qquad v_z = 0, \tag{7.12}$$

equation (7.11) reduces to the form given in Chapter 4 (eqn (4.116)),

$$\sigma_{xy}(t) = \int_{-\infty}^{t} \mathrm{d}t' G(t-t')\kappa(t'). \tag{7.13}$$

Let $\gamma(t)$ be the shear strain measured from the state at $t = 0$:

$$\gamma(t) = \int_{0}^{t} \mathrm{d}t' \kappa(t') \quad (-\infty < t < +\infty). \tag{7.14}$$

Equation (7.13) is then rewritten by integration by parts as

$$\sigma_{xy}(t) = \int_{-\infty}^{t} \mathrm{d}t' G(t-t') \frac{\partial \gamma(t')}{\partial t'} \tag{7.15}$$

$$= [G(t-t')\gamma(t')]_{-\infty}^{t} - \int_{-\infty}^{t} \mathrm{d}t' \frac{\partial G(t-t')}{\partial t'} \gamma(t')$$

$$= \int_{\infty}^{t} \mathrm{d}t' \frac{\partial G(t-t')}{\partial t'} (\gamma(t) - \gamma(t')). \tag{7.16}$$

Let us consider three typical situations shown in Fig. 7.1.

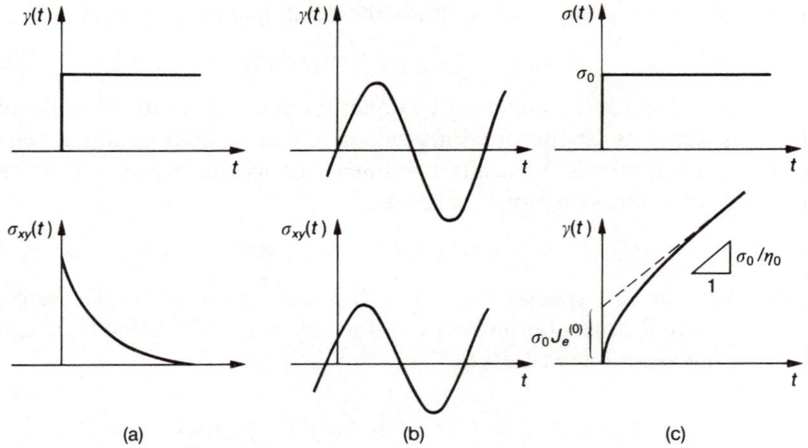

Fig. 7.1. (*a*) Stepwise shear, (*b*) oscillatory shear and (*c*) creep.

(a) Stepwise shear (Fig. 7.1*a*). The sample is deformed instantaneously at time $t = 0$. The shear strain $\gamma(t)$ is given by

$$\gamma(t) = \begin{cases} 0 & t < 0, \\ \gamma_0 & t > 0, \end{cases} \tag{7.17}$$

for which, eqn (7.16) gives

$$\sigma_{xy}(t) = \gamma_0 G(t). \tag{7.18}$$

This provides a direct determination of $G(t)$.

(b) Oscillatory shear (Fig. 7.1*b*). This has already been discussed in Chapter 4. The shear strain is give by

$$\gamma(t) = \gamma_0 \cos(\omega t) = \gamma_0 \operatorname{Re}(e^{i\omega t}). \tag{7.19}$$

The response defines the storage modulus $G'(\omega)$, loss modulus $G''(\omega)$, and the complex modulus $G^*(\omega)$,

$$\begin{aligned} \sigma_{xy}(t) &= \gamma_0(G' \cos(\omega t) - G''(\omega)\sin(\omega t)) \\ &\equiv \gamma_0 \operatorname{Re}(G^*(\omega)e^{i\omega t}) \end{aligned} \tag{7.20}$$

where

$$G^*(\omega) = G'(\omega) + iG''(\omega). \tag{7.21}$$

It is easy to show that eqn (7.16) gives

$$G^*(\omega) = i\omega \int_0^\infty dt e^{-i\omega t} G(t) \tag{7.22}$$

and

$$G'(\omega) = \omega \int_0^\infty \mathrm{d}t \sin(\omega t) G(t) \qquad G''(\omega) = \omega \int_0^\infty \mathrm{d}t \cos(\omega t) G(t). \quad (7.23)$$

(c) Creep (Fig. 7.1*c*). When a constant shear stress σ_0 is applied to the system at equilibrium, the system starts to flow. The shear strain $\gamma(t)$ is obtained from the integral equation:

$$\sigma_0 = \int_0^t \mathrm{d}t' G(t - t') \frac{\mathrm{d}\gamma(t')}{\mathrm{d}t'}. \quad (7.24)$$

This can be solved by the Fourier–Laplace transform:

$$\gamma(t) = \sigma_0 \int_{-\mathrm{i}\delta-\infty}^{-\mathrm{i}\delta+\infty} \frac{\mathrm{d}\omega}{2\pi} \frac{\mathrm{e}^{\mathrm{i}\omega t}}{\omega G^*(\omega)} \quad (7.25)$$

where δ is an arbitrary positive constant. The behaviour of $\gamma(t)$ for $t \to \infty$ is governed by the contribution from the pole at $\omega = 0$. Since for small ω, $G^*(\omega)$ is given by

$$G^*(\omega) = \mathrm{i}\omega \int_0^\infty \mathrm{d}t (1 - \mathrm{i}\omega t + \ldots) G(t) = \mathrm{i}\omega g_0 + \omega^2 g_1 + \ldots \quad (7.26)$$

with

$$g_0 = \int_0^\infty \mathrm{d}t G(t) \quad \text{and} \quad g_1 = \int_0^\infty \mathrm{d}t G(t) t, \quad (7.27)$$

the contribution from the pole at $\omega = 0$ is given by

$$\gamma(t) = \sigma_0 \left(\frac{t}{g_0} + \frac{g_1}{g_0^2} \right). \quad (7.28)$$

For large t, eqn (7.28) gives $\sigma_0 = g_0 \, \mathrm{d}\gamma/\mathrm{d}t$, whence g_0 can be identified with the steady state viscosity η_0. The constant term g_1/g_0^2 is called the steady-state compliance and written as $J_e^{(0)}$. Thus

$$\eta_0 = \int_0^\infty \mathrm{d}t G(t) = \lim_{\omega \to 0} \frac{G''(\omega)}{\omega}, \quad (7.29)$$

$$J_e^{(0)} = \int_0^\infty \mathrm{d}t G(t) t \Big/ \left[\int_0^\infty \mathrm{d}t G(t) \right]^2 = \lim_{\omega \to 0} \frac{G'(\omega)}{G''(\omega)^2}. \quad (7.30)$$

7.3.2 Calculation by Rouse model

We now study the linear viscoelasticity from the molecular viewpoint. First we consider Rouse dynamics. This corresponds to the case of short polymers in melts. The basic equations and their development are precisely the same as in Sections 4.5.3 and 4.5.4. The only distinction is that the viscous stress is now negligibly small, so that $G^{(p)}(t)$ in eqn (4.158), which corresponds to the polymer contribution to the relaxation modulus in Chapter 4, is now regarded as the relaxation modulus of the system. Thus

$$G(t) = \frac{c}{N} k_B T \sum_{p=1}^{\infty} \exp(-2tp^2/\tau_R) \tag{7.31}$$

where

$$\tau_R = \frac{\zeta N^2 b^2}{3\pi^2 k_B T}. \tag{7.32}$$

The viscosity and the steady-state compliance are calculated from $G(t)$,

$$\eta_0 = \int_0^{\infty} \mathrm{d}t G(t) = \frac{c}{N} k_B T \tfrac{1}{2} \tau_R \sum_{p=1}^{\infty} p^{-2} = \frac{\pi^2}{12}\left(\frac{c k_B T}{N}\right)\tau_R = \frac{c\zeta}{36} N b^2 \tag{7.33}$$

and

$$J_e^{(0)} = \frac{1}{\eta_0^2} \int_0^{\infty} \mathrm{d}t G(t) t = \frac{N}{c k_B T}\left(\sum_{p=1}^{\infty} p^{-4}\right)\left(\sum_{p=1}^{\infty} p^{-2}\right)^{-2} = \frac{2N}{5 c k_B T}. \tag{7.34}$$

These are written in terms of the molecular weight M, the weight of polymers in unit volume ρ ($\equiv cM/NN_A$), and the gas constant R ($\equiv N_A k_B$) as

$$\tau_R \propto M^2, \tag{7.35}$$

$$\eta_0 = \frac{\pi^2}{12}\left(\frac{\rho R T}{M}\right)\tau_R \propto \rho M, \tag{7.36}$$

$$J_e^{(0)} = \frac{2M}{5\rho R T}. \tag{7.37}$$

These results are confirmed for polymer melts with low molecular weight.[2,3]

7.3.3 *Calculation by reptation model*

Next we consider a polymer melt of high molecular weight in which entanglement is very important. To calculate $G(t)$, it is convenient to consider the stress relaxation after a step strain. Suppose at $t = 0$ a shear strain γ is applied to the system in equilibrium. The strain causes the deformation of the molecular conformation, and creates the stress, which relaxes with time as the conformation of polymers goes back to equilibrium. Though description of this process in the general case is slightly involved (see Section 7.5), the relaxation modulus in the linear regime can be obtained by a simple argument.

(A) For small t ($t \lesssim \tau_e$), the dynamics is described by the Rouse model and the relaxation modulus is given by eqn (7.31). Since $\tau_e \ll \tau_R$, eqn (7.31) is approximated as

$$G(t) = \frac{c}{N} k_B T \int_0^\infty \mathrm{d}p \, \exp(-2tp^2/\tau_R) = \frac{c}{2\sqrt{2}N} k_B T \left(\frac{\tau_R}{t}\right)^{1/2} \quad (t < \tau_e). \qquad (7.38)$$

(B) For $t \gtrsim \tau_e$, the Rouse behaviour is stopped by the tube constraints, and the reptation behaviour starts. According to reptation dynamics, the relaxation will generally involve two processes, the relaxation of the contour length, and the disengagement from the deformed tube, each being characterized by the time τ_R and τ_d respectively. In the linear regime (of small γ), however, the first process does not appear since the change in the contour length by the shear strain is an even function of γ and can be neglected to the first order in γ. Thus the relaxation for $t > \tau_e$ is only due to the disengagement. This can be evaluated as follows (see Fig. 7.2).

At $t \simeq \tau_e$, the whole polymer is confined in a deformed tube. As time passes, the polymer reptates, and at time t the parts of the polymer near the ends have disengaged from the deformed tube, while the part in the middle is still confined in the tube. Since only the segments in the deformed tube are oriented and contribute to the stress, the stress is proportional to the fraction $\psi(t)$ of the polymers still confined in the

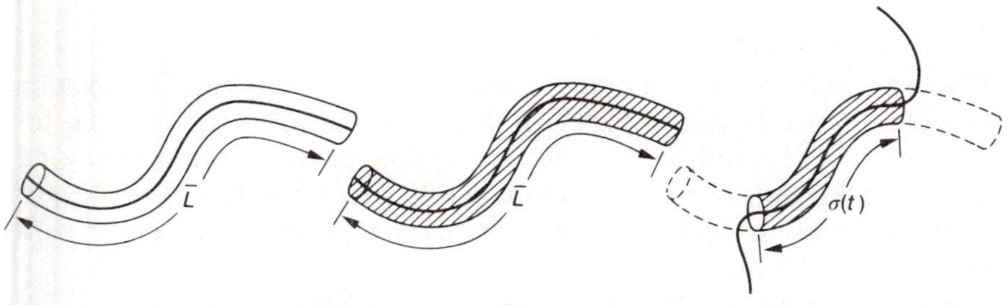

Fig. 7.2. Explanation of the stress relaxation after small step strain. (a) Before deformation, the conformation of the tube is in equilibrium. (b) Immediately after the deformation, the whole tube is deformed. The deformed part is indicated by the oblique lines. For small strain, the contour length of the tube is unchanged. (c) At a later time t, the chain is partly confined in a deformed tube. The average of the contour length $\sigma(t)$ of this part is equal to $\bar{L}\psi(t)$.

deformed tube, i.e.,

$$G(t) = G_N^{(0)}\psi(t) \quad (t \gtrsim \tau_e) \tag{7.39}$$

where $G_N^{(0)}$ is a certain constant and $\psi(t)$ has already been calculated in Section 6.2.2 (see eqn (6.16)),

$$\psi(t) = \sum_{p;\,\mathrm{odd}} \frac{8}{p^2\pi^2} \exp(-p^2 t/\tau_d) \tag{7.40}$$

with

$$\tau_d = \frac{\zeta N^3 b^2}{\pi^2 k_B T}\left(\frac{b}{a}\right)^2. \tag{7.41}$$

To obtain $G_N^{(0)}$, we utilize the fact that at $t \simeq \tau_e$, the Rouse-like behaviour (eqn (7.38)) smoothly crosses over to the reptation behaviour (eqn (7.39)), i.e.,

$$G_N^{(0)} \simeq G(\tau_e) \simeq \frac{c}{N} k_B T \left(\frac{\tau_R}{\tau_e}\right)^{1/2} \tag{7.42}$$

or using eqn (6.106)

$$G_N^{(0)} \simeq \frac{cb^2}{a^2} k_B T. \tag{7.43}$$

Equation (7.38) is then written as

$$G(t) \simeq G_N^{(0)}\left(\frac{\tau_e}{t}\right)^{1/2} \quad \text{for} \quad t \lesssim \tau_e. \tag{7.44}$$

Equations (7.39) and (7.44) give the relaxation modulus in the highly entangled state. We shall now compare this result with experiments.

7.3.4 *Comparison with experiments*

Overall shape of the relaxation modulus. The theoretical relaxation modulus $G(t)$ is shown in Fig. 7.3*a*. It is seen that $G(t)$ is nearly flat for $\tau_e < t < \tau_d$. The width of the plateau region increases with increasing molecular weight since τ_d is proportional to M^3, while τ_e and the height of the plateau are independent of M. This is in good agreement with experimental results.[30]

Figure 7.3*b* shows the storage modulus calculated by eqns (7.23), (7.39), and (7.44),

$$\begin{aligned} G'(\omega) &= G_N^{(0)}\left(\frac{\pi}{2}\omega\tau_e\right)^{1/2} \quad \text{for} \quad \omega\tau_e \gtrsim 1 \\ &= G_N^{(0)} \sum_{p;\,\mathrm{odd}} \frac{8}{\pi^2}\frac{1}{p^2}\frac{(\omega\tau_d/p^2)^2}{1+(\omega\tau_d/p^2)^2} \quad \text{for} \quad \omega\tau_e \lesssim 1. \end{aligned} \tag{7.45}$$

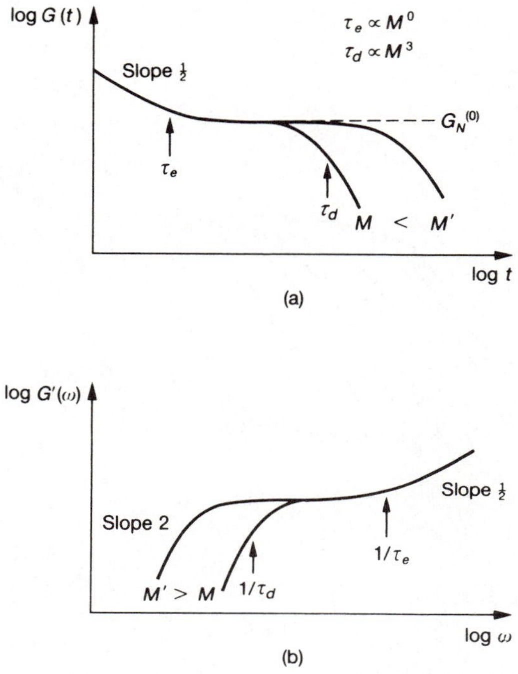

Fig. 7.3. Theoretical results of (a) $G(t)$ and (b) $G'(\omega)$.

The same characteristics as of $G(t)$ are observed here: $G'(\omega)$ has a plateau over a wide-frequency region $1/\tau_d < \omega < 1/\tau_e$. With increasing molecular weight, the low-frequency end of the plateau region decreases in proportion to M^{-3}, while the high-frequency end and the height of the plateau remain constant. This behaviour is strikingly demonstrated by the experiment of Onogi *et al.*[31] (see Fig. 7.4).

Maximum relaxation time, viscosity, and steady-state compliance. The longest relaxation time $\tau_{\max}$ of $G(t)$ is τ_d:

$$\tau_{\max} = \tau_d = \frac{\zeta N^3 b^2}{\pi^2 k_B T}\left(\frac{b}{a}\right)^2 = 3\frac{Nb^2}{a^2}\tau_R. \qquad (7.46)$$

The viscosity and the steady-state compliance are calculated from eqns (7.29) and (7.30). Since the contribution to the integral from the region $t < \tau_e$ is very small, η_0 and $J_e^{(0)}$ are given by

$$\eta_0 = G_N^{(0)}\int_0^\infty \mathrm{d}t\,\psi(t) = \frac{\pi^2}{12}G_N^{(0)}\tau_d \qquad (7.47)$$

Fig. 7.4. Storage modulus vs. frequency for narrow distribution polystyrene melts. Molecular weight ranges form $M_w = 8.9 \times 10^3$ (L9) to $M_w = 5.8 \times 10^5$ (L18). Reproduced from ref. 31.

and

$$J_e^{(0)} = \frac{1}{G_N^{(0)}} \left[\int_0^\infty \mathrm{d}t t\psi(t) \right] \Big/ \left[\int_0^\infty \mathrm{d}t\psi(t) \right]^2 = \frac{6}{5G_N^{(0)}}. \tag{7.48}$$

Thus $\tau_{\max}$, η_0 and $J_e^{(0)}$ depend on the molecular weight M as[32]

$$\tau_{\max} \sim M^3, \ \eta_0 \sim M^3, \ J_e^{(0)} \sim M^0. \tag{7.49}$$

That the steady-state compliance is independent of the molecular weight is in agreement with experimental results[2,3] (Fig. 7.5). On the other hand, the experimental exponent in the molecular weight dependence of $\tau_{\max}$ and η_0 is slightly larger than 3, ranging from 3 to 3.7.[33–35] An example of the viscosity in melts is given in Fig. 7.6. The reason for the discrepancy will be discussed later.

7.3.5 Tube diameter in melts

Given the general agreement in the shape of the relaxation modulus, it is possible to determine the step length a of the primitive chain. Though various ways are conceivable, a direct way is to use the plateau modulus

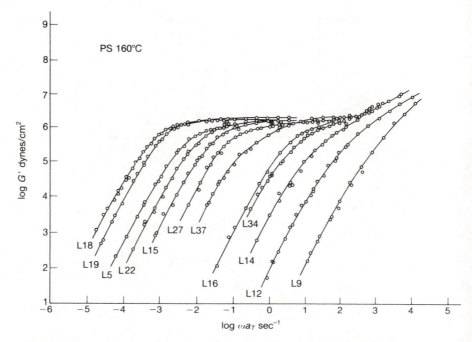

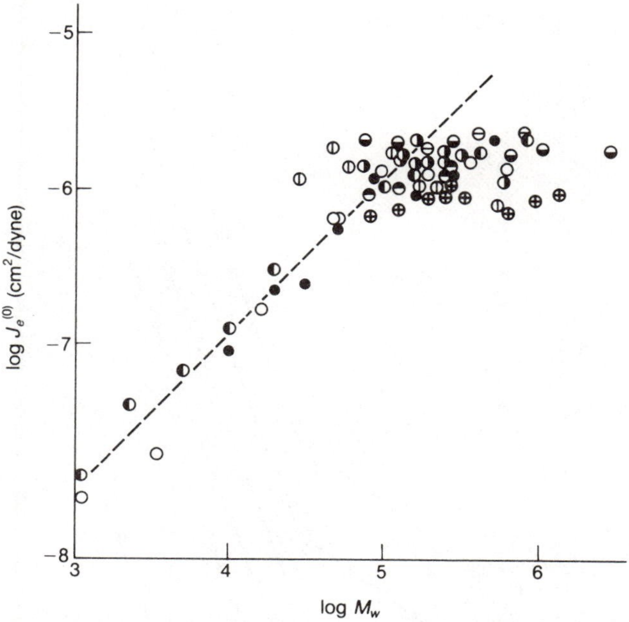

Fig. 7.5. Steady-state compliance of polystyrene at 160°C. Dotted line represents the result of the Rouse model.

○ Plazek–O'Rourke ● Mills–Nevin ◐ O'Reilly–Prest
◑ Murakami *et al.* ◓ Tobolsky *et al.* ◒ Nemoto
⊕ Akovali ⦶ Onogi *et al.* ⊖ Mieras–Rijn

Reproduced from ref. 35.

$G_N^{(0)}$, which is related to a by eqn (7.43). Since

$$cb^2 = \frac{c}{N} Nb^2 = \frac{\rho N_A}{M} Nb^2, \tag{7.50}$$

eqn (7.43) is written as

$$G_N^{(0)} \simeq \frac{\rho RT}{M} \frac{Nb^2}{a^2}. \tag{7.51}$$

Experimentally $G_N^{(0)}$ is often expressed by a characteristic molecular weight M_e, called the molecular weight between entanglements, which is defined by

$$M_e = \frac{\rho RT}{G_N^{(0)}}. \tag{7.52}$$

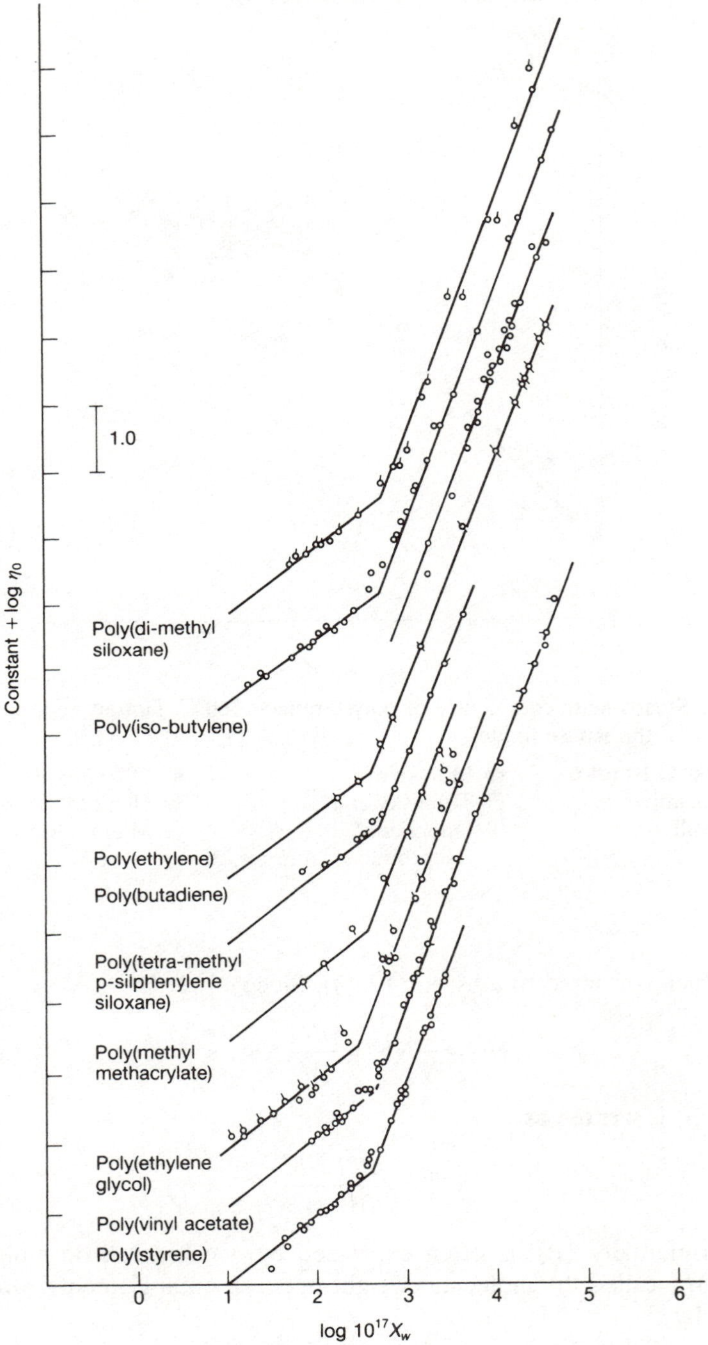

Fig. 7.6. Steady-state viscosity of various polymers in melt, where X_w is a parameter proportional to M_w. The curves are shifted vertically so as not to overlap each other. Reproduced from ref. 33.

It follows from eqns (7.51) and (7.52) that

$$a \simeq \left(\frac{M_e}{M} N b^2\right)^{1/2} \equiv \bar{R}_{M_e} \tag{7.53}$$

where $\bar{R}_{M_e}$ is the root mean square of the end-to-end distance of the polymer with the molecular weight M_e.

The precise numerical coefficient in eqn (7.53) is not given by the simple argument given in Section 7.3.3. To determine the coefficient, a further assumption is needed about the deformation of the tube under strain. A specific model described in the next section gives

$$a^2 = \frac{4}{5}\frac{M_e}{M} N b^2 = 0.8\bar{R}^2_{M_e} \tag{7.54}$$

which we shall now use.

The value of M_e in polymer melts is tabulated in the literature,[2] from which a is found to be 82 A for polystyrene and 34 A for polyethylene.[36] This distance is much larger than the correlation length measured by neutron scattering or X-ray scattering. At present, it is not fully understood what factors determine a, but various semiempirical relations are available for M_e.[33,37,38] Here we shall proceed regarding a as an adjustable parameter.

Graessley[36] showed that the above estimation of a is consistent with the results of diffusion experiments. According to the reptation theory, the self diffusion constant D_G is given by (see eqn (6.40))

$$D_G = \frac{a^2}{3Nb^2}\frac{k_B T}{N\zeta}. \tag{7.55}$$

To eliminate ζ, it is convenient to consider the zero shear rate viscosity calculated by the Rouse dynamics

$$\eta_0^{(R)}(M) \equiv \frac{1}{36} cN\zeta b^2 = \frac{\rho N_A}{36M} N\zeta \bar{R}^2_M \tag{7.56}$$

where $\bar{R}^2_M$ is the mean square end-to-end distance of the polymer of molecular weight M. Experimentally, $\eta_0^{(R)}(M)$ can be obtained by extracting the viscosity for low molecular weight according to the following equation (see Fig. 7.7)

$$\eta_0^{(R)}(M) = \eta_0(M_{ref})\frac{M}{M_{ref}} \tag{7.57}$$

where M_{ref} is an arbitrary molecular weight in the Rouse regime.

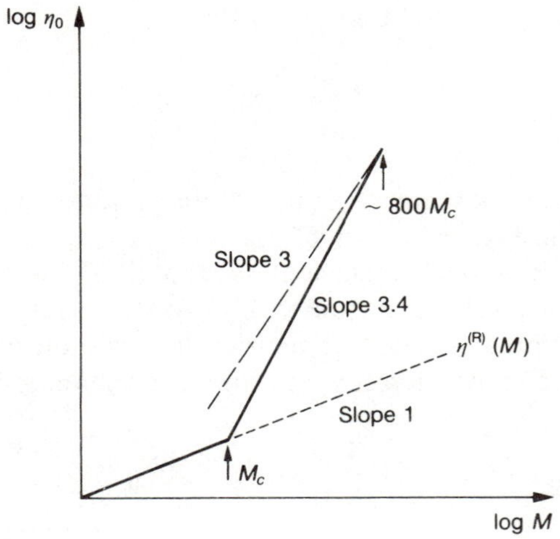

Fig. 7.7. Viscosity curve. Solid line: experimental formula eqns (7.71) and (7.72). Dashed line: theoretical curve, eqn (7.73).

From eqns (7.54)–(7.56), it follows that

$$D_G = \frac{1}{135}\left(\frac{\bar{R}_M^2}{M}\right)\frac{M_e}{M}\frac{\rho RT}{\eta_0^{(R)}(M)}. \tag{7.58}$$

According to Graessley,[36] eqn (7.58) gives values in reasonable agreement with experimental results[8,39] (e.g. for the case of polyethylene the error being only about 30%). Considering the different nature of the experiments, the agreement is remarkable.

That the length a is rather large seems to be consistent with the fact that both neutron scattering[6] and computer simulation[7,40,41] do not find it easy to detect reptation: the characteristic time-scale in these experiments is often shorter than τ_e. Recently, however, it has been reported that indications of reptation are observed in some situations.[41–43]

7.3.6 Semidilute and concentrated solutions

So far we have been considering polymer melts. It is expected that the same picture will hold in semidilute and concentrated solutions. Though this is generally believed to be the case, a few remarks must be made.

(i) In semidilute solutions, the excluded volume effect and the hydrodynamic interaction become important for dynamics on a length-scale shorter than the correlation length ξ (or the hydrodynamic screening length ξ_H). The problem of how this affects the reptation picture is delicate. However, in the case of a good solvent, certain conclusions can be drawn from scaling arguments.[44]†

Consider for example the maximum relaxation time $\tau_{\max}$, the steady state viscosity η_0, and the plateau modulus $G_N^{(0)}$. If one imposes the condition that they are invariant under the scaling transformation given by eqn (5.138), one can show that

$$\tau_{\max} = \frac{\eta_s N^{3\nu} b^3}{k_B T} f_1(c/c^*), \tag{7.59}$$

$$\eta_0 = \eta_s f_2(c/c^*), \tag{7.60}$$

and

$$G_N^{(0)} = \frac{c}{N} k_B T f_3(c/c^*), \tag{7.61}$$

where c^* is the overlap concentration

$$c^* \simeq \frac{N^{1-3\nu}}{b^3}. \tag{7.62}$$

The concentration dependence can be determined if one imposes a further condition that eqns (7.59)–(7.61) must be consistent with the reptation prediction:[44]

$$\tau_{\max} \propto N^3, \ \eta_0 \propto N^3, \ G_N^{(0)} \propto N^0, \tag{7.63}$$

then

$$\tau_{\max} \simeq \frac{\eta_s N^{3\nu} b^3}{k_B T} \left(\frac{c}{c^*}\right)^{(3-3\nu)/(3\nu-1)} \propto M^3 \rho^{3/2}, \tag{7.64}$$

$$\eta_0 \simeq \eta_s \left(\frac{c}{c^*}\right)^{3/(3\nu-1)} \propto M^3 \rho^{15/4}, \tag{7.65}$$

and

$$G_N^{(0)} = \frac{c}{N} k_B T \left(\frac{c}{c^*}\right)^{1/(3\nu-1)} \propto M^0 \rho^{9/4}. \tag{7.66}$$

(The last expressions indicate the result when ν is equal to 3/5). Experimentally, eqns (7.64)–(7.66) seem to hold at least approximately in the semidilute regime,[45–47] though the exponents in the molecular weight dependence of $\tau_{\max}$ and η_0 are slightly larger than 3, and the exponent in the concentration dependence of $G_N^{(0)}$ is somewhat higher than 9/4 depending on the quality of the solvent.

† It must be mentioned that whether dynamical scaling holds for systems dominated by the topological interaction is still a matter of debate, though eqns (7.64)–(7.66) seem to be in agreement with experimental results.

(ii) At higher concentration, the effect of the excluded volume and the hydrodynamic interaction becomes less important. If we disregard these effects, the dynamics is described by the same model as in melts except that a and ζ now depend on concentration. Experimentally it has been found[3,48,49] that

$$G_N^{(0)} \propto \rho^2 \tag{7.67}$$

which implies

$$a \propto \rho^{-1/2}. \tag{7.68}$$

The concentration dependence of $\tau_{\max}$ and η_0 is delicate since the friction constant ζ depends on the concentration in a nontrivial way.[2]†

7.4 Other relaxation modes

7.4.1 Discrepancy between the theory and experiments

Although the theory given in the previous section is largely in agreement with experiments on linear viscoelasticity, there remain certain discrepancies.

(i) The theoretical molecular weight dependence $\eta_0 \propto M^3$ and $\tau_{\max} \propto M^3$ is weaker than the experimental one; the measured exponent is higher than 3, ranging from 3 to 3.7.[34,35]

(ii) The theoretical relaxation modulus $G(t)$ is too close to a single exponential compared to the experimental modulus.[50] For example, the experimental value of $J_e^{(0)}G_N^{(0)}$, which measures the deviation from the single exponential behaviour of $G(t)$, is between 2 and 3[3,51] as against 6/5 for the theoretical result of eqn (7.48).

Part of these discrepancies can be attributed to the molecular weight distribution, which seriously affects the value of $J_e^{(0)}$.[2,3] On the other hand detailed comparison with experimental results indicates that not all the discrepancy can be resolved by the molecular weight distribution.[52]

The discrepancy in the exponent of the viscosity has been a matter of debate which is not yet settled. Various modifications of the reptation picture have been proposed. For example, Wendel and Noolandi[53] argued that if the polymers are trapped by some tight knots with an extremely long lifetime, the diffusion along the tube becomes non-Fickian and this gives a higher exponent in $\eta_0 \propto M^x$. However, such tight knots, if they exist, would create a rubbery plateau with extremely long

† A similar problem exists in the melt, where it is often observed that the viscosity does not follow the Rouse behaviour at small molecular weights much less than M_e. This is attributed to the fact that for short polymers the segmental friction constant ζ depends on the molecular weight.[33]

relaxation time, which no viscoelastic data seem to support. Curtiss and Bird[54] suggested that the segmental friction constant ζ may depend on the molecular weight, but this hypothesis seems to contradict the result of the diffusion experiment. Another possibility suggested by Ball[55] takes into account the long-range correlation in the motion of vacancies needed for the reptation motion to take place. This has been studied by computer simulation,[56] but the result is not yet conclusive.

At present, a more consistent explanation seems to be that given by Graessley,[36] who pointed out that the observed viscosity and the relaxation time are *smaller* than the calculated ones. Using eqns (7.32), (7.33), (7.46), and (7.47), one can show that

$$\eta_0(M) = \frac{15}{4}\eta_0(M_e)\left(\frac{M}{M_e}\right)^3 = \frac{15}{4}\eta_0^{(R)}(M)\left(\frac{M}{M_e}\right)^2 \tag{7.69}$$

and

$$\tau_{\max} = \frac{45}{\pi^2}\frac{M_e}{\rho RT}\eta_0(M_e)\left(\frac{M}{M_e}\right)^3 = \frac{15}{4}\tau_R(M)\frac{M}{M_e}. \tag{7.70}$$

On the other hand, a widely accepted empirical formula is

$$\eta_0^{(\exp)}(M) = \begin{cases} \eta_0(M_c)\dfrac{M}{M_c} & \text{for} \quad M < M_c, & (7.71) \\ \eta_0(M_c)\left(\dfrac{M}{M_c}\right)^{3.4} & \text{for} \quad M > M_c, & (7.72) \end{cases}$$

where M_c is a certain molecular weight which is two or three times larger than M_e. Using $M_c = 2M_e$, and $\eta_0(M_e) = \eta_0(M_c)(M_e/M_c)$, eqn (7.69) is written as

$$\eta_0^{(\text{theo})}(M) = 15\eta_0(M_c)\left(\frac{M}{M_c}\right)^3. \tag{7.73}$$

$\eta_0^{(\text{theo})}(M)$ is about 15 times larger than $\eta_0^{(\exp)}(M)$ at M_c, but the discrepancy decreases with increasing M and diminishes at $M = (15)^{1/0.4}M_c \simeq 800M_c$ (see Fig. 7.7). Graessley thus conjectured that although the pure reptation behaviour will be observed for very large molecular weight, there is a large cross-over region in the viscosity from the Rouse-like behaviour to the pure reptation behaviour, which gives an apparent exponent larger than 3.

Unfortunately since no data are available for molecular weights higher than $800M_c$, which is 3×10^7 for a polystyrene melt, the crucial test of Graessley's conjecture has not been given. However, it is obvious that any relaxation process which occurs concurrently with reptation decreases the viscosity and hence reduces the discrepancy between the theory and experiment.

7.4.2 Contour length fluctuation and tube reorganization

Two relaxation processes have been suggested to alter the pure reptation behaviour discussed in Section 7.3.3.

Contour length fluctuation. As was shown in the previous chapter, the contour length fluctuation reduces the disengagement time τ_d significantly. From eqns (6.20), (6.101), and (7.54), the disengagement time $\tau_d^{(F)}$ of a chain with fluctuation is given by

$$\tau_d^{(F)} = \tau_d^{(NF)}\left(1 - X'\left(\frac{M_e}{M}\right)^{1/2}\right)^2 \tag{7.74}$$

where X' is estimated as $1.47 * \sqrt{(4/5)} \simeq 1.3$.[57] For $M/M_e = 50$ the ratio between $\tau_d^{(F)}$ and $\tau_d^{(NF)}$ is about 0.67 which displays the considerable effect of the contour length fluctuation in the region where the polymers are usually regarded as 'fully entangled'. A crude calculation[57,58] indicates that the discrepancy in the viscosity and the steady state compliance is significantly improved if the contour length fluctuation is taken into account.

Tube reorganization. So far, it has been assumed that the tube is fixed in the material and its conformational change occurs only at the ends. It is conceivable that the conformational change of the tube can occur in the middle.[44] For example:

(i) Constraint release: The topological constraints for a polymer can be released (or created) by the reptation of the surrounding polymers as shown in Fig. 7.8. This will cause the conformational change of the tube in the middle. A model describing this process is to regard the conformational change as a local jump of the primitive chain. Since the jump rate is of the order $1/\tau_d$, this process has a negligible effect on the longest relaxation time.[59,60] However, the process gives an additional relaxation to $G(t)$ in the plateau region,[61] and improves the value of $J_e^{(0)}G_N^{(0)}$.

(ii) Tube deformation: If a polymer is in a strained conformation, it will tend to relax the strain by creating the deformation of the surrounding polymers. This effect may be handled by considering the deformation of a strained chain placed in a viscoelastic medium. So far no quantitative estimation of this effect has been done.

Though these processes are conceivable, estimation of their effect is

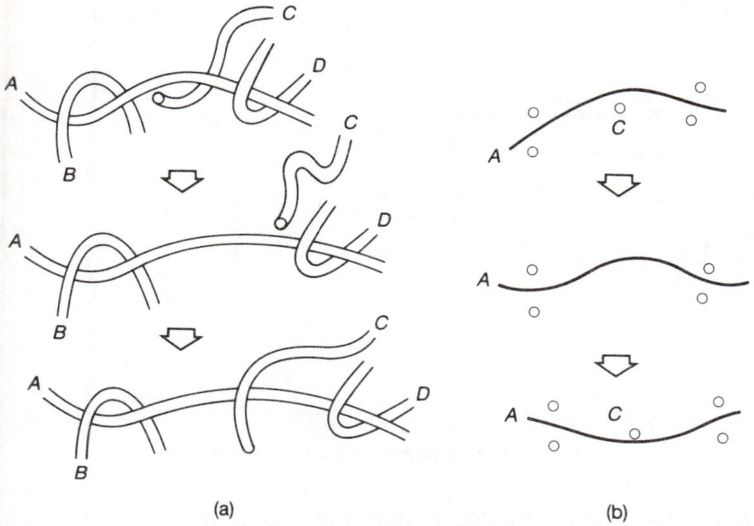

Fig. 7.8. Release and creation of the topological constraints. (*a*) The topological constraints imposed on the chain *A* by *C* is released and recreated by the motion of *C*. (*b*) In the two-dimensional representation, this process can be represented by the disappearance and reappearance of the obstacle *C*. The process causes the deformation of the tube in the middle.

still at the level of conjecture, and there are other theoretical treatments.[62,63] Experimentally, in linear polymers with narrow molecular weight distribution, it seems that the major difficulty of the theory can be resolved by including the contour length fluctuation. On the other hand the tube reorganization is believed to be important for polymers with broader molecular weight distribution,[60] or long branches, which will be discussed later.

7.5 Stress relaxation after large step strain

7.5.1 Experimental setup

Having seen the characteristic features of the linear viscoelasticity, we shall now study the nonlinear viscoelasticity. Before studying the general situation, we shall first consider a simple case, the stress relaxation after stepwise deformation.[14,64] Suppose that at time $t = 0$, a polymeric liquid is suddenly deformed homogeneously. The deformation creates a stress which gradually relaxes with time. Our problem is to find how this relaxation takes place.

For a homogeneous deformation, we may assume without loss of

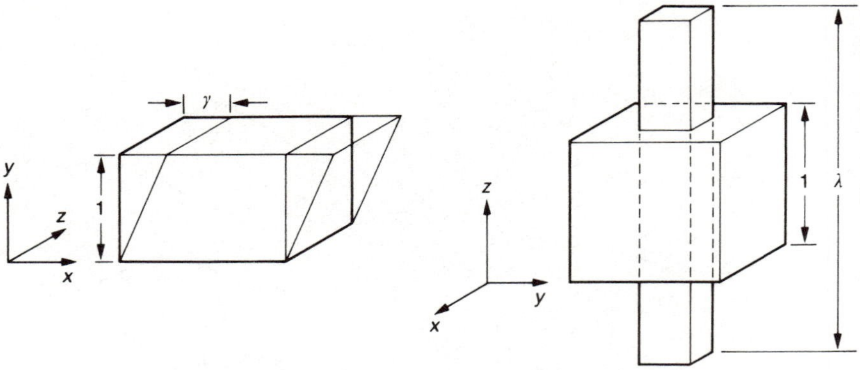

Fig. 7.9. (*a*) Shear and (*b*) elongation.

generality that a point $\boldsymbol{r}$ in the material is displaced to

$$\boldsymbol{r} \rightarrow \boldsymbol{r}' = \boldsymbol{E} \cdot \boldsymbol{r}. \tag{7.75}$$

The tensor $\boldsymbol{E}$ is called the deformation gradient.† Two particular cases are often studied experimentally.

(i) Shear (Fig. 7.9*a*), for which the material is deformed to

$$r'_x = r_x + \gamma r_y, \; r'_y = r_y, \; r'_z = r_z. \tag{7.76}$$

This deformation is characterized by a single parameter γ, the shear strain. Since the deformation has a reflection symmetry with respect to the xy plane, the stress components $\sigma_{zy}(=\sigma_{yz})$ and $\sigma_{xz}(=\sigma_{zx})$ vanish identically. However, the shear stress $\sigma_{xy} = \sigma_{yx}$ and the diagonal stresses σ_{xx}, σ_{yy} and σ_{zz} generally do not vanish. Since the isotropic part of the stress has no significance, only two of the diagonal components have meaning. The stresses

$$N_1 = \sigma_{xx} - \sigma_{yy} \tag{7.77}$$

and

$$N_2 = \sigma_{yy} - \sigma_{zz} \tag{7.78}$$

are called the first and the second normal stress differences. Thus the response of the shear deformation is characterized by three stress components σ_{xy}, N_1 and N_2, each of which are nonlinear functions of γ and t.

(ii) Uniaxial elongation (Fig. 7.9*b*). Here the sample is stretched in the

† A general deformation is described by the function $\boldsymbol{r}' = \boldsymbol{r}'(\boldsymbol{r})$ which connects the position vectors $\boldsymbol{r}$ and $\boldsymbol{r}'$ before and after the deformation. In such a case $E_{\alpha\beta}$ is given by $\partial r'_\alpha / \partial r_\beta$.

z direction by factor λ. Since the volume is unchanged, this causes a contraction in the x and y directions by a factor $1/\sqrt{\lambda}$. Hence, the deformation is described by

$$r'_x = \frac{1}{\sqrt{\lambda}} r_x, \qquad r'_y = \frac{1}{\sqrt{\lambda}} r_y, \qquad r'_z = \lambda r_z. \tag{7.79}$$

In this deformation, the off-diagonal components of the stress vanish and two of the diagonal components σ_{xx} and σ_{yy} are equal to each other. Thus the response is characterized by a component

$$\sigma_T = \sigma_{zz} - \sigma_{xx} \tag{7.80}$$

which is called the tensile stress.

7.5.2 Calculation by Rouse model

First we calculate the stress relaxation using the Rouse model. The stress tensor (eqn 7.4) is expressed by the normal coordinates X_p of the Rouse model (see Section 4.5.2, eqn (4.137))

$$\sigma_{\alpha\beta} = \frac{c}{N} \sum_{p=1}^{\infty} k_p \langle X_{p\alpha}(t) X_{p\beta}(t) \rangle \tag{7.81}$$

where

$$k_p = \frac{6\pi^2 k_B T}{Nb^2} p^2 \quad (p = 1, 2, \ldots). \tag{7.82}$$

According to Rouse dynamics, the position of the segments are changed in the same way as the macroscopic point;† thus if $\boldsymbol{R}_n(-0)$ and $\boldsymbol{R}_n(+0)$ are the positions of the segment before and after the deformation,

$$\boldsymbol{R}_n(+0) = \mathbf{E} \cdot \boldsymbol{R}_n(-0) \tag{7.83}$$

or

$$\boldsymbol{X}_p(+0) = \mathbf{E} \cdot \boldsymbol{X}_p(-0). \tag{7.84}$$

Since the system is in equilibrium for $t < 0$

$$\langle X_{p\alpha}(-0) X_{p\beta}(-0) \rangle = \frac{k_B T}{k_p} \delta_{\alpha\beta}. \tag{7.85}$$

† This is often called the affine deformation assumption, but it is actually derived from the Langevin equation for the Rouse model (eqn (4.139)). For an instantaneous deformation, the velocity gradient $\kappa_{\alpha\beta}(t)$ becomes so large that it dominates the other terms on the right-hand side of eqn (4.139). The equation for X_p then becomes the same as that for the macroscopic point (see eqn (7.151)), and therefore the change of X_p becomes affine.

From eqns (7.84) and (7.85), it follows that

$$\langle X_{p\alpha}(+0)X_{p\beta}(+0)\rangle = E_{\alpha\mu}E_{\beta\nu}\langle X_{p\mu}(-0)X_{p\nu}(-0)\rangle = E_{\alpha\mu}E_{\beta\nu}\delta_{\mu\nu}\frac{k_BT}{k_p}$$

$$= B_{\alpha\beta}(\mathbf{E})\frac{k_BT}{k_p} \tag{7.86}$$

where

$$B_{\alpha\beta}(\mathbf{E}) = E_{\alpha\mu}E_{\beta\mu} \tag{7.87}$$

which is called the Finger strain. For $t>0$, $\langle X_{p\alpha}(t)X_{p\beta}(t)\rangle$ satisfies eqn (4.141) with $\kappa_{\alpha\beta}=0$:

$$\frac{\partial}{\partial t}\langle X_{p\alpha}(t)X_{p\beta}(t)\rangle = -\frac{2p^2}{\tau_R}\left(\langle X_{p\alpha}(t)X_{p\beta}(t)\rangle - \delta_{\alpha\beta}\frac{k_BT}{k_p}\right). \tag{7.88}$$

Equation (7.88) is solved with the initial condition (7.86) by

$$\langle X_{p\alpha}(t)X_{p\beta}(t)\rangle = \frac{k_BT}{k_p}[B_{\alpha\beta}(\mathbf{E})\exp(-2p^2t/\tau_R)$$

$$+ \delta_{\alpha\beta}(1-\exp(-2p^2t/\tau_R))]. \tag{7.89}$$

Substituting eqn (7.89) into eqn (7.81) and dropping the isotropic term we have

$$\sigma_{\alpha\beta}(t) = \frac{c}{N}k_BTB_{\alpha\beta}(\mathbf{E})\sum_{p=1}^{\infty}\exp(-2p^2t/\tau_R) \tag{7.90}$$

$$= B_{\alpha\beta}(\mathbf{E})G(t) \tag{7.91}$$

where $G(t)$ is the linear relaxation modulus given by eqn (7.31).

For a shear deformation, $\mathbf{E}$ is given by

$$\mathbf{E} = \begin{bmatrix} 1 & \gamma & 0 \\ 0 & 1 & 0 \\ 0 & 0 & 1 \end{bmatrix} \tag{7.92}$$

so that

$$\mathbf{B} = \begin{bmatrix} 1+\gamma^2 & \gamma & 0 \\ \gamma & 1 & 0 \\ 0 & 0 & 1 \end{bmatrix}$$

and

$$\sigma_{xy} = \gamma G(t), \tag{7.93}$$

$$N_1 = \gamma^2 G(t), \tag{7.94}$$

$$N_2 = 0. \tag{7.95}$$

Note that the shear stress is a linear function of γ even if γ is large.

For the uniaxial elongation, the tensile stress is given by

$$\sigma_T = \left(\lambda^2 - \frac{1}{\lambda}\right)G(t). \tag{7.96}$$

The Rouse-like behaviour is expected to be seen in the initial stage of the relaxation ($t < \tau_e$).

7.5.3 *Calculation by reptation model*

Now we consider the behaviour for $t > \tau_e$ using the reptation model. To calculate the stress, we have to know

(i) How the conformation $\boldsymbol{R}(s, t)$ of the primitive chain is changed by the macroscopic deformation, and

(ii) How the stress is calculated for a given conformation $\boldsymbol{R}(s, t)$.

We shall discuss these problems separately.

Expression for the stress tensor. The microscopic expression for the stress tensor can be obtained by taking the average of eqn (7.4) for a given conformation of the primitive chain.[64] Alternatively, it can be derived by an elementary argument explained in Fig. 7.10. In both cases the result is

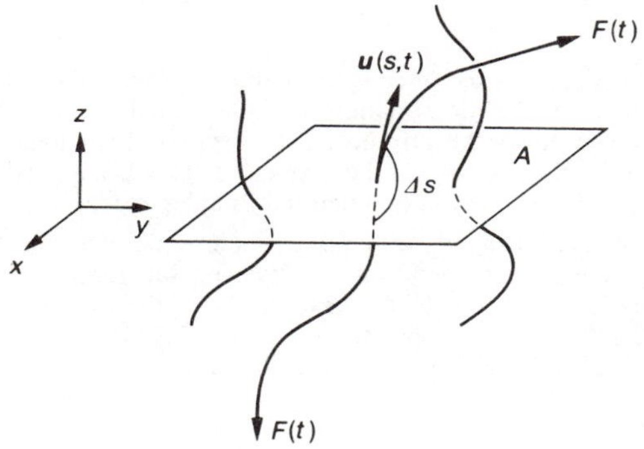

Fig. 7.10. Consider a plane of area A, normal to the z axis. The stress component $\sigma_{\alpha z}$ is given by the force (per area) S_α/A acting through the plane. Consider a part of the primitive chain between s and $s + \Delta s$. If this part is within the distance $u_z(s, t)\Delta s$ from the plane, it penetrates the plane and gives a contribution of $F_\alpha(s, t) = F(t)u_\alpha(s, t)$ to S_α, where $F(t)$ is the tensile force acting along the primitive chain. Since the number of the primitive chain in unit volume is (c/N), the number of such part is $(c/N)Au_z(s, t)\Delta s$, whence

$$S_\alpha = \sum_{\Delta s}\left\langle \frac{c}{N} Au_z(s, t)\Delta s F_\alpha(s, t)\right\rangle = \frac{c}{N} A\left\langle \int_0^L \mathrm{d}s F(t)u_\alpha(s, t)u_z(s, t)\right\rangle.$$

Thus the force per area S_α/A gives eqn (7.97).

written as

$$\sigma_{\alpha\beta}(t)=\frac{c}{N}\left\langle\int_0^L \mathrm{d}s F(t)u_\alpha(s,t)u_\beta(s,t)\right\rangle \tag{7.97}$$

where $\boldsymbol{u}(s,t)\equiv\partial\boldsymbol{R}(s,t)/\partial s$ is the unit vector tangent to the primitive chain and $F(t)$ is the tensile force acting along the primitive chain. In the equilibrium state $F(t)$ is given by (see eqn (6.83))

$$F(t)=\frac{3k_BT}{Nb^2}\bar{L}, \tag{7.98}$$

while in the non-equilibrium state in which the contour length is $L(t)$, $F(t)$ is given by†

$$F(t)=\frac{3k_BT}{Nb^2}L(t). \tag{7.99}$$

From eqns (7.97) and (7.99)

$$\sigma_{\alpha\beta}(t)=\frac{c}{N}\frac{3k_BT}{Nb^2}\left\langle\int_0^{L(t)} \mathrm{d}s L(t)(u_\alpha(s,t)u_\beta(s,t)-\tfrac{1}{3}\delta_{\alpha\beta})\right\rangle. \qquad (7.100)‡$$

This formula shows that the stress is determined by two quantities, the contour length $L(t)$ and the orientation of $\boldsymbol{u}(s,t)$.

† Here it is assumed that the tensile force is constant along the chain. This assumption is not correct immediately after the deformation because initially the Rouse segments are stretched or compressed along the primitive path depending on their direction (see Fig. 7.11). However, such local imbalance in the segment density is adjusted in the time τ_e, and for $t>\tau_e$, the tensile force $F(s,t)$ can be regarded as independent of s.

‡ Curtiss and Bird[54] derived a slightly different stress formula, which includes an adjustable parameter called the link tension coefficient ε. However, this formula is not consistent with the stress optical law unless $\varepsilon=0$. In the case of $\varepsilon=0$, the formula becomes essentially equivalent to eqn (7.120) which is a special case of eqn (7.100).

Equation (7.100) is also derived from the principle of virtual work. The free energy per unit volume is

$$\mathscr{A}=\frac{c}{N}\frac{3k_BT\langle L(t)^2\rangle}{2Nb^2}.$$

Under a virtual deformation $\delta\varepsilon_{\alpha\beta}$, $L(t)$ changes by

$$\delta L=\int_0^L \mathrm{d}s\,\delta\varepsilon_{\alpha\beta}u_\alpha(s,t)u_\beta(s,t)$$

whence

$$\delta\mathscr{A}=\frac{c}{N}\frac{3k_BT}{Nb^2}\langle L(t)\delta L(t)\rangle=\delta\varepsilon_{\alpha\beta}\frac{c}{N}\frac{3k_BT}{Nb^2}\left\langle L(t)\int_0^{L(t)}\mathrm{d}s u_\alpha(s,t)u_\beta(s,t)\right\rangle$$

which gives eqn (7.100).

Deformation of the primitive path. Our next question is how the conformation $\boldsymbol{R}(s, t)$ of the primitive chain is deformed by the macroscopic strain. The simplest assumption is that the deformation is affine, i.e., that the primitive chain (or the central axis of the tube) is deformed in the same way as the macroscopic deformation. Thus the point $\boldsymbol{R}(s, -0)$ on the primitive chain is displaced as

$$\boldsymbol{R}(s, -0) \rightarrow \boldsymbol{E} \cdot \boldsymbol{R}(s, -0). \tag{7.101}$$

Let us now study how this changes the contour length $L(t)$ and the orientation of $\boldsymbol{u}(s, t)$. This is illustrated in Fig. 7.11.

(i) The change of the contour length: The transformation (7.101) changes the length Δs of a line segment on the primitive chain to

$$\Delta\bar{s} = \Delta s \, |\boldsymbol{E} \cdot \boldsymbol{u}(s, -0)|. \tag{7.102}$$

The length $\Delta\bar{s}$ can be larger or smaller than the original length Δs depending on $\boldsymbol{u}(s, -0)$. Since the distribution of $\boldsymbol{u}(s, -0)$ is isotropic, the average ratio between $\Delta\bar{s}$ and Δs is given by

$$\alpha(\boldsymbol{E}) = \langle |\boldsymbol{E} \cdot \boldsymbol{u}| \rangle_0 \tag{7.103}$$

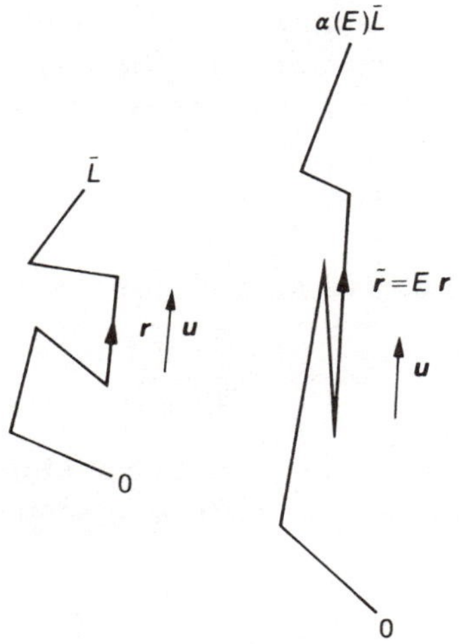

Fig. 7.11. Deformation of the primitive chain by macroscopic strain. Here for the purpose of explanation, the primitive chain is represented by randomly connected line segments. According to the affine deformation assumption, a line segment represented by $\boldsymbol{r}$ is transformed to $\boldsymbol{E} \cdot \boldsymbol{r}$. Thus the length $\Delta s = |\boldsymbol{r}|$ and the direction $\boldsymbol{u} = \boldsymbol{r}/|\boldsymbol{r}|$ are changed as $\Delta s \rightarrow |\boldsymbol{E} \cdot \boldsymbol{u}| \, \Delta s$ and $\boldsymbol{u} \rightarrow \boldsymbol{E} \cdot \boldsymbol{u}/|\boldsymbol{E} \cdot \boldsymbol{u}|$.

where $\langle \ldots \rangle_0$ denotes the average of $\boldsymbol{u}$ over the isotropic state,

$$\langle \ldots \rangle_0 = \int \frac{\mathrm{d}\boldsymbol{u}}{4\pi} \cdots . \tag{7.104}$$

For any deformation which conserves the volume (det $|\boldsymbol{E}| = 1$), it can be shown that[13]

$$\alpha(\boldsymbol{E}) \geqslant 1. \tag{7.105}$$

Thus the contour length of the primitive chain immediately after the deformation is given by

$$\langle L(+0) \rangle = \alpha(\boldsymbol{E}) \bar{L} \tag{7.106}$$

which is always larger than $\bar{L}$.

(ii) The change of the orientation: To denote the orientation of the primitive path, we define the orientational tensor

$$S_{\alpha\beta}(s, t) = \langle u_\alpha(s, t) u_\beta(s, t) - \tfrac{1}{3}\delta_{\alpha\beta} \rangle \tag{7.107}$$

which vanishes in the isotropic state, but does not vanish in the oriented state. Since the distribution of $\boldsymbol{u}(s, -0)$ is independent of s, $S_{\alpha\beta}(s, +0)$ will also be independent of s and can be written as

$$S_{\alpha\beta}(s, +0) = Q_{\alpha\beta}(\boldsymbol{E}). \tag{7.108}$$

To calculate $Q_{\alpha\beta}(\boldsymbol{E})$, let us consider the probability distribution function $f(\boldsymbol{u}, s, t)$ for the tangent vector $\boldsymbol{u}(s, t)$. ($f(\boldsymbol{u}, s, t)$ is the probability that the tangent vector at s and t is in the direction $\boldsymbol{u}$.) Obviously

$$f(\boldsymbol{u}, s, -0) = \frac{1}{4\pi}. \tag{7.109}$$

By the deformation, the unit vector $\boldsymbol{u}$ changes as

$$\boldsymbol{u} \to \tilde{\boldsymbol{u}} = \frac{\boldsymbol{E} \cdot \boldsymbol{u}}{|\boldsymbol{E} \cdot \boldsymbol{u}|}. \tag{7.110}$$

The probability that an arbitrarily chosen point of the deformed primitive chain is in the direction of $\boldsymbol{u}' = \boldsymbol{E} \cdot \boldsymbol{u} / |\boldsymbol{E} \cdot \boldsymbol{u}|$ is proportional to $|\boldsymbol{E} \cdot \boldsymbol{u}|$, the length of such a part. Thus

$$f(\boldsymbol{u}', s, +0) = C \int \mathrm{d}\boldsymbol{u}\, |\boldsymbol{E} \cdot \boldsymbol{u}|\, \delta\!\left(\boldsymbol{u}' - \frac{\boldsymbol{E} \cdot \boldsymbol{u}}{|\boldsymbol{E} \cdot \boldsymbol{u}|}\right) f(\boldsymbol{u}, s, -0) \tag{7.111}$$

where C is the normalization constant which is determined from the condition

$$1 = C \int \mathrm{d}\boldsymbol{u}' f(\boldsymbol{u}', s, t) \tag{7.112}$$

to be

$$C^{-1} = (4\pi)^{-1} \int d\boldsymbol{u}\, |\mathbf{E} \cdot \boldsymbol{u}| = \langle |\mathbf{E} \cdot \boldsymbol{u}| \rangle_0. \tag{7.113}$$

Then

$$f(\boldsymbol{u}, s, +0) = \frac{1}{4\pi \langle |\mathbf{E} \cdot \boldsymbol{u}| \rangle_0} \int d\boldsymbol{u}'\, |\mathbf{E} \cdot \boldsymbol{u}'|\, \delta\left(\boldsymbol{u} - \frac{\mathbf{E} \cdot \boldsymbol{u}'}{|\mathbf{E} \cdot \boldsymbol{u}'|}\right) \tag{7.114}$$

so that

$$\begin{aligned} Q_{\alpha\beta}(\mathbf{E}) &= \int d\boldsymbol{u} u_\alpha u_\beta f(\boldsymbol{u}, s, +0) - \tfrac{1}{3}\delta_{\alpha\beta} \\ &= \left\langle \frac{(\mathbf{E} \cdot \boldsymbol{u})_\alpha (\mathbf{E} \cdot \boldsymbol{u})_\beta}{|\mathbf{E} \cdot \boldsymbol{u}|} \right\rangle_0 \Big/ \langle |\mathbf{E} \cdot \boldsymbol{u}| \rangle_0 - \tfrac{1}{3}\delta_{\alpha\beta}. \end{aligned} \tag{7.115}$$

Stress relaxation. Now it is easy to calculate the stress relaxation. According to the model described in Section 6.4, the relaxation of $L(t)$ occurs on the time-scale of τ_R, while that of orientation occurs on the time-scale τ_d. Thus the stress relaxation for $t > \tau_e$ occurs in two steps (see Fig. 7.12).

(i) Contour length relaxation: In the time-scale of τ_R, $S_{\alpha\beta}(s, t)$ can be regarded as equal to the initial value $Q_{\alpha\beta}(\mathbf{E})$, so that eqn (7.100) is

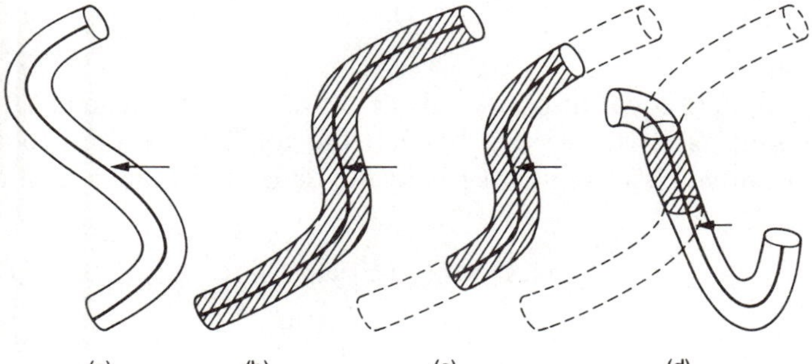

Fig. 7.12. Explanation of the stress relaxation after large step strain. (*a*) Before deformation the conformation of the primitive chain is in equilibrium ($t = -0$). (*b*) Immediately after deformation, the primitive chain is in the affinely deformed conformation ($t = +0$). (*c*) After time τ_R, the primitive chain contracts along the tube and recovers the equilibrium contour length ($t \simeq \tau_R$). (*d*) After the time τ_d, the primitive chain leaves the deformed tube by reptation ($t \simeq \tau_d$). The oblique lines indicates the deformed part of the tube. Reproduced from ref. 107.

written as

$$\sigma_{\alpha\beta}(t) = \frac{c}{N}\frac{3k_BT}{Nb^2}\langle L(t)\rangle^2 Q_{\alpha\beta}(\mathbf{E}) \quad \text{for} \quad \tau_e \lesssim t \lesssim \tau_R. \tag{7.116}$$

Here $\langle L(t)^2\rangle$ is replaced by $\langle L(t)\rangle^2$ since the distribution of $L(t)$ is quite narrow (the error $\langle \Delta L^2\rangle/\langle L^2\rangle$ being of the order of $a/L \simeq M_e/M$.). The relaxation of $\langle L(t)\rangle$ can be calculated by the model described in Section 6.4.2.[64] However, for simplicity, we shall use here

$$\langle L(t)\rangle = \bar{L}[1 + (\alpha(\mathbf{E}) - 1)\exp(-t/\tau_R)] \tag{7.117}$$

which simply states that $\langle L(t)\rangle$ decreases from $\alpha(\mathbf{E})\bar{L}$ to $\bar{L}$ with the relaxation time τ_R. From eqns (7.116) and (7.117), it follows that

$$\sigma_{\alpha\beta}(t) = G_e(1 + (\alpha(\mathbf{E}) - 1)\exp(-t/\tau_R))^2 Q_{\alpha\beta}(\mathbf{E}) \quad \text{for} \quad \tau_e \lesssim \tau \lesssim \tau_R \tag{7.118}$$

where

$$G_e = \frac{c}{N}\frac{3k_BT}{Nb^2}\bar{L}^2 = 3k_BT\frac{cb^2}{a^2}. \tag{7.119}$$

(ii) Disengagement: For $t > \tau_R$, $L(t)$ is at the equilibrium value $\bar{L}$ so that eqn (7.100) is written as

$$\sigma_{\alpha\beta}(t) = \frac{c}{N}\frac{3k_BT}{Nb^2}\bar{L}\int_0^{\bar{L}} ds\langle u_\alpha(s, t)u_\beta(s, t) - \tfrac{1}{3}\delta_{\alpha\beta}\rangle$$

$$= G_e\frac{1}{\bar{L}}\int_0^{\bar{L}} ds S_{\alpha\beta}(s, t). \tag{7.120}$$

Now $S_{\alpha\beta}(s, t)$ is equal to $Q_{\alpha\beta}(\mathbf{E})$ if the primitive chain segment s in the deformed tube, and is zero if it has left the tube. Since the probability that the primitive chain segment s is in the deformed tube is $\psi(s, t)$ (see Section 6.2.2)

$$S_{\alpha\beta}(s, t) = Q_{\alpha\beta}(\mathbf{E})\psi(s, t). \tag{7.121}$$

Hence

$$\sigma_{\alpha\beta}(t) = G_e Q_{\alpha\beta}(\mathbf{E})\frac{1}{\bar{L}}\int_0^{\bar{L}} ds\psi(s, t)$$

$$= G_e Q_{\alpha\beta}(\mathbf{E})\psi(t) \qquad t \gtrsim \tau_R. \tag{7.122}$$

Here $\psi(t)$ is given by eqn (7.40). Combining eqns (7.118) and (7.122) we finally have

$$\sigma_{\alpha\beta}(t) = G_e Q_{\alpha\beta}(\mathbf{E})(1 + (\alpha(\mathbf{E}) - 1)\exp(-t/\tau_R))^2\psi(t) \quad (t \gtrsim \tau_e). \tag{7.123}$$

In the case of a shear deformation, we may write $\alpha(\boldsymbol{E})$ and $Q_{\alpha\beta}(\boldsymbol{E})$ as $\alpha(\gamma)$ and $Q_{\alpha\beta}(\gamma)$, respectively. For small γ, $Q_{xy}(\gamma)$ and $\alpha(\gamma)$ are easily calculated.

$$\begin{aligned}\alpha(\gamma) &= \langle(1+2\gamma u_x u_y + \gamma^2 u_y^2)^{1/2}\rangle_0 \\ &= \langle 1+\gamma u_x u_y - \tfrac{1}{2}\gamma^2 u_x^2 u_y^2 + \tfrac{1}{2}\gamma^2 u_y^2\rangle_0 = 1+\tfrac{2}{15}\gamma^2 + O(\gamma^4) \qquad (7.124)\end{aligned}$$

and

$$\begin{aligned}Q_{xy}(\gamma) &= \frac{1}{\alpha(\gamma)}\left\langle\frac{(u_x+\gamma u_y)u_y}{(1+2\gamma u_x u_y+\gamma^2 u_y^2)^{1/2}}\right\rangle_0 \\ &= \gamma\langle u_y^2 - u_x^2 u_y^2\rangle_0 + O(\gamma^3) = \tfrac{4}{15}\gamma + O(\gamma^3). \qquad (7.125)\end{aligned}$$

Thus eqn (7.123) becomes

$$\sigma_{xy}(t) = \tfrac{4}{15}\gamma G_e \psi(t) + O(\gamma^3). \qquad (7.126)$$

Hence, the relaxation modulus in the linear viscoelasticity is given by

$$G(t) = \tfrac{4}{15} G_e \psi(t). \qquad (7.127)$$

This determines the plateau modulus to be

$$G_N^{(0)} = \tfrac{4}{15} G_e = \frac{4}{5}\frac{cb^2}{a^2} k_B T \qquad (7.128)$$

which gives eqn (7.54).

We shall now compare the results for large strain with experimental results.

7.5.4 *Comparison with experimental results*

Shear. Extensive experiments on the stress relaxation for shear deformation have been done by Osaki *et al.*[65–69] For convenience of comparison, we shall represent the relaxation of the shear stress by the nonlinear relaxation modulus defined by

$$G(t,\gamma) = \frac{1}{\gamma}\sigma_{xy}(t,\gamma). \qquad (7.129)$$

In the limit of $\gamma \to 0$, this reduces to the relaxation modulus of linear viscoelasticity.

Equation (7.123) gives

$$\begin{aligned}G(t,\gamma) &= G_e \frac{Q_{xy}(\gamma)}{\gamma}(1+(\alpha(\gamma)-1)\exp(-t/\tau_R))^2\psi(t) \\ &= h(\gamma)G(t)(1+(\alpha(\gamma)-1)\exp(-t/\tau_R))^2 \qquad (7.130)\end{aligned}$$

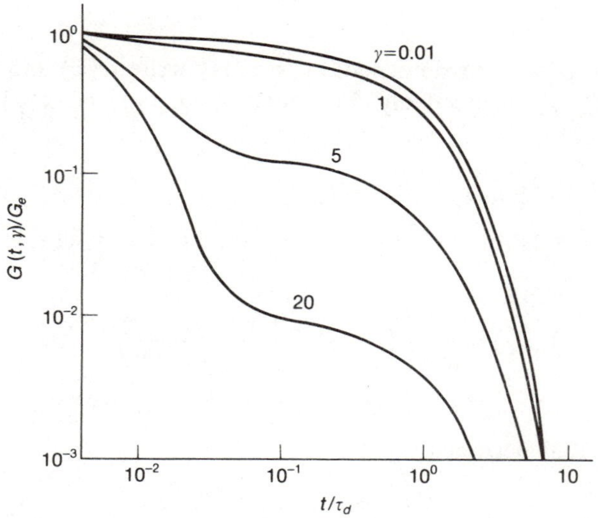

Fig. 7.13. Theoretical curve of the nonlinear relaxation modulus $G(t, \gamma)$. The case of $\tau_d/\tau_R = 100$ is shown. Reproduced from ref. 64.

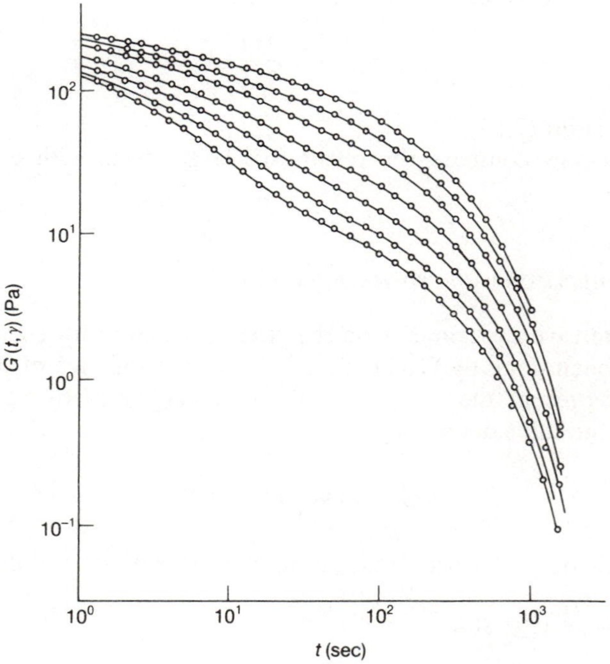

Fig. 7.14. Nonlinear relaxation modulus $G(t, \gamma)$ for polystyrene solution of chlorinated biphenyl at 30°C. The molecular weight of the polymer is 8.42×10^6 and the concentration is $0.06\ \mathrm{g/cm^3}$. Magnitudes of shear γ are <0.57, 1.25, 2.06, 3.04, 4.0, 5.3, and 6.1, from top to bottom. Reproduced from ref. 69.

where $G(t)$ is the relaxation modulus of linear viscoelasticity and

$$h(\gamma) = Q_{xy}(\gamma)/(\tfrac{4}{15}\gamma). \tag{7.131}$$

Equation (7.130) is plotted in Fig. 7.13. For small γ, $G(t, \gamma)$ decays roughly in a single exponential manner with relaxation time τ_d. For large γ, $G(t, \gamma)$ shows another relaxation characterized by τ_R corresponding to the relaxation of the contour length. Such behaviour has actually been observed experimentally[65,69] as shown in Fig. 7.14. Detailed comparison reveals good agreement between the theory and experiment.

(i) Osaki *et al*[65] found that at large t, curves in Fig. 7.14 for various γ can be superimposed by a vertical shift (see Fig. 7.15). This implies that, for large t, $G(t, \gamma)$ can be written as a product of two functions, one depending on time and the other on strain. This agrees with eqn (7.130), which is written for $t > \tau_R$ as

$$G(t, \gamma) = h(\gamma)G(t). \tag{7.132}$$

The function $h(\gamma)$, called the damping function, is found to be independent of the molecular weight and concentration over a wide range.[68–70] Figure 7.16 shows the comparison between the theoretical damping function and the experimental one. The agreement is very good considering that $h(\gamma)$ includes no adjustable parameters.

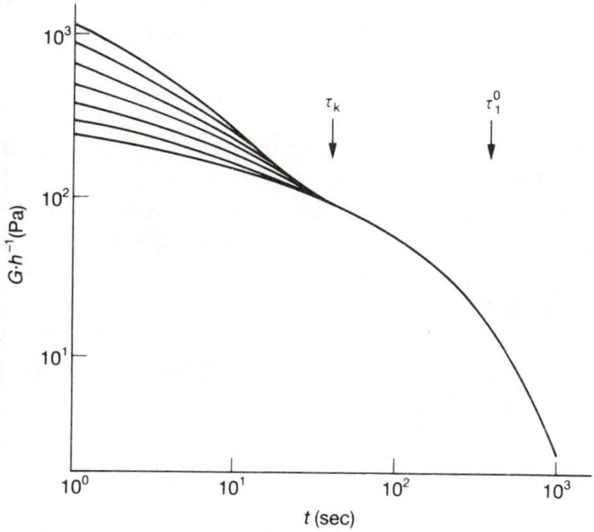

Fig. 7.15. Reduced relaxation modulus $G(t, \gamma)/h(\gamma)$ derived from Fig. 7.14. Each curve for $\gamma > 1.25$ in Fig. 7.14 is shifted vertically by an amount $-\log h(\gamma)$ so that it superposes on the top curve in the long time region. τ_1^0 indicates the longest relaxation time, and τ_k the characteristic time below which the superposition is not possible. Reproduced from ref. 69.

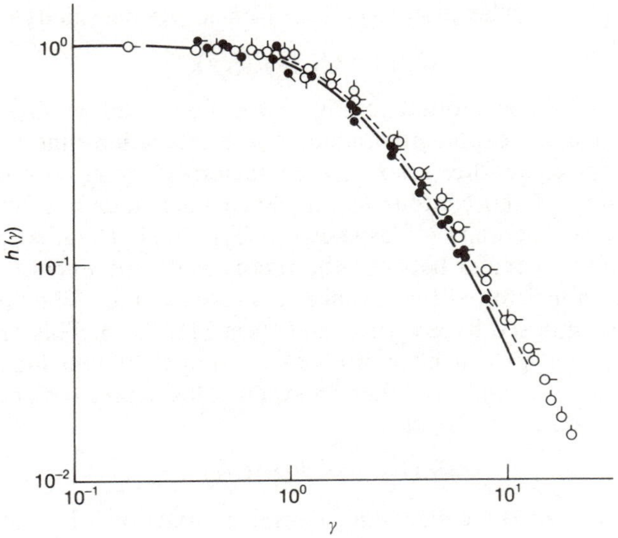

Fig. 7.16. $h(\gamma)$ determined from the procedure explained in Fig. 7.15. Filled circles represent polystyrene of molecular weight 8.42×10^6 and the unfilled circles of 4.48×10^6. Directions of pips indicate concentrations which range from $0.02\ \mathrm{g\ cm^3}$ to $0.08\ \mathrm{g\ cm^3}$. The solid curve represents the theoretical value (eqn (7.131)), and the dashed curve the result of the independent alignment approximation (eqn 7.187). Reproduced from ref. 69.

(ii) Experimentally, the first relaxation can be characterized by the time τ_k below which the factorization of $G(t, \gamma)$ is not possible (see Fig. 7.15). Osaki *et al.*[69] found that the ratio between τ_k and the Rouse relaxation time τ_R is about 4.5 and essentially independent of the molecular weight and concentration.

(iii) The relaxation of the other stress components $N_1(t, \gamma)$ and $N_2(t, \gamma)$, measured by birefringence, have precisely the same time dependence as $\sigma_{xy}(t, \gamma)$, and their ratio depends only on γ. This agrees with eqn (7.123), according to which

$$N_1(t, \gamma) = \frac{Q_{xx}(\gamma) - Q_{yy}(\gamma)}{Q_{xy}(\gamma)} \sigma_{xy}(t, \gamma) \tag{7.133}$$

$$N_2(t, \gamma) = \frac{Q_{yy}(\gamma) - Q_{zz}(\gamma)}{Q_{xy}(\gamma)} \sigma_{xy}(t, \gamma). \tag{7.134}$$

Since it can be proved[14] that

$$Q_{xx}(\gamma) - Q_{yy}(\gamma) = \gamma Q_{xy}(\gamma), \tag{7.135}$$

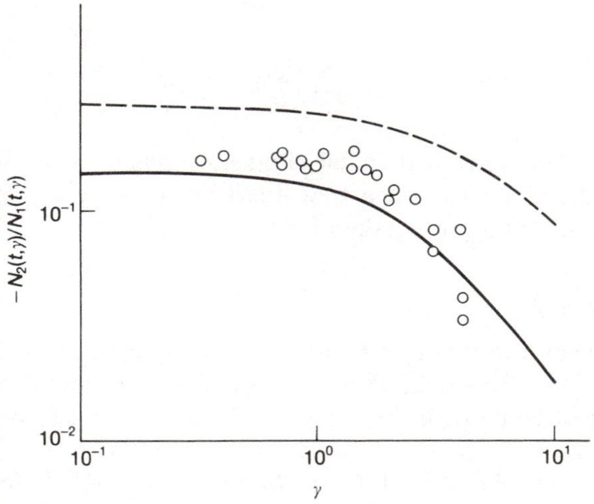

Fig. 7.17. Quantity $-N_2/N_1$ is plotted against magnitude of shear γ. Sample: polystyrene solution in chlorinated biphenyl ($M_w = 6.7 \times 10^5$, $\rho = 0.40\,\text{g cm}^{-3}$). The number of entanglements Z corresponds to about 14. The solid line is the theoretical value, $-(Q_{yy}(\gamma) - Q_{zz}(\gamma))/(Q_{xx}(\gamma) - Q_{yy}(\gamma))$. The dashed line is the result of the independent alignment approximation. Reproduced from ref. 67.

eqn (7.133) is written as

$$N_1(t, \gamma) = \gamma\sigma_{xy}(t, \gamma). \tag{7.136}$$

This relation, first found by Lodge and Meissner[71] using a phenomenological argument, has been well confirmed.[66,67] The ratio between the second normal stress difference $N_2(t, \gamma)$ and the first normal stress difference $N_1(t, \gamma)$ is shown in Fig. 7.17. The experimental values are again in reasonable agreement with the theory.

Uniaxial elongation. For uniaxial elongation, the tensile stress is given by

$$\sigma_T(t, \lambda) = [1 + (\alpha(\lambda) - 1)\exp(-t/\tau_R)]^2 f(\lambda)G(t). \tag{7.137}$$

Here $\alpha(\lambda)$ denotes $\alpha(\boldsymbol{E})$ of uniaxial elongation and

$$f(\lambda) = \tfrac{15}{4}(Q_{zz}(\lambda) - Q_{xx}(\lambda)). \tag{7.138}$$

Explicit formulae for $\alpha(\lambda)$ and $f(\lambda)$ can be calculated analytically,[72]

$$\alpha(\lambda) = \tfrac{1}{2}\lambda(1 + A(\lambda)) \tag{7.139}$$

and

$$f(\lambda) = \frac{15(\lambda^3 + 1/2)}{4(\lambda^3 - 1)} \frac{1}{1 + A(\lambda)}\left(1 - \frac{4\lambda^3 - 1}{2\lambda^3 + 1} A(\lambda)\right) \tag{7.140}$$

where

$$A(\lambda) = \frac{\sinh^{-1}[(\lambda^3 - 1)^{1/2}]}{[\lambda^3(\lambda^3 - 1)]^{1/2}}. \tag{7.141}$$

The stress relaxation for uniaxial elongation has been studied by Ferry *et al.*,[73] and the result has been well fitted by eqn (7.137) if the effect of the molecular distribution is taken into account.[2]

7.5.5 *Discussion*

As we have seen the theory has predicted many aspects of the nonlinear stress relaxation. However, there are some experimental results which are not in accordance with the theory and need some discussion.

Anomalous stress relaxation in shear flow. Osaki *et al.*[65,68] found that the nonlinear relaxation modulus $G(t, \gamma)$ of polystyrene solutions does not agree with the theory for very-high-molecular-weight samples for which

$$M\rho > 10^6 \text{ g/cm}^3. \tag{7.142}$$

For these samples, the stress near τ_k decreases much more steeply than predicted by the theory, and shows complex M, ρ dependence. A similar anomaly is also reported by Vrentas *et al.*[74] This result is puzzling since the theory should become most valid in the high molecular weight limit. A possible explanation, however, was given by Marrucci and Grizzuti,[75] who pointed out that the theoretical damping function $h(\gamma)$ has a region where the differential rigidity modulus $\partial(\gamma h(\gamma))/\partial\gamma$ is negative. In this region, the elastic energy can be lowered by microscopic phase separation, each phase having different local shear strains. Indeed the observed anomalous behaviour can be reproduced with suitable assumptions.[75,76] Experimental evidence of such microphase separation is, however, lacking and further study is expected.

Tube reorganization. The theory described in Section 7.5 includes two essential assumptions; (i) the conformation of the tube remains unchanged and (ii) the contour length of the tube returns to the equilibrium value $\bar{L}$ even if the environment is not in equilibrium. The validity of those assumptions is not established and it is worthwhile to study the consequences of the theory based on other assumptions.

Marrucci *et al.*[72,77] assumed that the volume of the tube remains constant by deformation, and derived a result which has the same time dependence as eqn (7.122) but different strain dependence. Though an experiment on PMMA[72] seems to fit with the modified formula, caution is needed in accepting the modification since critical experiments need data over a large time-scale and for a monodisperse sample, but the quoted experiment does not meet these conditions.

Viovy *et al.*[78] argued that as the contour length of the surrounding primitive chain contracts, there will be an extra relaxation by the release of the topological constraints. They proposed a theory which gives a slightly different relaxation behaviour for $t \lesssim \tau_R$. Though this proposal seems plausible, precise experimental evidence which supports the improved formula has not yet been given.

As in the case of linear viscoelasticity, the effect of the tube reorganization will play an important role in the molecular weight distribution. The problem of how it affects the nonlinear behaviour is interesting but unsolved.

Neutron scattering. In a series of experiments, Boué *et al.*[79] have studied by neutron scattering the conformational relaxation of the labelled chain after the stepwise deformation. In the short time-scale, the observed relaxation is well described by the Rouse dynamics. In the long time-scale (near τ_R), no clear indication has been found so far for the contraction of the contour length. Various reasons for this behaviour are conceivable such as the limited range of the scattering wave vector or polydispersity of the sample. On the other hand the results may indicate the importance of the tube deformation in the nonequilibrium state.[80]

Theoretical calculations of the scattering intensity based on the reptation dynamics are given in refs 81–83.

7.6 Nonlinear viscoelasticity

7.6.1 Phenomena of nonlinear viscoelasticity

We shall now consider the nonlinear viscoelasticity in the general situation. Experimentally a variety of interesting phenomena have been found in the nonlinear region. Leaving detailed description of them to the literature of rheology,[16,18,19] we shall limit ourselves here to a few typical aspects of nonlinear properties.

Shear flow. The general shear flow is characterized by the time-dependent shear rate $\kappa(t)$:

$$v_x = \kappa(t)y, \qquad v_y = 0, \qquad v_z = 0. \tag{7.143}$$

By symmetry, the stress components σ_{xz} and σ_{yz} again vanish identically, so that the relevant stress components are the shear stress σ_{xy} and the normal stresses $N_1 = \sigma_{xx} - \sigma_{yy}$, and $N_2 = \sigma_{yy} - \sigma_{zz}$, which are nonlinear functionals of $\kappa(t)$. Typical flows are:

(i) Steady shear flow. In the steady state, the stress depends only on κ. The steady state viscosity $\eta(\kappa)$ is defined by

$$\eta(\kappa) = \sigma_{xy}(\kappa)/\kappa. \tag{7.144}$$

The normal stresses N_1 and N_2 are usually expressed by

$$\Psi_1(\kappa) = N_1(\kappa)/\kappa^2 \quad \text{(first normal stress coefficient)} \tag{7.145}$$

and

$$\Psi_2(\kappa) = N_2(\kappa)/\kappa^2 \quad \text{(second normal stress coefficient)} \tag{7.146}$$

An example of $\eta(\kappa)$, $\Psi_1(\kappa)$ is shown in Fig. 7.18. Both quantities decreases significantly with increasing shear rate. For large shear rate, N_1 becomes much larger than σ_{xy}, and this causes many interesting phenomena known collectively as the Weissenberg effect.[18]

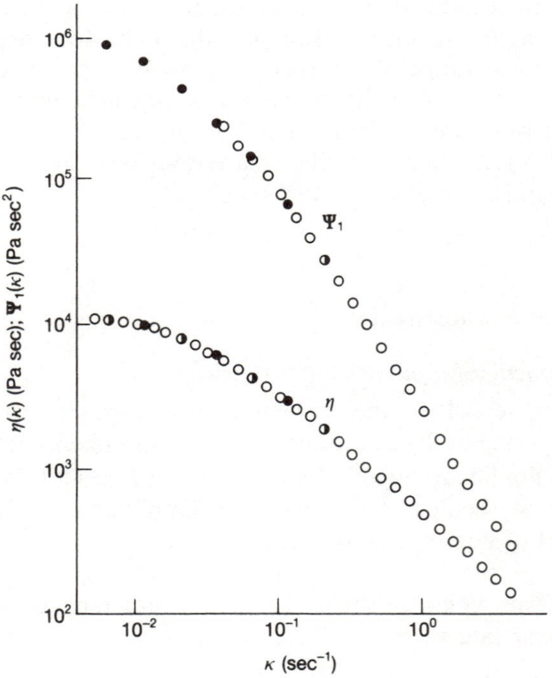

Fig. 7.18. Steady-state viscosity $\eta(\kappa)$ and the first normal stress coefficient $\Psi_1(\kappa)$. Sample: 8% solution of polystyrene ($M = 3.0 \times 10^6$) in chlorinated biphenyl. Open circles represent results from Weissenberg Rheogoniometry and closed circles results from the birefringence method. Reproduced from ref. 84.

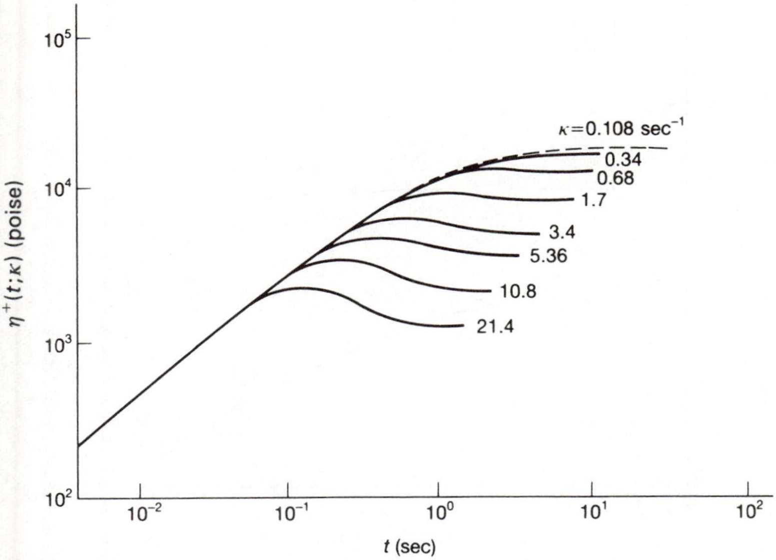

Fig. 7.19. Shear stress growth function at several shear rates for solution of polybutadiene ($M = 3.5 \times 10^5$). In this sample τ_d is about 1 sec. Reproduced from ref. 102.

(ii) Stress growth. Figure 7.19 shows the growth of the shear stress when a shear flow of constant shear rate is started (i.e., $\kappa(t) = \kappa\Theta(t)$). Here $\eta^+(t;\kappa)$ is defined by

$$\eta^+(t;\kappa) = \sigma_{xy}(t;\kappa)/\kappa. \tag{7.147}$$

It is seen that for large κ the shear stress reaches a maximum before reaching the steady state value. This phenomena is called the stress overshoot. Sometimes the overshoot is observed for the first normal stress difference at higher shear rate.

Elongational flow. The elongational flow is given by

$$v_x = -\tfrac{1}{2}\dot{\varepsilon}(t)x, \qquad v_y = -\tfrac{1}{2}\dot{\varepsilon}(t)y, \qquad v_z = \dot{\varepsilon}(t)z. \tag{7.148}$$

When the flow of constant elongational rate $\dot{\varepsilon}$ is started, the tensile stress $\sigma_{zz} - \sigma_{xx}$ increases with time and reaches the steady state. The steady state elongational viscosity is defined by

$$\eta_E(\dot{\varepsilon}) = \frac{\sigma_{zz} - \sigma_{xx}}{\dot{\varepsilon}}. \tag{7.149}$$

For small $\dot{\varepsilon}$, it follows from the linear constitutive equation (7.11) that $\eta_E(\dot{\varepsilon})$ approaches $3\eta(0)$. An example of $\eta_E(\dot{\varepsilon})$ is shown in Fig. 7.20.

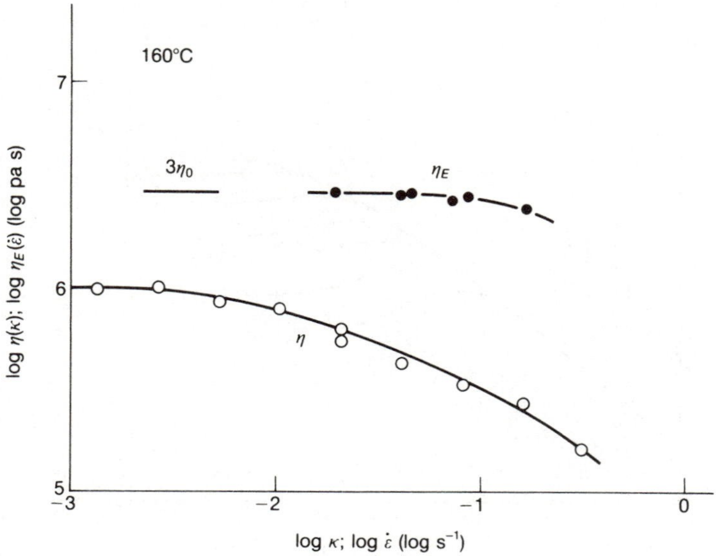

Fig. 7.20. Shear viscosity $\eta(\kappa)$ and elongational viscosity $\eta_E(\dot{\varepsilon})$ in a steady flow of monodisperse polystyrene ($M_w = 2.54 \times 10^5$). Reproduced from ref. 85.

7.6.2 Deformation gradient tensor

For many purposes it is convenient to describe the history of the velocity gradient by another quantity. Consider the motion of a point $r_\alpha(t)$ fixed on the material. In the homogeneous flow in which the velocity field is given by

$$v_\alpha(r, t) = \kappa_{\alpha\beta}(t) r_\beta \tag{7.150}$$

the point moves as

$$\frac{\mathrm{d}}{\mathrm{d}t} r_\alpha(t) = \kappa_{\alpha\beta}(t) r_\beta(t). \tag{7.151}$$

Since this is a linear equation, its solution is written as

$$r_\alpha(t) = E_{\alpha\beta}(t, t') r_\beta(t'). \tag{7.152}$$

This is analogous to eqn (7.75). The tensor $\boldsymbol{E}(t, t')$ denotes the deformation gradient at time t referred to the state at time t'. From eqn (7.152) it follows that

$$E_{\alpha\beta}(t, t') = E_{\alpha\gamma}(t, t'') E_{\gamma\beta}(t'', t'). \tag{7.153}$$

Note that eqn (7.153) holds for arbitrary t'' from $-\infty$ to $+\infty$.

From eqns (7.151) and (7.152), $E_{\alpha\beta}(t, t')$ satisfies

$$\frac{\partial}{\partial t} E_{\alpha\beta}(t, t') = \kappa_{\alpha\gamma}(t) E_{\gamma\beta}(t, t'). \tag{7.154}$$

This equation, together with the initial condition

$$E_{\alpha\beta}(t, t) = \delta_{\alpha\beta} \tag{7.155}$$

determines $\mathbf{E}(t, t')$ for given $\boldsymbol{\kappa}(t)$. Conversely, $\boldsymbol{\kappa}(t)$ is obtained from $\mathbf{E}(t, t')$ by eqn (7.154).

For the shear flow, $\mathbf{E}(t, t')$ is given as

$$\mathbf{E}(t, t') = \begin{bmatrix} 1 & \gamma(t, t') & 0 \\ 0 & 1 & 0 \\ 0 & 0 & 1 \end{bmatrix} \tag{7.156}$$

where

$$\gamma(t, t') = \int_{t'}^{t} dt'' \kappa(t''). \tag{7.157}$$

For the elongational flow

$$\mathbf{E}(t, t') = \begin{bmatrix} \lambda(t, t')^{-1/2} & 0 & 0 \\ 0 & \lambda(t, t')^{-1/2} & 0 \\ 0 & 0 & \lambda(t, t') \end{bmatrix} \tag{7.158}$$

where

$$\lambda(t, t') = \exp\left(\int_{t'}^{t} dt'' \dot{\varepsilon}(t'')\right). \tag{7.159}$$

7.6.3 Constitutive equation derived from Rouse model

For the Rouse model, the constitutive equation is obtained in a simple closed form. To calculate the stress given by eqn (7.81), we solve the time evolution equation for $\langle X_{p\alpha}(t) X_{p\beta}(t) \rangle$ (see eqn (4.141)):

$$\frac{\partial}{\partial t} \langle X_{p\alpha}(t) X_{p\beta}(t) \rangle = \frac{1}{\zeta_p} [2k_B T \delta_{\alpha\beta} - 2k_p \langle X_{p\alpha}(t) X_{p\beta}(t) \rangle] + \kappa_{\alpha\mu}(t) \langle X_{p\mu}(t) X_{p\beta}(t) \rangle + \kappa_{\beta\mu}(t) \langle X_{p\mu}(t) X_{p\alpha}(t) \rangle. \tag{7.160}$$

Since this is a linear differential equation of the first order, it can be solved by the standard method of the variation of constants (see ref. 19

Chapter 10, for example). The result is

$$\langle X_{p\alpha}(t)X_{p\beta}(t)\rangle = \int_{-\infty}^{t} \mathrm{d}t' \frac{2k_BT}{\zeta_p} E_{\alpha\mu}(t, t')E_{\beta\mu}(t, t')\exp[-2p^2(t-t')/\tau_R]$$
$$= \frac{1}{k_p}\int_{-\infty}^{t} \mathrm{d}t' B_{\alpha\beta}(\mathbf{E}(t, t'))\frac{\partial}{\partial t'}\exp[-2p^2(t-t')/\tau_R]. \quad (7.161)$$

Substituting this in eqn (7.81), we have

$$\sigma_{\alpha\beta}(t) = \int_{-\infty}^{t} \mathrm{d}t' \frac{\partial G(t-t')}{\partial t'} B_{\alpha\beta}(\mathbf{E}(t, t')) \quad (7.162)$$

where $G(t)$ is the linear relaxation modulus of the Rouse model (eqn (7.31)). It is easy to check that eqn (7.162) gives eqn (7.91) for the case of step deformation.

In the case of steady shear flow, $B_{\alpha\beta}(\mathbf{E}(t, t'))$ is given by

$$B_{xy}(\gamma(t, t')) = (t-t')\kappa, \quad (7.163)$$
$$B_{xx}(\gamma(t, t')) - B_{yy}(\gamma(t, t')) = (t-t')^2\kappa^2, \quad (7.164)$$
$$B_{yy}(\gamma(t, t')) - B_{zz}(\gamma(t, t')) = 0. \quad (7.165)$$

Hence

$$\sigma_{xy}(\kappa) = \kappa\int_{-\infty}^{t} \mathrm{d}t'(t-t')\frac{\partial}{\partial t'}G(t-t') = \kappa\int_{0}^{\infty} \mathrm{d}t' G(t') = \eta_0\kappa, \quad (7.166)$$

$$N_1(\kappa) = \int_{-\infty}^{t} \mathrm{d}t' \kappa^2(t-t')^2\frac{\partial}{\partial t'}G(t-t') = 2J_e^{(0)}\eta_0^2\kappa^2, \quad (7.167)$$

and

$$N_2(\kappa) = 0, \quad (7.168)$$

where η_0 and $J_e^{(0)}$ are given by eqns (7.33) and (7.34). Thus the shear viscosity $\eta(\kappa)$ and the first normal stress coefficient $\Psi_1(\kappa)$ are independent of the shear rate.

7.7 Approximate constitutive equation for reptation model

We shall now derive the constitutive equation for reptation dynamics. To simplify the analysis, we assume that the contour length of the primitive chain remains at the equilibrium value $\bar{L}$ under macroscopic deformation (inextensible primitive chain). This assumption is valid if the characteristic magnitude of the velocity gradient is much less than $1/\tau_R$, i.e,

$$\kappa\tau_R < 1. \quad (7.169)$$

Since nonlinear behaviour starts to be observed when the characteristic magnitude of the velocity gradient becomes of the order of $1/\tau_d$, the assumption is not restrictive for a polymer of $\tau_R/\tau_d \ll 1$, i.e., $M \gg M_e$. In practice, the condition (7.169) is not always satisfied, and the elongation of the contour length can be important, but this will not be considered here.

For the sake of simplicity, we shall use a slightly different notation in this and the following two sections: the equilibrium contour length will be denoted by L (because L and $\bar{L}$ need not be distinguished for the inextensible model), and the segments of the primitive chain are labelled from $-L/2$ to $L/2$. (Thus the segment 0 corresponds to the middle of the chain.)

7.7.1 Deformation of the primitive chain

First we express the transformation rule of the inextensible primitive chain in mathematical terms. Let $\boldsymbol{R}(s)$ and $\tilde{\boldsymbol{R}}(s)$ be the conformations of the primitive chain before and after the deformation. The transformation rule is explained in Fig. 7.21, i.e.,

(a) The segment in the middle changes its position affinely, i.e.,

$$\tilde{\boldsymbol{R}}(0) = \boldsymbol{E} \cdot \boldsymbol{R}(0). \tag{7.170}$$

(b) The segment $\tilde{s}$ lies on the curve $\boldsymbol{E} \cdot \boldsymbol{R}(s)$, so that

$$\tilde{\boldsymbol{R}}(\tilde{s}) = \boldsymbol{E} \cdot \boldsymbol{R}(s) \tag{7.171}$$

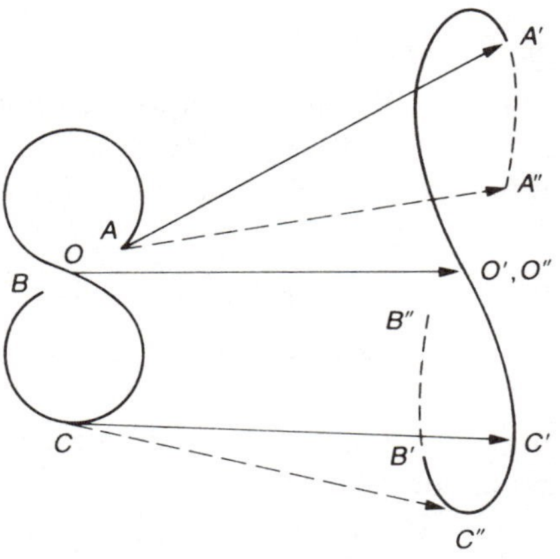

Fig. 7.21. The deformation of an inextensible primitive chain by a macroscopic strain. The new conformation $A'O'B'$ is on the curve $A''O''B''$, which is the affine transformation of AOB. The new position of a segment, say C, is determined from the condition that the contour length $O'C'$ is equal to OC, where O' is the affine transformation of O and coincides with O''. Reproduced from ref. 108.

where $\tilde{s}$ is the contour length along the curve $\boldsymbol{E}\cdot\boldsymbol{R}(s')$ from $s'=0$ to $s'=s$, i.e.,

$$\tilde{s}=\int_0^s \mathrm{d}s'\,|\boldsymbol{E}\cdot\boldsymbol{u}(s')|. \tag{7.172}$$

From eqn (7.171), the transformation rule for the direction $\boldsymbol{u}(s)$ is obtained as

$$\tilde{\boldsymbol{u}}(\tilde{s})=\frac{\partial}{\partial\tilde{s}}\tilde{\boldsymbol{R}}(\tilde{s})=\frac{\partial}{\partial\tilde{s}}\boldsymbol{E}\cdot\boldsymbol{R}(s)=\frac{\partial s}{\partial\tilde{s}}\frac{\partial}{\partial s}\boldsymbol{E}\cdot\boldsymbol{R}(s) \tag{7.173}$$

$$=\frac{\partial s}{\partial\tilde{s}}\boldsymbol{E}\cdot\boldsymbol{u}(s). \tag{7.174}$$

Equation (7.172) gives

$$1=\frac{\partial s}{\partial\tilde{s}}|\boldsymbol{E}\cdot\boldsymbol{u}(s)|. \tag{7.175}$$

From eqns (7.174) and (7.175) it follows that

$$\tilde{\boldsymbol{u}}(\tilde{s})=\frac{\boldsymbol{E}\cdot\boldsymbol{u}(s)}{|\boldsymbol{E}\cdot\boldsymbol{u}(s)|}. \tag{7.176}$$

This is equivalent to the transformation given in Fig. 7.11.

7.7.2 *Independent alignment approximation*

According to eqn (7.172) s and $\tilde{s}$ are not equal to each other. This leads to a constitutive equation of a rather complicated form (see Section 7.9). If we disregard the difference between s and $\tilde{s}$ and assume the following transformation rule

$$\tilde{\boldsymbol{u}}(s)=\frac{\boldsymbol{E}\cdot\boldsymbol{u}(s)}{|\boldsymbol{E}\cdot\boldsymbol{u}(s)|} \tag{7.177}$$

the constitutive equation is obtained in a simple form. This prompts the study of the approximation (7.177), which we call the independent alignment approximation (IA approximation).[14,15] Though the physical justification of this approximation is not clear, its error usually turns out to be small except for a few cases which will be discussed later. Therefore we shall first proceed using this approximation.

According to eqn (7.100) the stress for the inextensible model is given by

$$\sigma_{\alpha\beta}(t) = \frac{c}{N}\frac{3k_BTL^2}{Nb^2}\frac{1}{L}\int_{-L/2}^{L/2} \mathrm{d}s\langle u_\alpha(s,t)u_\beta(s,t) - \tfrac{1}{3}\delta_{\alpha\beta}\rangle$$
$$= G_e\frac{1}{L}\int_{-L/2}^{L/2} \mathrm{d}s\langle u_\alpha(s,t)u_\beta(s,t) - \tfrac{1}{3}\delta_{\alpha\beta}\rangle$$
$$= G_e\frac{1}{L}\int_{-L/2}^{L/2} \mathrm{d}sS_{\alpha\beta}(s,t). \tag{7.178}$$

To calculate $S_{\alpha\beta}(s, t)$ we need to obtain the probability distribution function $f(\boldsymbol{u}, s, t)$ that the tangent vector at the segment s is in the direction $\boldsymbol{u}$ at time t. The time evolution equation for $f(\boldsymbol{u}, s, t)$ is obtained in the same way as in Section 6.3.

Suppose that in the time interval between t and $t+\Delta t$, the chain segment s moves to the position at which the chain segment $s+\Delta\xi$ was located at time t, then $\boldsymbol{u}(s, t+\Delta t)$ is given by

$$\boldsymbol{u}(s, t+\Delta t) = \frac{\boldsymbol{E}(t+\Delta t, t)\cdot \boldsymbol{u}(s+\Delta\xi, t)}{|\boldsymbol{E}(t+\Delta t, t)\cdot \boldsymbol{u}(s+\Delta\xi, t)|}. \tag{7.179}$$

Since the distribution function of $\boldsymbol{u}(s+\Delta\xi, t)$ and $\Delta\xi$ are given by $f(\boldsymbol{u}, s+\Delta\xi, t)$ and $\Psi(\Delta\xi)$, respectively (see eqn (6.24)), the distribution function of $\boldsymbol{u}(s, t+\Delta t)$ is given by

$$f(\boldsymbol{u}, s, t+\Delta t) = \int \mathrm{d}\Delta\xi\Psi(\Delta\xi)\int \mathrm{d}\boldsymbol{u}'\delta\left(\boldsymbol{u} - \frac{\boldsymbol{E}(t+\Delta t, t)\cdot\boldsymbol{u}'}{|\boldsymbol{E}(t+\Delta t, t)\cdot\boldsymbol{u}'|}\right)$$
$$\times f(\boldsymbol{u}', s+\Delta\xi, t). \tag{7.180}$$

To assess the accuracy of the IA approximation, let us consider the stepwise deformation $\boldsymbol{E}$ imposed at $t=0$. In this case the orientational distribution before the deformation is

$$f(\boldsymbol{u}, s, -0) = \frac{1}{4\pi}. \tag{7.181}$$

Since $\Psi(\Delta\xi)$ becomes $\delta(\Delta\xi)$ for an infinitesimally small time-interval Δt, eqn (7.180) gives

$$f(\boldsymbol{u}, s, +0) = \int \frac{\mathrm{d}\boldsymbol{u}'}{4\pi}\delta\left(\boldsymbol{u} - \frac{\boldsymbol{E}\cdot\boldsymbol{u}'}{|\boldsymbol{E}\cdot\boldsymbol{u}'|}\right). \tag{7.182}$$

Hence $S_{\alpha\beta}(s, +0)$ is given by

$$\begin{aligned}S_{\alpha\beta}(s, +0) &= \int d\boldsymbol{u}(u_\alpha u_\beta - \tfrac{1}{3}\delta_{\alpha\beta})f(\boldsymbol{u}, s, +0)\\ &= \int d\boldsymbol{u} \int \frac{d\boldsymbol{u}'}{4\pi}\left(u_\alpha u_\beta - \tfrac{1}{3}\delta_{\alpha\beta}\right)\delta\left(\boldsymbol{u} - \frac{\boldsymbol{E}\cdot\boldsymbol{u}'}{|\boldsymbol{E}\cdot\boldsymbol{u}'|}\right)\\ &= \int \frac{d\boldsymbol{u}}{4\pi}\frac{(\boldsymbol{E}\cdot\boldsymbol{u})_\alpha(\boldsymbol{E}\cdot\boldsymbol{u})_\beta}{|\boldsymbol{E}\cdot\boldsymbol{u}|^2} - \tfrac{1}{3}\delta_{\alpha\beta}\\ &= \left\langle \frac{(\boldsymbol{E}\cdot\boldsymbol{u})_\alpha(\boldsymbol{E}\cdot\boldsymbol{u})_\beta}{|\boldsymbol{E}\cdot\boldsymbol{u}|^2} - \tfrac{1}{3}\delta_{\alpha\beta}\right\rangle_0\\ &\equiv Q^{(\mathrm{IA})}_{\alpha\beta}(\boldsymbol{E}).\end{aligned} \tag{7.183}$$

On the other hand the correct value of $S_{\alpha\beta}(s, +0)$ is $Q_{\alpha\beta}(\boldsymbol{E})$ given by eqn (7.115). Thus in the case of stepwise deformation, the IA approximation amounts to a decoupling approximation

$$\left\langle \frac{(\boldsymbol{E}\cdot\boldsymbol{u})_\alpha(\boldsymbol{E}\cdot\boldsymbol{u})_\beta}{|\boldsymbol{E}\cdot\boldsymbol{u}|}\right\rangle_0 \frac{1}{\langle|\boldsymbol{E}\cdot\boldsymbol{u}|\rangle_0} \simeq \left\langle \frac{(\boldsymbol{E}\cdot\boldsymbol{u})_\alpha(\boldsymbol{E}\cdot\boldsymbol{u})_\beta}{|\boldsymbol{E}\cdot\boldsymbol{u}|^2}\right\rangle_0. \tag{7.184}$$

It can be seen that the error of this approximation will not be large for any form of $\boldsymbol{E}$. Indeed for the case of shear deformation, the IA approximation gives

$$Q^{(\mathrm{IA})}_{xy}(\gamma) = \tfrac{1}{5}\gamma \qquad (\gamma \ll 1), \tag{7.185}$$

$$Q^{(\mathrm{IA})}_{yy}(\gamma) - Q^{(\mathrm{IA})}_{zz}(\gamma) = -\tfrac{2}{35}\gamma^2(\gamma \ll 1). \tag{7.186}$$

The damping function is thus given by

$$h^{(\mathrm{IA})}(\gamma) = Q^{(\mathrm{IA})}_{xy}(\gamma)\Big/\left(\frac{\gamma}{5}\right). \tag{7.187}$$

This is shown by the dashed line in Fig. 7.16. It is seen that the error of the IA approximation is small over a wide range of γ. A useful formula for $Q^{(\mathrm{IA})}_{\alpha\beta}(\boldsymbol{E})$ for general $\boldsymbol{E}$ is given in ref. 87.

7.7.3 *Constitutive equation*

To obtain $f(\boldsymbol{u}, s, t)$ in the general case, we rewrite eqn (7.180) in a differential form. For small Δt, $\boldsymbol{E}(t+\Delta t, t)$ is written as $\boldsymbol{I} + \boldsymbol{\kappa}(t)\Delta t$. Hence

$$\begin{aligned}\frac{\boldsymbol{E}(t+\Delta t, t)\cdot\boldsymbol{u}}{|\boldsymbol{E}(t+\Delta t, t)\cdot\boldsymbol{u}|} &= \frac{\boldsymbol{u} + \boldsymbol{\kappa}(t)\cdot\boldsymbol{u}\Delta t}{|\boldsymbol{u} + \boldsymbol{\kappa}(t)\cdot\boldsymbol{u}\Delta t|}\\ &= \boldsymbol{u} + (\boldsymbol{\kappa}(t)\cdot\boldsymbol{u} - (\boldsymbol{uu}:\boldsymbol{\kappa}(t))\boldsymbol{u})\Delta t = \boldsymbol{u} + \Delta t\boldsymbol{\Gamma}(\boldsymbol{u}, t)\end{aligned} \tag{7.188}$$

where

$$\boldsymbol{\Gamma}(\boldsymbol{u}, t) \equiv \boldsymbol{\kappa}(t)\cdot\boldsymbol{u} - (\boldsymbol{uu}:\boldsymbol{\kappa}(t))\boldsymbol{u}. \tag{7.189}$$

Thus if we neglect the terms of order $(\Delta t)^2$, eqn (7.180) is rewritten as

$$
\begin{aligned}
f(\boldsymbol{u}, s, t+\Delta t) &= \int \mathrm{d}\Delta\xi \Psi(\Delta\xi) \int \mathrm{d}\boldsymbol{u}' \delta(\boldsymbol{u}-\boldsymbol{u}'-\Delta t \boldsymbol{\Gamma}(\boldsymbol{u}', t)) f(\boldsymbol{u}', s+\Delta\xi, t) \\
&= \int \mathrm{d}\Delta\xi \Psi(\Delta\xi)\left(1-\Delta t \frac{\partial}{\partial \boldsymbol{u}} \cdot \boldsymbol{\Gamma}(\boldsymbol{u}, t)\right) f(\boldsymbol{u}, s+\Delta\xi, t) \\
&= \int \mathrm{d}\Delta\xi \Psi(\Delta\xi)\left(1+\Delta\xi \frac{\partial}{\partial s}+\tfrac{1}{2}(\Delta\xi)^2 \frac{\partial^2}{\partial s^2}\right)\left(1-\Delta t \frac{\partial}{\partial \boldsymbol{u}} \cdot \boldsymbol{\Gamma}(\boldsymbol{u}, t)\right) f(\boldsymbol{u}, s, t) \\
&= \left(1+\Delta t D_c \frac{\partial^2}{\partial s^2}\right)\left(1-\Delta t \frac{\partial}{\partial \boldsymbol{u}} \cdot \boldsymbol{\Gamma}(\boldsymbol{u}, t)\right) f(\boldsymbol{u}, s, t).
\end{aligned} \tag{7.190}
$$

Comparing the terms of order Δt, we have

$$
\frac{\partial}{\partial t} f(\boldsymbol{u}, s, t) = D_c \frac{\partial^2}{\partial s^2} f(\boldsymbol{u}, s, t) - \frac{\partial}{\partial \boldsymbol{u}} \cdot \boldsymbol{\Gamma}(\boldsymbol{u}, t) f(\boldsymbol{u}, s, t). \tag{7.191}
$$

The boundary condition is that the tangent vector at the chain end is isotropic,

$$
f(\boldsymbol{u}, s, t) = \frac{1}{4\pi} \quad \text{at} \quad s = \pm L/2. \tag{7.192}
$$

Equation (7.191) can be rigorously solved to give[15]

$$
f(\boldsymbol{u}, s, t) = \int_{-\infty}^{t} \mathrm{d}t' \left(\frac{\partial}{\partial t'} \psi(s, t-t')\right) \int \frac{\mathrm{d}\boldsymbol{u}'}{4\pi} \delta\left(\boldsymbol{u} - \frac{\boldsymbol{E}(t, t') \cdot \boldsymbol{u}'}{|\boldsymbol{E}(t, t') \cdot \boldsymbol{u}'|}\right) \tag{7.193}
$$

where $\psi(s, t)$ is given by eqn (6.14). This solution could also have been arrived at by physical argument. Suppose that a tube segment is created in the direction $\boldsymbol{u}'$ at either of the chain ends between time t' and $t' + \mathrm{d}t'$. If this still survives at time t and is now occupied by the primitive chain segment s, $\boldsymbol{u}(s, t)$ must be $\boldsymbol{E}(t, t') \cdot \boldsymbol{u}' / |\boldsymbol{E}(t, t') \cdot \boldsymbol{u}'|$ (this is the result of the IA approximation). Since the probability that this happens is $[\partial\psi(s, t-t')/\partial t']\,\mathrm{d}t'$ and the distribution of $\boldsymbol{u}'$ is $1/4\pi$, we get eqn (7.193).

From eqn (7.193), the orientation of the primitive chain segment is given by

$$
\begin{aligned}
S_{\alpha\beta}(s, t) &= \int_{-\infty}^{t} \mathrm{d}t' \left(\frac{\partial}{\partial t'} \psi(s, t-t')\right) \int \mathrm{d}\boldsymbol{u} \int \frac{\mathrm{d}\boldsymbol{u}'}{4\pi} \\
&\quad \times (u_\alpha u_\beta - \tfrac{1}{3}\delta_{\alpha\beta}) \delta\left(\boldsymbol{u} - \frac{\boldsymbol{E}(t, t') \cdot \boldsymbol{u}'}{|\boldsymbol{E}(t, t') \cdot \boldsymbol{u}'|}\right) \\
&= \int_{-\infty}^{t} \mathrm{d}t' \left(\frac{\partial}{\partial t'} \psi(s, t-t')\right) Q_{\alpha\beta}^{(\mathrm{IA})}(\boldsymbol{E}(t, t')).
\end{aligned} \tag{7.194}
$$

Substituting this into eqn (7.178), we finally have

$$\sigma_{\alpha\beta}(t) = G_e \int_{-\infty}^{t} \mathrm{d}t' \left(\frac{\partial}{\partial t'} \psi(t-t') \right) Q_{\alpha\beta}^{(\mathrm{IA})}(\boldsymbol{E}(t, t')) \tag{7.195}$$

where $\psi(t)$ is given by eqn (7.40). Equation (7.195) is the constitutive equation which comes out of the IA approximation. We shall now compare this equation with experimental results.

7.7.4 *Comparison with experiments*

General feature. Equation (7.195) can be written in a more convenient form. Consider that a stepwise strain $\boldsymbol{E}$ is applied at time 0, then $\boldsymbol{E}(t, t')$ is given by

$$\boldsymbol{E}(t, t') = \begin{cases} \boldsymbol{E} & \text{if } t > 0 \text{ and } t' < 0, \\ \boldsymbol{I} & \text{otherwise.} \end{cases} \tag{7.196}$$

Let $\phi_{\alpha\beta}(t, \boldsymbol{E})$ be the stress caused by this deformation at a positive time t. Equations (7.195) and (7.196) give

$$\begin{aligned} \phi_{\alpha\beta}(t, \boldsymbol{E}) &= G_e \int_{-\infty}^{0} \mathrm{d}t' \left(\frac{\partial}{\partial t'} \psi(t-t') \right) Q_{\alpha\beta}^{(\mathrm{IA})}(\boldsymbol{E}) \\ &= G_e \psi(t) Q_{\alpha\beta}^{(\mathrm{IA})}(\boldsymbol{E}). \end{aligned} \tag{7.197}$$

This is very similar to eqn (7.122) except that $Q_{\alpha\beta}(\boldsymbol{E})$ is now replaced by $Q_{\alpha\beta}^{(\mathrm{IA})}(\boldsymbol{E})$. In fact eqn (7.197) can be derived much more simply by the reasoning given in Section 7.5.3, if it is noted that $S_{\alpha\beta}(s, +0)$ is given by $Q_{\alpha\beta}^{(\mathrm{IA})}(\boldsymbol{E})$ in the independent alignment approximation.

Using $\phi_{\alpha\beta}(t, \boldsymbol{E})$, the stress for an arbitrary flow history is given by

$$\sigma_{\alpha\beta}(t) = \int_{-\infty}^{t} \mathrm{d}t' \left(\frac{\partial}{\partial t'} \phi_{\alpha\beta}(t-t', \boldsymbol{E}) \right)_{\boldsymbol{E}=\boldsymbol{E}(t,t')}. \tag{7.198}$$

Equation (7.198) agrees with the empirical equation proposed by Bernstein, Kearseley, and Zapas (BKZ),[88] who found that the stress response for various flow histories can be predicted by eqn (7.198) using the stress relaxation function $\phi_{\alpha\beta}(t, \boldsymbol{E})$ determined experimentally. Subsequent experiments done by many authors[89–93] revealed that the BKZ equation is one of the most successful empirical constitutive equations.

As was shown in Section 7.5, the empirical stress relaxation function $\phi_{\alpha\beta}(t, \boldsymbol{E})$ is in good agreement with the reptation theory for linear polymers of narrow molecular weight distribution. This, together with the success of the BKZ equation, indicates that the constitutive equation derived by the reptation theory works well for general flow histories. Indeed eqn (7.198) reproduces many characteristic features of the nonlinear viscoelasticity. Leaving detailed comparison to refs 51 and 94–97, we shall study the main features briefly.

Nonlinear viscoelastic behaviour. To see the characteristic features of the constitutive equation (7.195), we approximate $\psi(t)$ by

$$\psi(t) \simeq \begin{cases} 1 & \text{for} \quad t < \tau_d, \\ 0 & \text{for} \quad t > \tau_d. \end{cases} \tag{7.199}$$

Then eqn (7.195) gives

$$\sigma_{\alpha\beta} \simeq G_e Q^{(\mathrm{IA})}_{\alpha\beta}(\boldsymbol{E}(t, t-\tau_d)) \tag{7.200}$$

which simply says that the stress is given by the elastic deformation caused between $t - \tau_d$ and t.

(i) Steady shear flow. In the steady shear flow, eqn (7.200) gives the shear stress

$$\sigma_{xy}(\kappa) \simeq G_e Q^{(\mathrm{IA})}_{xy}(\kappa\tau_d) \simeq G_e \kappa\tau_d h^{(\mathrm{IA})}(\kappa\tau_d). \tag{7.201}$$

Thus the viscosity becomes

$$\eta(\kappa) = \eta(0) h^{(\mathrm{IA})}(\kappa\tau_d). \tag{7.202}$$

Equation (7.202) indicates the shear thinning occurs at $\kappa \simeq 1/\tau_d$, which becomes very small for large molecules. This explains why the nonlinear response is important in polymeric liquids.

The first normal stress coefficient is also estimated as

$$\Psi_1(\kappa) \simeq G_e \tau_d^2 h^{(\mathrm{IA})}(\kappa\tau_d) \tag{7.203}$$

which again decreases with the shear rate as shown in Fig. 7.18. The second normal stress coefficient

$$\Psi_2(\kappa) \simeq -G_e \tau_d^2 h_2^{(\mathrm{IA})}(\kappa\tau_d),$$

where

$$h_2^{(\mathrm{IA})}(\gamma) = -\frac{35}{2\gamma^2}\left(Q^{(\mathrm{IA})}_{yy}(\gamma) - Q^{(\mathrm{IA})}_{zz}(\gamma)\right), \tag{7.204}$$

is negative and again decreases with κ, in agreement with experiments. The ratio $\Psi_2(0)/\Psi_1(0)$ is precisely evaluated as[94]†

$$\frac{\Psi_2(0)}{\Psi_1(0)} = \left.\frac{Q^{(\mathrm{IA})}_{yy}(\gamma) - Q^{(\mathrm{IA})}_{zz}(\gamma)}{Q^{(\mathrm{IA})}_{xx}(\gamma) - Q^{(\mathrm{IA})}_{yy}(\gamma)}\right|_{\gamma=0} = -\tfrac{2}{7} \simeq -0.3. \tag{7.205}$$

The experimental value is between -0.1 and -0.3.[98,99]

† If the IA approximation is not used, the precise value of $\Psi_2(0)/\Psi_1(0)$ becomes $-1/7 = -0.14$.

(ii) Stress growth in shear flow. When a shear flow is started with constant shear rate κ at time $t=0$, eqn (7.200) gives

$$\eta^+(t;\kappa) \simeq \frac{1}{\kappa} G_e \gamma(t, t-\tau_d) h^{(\mathrm{IA})}(\gamma(t, t-\tau_d)). \tag{7.206}$$

Since

$$\gamma(t, t-\tau_d) = \begin{cases} \kappa t & \text{for} \quad t < \tau_d, \\ \kappa\tau_d & \text{for} \quad t > \tau_d, \end{cases} \tag{7.207}$$

therefore

$$\eta^+(t;\kappa) \simeq \begin{cases} G_e t h^{(\mathrm{IA})}(\kappa t) & \text{for} \quad t < \tau_d \\ G_e \tau_d h^{(\mathrm{IA})}(\kappa\tau_d) & \text{for} \quad t > \tau_d. \end{cases} \tag{7.208}$$

Since $\gamma h(\gamma)$ has a maximum at $\gamma \simeq 2$, $\eta^+(t;\kappa)$ shows a maximum at $t \simeq 2/\kappa$ provided $\kappa\tau_d \gtrsim 2$. The height of the maximum decreases as $1/\kappa$. These features are in agreement with the experimental results shown in Fig. 7.19.

(iii) Steady elongational flow. The steady elongational viscosity is given as

$$\eta_E(\dot{\varepsilon}) \simeq \frac{G_e}{\dot{\varepsilon}} (Q_{zz}^{(\mathrm{IA})}(\dot{\varepsilon}\tau_d) - Q_{xx}^{(\mathrm{IA})}(\dot{\varepsilon}\tau_d)) \tag{7.209}$$

which first increases slightly with $\dot{\varepsilon}$ and then decreases with $\dot{\varepsilon}$. This was in contradiction with earlier data[86] for low density polyethylene, which indicated a sharp rise of $\eta_E(\dot{\varepsilon})$, but recent data for monodisperse linear polymers[85,100] are consistent with the theory.

7.7.5 Discussion

Though the theoretical constitutive equation (7.195) explains many features of nonlinear viscoelasticity, there are some discrepancies which are worth discussing.

Steady shear flow. The predicted steady state viscosity $\eta(\kappa)$ depends on the shear rate κ too strongly. In fact eqn (7.195) predicts that at high shear rate of $\kappa\tau_d \gg 1$[94]

$$\eta(\kappa) \simeq \eta(0)(\kappa\tau_d)^{-3/2} \tag{7.210}$$

i.e., the shear stress $\sigma_{xy}(\kappa) = \eta(\kappa)\kappa$ *decreases* with increasing shear rate, which means that the shear flow is not stable at high shear rate. At first sight this conclusion may seem to contradict the many experiments which show stable shear flow up to very high shear rate. However, the theoretical prediction is not entirely ruled out for various reasons:

(i) Equation (7.195) indicates that the form of $\eta(\kappa)$ is sensitive to the relaxation spectra of the linear relaxation modulus $G(t)$: the broader the relaxation spectra is, the smaller the exponent x in $\eta(\kappa) \propto \kappa^{-x}$ becomes. If the sample has broader molecular weight distribution, the relaxation spectra of $G(t)$ becomes broad and the anomalous behaviour of $\eta(\kappa)$ disappears. Also, even for the monodisperse sample, the various relaxation processes discussed in Section 7.4.2 broaden the relaxation spectra and weaken the shear rate dependence of $\eta(\kappa)$.

(ii) Equation (7.210) is derived under the condition

$$1/\tau_d \ll \kappa \ll 1/\tau_R. \tag{7.211}$$

If $\kappa\tau_R$ becomes of the order of unity, the contour length $L(\kappa)$ increases with the shear rate, and the stress starts to increase according to

$$\sigma_{xy} \propto L(\kappa)^2 \tag{7.212}$$

(see eqn (7.100)). Therefore if $\tau_d/\tau_R \simeq M/M_e$ is not sufficiently large, which is the case in many experiments with monodisperse samples, the minimum of $\sigma_{xy}(\kappa)$ will not be observed and the flow will be stable.

(iii) On the other hand if the system is monodisperse and if M/M_e is large enough, the theory predicts the shear stress shown in Fig. 7.22. Such behaviour has indeed been proposed by Vinogradov[101] and by Ball and McLeish,[76] to interpret the finding that the flow rate of polymers through pipes changes abruptly as the shear rate is raised if the molecular weight of the polymer is high and has a narrow distribution. Though the

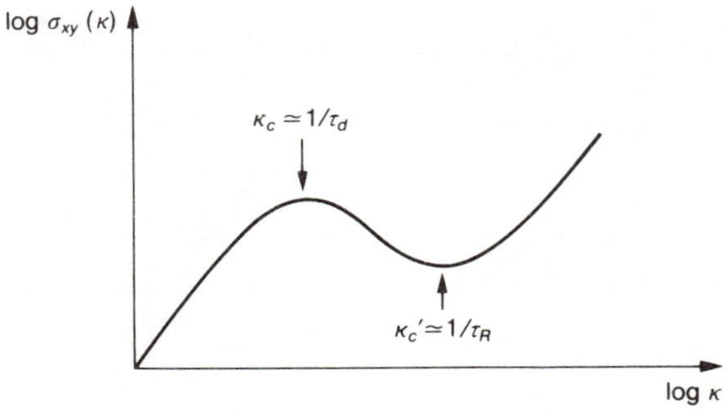

Fig. 7.22. Shear stress predicted by the theory for monodisperse systems with M/M_e large.

detailed analysis has not been done, it should be worthwhile to study the phenomenon under well-controlled conditions.

Stress overshoot. According to eqn (7.195), the stress maximum at the start of the shear flow appears in the shear stress, but not in the first normal stress difference $N_1(t, \kappa) = \sigma_{xx}(t; \kappa) - \sigma_{yy}(t; \kappa)$,[94] whilst experimentally the maximum is often observed in $N_1(t; \kappa)$. This is possibly due to the elongation of the contour length. Indeed the overshoot in the normal stress appears at a higher shear rate than in the shear stress.[102,103]

That the elongation of the contour length is important under usual flow conditions is indicated by the stress relaxation after the steady shear flow. It has been observed[104] that when the shear rate becomes larger than $1/\tau_R$, the relaxation curves begin to show a short-time component which corresponds to the relaxation in the contour length.

7.8 Stress relaxation after double step strain

Though the BKZ-type constitutive equation has been quite successful in many phenomena, it has been reported that under certain flow history, the equation gives unsatisfactory predictions. One such experiment is the stress relaxation after application of double step strain.[105,106] The flow history of this experiment is illustrated in Fig. 7.23.

Two step shears γ_1 and γ_2 are applied with time interval t_1, one at time $-t_1$ and the other at time 0. The BKZ equation (7.198) predicts

$$\sigma_{\alpha\beta}(t) = \phi_{\alpha\beta}(t, \gamma_2) + \phi_{\alpha\beta}(t + t_1, \gamma_2 + \gamma_1) - \phi_{\alpha\beta}(t + t_1, \gamma_2). \quad (7.213)$$

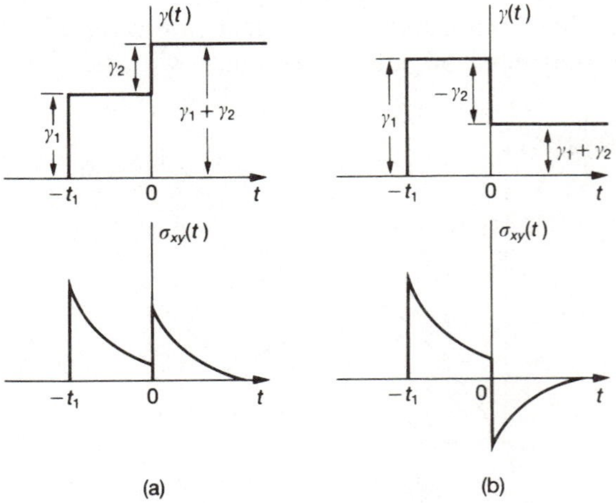

Fig. 7.23. Double step strain experiments.

According to Osaki *et al.*,[106] eqn (7.213) predicts the stress with reasonable accuracy when $\gamma_1\gamma_2 > 0$, i.e, when the sense of the two shears are the same, while when $\gamma_1\gamma_2 < 0$, a large discrepancy is found. It has been shown[107] that this discrepancy is caused by the IA approximation and the rigorous analysis of the model gives good agreement with experiments.

We consider the inextensible chain model. Figure 7.24 explains the change of polymer conformation under the double step strain. Figure 7.24*a* shows the undeformed state just before the first deformation. Figure 7.24*b* represents the state immediately after the deformation: the primitive chain is deformed by the shear γ_1. Figure 7.24*c* indicates the state just before the second deformation; the inner part AB still remains in the deformed tube, while the outer parts are in the undeformed tube. Now when the second deformation is applied, the inner part AB is deformed by the shear $\gamma_1 + \gamma_2$ from the equilibrium state, while the outer part is deformed by the shear γ_2. It is important to note that the second shear stretches the contour length of the outer part by the factor $\alpha(\gamma_2)$, but that of the inner part by the factor

$$\beta = \alpha(\gamma_1 + \gamma_2)/\alpha(\gamma_1), \tag{7.214}$$

since the inner part is already stretched by the factor $\alpha(\gamma_1)$. If the coordinates of A and B are s_1 and s_2, respectively, the coordinates of A' and B' are βs_1 and βs_2. Hence the probability that a primitive chain segment s is between $A'B'$ is equal to the probability that it is between

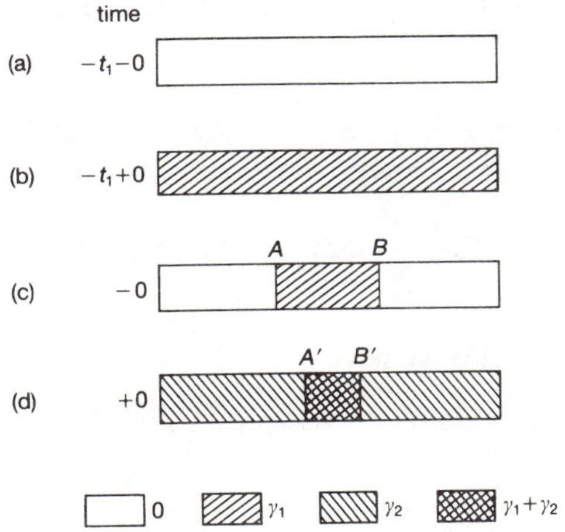

Fig. 7.24. Microscopic process in the relaxation of double step strain. The deformation of the primitive chain at various times are shown. The deformation of the primitive chain relative to the equilibrium state is shown by oblique lines.

AB at time $t=-0$, and is given by $\psi(s/\beta, t_1)$. Hence the average of $S_{\alpha\beta}(s, +0)$ is given by†

$$S_{\alpha\beta}(s, +0) = \begin{cases} \psi(s/\beta, t_1)Q_{\alpha\beta}(\gamma_1+\gamma_2) + [1-\psi(s/\beta, t_1)]Q_{\alpha\beta}(\gamma_2) & |s/\beta| < L/2, \\ Q_{\alpha\beta}(\gamma_2), & |s/\beta| > L/2. \end{cases} \tag{7.215}$$

For $t>0$, $S_{\alpha\beta}(s, t)$ satisfies

$$\frac{\partial}{\partial t} S_{\alpha\beta}(s, t) = D_c \frac{\partial^2}{\partial s^2} S_{\alpha\beta}(s, t) \tag{7.216}$$

and the boundary condition

$$S_{\alpha\beta}(s, t) = 0 \quad \text{at} \quad s = \pm\frac{L}{2}. \tag{7.217}$$

Hence $S_{\alpha\beta}(s, t)$ is given by

$$S_{\alpha\beta}(s, t) = \int_{-L/2}^{L/2} ds' G(s, s', t) S_{\alpha\beta}(s', +0) \tag{7.218}$$

where

$$G(s, s', t) = \frac{2}{L}\sum_{p=1}^{\infty} \sin\left(\frac{p\pi}{L}\left(s+\frac{L}{2}\right)\right)\sin\left(\frac{p\pi}{L}\left(s'+\frac{L}{2}\right)\right)\exp(-p^2 t/\tau_d). \tag{7.219}$$

From eqns (7.178), (7.218), and (7.219), the stress at time $t(t>0)$ is given by

$$\begin{aligned} \sigma_{\alpha\beta}(t) &= G_e \frac{1}{L}\int_{-L/2}^{L/2} ds' \int_{-L/2}^{L/2} ds G(s, s', t) S_{\alpha\beta}(s', +0) \\ &= G_e \frac{1}{L}\int_{-L/2}^{L/2} ds' \psi(s', t) S_{\alpha\beta}(s', +0). \end{aligned} \tag{7.220}$$

Substituting eqn (7.215) we have

$$\sigma_{\alpha\beta}(t) = G_e Q_{\alpha\beta}(\gamma_2)\psi(t) + G_e[Q_{\alpha\beta}(\gamma_1+\gamma_2) - Q_{\alpha\beta}(\gamma_2)]\psi'(t, t_1, \beta) \tag{7.221}$$

† Here it is assumed that if $|s/\beta|$ is larger than $L/2$, the primitive chain segment s is deformed by γ_2. Strictly speaking this is only approximately correct. Actually there is a small correction term to eqn (7.215),[107] which is neglected here because it is numerically insignificant.

where

$$\psi'(t, t_1, \beta) = \begin{cases} \dfrac{1}{L}\displaystyle\int_{L/2}^{\beta L/2} ds\,\psi(s, t)\psi(s/\beta, t_1) & \text{for } \beta < 1, \\ \dfrac{1}{L}\displaystyle\int_{-L/2}^{-\beta L/2} ds\,\psi(s, t)\psi(s/\beta, t_1) & \text{for } \beta > 1. \end{cases} \quad (7.222)$$

Using eqn (6.14) and doing the integral, we get

$$\psi'(t, t_1, \beta) = \begin{cases} \dfrac{32}{\pi^3}\beta \displaystyle\sum_{p,q:\text{odd}} (-)^{(p+q-2)/2} \frac{\sin[(\pi/2)(q - p\beta)]}{p(q^2 - p^2\beta^2)} \\ \qquad \times \exp[-(p^2 t + q^2 t_1)/\tau_d] \quad \text{for } \beta < 1, \\ \dfrac{32}{\pi^3} \displaystyle\sum_{p,q:\text{odd}} (-)^{(p+q-2)/2} \frac{\sin[(\pi/2)(p - q/\beta)]}{q(p^2 - q^2/\beta^2)} \\ \qquad \times \exp[-(p^2 t + q^2 t_1)/\tau_d] \quad \text{for } \beta > 1. \end{cases} \quad (7.223)$$

To simplify the equation we consider the case of large t and t_1. If $t > \tau_d$ and $t_1 > \tau_d$, only the first term in the sum of eqn (7.223) is important and $\psi'(t, t_1, \beta)$ is approximated by

$$\psi'(t, t_1, \beta) = \begin{cases} \dfrac{32}{\pi^3} \dfrac{\beta \cos(\pi\beta/2)}{1 - \beta^2} \exp[-(t + t_1)/\tau_d] & \text{for } \beta < 1 \\ \dfrac{32}{\pi^3} \dfrac{\cos(\pi/2\beta)}{1 - \beta^{-2}} \exp[-(t + t_1)/\tau_d] & \text{for } \beta > 1 \end{cases} \quad (7.224)$$

or it may be written as

$$\psi'(t, t_1, \beta) = A(\beta)\psi(t + t_1) \quad (7.225)$$

where

$$A(\beta) = \begin{cases} \dfrac{4}{\pi} \dfrac{\beta \cos(\pi\beta/2)}{1 - \beta^2} & \text{for } \beta < 1, \\ \dfrac{4}{\pi} \dfrac{\cos(\pi/2\beta)}{1 - \beta^{-2}} & \text{for } \beta > 1. \end{cases} \quad (7.226)$$

Although eqn (7.225) is obtained under the condition $t > \tau_d$ and $t_1 > \tau_d$, it turns out that eqn (7.225) is actually a good approximation for the entire regime of t and t_1.[107] If eqn (7.225) is used, eqn (7.221) is written as

$$\sigma_{\alpha\beta}(t) = \phi_{\alpha\beta}(t, \gamma_2) + A(\beta)[\phi_{\alpha\beta}(t + t_1, \gamma_1 + \gamma_2) - \phi_{\alpha\beta}(t + t_1, \gamma_2)]. \quad (7.227)$$

Equation (7.227) has been thoroughly checked by Osaki *et al.*[68,106] An example is given in Fig. 7.25. This indicates that the theory described in

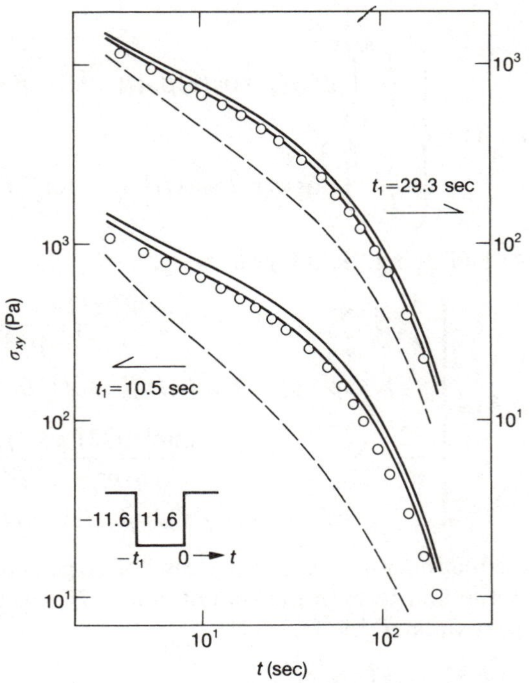

Fig. 7.25. Shear stresses for double step shear deformation. $-\gamma_1 = \gamma_2 = 11.6$, and t_1 is indicated in the figure. Sample polystyrene solution in diethyl phthalate $M = 3.10 \times 10^6$ and $\rho = 0.221\ \mathrm{g\,cm^{-3}}$. The heavy lines represent stress for single step deformation. The light solid line represents eqn (7.227) and the light broken lines the result of the BKZ equation (eqn (7.213)). Reproduced from ref. 68.

Section 7.7 correctly reflects the reality of polymer dynamics in an entangled state.

7.9 Rigorous constitutive equation for reptation model

Having seen that the IA approximation causes a serious error in certain situations, we now derive a constitutive equation without using the IA approximation.[108]

In a small time-interval Δt, $\mathbf{E}(t + \Delta t, t)$ is given by

$$\mathbf{E}(t + \Delta t, t) = \mathbf{I} + \boldsymbol{\kappa}(t)\Delta t. \tag{7.228}$$

Thus the transformation rule described by eqn (7.171) is written as

$$\mathbf{R}(\tilde{s}, t + \Delta t) = \mathbf{R}(s, t) + \boldsymbol{\kappa}(t) \cdot \mathbf{R}(s, t)\Delta t = \mathbf{R}(s, t) + \boldsymbol{\kappa}(t) \cdot \mathbf{R}(\tilde{s}, t)\Delta t. \tag{7.229}$$

The second equality holds since $s - \bar{s}$ is of order Δt. Similarly eqn (7.172) becomes

$$\begin{aligned}\bar{s} &= \int_0^s \mathrm{d}s' \,|\boldsymbol{u}(s', t) + \boldsymbol{\kappa}(t) \cdot \boldsymbol{u}(s', t)\Delta t| \\ &= \int_0^s \mathrm{d}s'(1 + \boldsymbol{\kappa}(t) : \boldsymbol{u}(s', t)\boldsymbol{u}(s', t)\Delta t) + O(\Delta t^2) \\ &= s + \Delta t \int_0^s \mathrm{d}s' \boldsymbol{\kappa}(t) : \boldsymbol{u}(s', t)\boldsymbol{u}(s', t).\end{aligned} \tag{7.230}$$

Therefore, to the order of Δt, s is expressed by $\bar{s}$ as

$$\begin{aligned}s &= \bar{s} - \Delta t \int_0^{\bar{s}} \mathrm{d}s' \boldsymbol{\kappa}(t) : \boldsymbol{u}(s', t)\boldsymbol{u}(s', t) \\ &= \bar{s} - \Delta t \dot{\xi}(\bar{s}, t)\end{aligned} \tag{7.231}$$

where

$$\dot{\xi}(s, t) = \int_0^s \mathrm{d}s' \boldsymbol{\kappa}(t) : \boldsymbol{u}(s', t)\boldsymbol{u}(s', t). \tag{7.232}$$

From eqns (7.229) and (7.230), the change in the tangent vector $\boldsymbol{u}(s, t)$ becomes

$$\begin{aligned}\boldsymbol{u}(\bar{s}, t + \Delta t) &= \frac{\partial}{\partial \bar{s}} \boldsymbol{R}(\bar{s}, t + \Delta t) = \frac{\partial}{\partial \bar{s}} (\boldsymbol{R}(s, t) + \boldsymbol{\kappa}(t) \cdot \boldsymbol{R}(\bar{s}, t)\Delta t) \\ &= \frac{\partial s}{\partial \bar{s}} \frac{\partial}{\partial s} \boldsymbol{R}(s, t) + \boldsymbol{\kappa}(t) \cdot \frac{\partial}{\partial \bar{s}} \boldsymbol{R}(\bar{s}, t)\Delta t \\ &= \left(1 - \Delta t \frac{\partial}{\partial \bar{s}} \dot{\xi}(\bar{s}, t)\right) \boldsymbol{u}(s, t) + \boldsymbol{\kappa}(t) \cdot \boldsymbol{u}(\bar{s}, t)\Delta t.\end{aligned} \tag{7.233}$$

Using eqn (7.232),

$$\begin{aligned}\boldsymbol{u}(\bar{s}, t + \Delta t) &= \boldsymbol{u}(s, t) - [\boldsymbol{\kappa}(t) : \boldsymbol{u}(\bar{s}, t)\boldsymbol{u}(\bar{s}, t)\boldsymbol{u}(s, t) - \boldsymbol{\kappa}(t) \cdot \boldsymbol{u}(\bar{s}, t)]\Delta t \\ &= \boldsymbol{u}(s, t) - [\boldsymbol{\kappa}(t) : \boldsymbol{u}(\bar{s}, t)\boldsymbol{u}(\bar{s}, t)\boldsymbol{u}(\bar{s}, t) - \boldsymbol{\kappa}(t) \cdot \boldsymbol{u}(\bar{s}, t)]\Delta t + O(\Delta t^2) \\ &= \boldsymbol{u}(s, t) + \boldsymbol{\Gamma}(\boldsymbol{u}(\bar{s}, t), t)\Delta t\end{aligned} \tag{7.234}$$

where $\boldsymbol{\Gamma}(\boldsymbol{u}, t)$ is given by eqn (7.189)

In eqn (7.234) the effect of Brownian motion was not taken into account. If this is included, the final equation becomes

$$\begin{aligned}\boldsymbol{u}(\bar{s}, t + \Delta t) &= \boldsymbol{u}(s + \Delta\xi, t) + \boldsymbol{\Gamma}(\boldsymbol{u}(\bar{s}, t), t)\Delta t \\ &= \boldsymbol{u}(\bar{s} - \dot{\xi}(\bar{s}, t)\Delta t + \Delta\xi, t) + \boldsymbol{\Gamma}(\boldsymbol{u}(\bar{s}, t), t)\Delta t,\end{aligned} \tag{7.235}$$

or replacing $\bar{s}$ by s

$$\boldsymbol{u}(s, t + \Delta t) = \boldsymbol{u}(s - \dot{\xi}(s, t)\Delta t + \Delta\xi, t) + \boldsymbol{\Gamma}(\boldsymbol{u}(s, t), t)\Delta t. \quad (7.236)$$

This is the time-evolution equation for the tangent vector $\boldsymbol{u}(s, t)$ of the inextensible primitive chain. Thus the equation for $f(\boldsymbol{u}, s, t)$ is

$$\begin{aligned} f(\boldsymbol{u}, s, t + \Delta t) &= \langle \delta[\boldsymbol{u} - \boldsymbol{u}(s, t + \Delta t)] \rangle \\ &= \langle \delta[\boldsymbol{u} - \boldsymbol{u}(s - \dot{\xi}(s, t)\Delta t + \Delta\xi(t), t) - \boldsymbol{\Gamma}(\boldsymbol{u}(s, t), t)\Delta t] \rangle \\ &= \langle \delta[\boldsymbol{u} - \boldsymbol{u}(s - \dot{\xi}(s, t)\Delta t + \Delta\xi(t), t)] \rangle \\ &\quad - \Delta t \frac{\partial}{\partial \boldsymbol{u}} \cdot (\boldsymbol{\Gamma}(\boldsymbol{u}, t) f(\boldsymbol{u}, s, t)). \end{aligned} \quad (7.237)$$

The first term is written

$$\begin{aligned} X &\equiv \langle \delta[\boldsymbol{u} - \boldsymbol{u}(s - \dot{\xi}(s, t)\Delta t + \Delta\xi(t), t)] \rangle \\ &= \left\langle \left[1 + \Delta\xi \frac{\partial}{\partial s} + \frac{\Delta\xi^2}{2} \frac{\partial^2}{\partial s^2} - \Delta t \dot{\xi} \frac{\partial}{\partial s} \right] \delta[\boldsymbol{u} - \boldsymbol{u}(s, t)] \right\rangle \\ &= \left[1 + D_c \Delta t \frac{\partial^2}{\partial s^2} \right] f(\boldsymbol{u}, \mathbf{s}, t) - \Delta t \underline{\left\langle \dot{\xi} \frac{\partial}{\partial s} \delta[\boldsymbol{u} - \boldsymbol{u}(s, t)] \right\rangle}. \end{aligned} \quad (7.238)$$

The underlined term is rewritten as

$$Y \equiv \left\langle \dot{\xi} \frac{\partial}{\partial s} \delta[\boldsymbol{u} - \boldsymbol{u}(s, t)] \right\rangle = \frac{\partial}{\partial s} \langle \dot{\xi} \delta[\boldsymbol{u} - \boldsymbol{u}(s, t)] \rangle - \left\langle \delta[\boldsymbol{u} - \boldsymbol{u}(s, t)] \frac{\partial \dot{\xi}}{\partial s} \right\rangle. \quad (7.239)$$

Since the correlation between $\boldsymbol{u}(s, t)$ and $\boldsymbol{u}(s', t)$ decreases quickly with an increase in $|s - s'|$, the first average in eqn (7.239) becomes

$$\begin{aligned} \langle \dot{\xi} \delta[\boldsymbol{u} - \boldsymbol{u}(s, t)] \rangle &= \int_0^s \mathrm{d}s' \boldsymbol{\kappa}(t) : \langle \boldsymbol{u}(s', t) \boldsymbol{u}(s', t) \delta[\boldsymbol{u} - \boldsymbol{u}(s, t)] \rangle \\ &\simeq \int_0^s \mathrm{d}s' \boldsymbol{\kappa}(t) : \langle \boldsymbol{u}(s', t) \boldsymbol{u}(s', t) \rangle \langle \delta[\boldsymbol{u} - \boldsymbol{u}(s, t)] \rangle \\ &= \int_0^s \mathrm{d}s' \boldsymbol{\kappa}(t) : \langle \boldsymbol{u}(s', t) \boldsymbol{u}(s', t) \rangle f(\boldsymbol{u}, s, t). \end{aligned} \quad (7.240)$$

From eqn (7.232) it follows that

$$\frac{\partial \dot{\xi}}{\partial s} = \boldsymbol{\kappa}(t) : \boldsymbol{u}(s, t) \boldsymbol{u}(s, t). \quad (7.241)$$

From eqns (7.239)–(7.241), one has

$$Y = \frac{\partial}{\partial s}\left[\int_0^s \mathrm{d}s' \boldsymbol{\kappa}(t) : \langle \boldsymbol{u}(s', t)\boldsymbol{u}(s', t)\rangle f(\boldsymbol{u}, s, t)\right]$$
$$- \langle \boldsymbol{\kappa}(t) : \boldsymbol{u}(s, t)\boldsymbol{u}(s, t)\delta[\boldsymbol{u} - \boldsymbol{u}(s, t)]\rangle$$
$$= \boldsymbol{\kappa}(t) : \langle \boldsymbol{u}(s, t)\boldsymbol{u}(s, t)\rangle f(\boldsymbol{u}, s, t) + \langle \dot{\xi}(s, t)\rangle \frac{\partial}{\partial s} f(\boldsymbol{u}, s, t)$$
$$- \boldsymbol{\kappa}(t) : \boldsymbol{u}\boldsymbol{u} f(\boldsymbol{u}, s, t). \tag{7.242}$$

Hence the time evolution equation for $f(\boldsymbol{u}, s, t)$ is obtained as†

$$\frac{\partial f}{\partial t} = \left(D_c \frac{\partial^2}{\partial s^2} - \langle \dot{\xi}(s, t)\rangle \frac{\partial}{\partial s}\right) f(\boldsymbol{u}, s, t) - \frac{\partial}{\partial \boldsymbol{u}} \cdot (\boldsymbol{\Gamma}(\boldsymbol{u}, t) f(\boldsymbol{u}, s, t))$$
$$+ \boldsymbol{\kappa}(t) : (\boldsymbol{u}\boldsymbol{u} - \langle \boldsymbol{u}(s, t)\boldsymbol{u}(s, t)\rangle) f(\boldsymbol{u}, s, t). \tag{7.243}$$

The average in eqn (7.243) can be expressed by $f(\boldsymbol{u}, s, t)$ as

$$\langle \dot{\xi}(s, t)\rangle = \int_0^s \mathrm{d}s' \int \mathrm{d}\boldsymbol{u} \boldsymbol{\kappa}(t) : \boldsymbol{u}\boldsymbol{u} f(\boldsymbol{u}, s', t) \tag{7.244}$$

and

$$\langle u_\alpha(s, t) u_\beta(s, t)\rangle = \int \mathrm{d}\boldsymbol{u}\, u_\alpha u_\beta f(\boldsymbol{u}, s, t). \tag{7.245}$$

Hence eqn (7.243) is a nonlinear integro-differential equation for $f(\boldsymbol{u}, s, t)$.

Equation (7.243) can be rewritten into more tractable form. By a similar technique described in ref. 108, eqn (7.243) can be transformed into a closed equation for $S_{\alpha\beta}(s, t)$:

$$S_{\alpha\beta}(s, t) = \int_{-\infty}^{t} \mathrm{d}t' \left(\frac{\partial}{\partial t'} K(s, t, t')\right) Q_{\alpha\beta}(\mathbf{E}(t, t')) \tag{7.246}$$

where $K(s, t, t')$ is the solution of the differential equation

$$\left(\frac{\partial}{\partial t} - D_c \frac{\partial^2}{\partial s^2} + \langle \dot{\xi}(s, t)\rangle \frac{\partial}{\partial s}\right) K(s, t, t') = 0 \tag{7.247}$$

with the initial condition

$$K(s, t, t') = 1 \quad \text{at} \quad t = t' \tag{7.248}$$

† In ref. 108 the terms in the last parenthesis are erroneously omitted. This gives a constitutive equation which includes $Q^{(\mathrm{IA})}(\mathbf{E})$ instead of $Q(\mathbf{E})$ in eqn (7.246). Since the difference in $Q^{(\mathrm{IA})}(\mathbf{E})$ and $Q(\mathbf{E})$ is small, the error caused by this is not serious.

and the boundary condition

$$K(s, t, t') = 0 \quad \text{at} \quad s = -L/2 \quad \text{and} \quad s = L/2. \tag{7.249}$$

Finally $\langle \dot{\xi}(s, t) \rangle$ is given by

$$\langle \dot{\xi}(s, t) \rangle = \int_0^s \mathrm{d}s' \kappa_{\alpha\beta}(t) S_{\alpha\beta}(s', t). \tag{7.250}$$

Equations (7.246)–(7.250) determine $S_{\alpha\beta}(s, t)$. Given $S_{\alpha\beta}(s, t)$, the stress can be calculated by eqn (7.178).†

In the special case of step strain, one can solve the set of equations (7.246)–(7.250) rigorously and obtain the results given in eqn (7.122). In the general case, the solution of the equation needs numerical calculation. It turns out that the difference between eqns (7.194) and (7.246) is not large for the usual flow history discussed in Section 7.6. For such flows, the simple constitutive equation will be useful.

The effect of the IA approximation has also been examined for large amplitude oscillatory shear deformation.[109] In this case the result of the constitutive equation without using the IA approximation is shown to be in better agreement with experimental results.[110]

It has been shown by Marrucci[110a] and Marrucci and Grizzuti[110b] that analysis at the level of accuracy of this section is required to derive Weissenberg effect correctly.

7.10 Further applications

Here we shall briefly discuss some pending problems which have not been discussed in the previous sections.

7.10.1 Branched polymers

As discussed in Section 6.4.5, reptation is severely suppressed if the polymer has long branches. Indeed it has been observed that the dynamical properties of branched polymers are quite distinct from those of linear polymers. So far studies have been done for branched polymers of the simplest type, the star-shaped polymer in which f chains are connected to a centre. The observed phenomena are:

(i) The diffusion constant D_G of a star polymer in a high molecular weight matrix is much smaller than that of a linear polymer of the same molecular weight,[111] and the molecular weight dependence of D_G is much stronger than that of linear polymers. This is consistent with the prediction of the reptation theory[112] (eqn (6.118)).

† The same constitutive equation has recently been derived by G. Marrucci (*J. Non-Newtonian Fluid Mech*, to appear) by a different method.

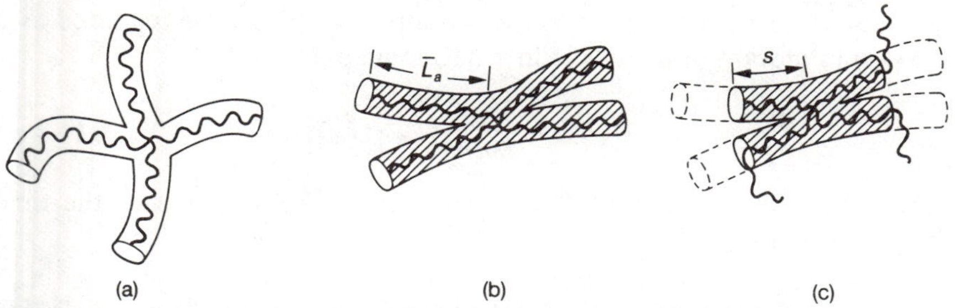

Fig. 7.26. Relaxation of a star polymer. Figures show the states (a) before the deformation, (b) immediately after the deformation, (c) at a later time t. The deformed part of the tube which contributes to the stress is denoted by oblique lines.

(ii) The zero shear rate viscosity of the melts of the star polymers increases more steeply with the molecular weight M than that of linear polymers.[113] On the other hand, the storage modulus $G'(\omega)$ shows no plateau region and the steady-state compliance $J_e^{(0)}$ is Rouse like (i.e., proportional to M) even if the molecular weight becomes quite high.[114]

The anomalous behaviour in the linear viscoelasticity has been explained by the tube model.[112,115,116] Figure 7.26 shows schematically how the stress relaxation takes place in star polymers. In the crude theory,[115] it is assumed that the centre of the star is fixed during the viscoelastic relaxation time and that the relaxation takes place only by the contour length fluctuation, i.e., by the process that the polymer retracts its arm down the tube and evacuates from the deformed tube as shown in Fig. 7.26.

Let $\psi(s, t)$ be the probability that the tube segment s which is separated from the centre by the contour length s still remains at time t, then the relaxation modulus is written as

$$G(t) = G_N^{(0)} \frac{1}{\bar{L}_a} \int_0^{\bar{L}_a} ds\, \psi(s, t) \tag{7.251}$$

where $\bar{L}_a$ is the equilibrium length of the tube for the arm of the star polymer, and the constant $G_N^{(0)}$ can be identified, in a first approximation, with the plateau modulus for linear polymers:

$$G_N^{(0)} = \frac{\rho RT}{M_e}. \tag{7.252}$$

A simple approximation for $\psi(s, t)$ is

$$\psi(s, t) = \exp(-t/\tau(s)) \tag{7.253}$$

where $\tau(s)$ is the average time at which the chain end first reaches the tube segment s, i.e., the contour length $L_a(t)$ first becomes equal to s. As was discussed in Section 6.4.5, the motion of $L_a(t)$ can be regarded as a Brownian motion of a particle in a harmonic potential

$$U(L_a) = \frac{3k_BT}{2N_ab^2}(L_a - \bar{L}_a)^2 \tag{7.254}$$

where N_a is the number of Rouse segments in the arm. Hence the time $\tau(s)$ is estimated as

$$\tau(s) \simeq \frac{(\bar{L}_a - s)^2}{D_c} \exp([U(L_a = s) - U(\bar{L}_a)]/k_BT). \tag{7.255}$$

The maximum relaxation time is given by

$$\tau_{\max} = \tau(s=0) \simeq \frac{\bar{L}_a^2}{D_c} \exp\left(\frac{3}{2N_ab^2}\bar{L}_a^2\right) \simeq \frac{\zeta N_a^3 b^4}{k_BT\,a^2} \exp\left[\tfrac{3}{2}N_a\left(\frac{b}{a}\right)^2\right]. \tag{7.256}$$

Equation (7.255) is then written as

$$\tau(s) = \tau_{\max}(1-\xi)^2 \exp\left(\tfrac{3}{2}N_a\left(\frac{b}{a}\right)^2(\xi^2 - 2\xi)\right) \tag{7.257}$$

where

$$\xi = s/\bar{L}_a. \tag{7.258}$$

Consider the case

$$\alpha \equiv \tfrac{3}{2}N_a\left(\frac{b}{a}\right)^2 \gg 1, \tag{7.259}$$

then the viscosity is evaluated as

$$\eta_0 = \int_0^\infty \mathrm{d}t G(t) = G_N^{(0)}\frac{1}{\bar{L}_a}\int_0^{\bar{L}_a} \mathrm{d}s\tau(s) = G_N^{(0)}\tau_{\max}\int_0^1 \mathrm{d}\xi(1-\xi)^2\exp(-2\alpha\xi + \alpha\xi^2)$$

$$\simeq G_N^{(0)}\tau_{\max}\int_0^\infty \mathrm{d}\xi\exp(-2\alpha\xi) = \frac{1}{2\alpha}G_N^{(0)}\tau_{\max}. \tag{7.260}$$

Similarly, the steady-state compliance is obtained as

$$J_e^{(0)} = \frac{1}{\eta_0^2}\int_0^\infty \mathrm{d}t G(t)t = \frac{\alpha}{G_N^{(0)}}. \tag{7.261}$$

Since the molecular weight of an arm is M/f, it follows from eqns (7.54) and (7.259)

$$\alpha = \frac{15M}{8fM_e}. \tag{7.262}$$

Thus eqns (7.260) and (7.261) are written as

$$\eta_0 \propto \left(\frac{M}{fM_e}\right)^2 \exp\left(\frac{15M}{8fM_e}\right) \tag{7.263}$$

and

$$J_e^{(0)} = \frac{15M}{8f\rho RT}. \tag{7.264}$$

The results of eqns (7.263) and (7.264) are in qualitative agreement with experimental results: the viscosity increases steeply because of the exponential factor, and the steady state compliance is proportional to M. However, the quantitative agreement is not satisfactory. The observed viscosity is smaller than the calculated one, and the best fit with experiments is obtained only when the numerical coefficient in the exponential of eqn (7.263) is replaced by a smaller number (about 1/2) instead of 15/8.[116] This suggests that relaxation mechanisms other than the contour length fluctuations are important for star polymers. Indeed it has been pointed out[61,97] that in the case of star polymers the constraint release, and perhaps other tube reorganization processes, are as important as the contour length fluctuation.

That the tube reorganization is important for star polymers is indicated by another experiment. Kan *et al.*[117] found that the relaxation time of a star polymer dispersed in a crosslinked system is by orders of magnitude larger than that in the melt, while for linear polymers the former is larger only by a factor of 2 or 3.

The constraint release or other mechanisms of the tube reorganization are supposed to be important in other branched polymers such as H-shaped polymers[118] or ring polymers.[119] Theoretical prediction for the rheological properties of these polymers is interesting and challenging.

7.10.2 Molecular weight distribution

Various experimental data suggest that the tube reorganization is important in linear polymers with molecular weight distribution.

(i) The diffusion constant of a polymer (of molecular weight M) in a matrix (of molecular weight P) has been found to be essentially independent of P if P is larger than a certain value P_c which is between M and M_e.[8,10] This indicates that the tube reorganization is weak in monodisperse systems ($M = P$). On the other hand, if P becomes smaller than P_c, the diffusion constant increases with decreasing P.[120,121]

(ii) The linear viscoelasticity of a mixture of two polymers of the same

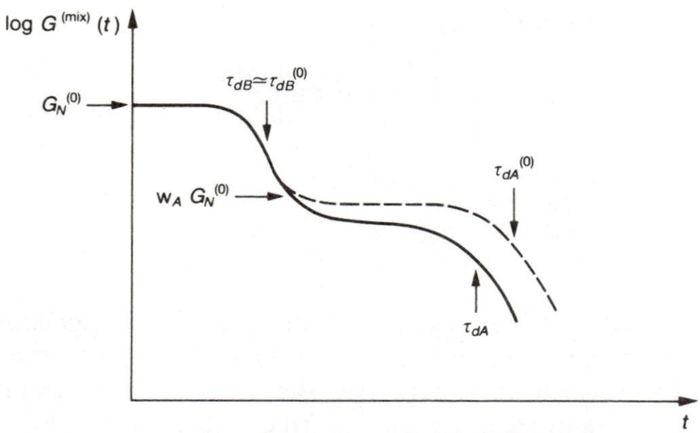

Fig. 7.27. Discrepancy between experimental results and eqn (7.265).

species, but different molecular weights M_A and M_B ($M_A > M_B$) is not explained by the model which includes only reptation. According to the fixed tube model, the relaxation modulus of the mixture is the weight average of that of the pure melts of individual components:[15]

$$G^{(\mathrm{mix})}(t) = w_A G_A(t) + w_B G_B(t) \tag{7.265}$$

where

$$w_A = \frac{\rho_A}{\rho_A + \rho_B}, \qquad w_B = \frac{\rho_B}{\rho_A + \rho_B}. \tag{7.266}$$

The discrepancy between the experimental results[122] and eqn (7.265) is schematically explained in Fig. 7.27: $G^{(\mathrm{mix})}(t)$ shows two characteristic relaxations, each corresponding to the disengagement of polymer A and B. Though this feature is in agreement with eqn (7.265), the relaxation time of the larger component τ_{dA} is shorter than that in the pure A component $\tau_{dA}^{(0)}$, and the plateau modulus for the larger polymer is lower than expected from eqn (7.265). Kurata[123] suggested that the experimental data can be fitted by

$$G^{(\mathrm{mix})}(t) = (1 - w_A^2) G_B(t) + w_A^2 G_A(t/w_A). \tag{7.267}$$

These results clearly indicate that the tube constraint for a polymer becomes weaker if it is made of shorter polymers. The weakening of the tube can be expressed either by an increase in the step length,[62,123] or by an increase in the constraint release process,[59,124] or both.[125,126] However, the interpretation seems to be still at a tentative level.

7.10.3 Future problems

The reptation model has been applied to various problems other than the problems of viscoelasticity and diffusion that have been discussed. These include dielectric relaxation,[127] spinodal decomposition,[128,129] polymer–polymer welding,[130,131] diffusion controlled reaction,[132,133] and crazing.[134,135] A concise review of various applications is given by de Gennes and Léger.[136]

On the whole the reptation model works well qualitatively, and for several problems it gives quantitatively successful predictions. However, many problems remain unsolved.

Perhaps the most important problem is the tube reorganization. We have seen that the tube reorganization is important in branched polymers and in linear polymers with polydispersity. It will also be important in a nonuniform system such as polymer mixtures. So far the reptation theory is based on the assumption that there is a tube which is characterized by a single parameter a, the step length of the tube. Though the outcome of this simple assumption is quite fruitful, one could ask: to what extent is this picture correct?

A complete answer to this question will be given when the tube is derived from more basic equations such as eqn (5.84) by a kind of mean field approximation. This will require a new development of statistical mechanics since the tube is a dynamical concept rather than static. (Notice that the mean force acting on the polymer vanishes if it is averaged over a time longer than τ_d, so that the average of the surrounding field must be taken over a finite time.) Perhaps the tube is better understood as representing the effect of dynamical correlation of the environment rather than the usual mean field.

A slightly different, but closely related, problem is rubber elasticity. Here the dynamical problem does not arise since the topological constraints are permanent. However, the correlation plays an essential role in the problem. Indeed it is the correlation in the topological structure between the undeformed state and the deformed state that gives rise to the rubber elasticity. In the modern theory of rubber elasticity, this correlation is neatly handled by the replica method.[137,138] Generalization of this method to dynamical problems might be quite useful.

On the other hand, apart from that purely theoretical approach, it will be quite promising to develop a theory by closely studying experimental results. Collaboration between experiment and theory will be essential for further progress.

References

1. Ferry, J. D., Landel, R. F. and Williams, M. L., *J. Appl. Phys.* **26,** 359 (1955).
2. Ferry, J. D., *Viscoelastic Properties of Polymers* (3rd edn). Wiley, New York (1980).
3. Graessley, W. W., *Adv. Polym. Sci.* **16,** 1 (1974).

4. Baur, M. E., and Stockmayer, W. H., *J. Chem. Phys.* **43,** 4319 (1965).
5. Higgins, J. S., Nicholson, L. K., and Hayter, J. B., *Polymer* **22,** 163 (1981).
6. Richter, D., Hayter, J. B., Mezei, F., and Ewen, B., *Phys. Rev. Lett.* **41,** 1484 (1978).
7. Baumgärtner, A., Kremer, K., and Binder, K., *Faraday Symp. Chem. Soc.* **18,** 37 (1983).
8. Klein, J., *Nature (London),* **271,** 143 (1978); *Phil. Mag.* **A43,** 771 (1981).
9. Klein, J., and Briscoe, B. J., *Proc. R. Soc. London* **A365,** 53 (1979).
10. Hervet, H., Léger, L., Rondelez, F., *Phys. Rev. Lett.* **42,** 1681 (1979); Léger, L., Hervet, H., and Rondelez, F., *Macromolecules* **14,** 1732 (1981).
11. Tanner, J. E., *Macromolecules* **4,** 748 (1971); Tanner, J. E., Liu, K. J., and Anderson, J. E., *Macromolecules* **4,** 586 (1971).
12. Bachus, R., and Kimmich, R., *Polymer* **24,** 964 (1983).
13. Tirrell, M., *Rubber Chem. Tech.* **57,** 523 (1984).
14. Doi, M., and Edwards, S. F., *J. C. S. Faraday Trans. 2* **74,** 1802 (1978).
15. Doi, M., and Edwards, S. F., *J. C. S. Faraday Trans. 2* **74,** 1818 (1978).
16. Walters, K., *Rheometry*. Chapman & Hall, London. New York (1975).
17. Astarita, G., and Marrucci, G., *Principles of Non-Newtonian Fluid Mechanics,* McGraw-Hill, London (1974).
18. Lodge, A. S., *Elastic Liquids*. Academic Press, London (1964).
19. Bird, R. B., Armstrong, R. C., Hassager, O., and Curtiss, C. F., *Dynamics of Polymeric Liquids,* Vols. 1, 2. Wiley, New York (1977).
20. See for example Treloar, L. R. G., *Rep. Prog. Phys.* **36,** 755 (1973); and Treloar, L. R. G., *The Physics of Rubber Elasticity,* (3rd edn). Clarendon Press, Oxford (1975).
21. Green, M. S., and Tobolsky, A. V., *J. Chem. Phys.* **14,** 80 (1946).
22. Yamamoto, M., *J. Phys. Soc. Jpn* **11,** 413 (1956), **12,** 1148 (1957); **13,** 1200 (1958). Lodge, A. S., *Rheol. Acta* **7,** 379 (1968).
23. Coleman, B. D., *Arch. Ratl. Mech.* **17,** 1 (1964), **17,** 230 (1964). Truesdell, C., and Noll, W., The nonlinear field theories of mechanics, in *Encyclopedia of Physics* III/3. Springer (1965).
24. Janeschitz-Kriegl, H., *Polymer Melt Rheology and Flow Birefringence.* Springer, New York (1983).
25. Janeschitz-Kriegl, H., *Adv. Polym. Sci.* **6,** 170 (1969).
26. Historical development of the stress optical law is described in refs 20 and 24. A recent study is given by Wales J. L. S., *The Application of Flow Birefringence to Rheological Studies of Polymer Melts.* Delft Univ. Press, Rotterdam (1976).
27. DiMarzio, E. A., *J. Chem. Phys.* **36,** 1563 (1962).
28. Fukuda, M., Wilkes, G. L., and Stein, R. S., *J. Polym. Sci.* **A2,** 9, 1417 (1971).
29. Jarry, J. P., and Monnerie, L., *Macromolecules* **12,** 316 (1979).
30. Tobolsky, A. V., *Properties and Structure of Polymers.* Wiley, New York (1960).
31. Onogi, S., Masuda, T., and Kitagawa, K., *Macromolecules* **3,** 109 (1970).
32. Doi, M., *Chem. Phys. Lett.* **26,** 269 (1974).
33. Berry, G. C., and Fox, T. G., *Adv. Polym. Sci.* **5,** 261 (1968).
34. Casale, A., Porter, R. S., and Johnson, J. F., *J. Macromol. Sci. Rev. Macromol. Chem.* **C5,** 387 (1971).

35. Odani, H., Nemoto, N., and Kurata, M., *Bull. Inst. Chem. Res. Kyoto Univ.* **50,** 117 (1972).
36. Graessley, W. W., *J. Polym. Sci.* **18,** 27 (1980).
37. van Krevelen, D. W., *Properties of Polymers,* p. 338. Elsevier, Amsterdam (1976).
38. Graessley, W. W., and Edwards, S. F., *Polymer* **22,** 1329 (1981).
39. Bartels, C. R., Crist, B., and Graessley, W. W., *Macromolecules* **17,** 2702 (1984).
40. Kremer, K., *Macromolecules* **16,** 1632 (1983).
41. A recent review on the computer simulation for polymer dynamics is given by Baumgärtner, A., *Ann. Rev. Phys. Chem.* **35,** 419 (1984).
42. Deutsch, J. M., *Phys. Rev. Lett.* **49,** 926 (1982).
43. Higgins, J. S., and Roots, J. E., *J. C. S. Faraday Trans II* **81,** 757 (1985).
44. de Gennes, P. G., *Macromolecules* **9,** 587, 594 (1976).
45. Adam, M., and Delsanti, M., *J. Phys.* (*Paris*) **44,** 1185 (1983).
46. Raju, V. R., Menezes, E. V., Marin, G., Graessley, W. W., and Fetters, L. J., *Macromolecules* **14,** 1668 (1981).
47. Onogi, S., Masuda, T., Miyanaga, N., and Kimura, Y., *J. Polym. Sci. Part A2* **5,** 899 (1967); Onogi, S., Kimura, S., Kato, T., Masuda, T., and Miyanaga, N., *J. Polym. Sci.* **C15,** 381 (1966).
48. Masuda, T., Toda, N., Aoto, Y., and Onogi, S., *Polymer J.* **3,** 315 (1972).
49. Nemoto, N., Ogawa, T., Odani, H., and Kurata, M., *Macromolecules* **5,** 641 (1972).
50. See for example, Osaki, K., Fukuda, M., and Kurata, M., *J. Polym. Sci. Phys. ed.* **13,** 775 (1975). Recent data for melt are given by Lin, Y. H., *Macromolecules* **17,** 2846 (1984); *J. Rheol.* **28,** 1 (1984).
51. Graessley, W. W., *Faraday Symp. Chem. Soc.* **18,** 7 (1983).
52. Bernard, D. A., and Noolandi, J., *Macromolecules* **15,** 1553 (1982); **16,** 548 (1983).
53. Wendel, H., and Noolandi, J., *Macromolecules* **15,** 1318 (1982); Wendel, H., *Colloid Polym. Sci.* **259,** 908 (1981).
54. Curtiss, C. F., and Bird, R. B., *J. Chem. Phys.* **74,** 2016, 2026 (1981).
55. Ball R., private communication.
56. Deutsch, J. M., *Phys. Rev. Lett.* **54,** 56 (1985).
57. Doi, M., *J. Polymer Sci.* **21,** 667 (1983); *J. Polym. Sci. Lett.* **19,** 265 (1981).
58. Lin, Y. H., *Macromolecules* **19,** 159, 168 (1986).
59. Klein, J., *Macromolecules* **11,** 852 (1978).
60. Daoud, M., and de Gennes, P. G., *J. Polym. Sci. Phys ed.* **17,** 1971 (1979).
61. Graessley, W. W., *Adv. Polym. Sci.* **47,** 67 (1982).
62. Marrucci, G., *J. Polym. Sci. Phys.* **23,** 159 (1985).
63. Viovy, J. L., *J. Physique Lett.* **46,** 847 (1985).
64. Doi, M., *J. Polym. Sci.* **18,** 1005 (1980).
65. Einaga, Y., Osaki, K., Kurata, M., Kimura, S., and Tamura, M., *Polymer. J.* **2,** 550 (1971); Fukuda, M., Osaki, K., and Kurata, M., *J. Polym. Sci. Phys.* **13,** 1563 (1975).
66. Osaki, K., Bessho, N., Kojimoto, T., and Kurata, M., *J. Rheol.* **23,** 617 (1979).
67. Osaki, K., Kimura, S., and Kurata, M., *J. Polym. Sci. Phys. ed.* **19,** 517 (1981).

68. Osaki, K., and Kurata, M., *Macromolecules* **13,** 671 (1980).
69. Osaki, K., Nishizawa, K., and Kurata, M., *Macromolecules* **15,** 1068 (1982).
70. Takahashi, M., Nakamura, H., Masuda, T., and Onogi, S., *Polymer Preprints Japan* **30,** 1970 (1981).
71. Lodge, A. S., and Meissner, J., *Rheol. Acta,* **11,** 351 (1972); Lodge, A. S., ibid. **14,** 664 (1975).
72. Marrucci, G., and de Cindio, B., *Rheol. Acta,* **19,** 68 (1980).
73. Taylor, C. R., Greco, R., Kramer, O., and Ferry, J. D., *Trans. Soc. Rheol.* **20,** 141 (1976); Noordermeer, J. M., and Ferry, J. D., *J. Polym. Sci.* **14,** 509 (1976).
74. Vrentas, C. M., and Graessley, W. W., *J. Rheol.* **26,** 359 (1982); Pearson, D. S., *IUPAC Proceedings,* 28th Macromolecular Symposium, July, p. 866 (1982).
75. Marrucci, G., and Grizzuti, N., *J. Rheol.* **27,** 433 (1983).
76. McLeish, T. C. B., and Ball, R. C., *J. Polym. Sci.,* to be published.
77. Marrucci, G., and Hermans, J. J., *Macromolecules* **13,** 380 (1980).
78. Viovy, J. L., Monnerie, L., and Tassin, J. F., *J. Polym. Sci. Phys. ed.* **21,** 2427 (1983); Viovy, J. L., *J. Polym. Sci. Phys. ed.* **23,** 2423 (1985).
79. Boué, F., *Adv. Polym. Sci.* to be published; Boué, F., Nierlich, M., and Osaki, K., *Faraday Symp. Chem. Soc.* **18,** 83 (1983): Boué, F., Nierlich, M., Jannink, G., and Ball, R. C., *J. Phys. (Paris)* **43,** 137 (1982); *J. Phys. Lett.* **43,** L585, L593 (1982).
80. Bastide, J., Herz, J., and Boué, F., *J. Physique* **46,** 1967 (1985).
81. Sekiya, M., and Doi, M., *J. Phys. Soc. Jpn* **51,** 3672 (1982).
82. Noolandi, J., and Hong, K. M., *J. Physique Lett.* **45,** L149 (1984).
83. Boué, F., Osaki, K., and Ball, R. C., *J. Polym. Sci.* **23,** 833 (1985).
84. Takahashi, M., Masuda, T., Bessho, N., and Osaki, K., *J. Rheol.* **24,** 517 (1980).
85. Takahashi, M., Masuda, T., Oono, H., and Onogi, S., *Polymer Preprints Japan* **33,** 871 (1984).
86. Meissner, J., *Rheol. Acta,* **10,** 230 (1971).
87. Currie, P. K., *J. Non-Newtonian Fluid Mech.* **11,** 53 (1982).
88. Bernstein, B., Kearsley, E. A., and Zapas, L. J., *Trans. Soc. Rheol.* **7,** 391 (1963).
89. Osaki, K., Ohta, S., Fukuda, M., and Kurata, M., *J. Polym. Sci.* **A14,** 1701 (1976).
90. Chang, W. V., Bloch, R., and Tschoegl, N. W., *Rheol. Acta* **15,** 367 (1976); *J. Polym. Sci.* **A15,** 923 (1977).
91. Wagner, M. H., *Rheol. Acta* **15,** 136 (1976); **16,** 43 (1977).
92. Phillips, M. C., *J. Non-Newtonian Fluid. Mech.* **2,** 109, 123, 139 (1977).
93. Osaki, K., *Proceeding of the 7th International Congress on Rheology* (eds C. Klason and J. Kubat). Chalmers University of Technology, Gothenberg, p. 104 (1976).
94. Doi, M., and Edwards, S. F., *J. C. S. Faraday Trans. 2* **75,** 38 (1979).
95. Bird, R. B., Saab, H. H., and Curtiss, C. F., *J. Phys. Chem.* **86,** 1102 (1982); *J. Chem. Phys.* **77,** 4747, 4758 (1982).
96. Osaki, K., and Doi, M., *Polym. Eng. Rev.* **4,** 35 (1984).
97. Marrucci, G., *Adv. Transport Processes* **5** (eds. A. S. Mujumdar and R. A. Mashelkar). New York (1985).
98. Ramachandran, S., Gao, H. W., and Christiansen, E. B., *J. Rheol.* **25,** 213 (1981).

99. Tanner, R. I., *Trans. Soc. Rheol.* **17,** 365 (1973).
100. Laun, H. M., and Munstedt, H., *Rheol. Acta* **18,** 427 (1979); Munstedt, H., *J. Rheol.* **23,** 421 (1979).
101. Vinogradov, G. V., *Rheol. Acta* **12,** 273 (1973); see also Lin, Y. H., *J. Rheol.* **29,** 65 (1985).
102. Menezes, E. V., and Graessley, W. W., *J. Polym. Sci. Phys.* **20,** 1817 (1982).
103. Takahashi, M., Masuda, T., and Onogi, S., *Polymer Preprints Japan* **29,** 1807 (1980).
104. Osaki, K., and Kurata, M., *J. Polym. Sci. Phys.* **18,** 2421 (1980).
105. Zapas, L. J., *Deformation and Fracture of High Polymers* (eds H. H. Kausch, J. A. Hassell, and R. I. Jaffe). Plenum Press, New York (1974); see also McKenna, G. B., and Zapas, L. J., *J. Rheol.* **23,** 151 (1979); **24,** 367 (1980).
106. Osaki, K., Kimura, S., and Kurata, M., *J. Rheol.* **25,** 549 (1981).
107. Doi, M., *J. Polym. Sci.* **18,** 1891 (1980).
108. Doi, M., *J. Polym. Sci.* **18,** 2055 (1980).
109. Helfand, E., and Pearson, D. S., *J. Polym. Sci.* **20,** 1249 (1982).
110. Pearson, D. S., and Rochefort, W. E., *J. Polym. Sci. Phys. ed.* **20,** 83 (1982).
110a. Marrucci, G., *J. Non-Newtonian Fluid Mech.* **21,** 329–36 (1986).
110b. Marrucci, G. and Grizzuti, N., *J. Non-Newtonian Fluid Mech.* **21,** 319–28 (1986).
111. Klein, J., Fletcher, D., and Fetters, L. J., *Faraday Symp. Chem. Soc.* **18,** 159 (1983).
112. de Gennes, P. G., *J. Phys.* (*Paris*) **36,** 1199 (1975).
113. Kraus, G., and Gruver, J. T., *J. Polym. Sci.* **A3,** 105 (1965); *J. Appl. Polym. Sci.* **9,** 739 (1965).
114. Graessley, W. W., Masuda, T., Roovers, J. E. L., and Hadjichristidis, N., *Macromolecules* **9,** 127 (1976); Graessley, W. W., and Roovers, J., *Macromolecules* **12,** 959 (1979); Raju, V. R., Menezes, E. V., Marin, G., Graessley, W. W., and Fetters, L. J., *Macromolecules* **14,** 1668 (1981).
115. Doi, M., and Kuzuu, N., *J. Polym. Sci. Lett.* **18,** 775 (1980).
116. Pearson, D. S., and Helfand, E., *Macromolecules* **17,** 888 (1984).
117. Kan, H. C., Ferry, J. D., and Fetters, L. J., *Macromolecules* **13,** 1571 (1980).
118. Roovers, J., *Macromolecules* **17,** 1196 (1984).
119. Roovers, J., *J. Polym. Sci.* **23,** 1117 (1985).
120. Smith, B. A., Samulski, E. T., Yu, L. P., and Winnik, M. A., *Phys. Rev. Lett.* **52,** 45 (1984).
121. Green, P. F., Mills, P. J., Palmstrøm, C. J., Mayer, J. W., and Kramer, E. J., *Phys. Rev. Lett.* **53,** 2145 (1984).
122. Masuda, T., Takahashi, M., and Onogi, S., *Appl. Polym. Symp.* **20,** 49 (1973); Bogue, D. C., Masuda, T., Einaga, Y., and Onogi, S., *Polym. J.* **1,** 563 (1970).
123. Kurata, M., *Macromolecules* **17,** 895 (1984).
124. Montfort, J. P., Marin, G., and Monge, P., *Macromolecules* **17,** 1551 (1984).
125. Watanabe, H., and Kotaka, T., *Macromolecules* **17,** 2316 (1984).
126. Masuda, T., Yoshimatsu, S., Takahashi, M., and Onogi, S., *Polymer Preprints Japan* **33,** 2699 (1984).

127. Adachi, K., and Kotaka, T., *Macromolecules* **17,** 120 (1984); **18,** 466 (1985).
128. de Gennes, P. G., *J. Chem. Phys.* **72,** 4756 (1980).
129. Pincus, P., *J. Chem. Phys.* **75,** 1996 (1981).
130. de Gennes, P. G., *C. R. Acad. Sci. Paris* **B291,** 219 (1980).
131. Prager, S., and Tirrell, M., *J. Chem. Phys.* **75,** 5194 (1981); Adolf, D., Tirrell, M., and Prager, S., *J. Polym. Sci.* **23,** 413 (1985).
132. Tulig, T. J., and Tirrell, M., *Macromolecules* **14,** 1501 (1981).
133. de Gennes, P. G., *J. Chem. Phys.* **76,** 3316, 3322 (1982).
134. Kramer, E. J., *Adv. Polym. Sci.* **52/53,** 1 (1983).
135. Evans, K. E., and Donald, A. M., *Polymer* **26,** 101 (1985).
136. de Gennes, P. G., and Léger, L., *Ann. Rev. Phys. Chem.* **33,** 49 (1982).
137. Deam, R. T., and Edwards S. F., *Phil. Trans. R. Soc.* **A280,** 317 (1976).
138. Ball, R. C., Doi, M., Edwards S. F., and Warner, M., *Polymer* **22,** 1010 (1981).

8

DILUTE SOLUTIONS OF RIGID RODLIKE POLYMERS

8.1 Rodlike polymers

Though many polymers are flexible and take a random coil structure, there is a large class of polymers which are not flexible and assume a rodlike structure. For example, some polypeptides or polynucleotides form a helix structure which can be regarded effectively as a rigid rod. If the chemical bonds in the backbone chain consist of double bonds or phenylene rings, the internal rotation of the polymer is severely restricted and the polymer takes an elongated form. These latter type of rodlike polymers are quite important in polymer technology because of their capability of creating very strong fibres, and an increasing amount of research is being done as a result.

The physical properties of the rodlike polymers differ from those of flexible polymers in many respects.

Firstly, an obvious distinction is that rodlike polymers are much larger than flexible polymers with the same molecular weight. If the polymer is a straight rod, its radius of gyration R_g is proportional to the contour length of the polymer, or the molecular weight M, as compared to the relation $R_g \propto M^{\nu}$ ($\nu \simeq 0.6$) for flexible polymers. The elongated form of the polymer is reflected in various dilute solution properties such as the larger intrinsic viscosity, larger relaxation time, or smaller diffusion constant as compared to those of flexible polymers.[1]

Secondly, due to the large molecular anisotropy, rodlike polymers are much more easily oriented by an external field and show large birefringence. This enables us to use electric or magnetic birefringence as a practical tool to study the rotational motion of these polymers.[2]

Thirdly, the distinction between rodlike polymers and flexible polymers becomes more pronounced as concentration increases. Due to their larger size, the interaction of the rodlike polymers becomes important at a much lower concentration than with flexible polymers, and, as we shall show later, the effect of the entanglement is much more remarkable.

Fourthly, but not least, when the concentration becomes sufficiently high, rodlike polymers spontaneously orient towards some direction, and form a liquid crystalline phase.[3] It is this capability of forming a highly ordered phase that produces strong fibres.

In this and the following two chapters, we shall discuss the physical properties of such polymers. Although real polymers have finite rigidity and can bend to some extent, we shall mainly consider the extreme situation, i.e. rigid rodlike polymers. The effect of the flexibility will be discussed only briefly.

Theoretical treatment of rodlike polymers is much easier than for flexible polymers since rodlike polymers can have only two kinds of motion, i.e., translation and rotation. Once the basic equation is set up, mathematical analysis is easy. However, important physics is included in an essential way in the problems of rodlike polymers. In particular, the importance of the orientational degrees of freedom and the peculiar nature of the topological constraints will be seen clearly in this system.

In this chapter we shall discuss the properties of dilute solutions. The properties at higher concentrations will be discussed in later chapters.

8.2 Rotational diffusion

8.2.1 Rotational Brownian motion

Rodlike polymers do two kinds of Brownian motion, translation and rotation. The translational Brownian motion is the random motion of the position vector $\boldsymbol{R}$ of the centre of mass, and the rotational Brownian motion is the random motion of the unit vector $\boldsymbol{u}$ which is parallel to the polymer.

To visualize the rotational Brownian motion we imagine the trajectory of $\boldsymbol{u}(t)$, which is on the surface of the sphere $|\boldsymbol{u}|=1$ (see Fig. 8.1). For short times, the random motion of $\boldsymbol{u}(t)$ can be regarded as Brownian motion on a two-dimensional flat surface, and the mean square displace-

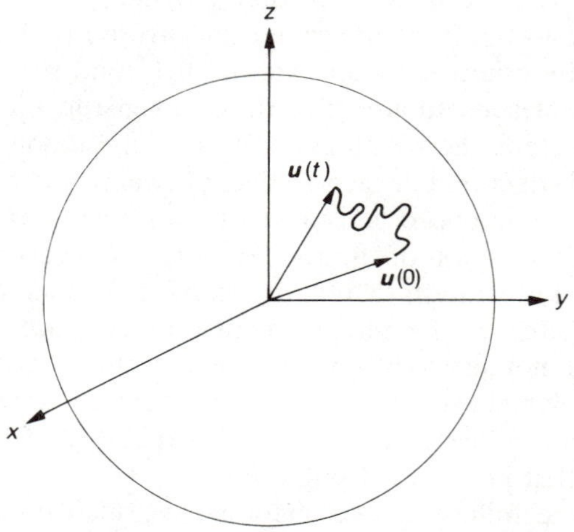

Fig. 8.1. Rotational diffusion.

ment of $\boldsymbol{u}(t)$ in time t is written as

$$\langle (\boldsymbol{u}(t) - \boldsymbol{u}(0))^2 \rangle = 4D_r t \quad (\text{for } D_r t \ll 1). \tag{8.1}$$

The coefficient D_r is called the rotational diffusion constant. Note that the dimension of D_r is $(\text{time})^{-1}$, and is not the same as that of the translational diffusion constant, which is $(\text{length})^2/(\text{time})$.

Equation (8.1) is correct only for $D_r t \ll 1$. To discuss the general case, we have to study the Smoluchowski equation for the rotational Brownian motion. This equation can be derived straightforwardly according to the Kirkwood theory[4] described in Section 3.8. Such a derivation is given in Appendix 8.I. Here we derive it by an elementary method to clarify the underlying physics.

8.2.2 *Hydrodynamics of rotational motion*

As was discussed in Chapter 3, the first step in deriving the Smoluchowski equation is to obtain the phenomenological relation between the force and flux by using the hydrodynamics of the problem.

Consider a rod placed in a quiescent viscous fluid. If an external field exerts a torque $\boldsymbol{N}$ on the rod, the rod will rotate with certain angular velocity $\boldsymbol{\omega}$. For thin rod, we may neglect the rotation around $\boldsymbol{u}$, and assume that both $\boldsymbol{\omega}$ and $\boldsymbol{N}$ are perpendicular to $\boldsymbol{u}$. If $\boldsymbol{N}$ is small, $\boldsymbol{\omega}$ is linear in $\boldsymbol{N}$, and by symmetry, parallel to $\boldsymbol{N}$.

$$\boldsymbol{\omega} = \frac{1}{\zeta_r} \boldsymbol{N}. \tag{8.2}$$

The coefficient ζ_r is called the rotational friction constant.

A simple estimation of ζ_r is done for the 'shish-kebab model' illustrated in Fig. 8.2: the rod is regarded as made up of $N = L/b$ 'beads', which are numbered from $-N/2$ to $N/2$. When the rod rotates with angular velocity $\boldsymbol{\omega}$, the bead n which is separated from the centre by the distance nb moves with velocity $\boldsymbol{V}_n = (\boldsymbol{\omega} \times nb\boldsymbol{u})$.

If the hydrodynamic interaction is neglected, the frictional force acting on the segment n is $-\zeta_0 \boldsymbol{V}_n$, where $\zeta_0 \equiv 3\pi\eta_s b$ is the translational friction constant of the bead. Thus the total torque due to the hydrodynamic friction is given by

$$\begin{aligned}
\boldsymbol{N}_{\text{friction}} &= -\sum_{n=-N/2}^{N/2} nb\boldsymbol{u} \times \zeta_0 \boldsymbol{V}_n \\
&= -\sum_{n=-N/2}^{N/2} nb\boldsymbol{u} \times (\zeta_0 \boldsymbol{\omega} \times nb\boldsymbol{u}) \\
&= -\zeta_0 \sum_{n=-N/2}^{N/2} n^2 b^2 \boldsymbol{\omega} = -(3\pi\eta_s b) b^2 \frac{2}{3} \left(\frac{N}{2}\right)^3 \boldsymbol{\omega} = -\eta_s \frac{\pi L^3}{4} \boldsymbol{\omega}
\end{aligned} \tag{8.3}$$

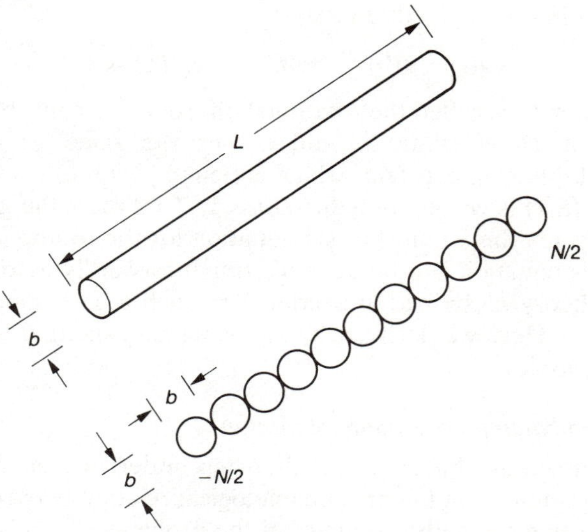

Fig. 8.2. Rodlike polymer and shish-kebab model, which consists of $N = L/b$ beads of diameter b placed along a straight line.

which must balance with the external torque $\boldsymbol{N}$. Hence

$$\zeta_r \simeq \frac{\pi \eta_s L^3}{4}. \tag{8.4}$$

If the hydrodynamic interaction among the beads is taken into account, ζ_r is shown to be (see Appendix 8.I):

$$\zeta_r = \frac{\pi \eta_s L^3}{3 \ln(L/2b)}. \tag{8.5}$$

More precise hydrodynamic calculation for the cylinder gives a correction γ to the denominator:†

$$\zeta_r = \frac{\pi \eta_s L^3}{3(\ln(L/b) - \gamma)}. \tag{8.6}$$

† For a prolate elipsoid, an exact calculation can be done and the result is[5,6,7]

$$\zeta_r = \frac{16\pi}{3} \eta_s a^3 \left(1 - \frac{1}{p^4}\right) \left[\frac{2p^2 - 1}{2p(p^2-1)^{1/2}} \ln\left(\frac{p + (p^2-1)^{1/2}}{p - (p^2-1)^{1/2}}\right) - 1\right]^{-1}$$

where $2a$ is the length of the long axis and p is the aspect ratio. For $p > 2$, the above equation is approximated as

$$\zeta_r = \frac{16\pi \eta_s a^3}{3[2 \ln(2p) - 1]}$$

which agrees with eqn (8.6)

The calculation based on the point force approximation[8] gives $\gamma = 0.8$, while more recent calculation[9,10] indicates that γ weakly depends on L/b.

The torque $\boldsymbol{N}$ is now expressed by the potential $U(\boldsymbol{u})$ of the external field. Consider a small rotation $\delta\boldsymbol{\psi}$ which changes $\boldsymbol{u}$ to $\boldsymbol{u} + \delta\boldsymbol{\psi} \times \boldsymbol{u}$. The work needed for this change is $-\boldsymbol{N} \cdot \delta\boldsymbol{\psi}$, which must be equal to the change in U, i.e.,

$$-\boldsymbol{N} \cdot \delta\boldsymbol{\psi} = U(\boldsymbol{u} + \delta\boldsymbol{\psi} \times \boldsymbol{u}) - U(\boldsymbol{u}) = (\delta\boldsymbol{\psi} \times \boldsymbol{u}) \cdot \frac{\partial}{\partial \boldsymbol{u}} U = \delta\boldsymbol{\psi} \cdot \left(\boldsymbol{u} \times \frac{\partial U}{\partial \boldsymbol{u}}\right) \tag{8.7}$$

Hence

$$\boldsymbol{N} = -\mathcal{R} U \tag{8.8}$$

where†

$$\mathcal{R} \equiv \boldsymbol{u} \times \frac{\partial}{\partial \boldsymbol{u}}. \tag{8.9}$$

The operator $\mathcal{R}$, called the rotational operator, plays the role of the gradient operator $\partial/\partial \boldsymbol{R}$ in translational diffusion.

An important property of $\mathcal{R}$ is the formula of integration by parts, i.e., for the integral over the entire surface of the sphere of $|\boldsymbol{u}| = 1$,

$$\int \mathrm{d}\boldsymbol{u} A(\boldsymbol{u}) \mathcal{R} B(\boldsymbol{u}) = -\int \mathrm{d}\boldsymbol{u} [\mathcal{R} A(\boldsymbol{u})] B(\boldsymbol{u}). \tag{8.10}$$

(In quantum mechanics, $-\mathrm{i}\mathcal{R}$ corresponds to the angular momentum operator so that eqn (8.10) is equivalent to the Hermitian property of this operator.)

Now if the fluid surrounding the rod is flowing with a certain velocity gradient, there will be an additional angular velocity $\boldsymbol{\omega}_0$ of the rod, which again can be calculated by hydrodynamics.‡ For a slender rod, $\boldsymbol{\omega}_0$ is obtained by a simple geometrical reasoning explained in Fig. 8.3:

$$\boldsymbol{\omega}_0 = \boldsymbol{u} \times \boldsymbol{\kappa} \cdot \boldsymbol{u}. \tag{8.11}$$

† In eqn (8.9) $\partial/\partial u_\gamma$ means the partial derivative in which u_x, u_y, u_z are regarded as independent variables. Since $\boldsymbol{u}$ is a unit vector, there are many ways to express U. For example, consider the following three quantities

$$F_1 = u_x, \qquad F_2 = \frac{u_x}{(u_x^2 + u_y^2 + u_z^2)^{1/2}}, \qquad F_3 = (1 - u_y^2 - u_z^2)^{1/2}.$$

All represent the same quantity for the unit vector $\boldsymbol{u}$. It is easily checked that, though $\partial F_1/\partial u_x$, $\partial F_2/\partial u_x$ and $\partial F_3/\partial u_x$ are not equal to each other, $\mathcal{R}F_1$, $\mathcal{R}F_2$, and $\mathcal{R}F_3$ are all equal. Thus the derivative $\mathcal{R}F$ has no ambiguity.

‡ For a spheroid of aspect ratio $p = a/b$, $\boldsymbol{\omega}_0$ is given by[5,6]

$$\boldsymbol{\omega}_0 = \boldsymbol{u} \times \left(\frac{p^2}{p^2+1} \boldsymbol{\kappa} \cdot \boldsymbol{u} - \frac{1}{p^2+1} \boldsymbol{\kappa}^+ \cdot \boldsymbol{u}\right).$$

In the limit of $p \to \infty$, this reduces to eqn (8.11).

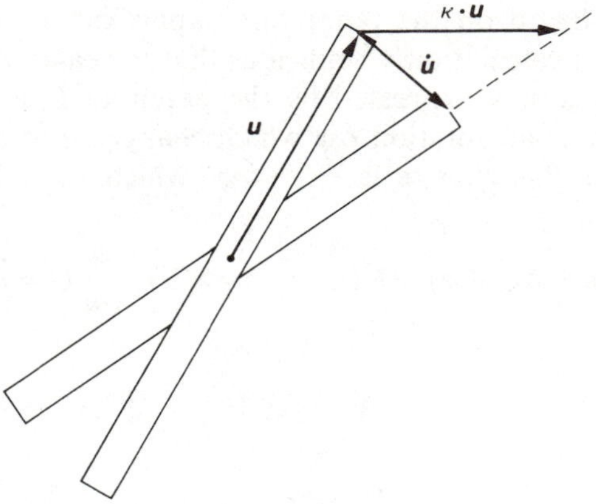

Fig. 8.3. Geometrical meaning of eqn (8.11). If the rod follows the macroscopic velocity gradient, its direction changes as $\dot{\boldsymbol{u}} = \boldsymbol{\kappa} \cdot \boldsymbol{u} - (\boldsymbol{uu}:\boldsymbol{\kappa})\boldsymbol{u}$. Hence the angular velocity $\boldsymbol{\omega}_0$ is given by $\boldsymbol{\omega}_0 = \boldsymbol{u} \times \dot{\boldsymbol{u}} = \boldsymbol{u} \times (\boldsymbol{\kappa} \cdot \boldsymbol{u})$.

Therefore the angular velocity of a rod immersed in a fluid with velocity gradient $\boldsymbol{\kappa}$ and subject to an external potential $U(\boldsymbol{u})$ is given as

$$\boldsymbol{\omega} = -\frac{1}{\zeta_r}\mathscr{R}U + \boldsymbol{u} \times \boldsymbol{\kappa} \cdot \boldsymbol{u}. \tag{8.12}$$

This consequence of the hydrodynamics corresponds to eqn (3.118) for translational motion.

8.2.3 *Smoluchowski equation for rotational motion*

Now it is easy to give an account of the Brownian motion. If $\Psi(\boldsymbol{u}; t)$ is the probability distribution function of $\boldsymbol{u}$, the Brownian motion is included by adding the 'Brownian potential' $k_BT \ln \Psi$ to U. The angular velocity $\boldsymbol{\omega}$ is now given by

$$\boldsymbol{\omega} = -\frac{1}{\zeta_r}\mathscr{R}(k_BT \ln \Psi + U) + \boldsymbol{u} \times \boldsymbol{\kappa} \cdot \boldsymbol{u}. \tag{8.13}$$

For given $\boldsymbol{\omega}$, $\boldsymbol{u}$ changes with the velocity $\boldsymbol{\omega} \times \boldsymbol{u}$, and the equation for the conservation of the probability becomes

$$\frac{\partial \Psi}{\partial t} = -\frac{\partial}{\partial \boldsymbol{u}} \cdot (\boldsymbol{\omega} \times \boldsymbol{u}\Psi) = -\left(\boldsymbol{u} \times \frac{\partial}{\partial \boldsymbol{u}}\right) \cdot \boldsymbol{\omega}\Psi = -\mathscr{R} \cdot (\boldsymbol{\omega}\Psi). \tag{8.14}$$

From eqns (8.13) and (8.14), we have the Smoluchowski equation for rotational diffusion

$$\frac{\partial\Psi}{\partial t}=\frac{1}{\zeta_r}\mathcal{R}\cdot[k_BT\mathcal{R}\Psi+\Psi\mathcal{R}U]-\mathcal{R}\cdot(\boldsymbol{u}\times\boldsymbol{\kappa}\cdot\boldsymbol{u}\Psi)$$

$$=D_r\mathcal{R}\cdot\left[\mathcal{R}\Psi+\frac{\Psi}{k_BT}\mathcal{R}U\right]-\mathcal{R}\cdot(\boldsymbol{u}\times\boldsymbol{\kappa}\cdot\boldsymbol{u}\Psi) \qquad (8.15)$$

where D_r is defined by

$$D_r=\frac{k_BT}{\zeta_r}=\frac{3k_BT(\ln(L/b)-\gamma)}{\pi\eta_sL^3}. \qquad (8.16)$$

We shall later show that D_r agrees with the rotational diffusion constant defined by eqn (8.1).

Note the formal similarity between the rotational diffusion equation and the usual translational diffusion equation: if the gradient operator $\partial/\partial\boldsymbol{R}$ in the translational diffusion equation is replaced by the operator $\mathcal{R}$, the rotational diffusion equation is obtained.

The rotational Brownian motion can also be described by the Langevin equation, but it is rarely used in the problem of rodlike polymers because it is less convenient for calculation than the Smoluchowski equation.

8.3 Translational diffusion

8.3.1 Hydrodynamics of translational motion

It is straightforward to include the translational motion into the Smoluchowski equation. Again the hydrodynamics is considered first. Suppose the rod is moving with the velocity $\boldsymbol{V}$ in a quiescent fluid (see Fig. 8.4). If the rod moves along $\boldsymbol{u}$, the rod will feel a hydrodynamic drag, which is parallel to $\boldsymbol{V}$ and is written as $\zeta_\parallel\boldsymbol{V}$. On the other hand if $\boldsymbol{V}$ is perpendicular to $\boldsymbol{u}$, the drag is again parallel to $\boldsymbol{V}$ and is written as $\zeta_\perp\boldsymbol{V}$.

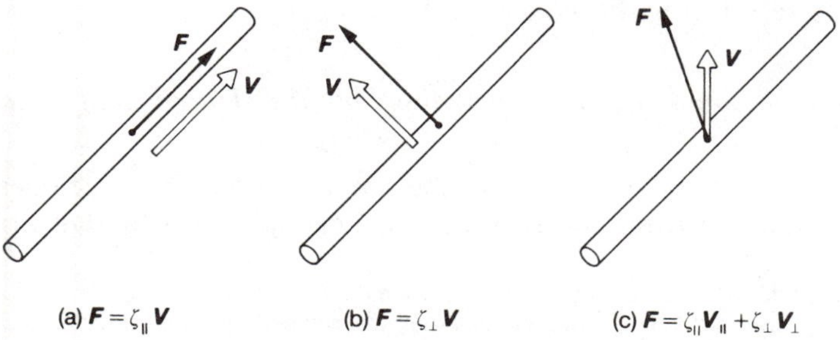

Fig. 8.4. Anisotropy in the translation friction constant. (*a*) $\boldsymbol{V}\parallel\boldsymbol{u}$, (*b*) $V\perp\boldsymbol{u}$, and (*c*) general direction $\boldsymbol{V}=\boldsymbol{V}_\parallel+\boldsymbol{V}_\perp$.

In general the coefficients $\zeta_{\parallel}$ and $\zeta_{\perp}$ are not equal to each other. They are called the parallel and perpendicular components of the translational friction constant, respectively.

Given $\zeta_{\parallel}$ and $\zeta_{\perp}$, the hydrodynamic drag for a translational motion in a general direction is obtained as follows. Since the Stokes equation (eqn (3.102)) is a linear equation, the hydrodynamic drag must be linear in $\boldsymbol{V}$. Thus if $\boldsymbol{V}_{\parallel}$ and $\boldsymbol{V}_{\perp}$ are the parallel and the perpendicular components of $\boldsymbol{V}$, the drag is written as

$$\boldsymbol{F} = \zeta_{\parallel}\boldsymbol{V}_{\parallel} + \zeta_{\perp}\boldsymbol{V}_{\perp}. \tag{8.17}$$

Substituting

$$\boldsymbol{V}_{\parallel} = (\boldsymbol{V}\cdot\boldsymbol{u})\boldsymbol{u} \quad \text{and} \quad \boldsymbol{V}_{\perp} = \boldsymbol{V} - \boldsymbol{V}_{\parallel} \tag{8.18}$$

we have

$$\boldsymbol{F} = \zeta_{\parallel}\boldsymbol{uu}\cdot\boldsymbol{V} + \zeta_{\perp}(\boldsymbol{I} - \boldsymbol{uu})\cdot\boldsymbol{V}. \tag{8.19}$$

The calculation of $\zeta_{\parallel}$ and $\zeta_{\perp}$ based on the Kirkwood theory is given in Appendix 8.I. The result is

$$\zeta_{\parallel} = \frac{2\pi\eta_s L}{\ln(L/b)} \tag{8.20}$$

$$\zeta_{\perp} = 2\zeta_{\parallel}. \tag{8.21}$$

Equation (8.19) is solved for $\boldsymbol{V}$ as

$$\boldsymbol{V} = \left[\frac{1}{\zeta_{\parallel}}\boldsymbol{uu} + \frac{1}{\zeta_{\perp}}(\boldsymbol{I} - \boldsymbol{uu})\right]\cdot\boldsymbol{F}. \tag{8.22}$$

If there is a macroscopic flow $\boldsymbol{v}(\boldsymbol{r}) = \boldsymbol{\kappa}\cdot\boldsymbol{r}$, there is an additional velocity $\boldsymbol{\kappa}\cdot\boldsymbol{R}$ for the rod at point $\boldsymbol{R}$, and eqn (8.22) becomes

$$\boldsymbol{V} = \left[\frac{1}{\zeta_{\parallel}}\boldsymbol{uu} + \frac{1}{\zeta_{\perp}}(\boldsymbol{I} - \boldsymbol{uu})\right]\cdot\boldsymbol{F} + \boldsymbol{\kappa}\cdot\boldsymbol{R}. \tag{8.23}$$

This is the result of hydrodynamic calculation.†

8.3.2 *Smoluchowski equation including both translational and rotational diffusion*

We can now write down the Smoluchowski equation which includes both the rotational and translational motions. Let $\Psi(\boldsymbol{R}, \boldsymbol{u}; t)$ be the probabil-

† Note that in the case of a rod, $\boldsymbol{V}$ is independent of the torque $\boldsymbol{N}$ acting on it. In the general case, this is not true. (Consider for example a screw: the torque turns the screw and causes a translational motion.) The formula for the general case is given in refs 6 and 11, and an example of its application to the Brownian dynamics is given in ref. 12

ity distribution function for the rod in the configuration $(\boldsymbol{R}, \boldsymbol{u})$. The velocity $\boldsymbol{V}$ is given by

$$\boldsymbol{V} = -\left[\frac{1}{\zeta_\parallel}\boldsymbol{uu} + \frac{1}{\zeta_\perp}(\boldsymbol{I} - \boldsymbol{uu})\right] \cdot \frac{\partial}{\partial \boldsymbol{R}}(k_B T \ln \Psi + U) + \boldsymbol{\kappa} \cdot \boldsymbol{R}. \quad (8.24)$$

The angular velocity is again given by eqn (8.13). Substituting this in the continuity equation

$$\frac{\partial \Psi}{\partial t} = -\frac{\partial}{\partial \boldsymbol{R}} \cdot (\boldsymbol{V}\Psi) - \mathcal{R} \cdot (\boldsymbol{\omega}\Psi) \quad (8.25)$$

we get

$$\frac{\partial \Psi}{\partial t} = \frac{\partial}{\partial \boldsymbol{R}} \cdot [D_\parallel \boldsymbol{uu} + D_\perp(\boldsymbol{I} - \boldsymbol{uu})] \cdot \left[\frac{\partial \Psi}{\partial \boldsymbol{R}} + \frac{\Psi}{k_B T}\frac{\partial U}{\partial \boldsymbol{R}}\right] - \frac{\partial}{\partial \boldsymbol{R}} \cdot (\boldsymbol{\kappa} \cdot \boldsymbol{R}\Psi)$$
$$+ D_r \mathcal{R} \cdot \left(\mathcal{R}\Psi + \frac{\Psi}{k_B T}\mathcal{R}U\right) - \mathcal{R} \cdot (\boldsymbol{u} \times \boldsymbol{\kappa} \cdot \boldsymbol{u}\Psi), \quad (8.26)$$

where

$$D_\parallel = \frac{k_B T}{\zeta_\parallel} = \frac{k_B T \ln(L/b)}{2\pi\eta_s L} \quad (8.27)$$

and

$$D_\perp = \frac{k_B T}{\zeta_\perp} = \frac{k_B T \ln(L/b)}{4\pi\eta_s L}. \quad (8.28)$$

The constants $D_\parallel$ and $D_\perp$ characterize the diffusion parallel and perpendicular to the rod axis: if the rod is along the z axis, then the displacement of $\boldsymbol{R}$ in a small time interval Δt is given as

$$\begin{aligned} \langle (R_x(\Delta t) - R_x(0))^2 \rangle &= \langle (R_y(\Delta t) - R_y(0))^2 \rangle = 2D_\perp \Delta t \\ \langle (R_z(\Delta t) - R_z(0))^2 \rangle &= 2D_\parallel \Delta t. \end{aligned} \quad (8.29)$$

Since $D_\parallel > D_\perp$, the rod can move more easily in the direction parallel to the axis than that perpendicular. Due to this anisotropy in the diffusion constant, translational and rotational motions of a rod are generally coupled with each other. For example, a concentration gradient of the rodlike polymer can induce an anisotropy in the orientational distribution. However, the reverse is not true: in a homogeneous system (in which the positional distribution is uniform), the translation–rotation coupling has no effect: if the system is homogeneous, it will remain homogeneous even if the orientational distribution is not isotropic. Thus in a homogeneous system, one can discuss the rotational diffusion using eqn (8.15) instead of the full Smoluchowski equation (8.26).

8.4 Brownian motion in the equilibrium state

Having obtained the Smoluchowski equation, we now study the characteristic features of the Brownian motion of a free polymer ($U=0$ and $\boldsymbol{\kappa}=0$).

8.4.1 *Vector correlation function* $\langle \boldsymbol{u}(t)\cdot\boldsymbol{u}(0)\rangle$

To see the rotational motion, let us consider the time correlation function $\langle \boldsymbol{u}(t)\cdot\boldsymbol{u}(0)\rangle$. According to the general prescription given in Chapter 3, this is calculated by

$$\langle \boldsymbol{u}(t)\cdot\boldsymbol{u}(0)\rangle = \int \mathrm{d}\boldsymbol{u}\,\mathrm{d}\boldsymbol{u}'\boldsymbol{u}\cdot\boldsymbol{u}'G(\boldsymbol{u},\boldsymbol{u}';t)\Psi_{\mathrm{eq}}(\boldsymbol{u}'), \tag{8.30}$$

where Ψ_{eq} is the equilibrium distribution function

$$\Psi_{\mathrm{eq}}(\boldsymbol{u}) = \frac{1}{4\pi} \tag{8.31}$$

and $G(\boldsymbol{u},\boldsymbol{u}';t)$ is the conditional probability that the polymer is in the direction of $\boldsymbol{u}$ at time t, given that it was in the direction $\boldsymbol{u}'$ at time $t=0$. This probability is the Green function of the diffusion equation

$$\frac{\partial}{\partial t}G(\boldsymbol{u},\boldsymbol{u}';t) = D_r\mathcal{R}^2G(\boldsymbol{u},\boldsymbol{u}';t) \tag{8.32}$$

with the initial condition

$$G(\boldsymbol{u},\boldsymbol{u}';t=0) = \delta(\boldsymbol{u}-\boldsymbol{u}'). \tag{8.33}$$

Though the explicit form of $G(\boldsymbol{u},\boldsymbol{u}';t)$ is available (see for example ref. 13 Chapter 7), the time correlation function can be calculated directly as before. The time derivative of $\langle \boldsymbol{u}(t)\cdot\boldsymbol{u}(0)\rangle$ becomes

$$\begin{aligned}\frac{\partial}{\partial t}\langle \boldsymbol{u}(t)\cdot\boldsymbol{u}(0)\rangle &= \int \mathrm{d}\boldsymbol{u}\,\mathrm{d}\boldsymbol{u}'\boldsymbol{u}\cdot\boldsymbol{u}'\left[\frac{\partial}{\partial t}G(\boldsymbol{u},\boldsymbol{u}';t)\right]\Psi_{\mathrm{eq}}(\boldsymbol{u}')\\ &= D_r\int \mathrm{d}\boldsymbol{u}\,\mathrm{d}\boldsymbol{u}'\boldsymbol{u}\cdot\boldsymbol{u}'[\mathcal{R}^2G(\boldsymbol{u},\boldsymbol{u}';t)]\Psi_{\mathrm{eq}}(\boldsymbol{u}').\end{aligned} \tag{8.34}$$

Using eqn (8.10) for the right-hand side, we get

$$\frac{\partial}{\partial t}\langle \boldsymbol{u}(t)\cdot\boldsymbol{u}(0)\rangle = D_r\int \mathrm{d}\boldsymbol{u}\,\mathrm{d}\boldsymbol{u}'[\mathcal{R}^2\boldsymbol{u}\cdot\boldsymbol{u}']G(\boldsymbol{u},\boldsymbol{u}';t)\Psi_{\mathrm{eq}}(\boldsymbol{u}'). \tag{8.35}$$

By a straightforward calculation†

$$\mathcal{R}_\alpha u_\beta = -e_{\alpha\beta\gamma}u_\gamma. \tag{8.36}$$

† Here $e_{\alpha\beta\gamma}$ is Levi Civita's symbol, i.e. $e_{\alpha\beta\gamma}\equiv \boldsymbol{e}_\alpha\cdot(\boldsymbol{e}_\beta\times\boldsymbol{e}_\gamma)$, where $\boldsymbol{e}_\alpha$ is the unit vector in the direction of the α axis.

Applying this twice, we have

$$\mathcal{R}_\alpha^2 u_\beta = -e_{\alpha\beta\gamma}\mathcal{R}_\alpha u_\gamma = e_{\alpha\beta\gamma}e_{\alpha\gamma\mu}u_\mu = -2u_\beta. \tag{8.37}$$

Hence eqn (8.35) is written as

$$\frac{\partial}{\partial t}\langle \boldsymbol{u}(t)\cdot\boldsymbol{u}(0)\rangle = -2D_r\int \mathrm{d}\boldsymbol{u}\,\mathrm{d}\boldsymbol{u}'\boldsymbol{u}\cdot\boldsymbol{u}'G(\boldsymbol{u},\boldsymbol{u}';t)\Psi_{\mathrm{eq}}(\boldsymbol{u}')$$
$$= -2D_r\langle \boldsymbol{u}(t)\cdot\boldsymbol{u}(0)\rangle. \tag{8.38}$$

Since $\langle \boldsymbol{u}(t)\cdot\boldsymbol{u}(0)\rangle$ is equal to 1 at time $t=0$, eqn (8.38) gives

$$\langle \boldsymbol{u}(t)\cdot\boldsymbol{u}(0)\rangle = \exp(-2D_r t). \tag{8.39}$$

The rotational correlation time τ_r is thus given by

$$\tau_r = 1/2D_r \tag{8.40}$$

From eqn (8.39), it follows that

$$\langle(\boldsymbol{u}(t)-\boldsymbol{u}(0))^2\rangle = 2-2\langle \boldsymbol{u}(t)\cdot\boldsymbol{u}(0)\rangle$$
$$= 2(1-\exp(-2D_r t)). \tag{8.41}$$

For $tD_r \ll 1$, eqn (8.41) reduces to eqn (8.1), which gives a clear physical meaning of D_r.

In the same way, one can show[13]

$$\langle \tfrac{3}{2}([\boldsymbol{u}(t)\cdot\boldsymbol{u}(0)]^2-\tfrac{1}{3})\rangle = \exp(-6D_r t) \tag{8.42}$$

or in general

$$\langle P_n(\boldsymbol{u}(t)\cdot\boldsymbol{u}(0))\rangle = \exp(-D_r n(n+1)t) \tag{8.43}$$

where $P_n(x)$ is the Legendre polynomial of n-th order

$$P_n(x) = \frac{1}{2^n n!}\frac{\mathrm{d}^n}{\mathrm{d}x^n}(x^2-1)^n. \tag{8.44}$$

8.4.2 *Translational diffusion*

Consider the mean square displacement of the centre of mass:

$$\phi(t) \equiv \langle(\boldsymbol{R}(t)-\boldsymbol{R}(0))^2\rangle \tag{8.45}$$

This is calculated by essentially the same method as before. Let $G(\boldsymbol{R},\boldsymbol{u},\boldsymbol{R}',\boldsymbol{u}';t)$ be the Green function for the configuration $(\boldsymbol{R},\boldsymbol{u})$,

$$\frac{\partial}{\partial t}G = \left[D_r\mathcal{R}^2 + \frac{\partial}{\partial \boldsymbol{R}}\cdot(D_\parallel \boldsymbol{uu} + D_\perp(\boldsymbol{I}-\boldsymbol{uu}))\cdot\frac{\partial}{\partial \boldsymbol{R}}\right]G \tag{8.46}$$

with the initial condition

$$G(\boldsymbol{R},\boldsymbol{u},\boldsymbol{R}',\boldsymbol{u}';t=0) = \delta(\boldsymbol{R}-\boldsymbol{R}')\delta(\boldsymbol{u}-\boldsymbol{u}'). \tag{8.47}$$

Then $\phi(t)$ is calculated from

$$\phi(t) = \int \mathrm{d}\boldsymbol{R}\, \mathrm{d}\boldsymbol{u}\, \mathrm{d}\boldsymbol{R}'\, \mathrm{d}\boldsymbol{u}' (\boldsymbol{R} - \boldsymbol{R}')^2 G(\boldsymbol{R}, \boldsymbol{u}, \boldsymbol{R}', \boldsymbol{u}'; t) \Psi_{\mathrm{eq}}(\boldsymbol{R}', \boldsymbol{u}'). \quad (8.48)$$

We again evaluate the time derivative of $\phi(t)$ using eqn (8.46). The resulting equation is written, after integration by parts, as

$$\frac{\partial}{\partial t}\phi(t) = \int \mathrm{d}\boldsymbol{R}\, \mathrm{d}\boldsymbol{u}\, \mathrm{d}\boldsymbol{R}'\, \mathrm{d}\boldsymbol{u}' G \Psi_{\mathrm{eq}}$$

$$\times \left[D_r \mathcal{R}^2 + \frac{\partial}{\partial \boldsymbol{R}} \cdot (D_\parallel \boldsymbol{uu} + D_\perp (\boldsymbol{I} - \boldsymbol{uu})) \cdot \frac{\partial}{\partial \boldsymbol{R}} \right] (\boldsymbol{R} - \boldsymbol{R}')^2$$

$$= \int \mathrm{d}\boldsymbol{R}\, \mathrm{d}\boldsymbol{u}\, \mathrm{d}\boldsymbol{R}'\, \mathrm{d}\boldsymbol{u}' G \Psi_{\mathrm{eq}} (2D_\parallel + 4D_\perp) = 2(D_\parallel + 2D_\perp). \quad (8.49)$$

Hence

$$\phi(t) = 2(D_\parallel + 2D_\perp)t. \quad (8.50)$$

Thus $\phi(t)$ increases linearly with t and the diffusion constant D_G defined by

$$D_G = \lim_{t \to \infty} \frac{1}{6t} \langle (\boldsymbol{R}(t) - \boldsymbol{R}(0))^2 \rangle \quad (8.51)$$

is given by

$$D_G = \frac{D_\parallel + 2D_\perp}{3} = \frac{\ln(L/b)}{3\pi\eta_s L} k_B T. \quad (8.52)$$

This formula can also be directly derived from the Kirkwood formula for the diffusion constant eqn (4.102).

It must be noted that although $\phi(t)$ increases linearly with time, the diffusion of $\boldsymbol{R}$ is not Fickian because of the translation–rotation coupling. Indeed as will be shown below, $G(\boldsymbol{R}, \boldsymbol{u}, \boldsymbol{R}', \boldsymbol{u}'; t)$ is not Gaussian in $\boldsymbol{R} - \boldsymbol{R}'$. The Fickian diffusion is recovered if the relevant length-scale is much larger than L.

8.4.3 Dynamic light scattering

As before, the Brownian motion of the polymer can be studied by dynamic light scattering.[13] If we take the shish-kebab model shown in Fig. 8.2, the dynamical structure factor is given by

$$g(\boldsymbol{k}, t) = \frac{1}{N^2} \sum_{n,m=-N/2}^{N/2} \langle \exp[\mathrm{i}\boldsymbol{k} \cdot (\boldsymbol{R}_n(t) - \boldsymbol{R}_m(0))] \rangle. \quad (8.53)$$

(Note the normalization of eqn (8.53) is chosen so that $g(0, 0) = 1$, which is different from that chosen for flexible polymers.) Since

$$\boldsymbol{R}_n = \boldsymbol{R} + nb\boldsymbol{u} \tag{8.54}$$

the sum over n is

$$\sum_{n=-N/2}^{N/2} \exp(\mathrm{i}\boldsymbol{k}\cdot\boldsymbol{R}_n) = \exp(\mathrm{i}\boldsymbol{k}\cdot\boldsymbol{R}) \int_{-N/2}^{N/2} \mathrm{d}n \exp(\mathrm{i}\boldsymbol{k}\cdot\boldsymbol{u}nb)$$

$$= 2\exp(\mathrm{i}\boldsymbol{k}\cdot\boldsymbol{R}) \frac{\sin(\boldsymbol{k}\cdot\boldsymbol{u}Nb/2)}{\boldsymbol{k}\cdot\boldsymbol{u}b} = N\exp(\mathrm{i}\boldsymbol{k}\cdot\boldsymbol{R}) \frac{\sin(\boldsymbol{k}\cdot\boldsymbol{u}L/2)}{\boldsymbol{k}\cdot\boldsymbol{u}L/2}. \tag{8.55}$$

Thus $g(\boldsymbol{k}, t)$ is expressed by $\boldsymbol{R}(t)$ and $\boldsymbol{u}(t)$ as

$$g(\boldsymbol{k}, t) = \left\langle \exp(\mathrm{i}\boldsymbol{k}\cdot[\boldsymbol{R}(t) - \boldsymbol{R}(0)]) \frac{\sin[\boldsymbol{K}\cdot\boldsymbol{u}(t)]}{\boldsymbol{K}\cdot\boldsymbol{u}(t)} \frac{\sin[\boldsymbol{K}\cdot\boldsymbol{u}(0)]}{\boldsymbol{K}\cdot\boldsymbol{u}(0)} \right\rangle \tag{8.56}$$

where

$$\boldsymbol{K} = \boldsymbol{k}L/2. \tag{8.57}$$

Let $G_{\boldsymbol{k}}(\boldsymbol{u}, \boldsymbol{u}'; t)$ be the Fourier transform of the Green function $G(\boldsymbol{R}, \boldsymbol{u}, \boldsymbol{R}', \boldsymbol{u}'; t)$, which depends only on $\boldsymbol{R} - \boldsymbol{R}'$,

$$G_{\boldsymbol{k}}(\boldsymbol{u}, \boldsymbol{u}'; t) \equiv \int \mathrm{d}\boldsymbol{R}\, \mathrm{e}^{\mathrm{i}\boldsymbol{k}\cdot\boldsymbol{R}} G(\boldsymbol{R}, \boldsymbol{u}, 0, \boldsymbol{u}'; t) \tag{8.58}$$

then

$$g(\boldsymbol{k}, t) = \int \mathrm{d}\boldsymbol{u} \int \mathrm{d}\boldsymbol{u}' G_{\boldsymbol{k}}(\boldsymbol{u}, \boldsymbol{u}'; t) \frac{\sin(\boldsymbol{K}\cdot\boldsymbol{u})}{\boldsymbol{K}\cdot\boldsymbol{u}} \frac{\sin(\boldsymbol{K}\cdot\boldsymbol{u}')}{\boldsymbol{K}\cdot\boldsymbol{u}'} \Psi_{\mathrm{eq}}(\boldsymbol{u}') \tag{8.59}$$

From eqns (8.46) and (8.58), $G_{\boldsymbol{k}}(\boldsymbol{u}, \boldsymbol{u}'; t)$ satisfies

$$\left(\frac{\partial}{\partial t} + \tilde{\Gamma}\right) G_{\boldsymbol{k}}(\boldsymbol{u}, \boldsymbol{u}'; t) = 0 \tag{8.60}$$

with

$$\tilde{\Gamma} = -D_r \mathcal{R}^2 + D_{\parallel}(\boldsymbol{k}\cdot\boldsymbol{u})^2 + D_{\perp}[\boldsymbol{k}^2 - (\boldsymbol{k}\cdot\boldsymbol{u})^2]. \tag{8.61}$$

The solution of eqn (8.60) is involved, so here we shall briefly describe its characteristic aspects. The limiting cases are:

(i) $|\boldsymbol{K}| \ll 1$, i.e., $|\boldsymbol{k}|\, L \ll 1$.

In this case, it is intuitively obvious that $g(\boldsymbol{k}, t)$ is described by the Fickian diffusion with the diffusion constant D_G, so that

$$g(\boldsymbol{k}, t) = \exp(-D_G \boldsymbol{k}^2 t). \tag{8.62}$$

A formal justification of this is made by considering the eigenfunction expansion of $G_{\boldsymbol{k}}$. Let ψ_p and λ_p be the eigenfunctions and the corresponding eigenvalues of $\tilde{\Gamma}$.

$$\tilde{\Gamma}\psi_p = \lambda_p \psi_p \tag{8.63}$$

Then

$$G_{\boldsymbol{k}} = \sum_p \exp(-\lambda_p t)\psi_p(\boldsymbol{u})\psi_p(\boldsymbol{u}'). \tag{8.64}$$

Substituting this into eqn (8.59) and using $K \ll 1$, we have

$$\begin{aligned} g(\boldsymbol{k}, t) &= \frac{1}{4\pi}\sum_p \exp(-\lambda_p t)\left(\int \mathrm{d}\boldsymbol{u}\psi_p(\boldsymbol{u})\frac{\sin(\boldsymbol{K}\cdot\boldsymbol{u})}{\boldsymbol{K}\cdot\boldsymbol{u}}\right)^2 \\ &\simeq \frac{1}{4\pi}\sum_p \exp(-\lambda_p t)\left(\int \mathrm{d}\boldsymbol{u}\psi_p(u)\right)^2. \end{aligned} \tag{8.65}$$

If $\boldsymbol{k} = 0$, the eigenfunctions are given by spherical harmonics $Y_{lm}(\boldsymbol{u})$ with the eigenvalues $\lambda_{lm}^{(0)} = D_r l(l+1)$ and only the term of the lowest eigenfunction $Y_{00}(u) = 1/\surd(4\pi)$ remains in the sum of eqn (8.65). If $\boldsymbol{k}$ is small, the relaxation is still dominated by the eigenfunction of the lowest eigenvalue

$$g(\boldsymbol{k},t) = \exp(-\lambda_0 t). \tag{8.66}$$

λ_0 is obtained by perturbation theory:

$$\lambda_0 = \int \mathrm{d}\boldsymbol{u} Y_{00}\tilde{\Gamma} Y_{00} \Big/ \int \mathrm{d}\boldsymbol{u} Y_{00}^2 = \frac{D_\parallel + 2D_\perp}{3}\boldsymbol{k}^2 = D_G \boldsymbol{k}^2 \tag{8.67}$$

which justifies eqn (8.62).

For the above perturbation calculation to be justified, λ_0 must be much smaller than the next smallest eigenvalue $\lambda_{2m}^{(0)} = 2D_r$, i.e.,

$$D_G \boldsymbol{k}^2 \ll 2D_r \tag{8.68}$$

which is rewritten using eqns (8.16) and (8.52) as

$$|\boldsymbol{k}|\, L \ll 1. \tag{8.69}$$

Equation (8.69) indicates that if the relevant length-scale is much larger than L, the diffusion can be regarded as Fickian with the diffusion constant D_G.

(ii) $|\boldsymbol{K}| \gg 1$, i.e., $|\boldsymbol{k}|\, L \gg 1$.

In this case, $g(\boldsymbol{k}, t)$ is not expressed by a single exponential. However, the initial decay rate is calculated easily:

$$\begin{aligned} \Gamma_{\boldsymbol{k}}^{(0)} &= -\frac{\mathrm{d}}{\mathrm{d}t}\ln(g(\boldsymbol{k}, t))|_{t=0} \\ &= \frac{1}{g(\boldsymbol{k}, 0)}\int \mathrm{d}\boldsymbol{u}\frac{\sin(\boldsymbol{K}\cdot\boldsymbol{u})}{\boldsymbol{K}\cdot\boldsymbol{u}}\tilde{\Gamma}\frac{\sin(\boldsymbol{K}\cdot\boldsymbol{u})}{\boldsymbol{K}\cdot\boldsymbol{u}}\Psi_{\mathrm{eq}}(\boldsymbol{u}) \\ &= \int_{-1}^{1} \mathrm{d}\xi \frac{\sin(K\xi)}{K\xi}\tilde{\Gamma}\frac{\sin(K\xi)}{K\xi} \Big/ \int_{-1}^{1} \mathrm{d}\xi\left(\frac{\sin(K\xi)}{K\xi}\right)^2 \end{aligned} \tag{8.70}$$

where

$$\xi = \boldsymbol{K} \cdot \boldsymbol{u}/|\boldsymbol{K}| \quad \text{and} \quad K = |\boldsymbol{K}|. \tag{8.71}$$

By using the relation

$$\mathcal{R}F(\xi) = (\mathcal{R}\xi)\frac{\partial F}{\partial \xi} = \frac{\boldsymbol{u} \times \boldsymbol{K}}{|\boldsymbol{K}|}\frac{\partial F}{\partial \xi} \tag{8.72}$$

we have, after some calculation,

$$\tilde{\Gamma}\left[\frac{\sin(K\xi)}{K\xi}\right] = \left[-D_r\frac{\partial}{\partial \xi}(1-\xi^2)\frac{\partial}{\partial \xi} + D_{\parallel}\boldsymbol{k}^2\xi^2 + D_{\perp}\boldsymbol{k}^2(1-\xi^2)\right]\left(\frac{\sin(K\xi)}{K\xi}\right). \tag{8.73}$$

The integral over ξ is carried out analytically by using the fact that K is large: for example,

$$\int_{-1}^{1} \mathrm{d}\xi\left(\frac{\sin(K\xi)}{K\xi}\right)^2 = \frac{1}{K}\int_{-K}^{K} \mathrm{d}\eta\left(\frac{\sin \eta}{\eta}\right)^2 = \frac{1}{K}\int_{-\infty}^{\infty} \mathrm{d}\eta\left(\frac{\sin \eta}{\eta}\right)^2 = \frac{\pi}{K}.$$

Straightforward calculation gives finally[16]

$$\Gamma_k^{(0)} = D_{\perp}\boldsymbol{k}^2 + \frac{L^2}{12}D_r\boldsymbol{k}^2. \tag{8.74}$$

More detailed studies are given in the literature.[13–18]

Since rodlike polymers have a large optical anisotropy, they have a significant depolarized light scattering, which is particularly suitable for studying rotational diffusion. In the small-angle regime $|\boldsymbol{k}|\,L \ll 1$, the dynamic structure factor is written as[13]

$$g_{\mathrm{dep}}(\boldsymbol{k}, t) \propto \exp(-D_G\boldsymbol{k}^2 t - 6D_r t), \tag{8.75}$$

the decay of which is mainly determined by D_r.

8.5 Orientation by an electric field

8.5.1 *The effect of an electric field*

An electric or magnetic field can orient the polymer, and measurement of this process gives information on the rotational motion of polymers in solution.[19,20] Using the Smoluchowski equation, we shall consider the orientation caused by an electric field.

Elementary electrostatics says that if an object with dipole moment $\boldsymbol{p}$ is placed in an electric field $\boldsymbol{E}(t)$, it feels a torque

$$\boldsymbol{N} = \boldsymbol{p} \times \boldsymbol{E}. \tag{8.76}$$

The dipole moment $\boldsymbol{p}$ consists of two parts, the permanent dipole $\boldsymbol{p}_p$ and the induced dipole $\boldsymbol{p}_i$. For thin, rodlike polymers the permanent dipole is always parallel to $\boldsymbol{u}$, the direction of the polymer, and is written as

$$\boldsymbol{p}_p = \mu \boldsymbol{u} \tag{8.77}$$

where μ is the magnitude of the permanent dipole moment. On the other hand, the induced dipole moment is written using the parallel and perpendicular polarizability $\alpha_\parallel$ and $\alpha_\perp$ as

$$\begin{aligned}\boldsymbol{p}_i &= \alpha_\parallel(\boldsymbol{u}\cdot\boldsymbol{E})\boldsymbol{u} + \alpha_\perp(\boldsymbol{E} - (\boldsymbol{u}\cdot\boldsymbol{E})\boldsymbol{u}) \\ &= \left[\frac{\alpha_\parallel + 2\alpha_\perp}{3}\boldsymbol{I} + (\alpha_\parallel - \alpha_\perp)\left(\boldsymbol{uu} - \frac{\boldsymbol{I}}{3}\right)\right]\cdot\boldsymbol{E} \equiv \boldsymbol{\alpha}\cdot\boldsymbol{E}.\end{aligned} \tag{8.78}$$

Hence the torque acting on the polymer is written as

$$\begin{aligned}\boldsymbol{N} &= (\boldsymbol{p}_p + \boldsymbol{p}_i)\times\boldsymbol{E} \\ &= \mu\boldsymbol{u}\times\boldsymbol{E} + \Delta\alpha(\boldsymbol{E}\cdot\boldsymbol{u})\boldsymbol{u}\times\boldsymbol{E}\end{aligned} \tag{8.79}$$

where

$$\Delta\alpha = \alpha_\parallel - \alpha_\perp. \tag{8.80}$$

The potential which gives such a torque is

$$U = -\mu\boldsymbol{u}\cdot\boldsymbol{E} - \tfrac{1}{2}\Delta\alpha(\boldsymbol{E}\cdot\boldsymbol{u})^2. \tag{8.81}$$

The orientation of the polymers can be studied by measurement of the dipole moment

$$\langle \boldsymbol{p} \rangle = \langle \mu\boldsymbol{u} + \boldsymbol{\alpha}\cdot\boldsymbol{E} \rangle \tag{8.82}$$

or the birefringence. The refractive index tensor $\hat{n}_{\alpha\beta}$ is written as

$$\hat{n}_{\alpha\beta} = n\delta_{\alpha\beta} + A\langle u_\alpha u_\beta - \tfrac{1}{3}\delta_{\alpha\beta}\rangle \tag{8.83}$$

where A is a constant which includes contributions from both the form birefringence and the intrinsic birefringence.

8.5.2 Dielectric relaxation

We now calculate the average dipole moment under a given electric field $\boldsymbol{E}(t)$. The diffusion equation to be solved is

$$\frac{\partial\Psi}{\partial t} = D_r\mathcal{R}\cdot\left[\mathcal{R}\Psi + \frac{\Psi}{k_BT}\mathcal{R}U\right]. \tag{8.84}$$

To obtain $\langle \boldsymbol{u} \rangle$, we multiply both sides of eqn (8.84) by $\boldsymbol{u}$ and integrate over $\boldsymbol{u}$.

$$\frac{\partial}{\partial t}\langle \boldsymbol{u} \rangle = D_r\int \mathrm{d}\boldsymbol{u}\boldsymbol{u}\left(\mathcal{R}^2\Psi + \mathcal{R}\frac{\Psi}{k_BT}\cdot\mathcal{R}U\right). \tag{8.85}$$

Integration by parts leads to

$$\frac{\partial}{\partial t}\langle \boldsymbol{u}\rangle = D_r\Big(\langle \mathcal{R}^2\boldsymbol{u}\rangle - \frac{1}{k_BT}\langle \mathcal{R}\boldsymbol{u}\cdot\mathcal{R}U\rangle\Big). \tag{8.86}$$

As before, $\mathcal{R}^2\boldsymbol{u}$ gives $-2\boldsymbol{u}$, and $\langle \mathcal{R}\boldsymbol{u}\cdot\mathcal{R}U\rangle$ is calculated from eqn (8.81). Hence

$$\frac{\partial}{\partial t}\langle \boldsymbol{u}\rangle = D_r\Big(-2\langle \boldsymbol{u}\rangle - \frac{\mu}{k_BT}\langle(\boldsymbol{u}\cdot\boldsymbol{E})\boldsymbol{u}-\boldsymbol{E}\rangle - \frac{\Delta\alpha}{k_BT}\langle(\boldsymbol{E}\cdot\boldsymbol{u})^2\boldsymbol{u}-(\boldsymbol{E}\cdot\boldsymbol{u})\boldsymbol{E}\rangle\Big). \tag{8.87}$$

We shall consider the linear response, in which case we can neglect the third term and evaluate the average in the second term for the isotropic distribution function of $\boldsymbol{u}$:

$$\langle(\boldsymbol{u}\cdot\boldsymbol{E})\boldsymbol{u}-\boldsymbol{E}\rangle_0 = \int\frac{\mathrm{d}\boldsymbol{u}}{4\pi}[(\boldsymbol{u}\cdot\boldsymbol{E})\boldsymbol{u}-\boldsymbol{E}] = -\tfrac{2}{3}\boldsymbol{E}. \tag{8.88}$$

Hence

$$\frac{\partial}{\partial t}\langle \boldsymbol{u}\rangle = -2D_r\langle \boldsymbol{u}\rangle + \frac{2D_r\mu}{3k_BT}\boldsymbol{E}. \tag{8.89}$$

The solution of this equation is

$$\langle \boldsymbol{u}\rangle = \int_{-\infty}^{t}\mathrm{d}t'\exp(-2D_r(t-t'))\frac{2D_r\mu}{3k_BT}\boldsymbol{E}(t'). \tag{8.90}$$

From eqns (8.82) and (8.90), the dipole moment is given by

$$\langle \boldsymbol{p}\rangle = \frac{2D_r\mu^2}{3k_BT}\int_{-\infty}^{t}\mathrm{d}t'\exp(-2D_r(t-t'))\boldsymbol{E}(t') + \frac{2\alpha_\perp+\alpha_\parallel}{3}\boldsymbol{E}(t). \tag{8.91}$$

For an oscillating electric field where

$$\boldsymbol{E}(t) = \mathrm{Re}[\boldsymbol{E}\exp(\mathrm{i}\omega t)], \tag{8.92}$$

the dipole moment is calculated as

$$\langle \boldsymbol{p}\rangle = \mathrm{Re}[\alpha^*(\omega)\boldsymbol{E}\exp(\mathrm{i}\omega t)] \tag{8.93}$$

where

$$\alpha^*(\omega) = \frac{\mu^2}{3k_BT}\frac{1}{1+\mathrm{i}\omega\tau} + \frac{\alpha_\parallel+2\alpha_\perp}{3} \tag{8.94}$$

with

$$\tau = 1/2D_r = \frac{\pi\eta_s L^3}{6k_BT(\ln(L/b)-\gamma)}. \tag{8.95}$$

$\alpha^*(\omega)$ is called the complex polarizability. Note that the induced dipole moment shows no frequency dependence. Dielectric relaxation experiments have been carried out for various polymers.[19,21–23]

8.5.3 *Electric birefringence*

Calculation of the electric birefringence is slightly more complicated because it is a second-order effect in the electric field.[20] Let us consider the case that a time-dependent, weak electric field $E(t)$ is applied in the z direction. Straightforward perturbation calculation gives[24–26]

$$\langle u_z^2 - u_x^2\rangle = \tfrac{2}{5}D_r\frac{\Delta\alpha}{k_BT}\int_{-\infty}^{t} dt_1 \exp[-6D_r(t-t_1)]E(t_1)^2$$
$$+\tfrac{4}{5}D_r^2\left(\frac{\mu}{k_BT}\right)^2\int_{-\infty}^{t} dt_1 \exp[-6D_r(t-t_1)]E(t_1)$$
$$\times\int_{\infty}^{t_1} dt_2 \exp[-2D_r(t_1-t_2)]E(t_2). \tag{8.96}$$

The difference in the refractive index $\Delta n = \hat{n}_{zz} - \hat{n}_{xx}$ is proportional to $\langle u_z^2 - u_x^2\rangle$. The response of Δn for various histories is immediately calculated from eqn (8.96).

(i) Steady state:
If the electric field is constant, eqn (8.96) gives

$$\langle u_z^2 - u_x^2\rangle = \tfrac{1}{15}\left[\frac{\Delta\alpha}{k_BT} + \left(\frac{\mu}{k_BT}\right)^2\right]E^2. \tag{8.97}$$

Thus

$$\Delta n = K_0E^2$$

with

$$K_0 = A\tfrac{1}{15}\left[\frac{\Delta\alpha}{k_BT} + \left(\frac{\mu}{k_BT}\right)^2\right] \tag{8.98}$$

which is called the Kerr constant. It consists of a permanent and an induced dipole term. If the polymer has a dipole moment, their ratio

$$R = \frac{\mu^2}{\Delta\alpha k_BT} \tag{8.99}$$

is much larger than unity since both μ and $\Delta\alpha$ increase in proportion to the molecular weight.

(ii) Decay: a constant field E is switched off at $t=0$, i.e., $E(t) = E\Theta(-t)$, and

$$\Delta n(t) = K_0E^2\exp(-6D_rt). \tag{8.100}$$

(iii) Rise: a constant field is applied at $t = 0$, i.e., $E(t) = E\Theta(t)$, and

$$\Delta n(t) = K_0 E^2\left(1 - \frac{3R}{2(R+1)}\exp(-2D_r t) + \frac{R-2}{2(R+1)}\exp(-6D_r t)\right). \tag{8.101}$$

(iv) Reversal: the direction of a constant field is changed at $t = 0$, i.e., $E = -E\Theta(-t) + E\Theta(t)$, and

$$\Delta n(t) = K_0 E^2\left[1 + \frac{3R}{R+1}(\exp(-6D_r t) - \exp(-2D_r t))\right]. \tag{8.102}$$

(v) Oscillation: $E(t) = E\cos(\omega t)$. The response for this field is written as

$$\Delta n(t) = E^2[K_{dc}(\omega) + \mathrm{Re}\, K_{ac}^*(\omega)\exp(2i\omega t)] \tag{8.103}$$

where

$$K_{dc} = \frac{K_0}{2(R+1)}\left(\frac{R}{1+(\omega\tau)^2} + 1\right) \tag{8.104}$$

and

$$K_{ac}(\omega) = \frac{K_0}{2(R+1)}\left[\frac{R}{(1+2i\omega\tau/3)(1+i\omega\tau)} + \frac{1}{(1+2i\omega\tau/3)}\right] \tag{8.105}$$

where τ is given by eqn (8.95).

8.6 Linear viscoelasticity

8.6.1 Expression for the stress tensor

We now consider the viscoelasticity of a solution of rodlike polymers. As was discussed in Chapter 3, the stress tensor consists of two terms, the elastic stress $\boldsymbol{\sigma}^{(E)}$ and the viscous stress $\boldsymbol{\sigma}^{(V)}$.

The elastic stress is related to the change in the free energy $\mathscr{A}$ (per volume) for a virtual deformation $\delta\varepsilon_{\alpha\beta}$ as

$$\delta\mathscr{A} = \sigma_{\alpha\beta}^{(E)}\delta\varepsilon_{\alpha\beta}. \tag{8.106}$$

Since the free energy is given by

$$\mathscr{A} = \nu\int d\boldsymbol{u}(k_B T\Psi \ln\Psi + \Psi U) \tag{8.107}$$

(where ν is the number of polymers in unit volume), the change in $\mathscr{A}$ is written as

$$\delta\mathscr{A} = \nu\int d\boldsymbol{u}(k_B T\delta\Psi \ln\Psi + k_B T\delta\Psi + \delta\Psi U). \tag{8.108}$$

The change $\delta\Psi$ is calculated using the Smoluchowski equation (8.15). For the instantaneous deformation, the velocity gradient $\kappa_{\alpha\beta} = \delta\varepsilon_{\alpha\beta}/\delta t$ (δt being the duration time of the deformation) dominates the time evolution of Ψ, so that

$$\frac{\partial\Psi}{\partial t} = -\mathcal{R}\cdot(\boldsymbol{u}\times\boldsymbol{\kappa}\cdot\boldsymbol{u}\Psi) \quad \text{during the deformation.} \tag{8.109}$$

Hence

$$\delta\Psi = -\mathcal{R}\cdot(\boldsymbol{u}\times\boldsymbol{\kappa}\cdot\boldsymbol{u}\Psi)\delta t = -\mathcal{R}\cdot(\boldsymbol{u}\times\delta\boldsymbol{\varepsilon}\cdot\boldsymbol{u}\Psi). \tag{8.110}$$

Substituting eqn (8.110) and integrating by parts, we have

$$\begin{aligned}\delta\mathcal{A} &= \nu\int d\boldsymbol{u}(k_BT(\boldsymbol{u}\times\delta\boldsymbol{\varepsilon}\cdot\boldsymbol{u})\cdot\mathcal{R}\Psi + (\boldsymbol{u}\times\delta\boldsymbol{\varepsilon}\cdot\boldsymbol{u}\Psi)\cdot\mathcal{R}U)\\ &= \nu\int d\boldsymbol{u}\Psi(-k_BT\underline{\mathcal{R}\cdot(\boldsymbol{u}\times\delta\boldsymbol{\varepsilon}\cdot\boldsymbol{u})} + (\boldsymbol{u}\times\delta\boldsymbol{\varepsilon}\cdot\boldsymbol{u})\cdot\mathcal{R}U).\end{aligned} \tag{8.111}$$

The underlined term can be calculated using eqn (8.36):

$$\mathcal{R}\cdot(\boldsymbol{u}\times\delta\boldsymbol{\varepsilon}\cdot\boldsymbol{u}) = -3\delta\varepsilon_{\alpha\beta}(u_\alpha u_\beta - \tfrac{1}{3}\delta_{\alpha\beta}). \tag{8.112}$$

Hence

$$\begin{aligned}\delta\mathcal{A} &= \nu\delta\varepsilon_{\alpha\beta}\int d\boldsymbol{u}\Psi(3k_BT(u_\alpha u_\beta - \tfrac{1}{3}\delta_{\alpha\beta}) - (\boldsymbol{u}\times\mathcal{R}U)_\alpha u_\beta)\\ &= \nu\delta\varepsilon_{\alpha\beta}(3k_BT\langle u_\alpha u_\beta - \tfrac{1}{3}\delta_{\alpha\beta}\rangle - \langle(\boldsymbol{u}\times\mathcal{R}U)_\alpha u_\beta\rangle).\end{aligned} \tag{8.113}$$

Thus

$$\sigma^{(E)}_{\alpha\beta} = 3\nu k_BT\langle u_\alpha u_\beta - \tfrac{1}{3}\delta_{\alpha\beta}\rangle - \nu\langle(\boldsymbol{u}\times\mathcal{R}U)_\alpha u_\beta\rangle. \tag{8.114}$$

If $U = 0$, the stress is given by

$$\sigma^{(E)}_{\alpha\beta} = 3\nu k_BTS_{\alpha\beta} \tag{8.115}$$

with

$$S_{\alpha\beta} = \langle u_\alpha u_\beta - \tfrac{1}{3}\delta_{\alpha\beta}\rangle, \tag{8.116}$$

which is called the orientational tensor. Note that this stress comes entirely from the Brownian potential. That the Brownian motion of a rod can produce a stress may be understood in the following way. Suppose a rod is placed at the origin along the z axis (see Fig. 8.5). If the rod rotates by Brownian motion, it will create flow of the surrounding fluid as in (a) or (b) of Fig. 8.5, depending on the direction of the rotation. Whichever direction the rod rotates, the fluid around the z axis comes towards the rod while the fluid in the $x-y$ plane goes away from the rod as shown in (c). This fluid motion is equivalent to the appearance of the stress $\sigma_{zz} - \sigma_{xx}$.

The viscous stress is related to the hydrodynamic energy dissipation W by (see Section 3.8.4)

$$W = \kappa_{\alpha\beta}\sigma^{(V)}_{\alpha\beta} \tag{8.117}$$

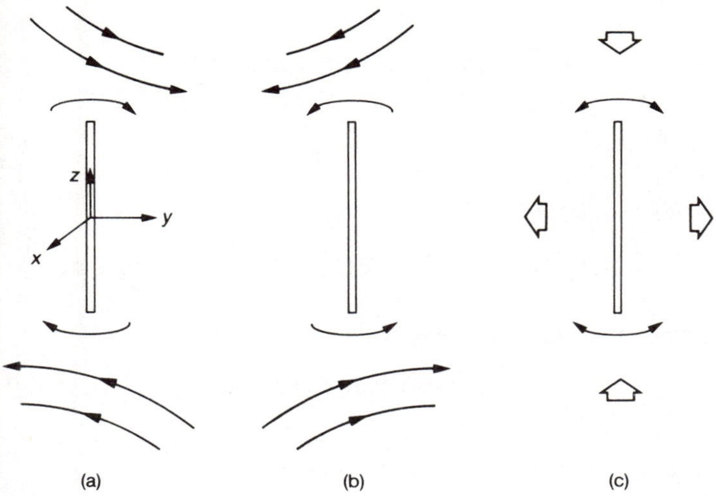

Fig. 8.5. Explanation of the stress expression (8.115). The rotation of the rod causes fluid flow as in (*a*) or (*b*). The direction of the rotation is random, but on average, the fluid moves as shown by thick arrows in (*c*).

A crude estimation of W is done easily again by neglecting the hydrodynamic interaction in the shish-kebab model. Under the velocity gradient $\boldsymbol{\kappa}$, the rod rotates with the angular velocity $\boldsymbol{\omega}_0 = \boldsymbol{u} \times (\boldsymbol{\kappa} \cdot \boldsymbol{u})$. Hence the velocity of the n-th bead relative to the fluid is

$$\boldsymbol{V}_{nr} = nb(\boldsymbol{\omega}_0 \times \boldsymbol{u} - \boldsymbol{\kappa} \cdot \boldsymbol{u}) = nb([\boldsymbol{u} \times (\boldsymbol{\kappa} \cdot \boldsymbol{u})] \times \boldsymbol{u} - \boldsymbol{\kappa} \cdot \boldsymbol{u}) = -nb\boldsymbol{u}(\boldsymbol{\kappa} : \boldsymbol{uu}). \tag{8.118}$$

The frictional force acting on the segment is $\boldsymbol{F}_n = \zeta_0 \boldsymbol{V}_{nr}$. Hence the work done by the frictional force in unit time and unit volume is

$$W = \nu \sum_n \langle \boldsymbol{F}_n \cdot \boldsymbol{V}_{nr} \rangle = \nu \sum_{n=-N/2}^{N/2} \zeta_0 n^2 b^2 \langle (\boldsymbol{\kappa} : \boldsymbol{uu})^2 \rangle = \nu \zeta_{\text{str}} \langle (\boldsymbol{\kappa} : \boldsymbol{uu})^2 \rangle \tag{8.119}$$

where

$$\zeta_{\text{str}} = \sum_{n=-N/2}^{N/2} \zeta_0 n^2 b^2 = \frac{\pi \eta_s L^3}{4}. \tag{8.120}$$

Hence

$$\sigma_{\alpha\beta}^{(\text{V})} = \nu \zeta_{\text{str}} \kappa_{\mu\nu} \langle u_\mu u_\nu u_\alpha u_\beta \rangle. \tag{8.121}$$

Note that in this case, ζ_{str} is equal to the rotational friction constant ζ_r (see eqn (8.4)). If the hydrodynamic interaction is taken into account, ζ_{str} is given by $\zeta_r/2$ (see Appendix 8.I). We shall consider only this case, whence

$$\sigma_{\alpha\beta}^{(\mathrm{V})} = \frac{\nu}{2}\zeta_r \kappa_{\mu\nu}\langle u_\alpha u_\beta u_\mu u_\nu \rangle. \tag{8.122}$$

Thus the stress due to the polymer is

$$\begin{aligned}\sigma_{\alpha\beta}^{(\mathrm{p})} &= \sigma_{\alpha\beta}^{(\mathrm{E})} + \sigma_{\alpha\beta}^{(\mathrm{V})} \\ &= 3\nu k_B T S_{\alpha\beta} - \nu\langle(\boldsymbol{u}\times \mathcal{R}U)_\alpha u_\beta\rangle + \frac{\nu}{2}\zeta_r\kappa_{\mu\nu}\langle u_\alpha u_\beta u_\mu u_\nu\rangle\end{aligned} \tag{8.123}$$

and the total macroscopic stress in the solution is

$$\sigma_{\alpha\beta} = \sigma_{\alpha\beta}^{(\mathrm{p})} + \eta_s(\kappa_{\alpha\beta} + \kappa_{\beta\alpha}). \tag{8.124}$$

8.6.2 *Calculation for weak velocity gradient*

Let us now calculate the stress tensor under a weak velocity gradient. If we consider the case $U=0$, then eqn (8.15) becomes

$$\frac{\partial \Psi}{\partial t} = D_r \mathcal{R}^2\Psi - \mathcal{R}\cdot(\boldsymbol{u}\times\boldsymbol{\kappa}\cdot\boldsymbol{u}\Psi). \tag{8.125}$$

To calculate $S_{\alpha\beta}$ we again multiply $u_\alpha u_\beta - \delta_{\alpha\beta}/3$ by eqn (8.125) and integrate over $\boldsymbol{u}$. The equation is then rewritten by integration by parts as

$$\frac{\partial}{\partial t}S_{\alpha\beta} = \int \mathrm{d}\boldsymbol{u}\Psi[D_r\mathcal{R}^2(u_\alpha u_\beta - \tfrac{1}{3}\delta_{\alpha\beta}) + \mathcal{R}(u_\alpha u_\beta - \tfrac{1}{3}\delta_{\alpha\beta})\cdot(\boldsymbol{u}\times\boldsymbol{\kappa}\cdot\boldsymbol{u})]. \tag{8.126}$$

The terms in the bracket is calculated directly giving

$$\begin{aligned}\frac{\partial}{\partial t}S_{\alpha\beta} &= -6D_r(u_\alpha u_\beta - \tfrac{1}{3}\delta_{\alpha\beta}) + \kappa_{\alpha\mu}\langle u_\mu u_\beta\rangle + \kappa_{\beta\mu}\langle u_\mu u_\alpha\rangle - 2\kappa_{\mu\nu}\langle u_\alpha u_\beta u_\mu u_\nu\rangle \\ &= -6D_r S_{\alpha\beta} + \tfrac{1}{3}(\kappa_{\alpha\beta}+\kappa_{\beta\alpha}) + \kappa_{\alpha\mu}S_{\beta\mu} + \kappa_{\beta\mu}S_{\alpha\mu} - 2\kappa_{\mu\nu}\langle u_\alpha u_\beta u_\mu u_\nu\rangle.\end{aligned} \tag{8.127}$$

To calculate the first order in κ, the coefficients of $\kappa_{\alpha\beta}$ can be replaced by their equilibrium values. Since at equilibrium

$$S_{\alpha\beta} = 0 \quad \text{and} \quad \langle u_\alpha u_\beta u_\mu u_\nu\rangle = \tfrac{1}{15}(\delta_{\alpha\beta}\delta_{\mu\nu} + \delta_{\alpha\mu}\delta_{\beta\nu} + \delta_{\alpha\nu}\delta_{\beta\mu}) \tag{8.128}$$

eqn (8.127) is approximated by

$$\begin{aligned}\frac{\partial}{\partial t}S_{\alpha\beta} &= -6D_r S_{\alpha\beta} + \tfrac{1}{3}(\kappa_{\alpha\beta}+\kappa_{\beta\alpha}) - \tfrac{2}{15}\kappa_{\mu\nu}(\delta_{\alpha\beta}\delta_{\mu\nu} + \delta_{\alpha\mu}\delta_{\beta\nu} + \delta_{\alpha\nu}\delta_{\beta\mu}) \\ &= -6D_r S_{\alpha\beta} + \tfrac{1}{5}(\kappa_{\alpha\beta}+\kappa_{\beta\alpha}).\end{aligned} \tag{8.129}$$

(Here the incompressible condition $\kappa_{\alpha\alpha} = 0$ has been used.) This is solved by

$$S_{\alpha\beta}(t) = \frac{1}{5}\int_{-\infty}^{t} dt' \exp[-6D_r(t-t')](\kappa_{\alpha\beta}(t') + \kappa_{\beta\alpha}(t')). \tag{8.130}$$

From eqns (8.123), (8.128), and (8.130), the stress is obtained, to the first order in $\kappa_{\alpha\beta}$, as

$$\sigma^{(p)}_{\alpha\beta}(t) = \tfrac{3}{5}\nu k_B T \int_{-\infty}^{t} dt' \exp[-6D_r(t-t')][\kappa_{\alpha\beta}(t') + \kappa_{\beta\alpha}(t')]$$
$$+ \frac{\nu}{30}\zeta_r(\kappa_{\alpha\beta}(t) + \kappa_{\beta\alpha}(t)). \tag{8.131}$$

For the shear flow

$$\kappa_{\alpha\beta} = \begin{cases} \kappa(t) & \text{for } \alpha = x \text{ and } \beta = y, \\ 0 & \text{otherwise.} \end{cases} \tag{8.132}$$

Equations (8.124) and (8.131) give

$$\sigma_{xy}(t) = \eta_s\kappa(t) + \tfrac{3}{5}\nu k_B T \int_{-\infty}^{t} dt' \exp[-6D_r(t-t')]\kappa(t') + \frac{\nu}{30}\zeta_r\kappa(t). \tag{8.133}$$

Let us consider two special cases.

(i) Steady shear flow: $\kappa(t)$ is constant, for which eqn (8.133) gives

$$\sigma_{xy} = \left(\eta_s + \frac{\nu k_B T}{10D_r} + \frac{\nu\zeta_r}{30}\right)\kappa = (\eta_s + \tfrac{2}{15}\nu\zeta_r)\kappa. \tag{8.134}$$

Thus the viscosity of the solution is

$$\eta = \eta_s + \tfrac{2}{15}\nu\zeta_r = \eta_s + \frac{2\pi\eta_s L^3}{45(\ln(L/b) - \gamma)}. \tag{8.135}$$

The intrinsic viscosity is defined by

$$[\eta] = \lim_{\rho\to 0}\frac{1}{\rho\eta_s}(\eta - \eta_s) \tag{8.136}$$

where ρ is the weight of polymer in unit volume, which is expressed in terms of ν and the molecular weight M of the polymer as

$$\rho = \frac{M}{N_A}\nu. \tag{8.137}$$

Thus

$$[\eta] = \frac{2\pi L^3}{45(\ln(L/b) - \gamma)} \frac{N_A}{M}. \tag{8.138}$$

(ii) Oscillatory flow: $\kappa(t) = \kappa_0 \operatorname{Re}(e^{i\omega t})$. The stress for this flow defines the complex viscosity $\eta^*(\omega)$ by

$$\sigma_{xy}(t) = \kappa_0 \operatorname{Re}(\eta^*(\omega) e^{i\omega t}). \tag{8.139}$$

(Note: The complex viscosity is related to the complex modulus by $\eta^*(\omega) = i\omega G^*(\omega)$.) From eqn (8.133), it is easy to show that

$$\eta^*(\omega) = \eta_s + \frac{\nu k_B T}{10 D_r}\left(\frac{1}{1 + i\omega\tau_v} + \frac{1}{3}\right) \tag{8.140}$$

where

$$\tau_v = \frac{1}{6D_r} = \frac{\pi \eta_s L^3}{18(\ln(L/b) - \gamma) k_B T}. \tag{8.141}$$

The intrinsic complex viscosity is defined by

$$[\eta^*(\omega)] = \lim_{\rho \to 0} \frac{1}{\rho \eta_s}(\eta^*(\omega) - \eta_s) \tag{8.142}$$

which is given by

$$[\eta^*(\omega)] = [\eta]\left(\frac{3}{4}\frac{1}{1 + i\omega\tau_v} + \frac{1}{4}\right). \tag{8.143}$$

Note that $[\eta^*(\omega)]$ has a finite value as $\omega \to \infty$, which arises from the rigid constraints of the rods. The relaxation time for $[\eta^*(\omega)]$ is one third of the relaxation time in the dielectric relaxation. Since L is proportional to M, $[\eta]$ and τ_v depend on the molecular weight,

$$[\eta] \propto M^2/(\ln M - K), \qquad \tau_v \propto M^3/(\ln M - K) \tag{8.144}$$

where K is a constant. These results have been well confirmed by experiment.[27–29]

8.7 Nonlinear viscoelasticity

8.7.1 Decoupling approximation

In Section 8.5 and 8.6, we considered the case that the perturbation due to the external field is small. This allowed us to replace the unknown fourth moment $\langle u_\alpha u_\beta u_\mu u_\nu \rangle$ by its equilibrium value. If the external field is large, this approximation becomes invalid.

The standard way of handling this nonlinear case is to expand the distribution function in spherical harmonics and solve the resulting equation for the expansion coefficients numerically. Although this exact numerical method is available,[30] it is valuable to see analytically where the characteristic features of the solution have their origin. To do this, a simple approximation is available which is to express the unknown quantity $\langle \boldsymbol{uuuu} \rangle : \boldsymbol{\kappa}$ in terms of $\langle \boldsymbol{uu} \rangle$. There are many ways of doing this (see the discussion in ref. 31). A simple one is to assume[26]

$$\langle u_\alpha u_\beta u_\mu u_\nu \rangle \kappa_{\mu\nu} = A \langle u_\alpha u_\beta \rangle \langle u_\mu u_\nu \rangle \kappa_{\mu\nu} + B \langle u_\alpha u_\mu \rangle \langle u_\beta u_\nu \rangle (\kappa_{\mu\nu} + \kappa_{\nu\mu}) \tag{8.145}$$

where A and B are constants to be determined. We impose the condition that eqn (8.145) holds rigorously when the traces of the second-order tensors of both sides are taken:

$$\langle u_\alpha u_\alpha u_\mu u_\nu \rangle \kappa_{\mu\nu} = A \langle u_\alpha u_\alpha \rangle \langle u_\mu u_\nu \rangle \kappa_{\mu\nu} + B \langle u_\alpha u_\mu \rangle \langle u_\alpha u_\nu \rangle \langle \kappa_{\mu\nu} + \kappa_{\nu\mu} \rangle \tag{8.146}$$

i.e.,

$$\langle u_\mu u_\nu \rangle \kappa_{\mu\nu} = A \langle u_\mu u_\nu \rangle \kappa_{\mu\nu} + B \langle u_\alpha u_\mu \rangle \langle u_\alpha u_\nu \rangle (\kappa_{\mu\nu} + \kappa_{\nu\mu}) \tag{8.147}$$

which gives $A = 1$ and $B = 0$. Thus the approximation gives

$$\langle u_\alpha u_\beta u_\mu u_\nu \rangle \kappa_{\mu\nu} = \langle u_\alpha u_\beta \rangle \langle u_\mu u_\nu \rangle \kappa_{\mu\nu}. \tag{8.148}$$

The advantage of this decoupling approximation is

(i) It preserves the symmetry and the trace of the original tensor $\langle \boldsymbol{uuuu} \rangle : \boldsymbol{\kappa}$. This property guarantees that the resulting equation for $S_{\alpha\beta}$ is symmetric ($S_{\alpha\beta} = S_{\beta\alpha}$) and traceless ($S_{\alpha\alpha} = 0$).

(ii) It becomes correct for the completely ordered state: $\Psi(\boldsymbol{u}) = \delta(\boldsymbol{u} - \boldsymbol{n})$ ($\boldsymbol{n}$ being a unit vector).

Other decoupling approximations are possible,[31] but here we shall only examine this approximation since it is mathematically simple.

For this approximation, eqn (8.127) is rewritten as

$$\begin{aligned} \frac{\partial}{\partial t} S_{\alpha\beta} = -6D_r S_{\alpha\beta} + \kappa_{\alpha\mu} S_{\beta\mu} + \kappa_{\beta\mu} S_{\alpha\mu} + \tfrac{1}{3}(\kappa_{\alpha\beta} + \kappa_{\beta\alpha}) \\ -2S_{\mu\nu}\kappa_{\mu\nu}(S_{\alpha\beta} + \tfrac{1}{3}\delta_{\alpha\beta}). \end{aligned} \tag{8.149}$$

8.7.2 *Elongational flow*

First we consider steady elongational flow,

$$\begin{aligned} &\kappa_{xx} = -\tfrac{1}{2}\dot{\varepsilon}, \qquad \kappa_{yy} = -\tfrac{1}{2}\dot{\varepsilon}, \qquad \kappa_{zz} = \dot{\varepsilon}, \\ &\kappa_{\alpha\beta} = 0 \quad \text{for other components.} \end{aligned} \tag{8.150}$$

In this flow, the diffusion equation can be solved rigorously, so that it is possible to check the accuracy of the decoupling approximation. Since $\boldsymbol{\kappa}$

is written as

$$\boldsymbol{\kappa} = \frac{\dot{\varepsilon}}{2}(3\boldsymbol{e}_z\boldsymbol{e}_z - \boldsymbol{I}),$$

eqn (8.125) becomes

$$\frac{\partial \Psi}{\partial t} = D_r \mathcal{R} \cdot \left(\mathcal{R}\Psi - \frac{3\dot{\varepsilon}}{2D_r} \boldsymbol{u} \times \boldsymbol{e}_z(\boldsymbol{u} \cdot \boldsymbol{e}_z)\Psi \right). \tag{8.151}$$

In the steady state, eqn (8.151) can be rigorously solved to give

$$\Psi = \text{const}\exp\left[\frac{3\dot{\varepsilon}}{4D_r}(\boldsymbol{u} \cdot \boldsymbol{e}_z)^2\right] = \text{const}\exp\left(\frac{3\dot{\varepsilon}}{4D_r} u_z^2\right). \tag{8.152}$$

Thus S_{zz} is calculated as

$$S_{zz} = \int d\boldsymbol{u}(u_z^2 - 1/3)\exp\left(\frac{3\dot{\varepsilon}}{4D_r} u_z^2\right) \Big/ \int d\boldsymbol{u} \exp\left(\frac{3\dot{\varepsilon}}{4D_r} u_z^2\right)$$

$$= \int_0^1 dt(t^2 - \tfrac{1}{3})\exp(\tfrac{3}{4}\xi t^2) \Big/ \int_0^1 dt \exp(\tfrac{3}{4}\xi t^2) \tag{8.153}$$

where $\xi \equiv \dot{\varepsilon}/D_r$.

On the other hand, eqn (8.149) becomes

$$\frac{\partial}{\partial t} S_{zz} = -6D_r S_{zz} + 2\dot{\varepsilon} S_{zz} + \tfrac{2}{3}\dot{\varepsilon} - 2S_{\mu\nu}\kappa_{\mu\nu}(S_{zz} + \tfrac{1}{3}). \tag{8.154}$$

Since $S_{xx} + S_{yy} + S_{zz} = 0$, $S_{\mu\nu}\kappa_{\mu\nu}$ is calculated to be

$$S_{\mu\nu}\kappa_{\mu\nu} = -\frac{\dot{\varepsilon}}{2}(S_{xx} + S_{yy}) + \dot{\varepsilon} S_{zz} = \tfrac{3}{2}\dot{\varepsilon} S_{zz}. \tag{8.155}$$

Hence eqn (8.154) is written as

$$\frac{\partial}{\partial t} S_{zz} = -D_r[6S_{zz} - 2\xi S_{zz} - \tfrac{2}{3}\xi + 3\xi S_{zz}(S_{zz} + \tfrac{1}{3})]. \tag{8.156}$$

The steady-state solution of eqn (8.156) is

$$S_{zz} = \frac{2}{3}\left[\frac{1}{4} - \frac{3}{2\xi} + \left(\frac{9}{16} - \frac{3}{4\xi} + \frac{9}{4\xi^2}\right)^{1/2}\right] \tag{8.157}$$

In Fig. 8.6, eqns (8.153) and (8.157) are compared. Though the decoupling approximation is not correct for small ξ, the difference becomes smaller as ξ becomes larger.

8.7.3 Shear flow

Next we study a simple shear for which

$$\kappa_{\alpha\beta} = \begin{cases} \kappa & \text{if } \alpha = x \text{ and } \beta = y, \\ 0 & \text{otherwise.} \end{cases} \tag{8.158}$$

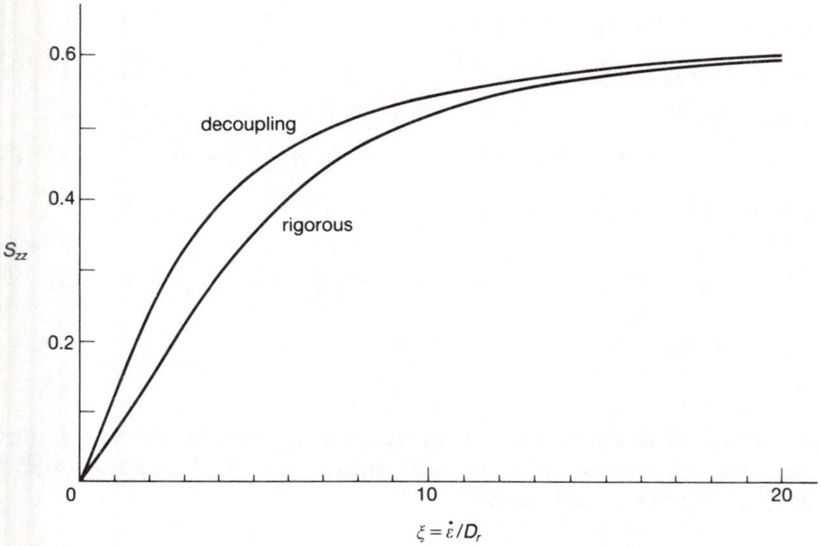

Fig. 8.6. Effect of the elongational flow on the orientational order parameter. For small ξ, the decoupling approximation gives $S_{zz} = \xi/9$, as compared to the rigorous results $S_{zz} = \xi/15$.

Therefore eqn (8.149) becomes

$$\frac{\partial}{\partial t} S_{xy} = -6D_r S_{xy} - 2\kappa S_{xy}^2 + \kappa S_{yy} + \tfrac{1}{3}\kappa, \tag{8.159}$$

$$\frac{\partial}{\partial t} S_{xx} = -6D_r S_{xx} - 2\kappa S_{xy}(S_{xx} + \tfrac{1}{3}) + 2\kappa S_{xy}, \tag{8.160}$$

$$\frac{\partial}{\partial t} S_{yy} = -6D_r S_{yy} - 2\kappa S_{xy}(S_{yy} + \tfrac{1}{3}), \tag{8.161}$$

$$\frac{\partial}{\partial t} S_{zz} = -6D_r S_{zz} - 2\kappa S_{xy}(S_{zz} + \tfrac{1}{3}). \tag{8.162}$$

In the steady state, these equations are rewritten as

$$S_{xy}(1 + \tfrac{1}{3}\xi S_{xy}) = \tfrac{1}{18}\xi + \tfrac{1}{6}\xi S_{yy}, \tag{8.163}$$

$$S_{xx} = \frac{2\xi S_{xy}}{9(1 + \tfrac{1}{3}\xi S_{xy})}, \tag{8.164}$$

$$S_{yy} = S_{zz} = -\tfrac{1}{2}S_{xx}, \tag{8.165}$$

where

$$\xi = \kappa/D_r. \tag{8.166}$$

For small ξ, these equations are solved by

$$S_{xy} = \tfrac{1}{18}\xi, \qquad S_{xx} = \tfrac{1}{81}\xi^2, \qquad S_{yy} = S_{zz} = -\tfrac{1}{162}\xi^2; \tag{8.167}$$

and for large ξ

$$S_{xy} = (2\xi)^{-1/3}, \qquad S_{xx} = \tfrac{2}{3} - 2(2/\xi^2)^{1/3}. \tag{8.168}$$

The shear stress is given by

$$\sigma_{xy}^{(\mathrm{p})} = 3\nu k_B T S_{xy} + \frac{\nu}{2}\zeta_r \kappa S_{xy}^2 = 3\nu k_B T S_{xy}\left(1 + \frac{\kappa}{6D_r} S_{xy}\right). \tag{8.169}$$

For small κ, eqn (8.167) and (8.169) gives

$$\sigma_{xy}^{(\mathrm{p})} = \tfrac{1}{6}\nu\zeta_r\kappa. \tag{8.170}$$

The viscosity calculated by the decoupling approximation is 5/4 times larger than the rigorous value of eqn (8.135). For large κ, $\sigma_{xy}^{(\mathrm{p})}$ is dominated by the viscous stress

$$\sigma_{xy}^{(\mathrm{p})} \simeq \frac{\nu}{2}\zeta_r \kappa S_{xy}^2 = \tfrac{1}{2}\nu k_B T\left(\frac{\kappa}{4D_r}\right)^{1/3}. \tag{8.171}$$

Thus the intrinsic viscosity decreases in proportion to $\kappa^{-2/3}$. On the other hand, rigorous asymptotic analysis[32] and the numerical calculation[33] show that the intrinsic viscosity decreases as $\kappa^{-1/3}$.

8.8 Effect of flexibility

So far we have been considering the dynamics of rigid rods. Real polymers have greater or lesser flexibility. This can be modelled by a rod which has an elastic energy for bending. For the polymer which has constant contour length, the simplest possible model will be the following. Let $\boldsymbol{R}(s)$ be the position of a point on the chain at the contour length s. The vector

$$\boldsymbol{u}(s) = \frac{\partial \boldsymbol{R}}{\partial s} \tag{8.172}$$

is a unit vector tangent to the chain. The straight rod corresponds to $\boldsymbol{u}(s) = \text{constant}$, or $\partial \boldsymbol{u}/\partial s = 0$. Thus the bending energy must be a quadratic of $\partial \boldsymbol{u}/\partial s$. Since $\boldsymbol{u}\cdot\partial\boldsymbol{u}/\partial s = 0$, the only quadratic form is

$$U_{\text{bend}} = \tfrac{1}{2}E\int_0^L \mathrm{d}s\left(\frac{\partial \boldsymbol{u}}{\partial s}\right)^2 \tag{8.173}$$

where E is a constant. The conformational distribution of the polymer is thus given by the Boltzmann distribution for this energy:

$$\Psi[\boldsymbol{u}] \propto \exp\left(-\frac{U_{\text{bend}}}{k_B T}\right) = \exp\left[-\frac{1}{4\lambda}\int_0^L \mathrm{d}s\left(\frac{\partial \boldsymbol{u}}{\partial s}\right)^2\right] \tag{8.174}$$

where

$$\lambda = \frac{k_B T}{2E}. \tag{8.175}$$

This is called the Kratky–Porod model,[34] and the length $(2\lambda)^{-1}$ is referred to as the persistence length.

Equation (8.173) indicates the analogy between the change of $\boldsymbol{u}(s)$ of the Kratky–Porod model and the time evolution of $\boldsymbol{u}(t)$ in rotational Brownian motion: both processes are Gaussian with the constraint $\boldsymbol{u}^2 = 1$.[35] From eqn (8.174) it can be shown that for small s

$$\langle(\boldsymbol{u}(s) - \boldsymbol{u}(0))^2\rangle = 4\lambda s \tag{8.176}$$

which indicates that λ plays the role of D_r in the rotational Brownian motion. Thus the correlation functions of $\boldsymbol{u}(s)$ are immediately obtained from the result of eqn (8.43). In particular

$$\langle \boldsymbol{u}(s) \cdot \boldsymbol{u}(0)\rangle = \exp(-2\lambda s) \tag{8.177}$$

from which the mean square end-to-end distance is calculated as

$$\begin{aligned}\bar{R}^2 &\equiv \langle(\boldsymbol{R}(L) - \boldsymbol{R}(0))^2\rangle = \int_0^L \mathrm{d}s \int_0^L \mathrm{d}s' \langle \boldsymbol{u}(s) \cdot \boldsymbol{u}(s')\rangle \\ &= 2\int_0^L \mathrm{d}s \int_0^s \mathrm{d}s' \exp(-2\lambda(s - s')) = \frac{L}{\lambda} - \frac{1}{2\lambda^2}[1 - \exp(-2\lambda L)]. \end{aligned} \tag{8.178}$$

The two limiting cases are:

(i) $L\lambda \gg 1$ (the random flight limit),

$$\bar{R}^2 = L/\lambda \tag{8.179}$$

(ii) $L\lambda \ll 1$ (the rigid rod limit),

$$\bar{R}^2 = L^2. \tag{8.180}$$

To develop a complete theory for the dynamical properties is difficult, but various approximate treatments have been proposed. Crudely speaking, the flexibility affects the dynamical properties in two respects.

Firstly, the flexibility changes the transport coefficients. As the size of the chain decreases with the increase in the flexibility, the diffusion constant D_G, which is roughly proportional to R_g^{-1}, increases, and the intrinsic viscosity $[\eta]$ ($\propto R_g^3$) decreases. The results of various approaches are summarized in refs 1 and 36.

Secondly, the flexibility gives a relaxation in the high-frequency region. For example, according to the rigid rod model, the complex intrinsic viscosity $[\eta^*(\omega)]$ approaches the finite value

$$[\eta_\infty] = \lim_{\omega \to \infty} [\eta^*(\omega)] = \tfrac{1}{4}[\eta]. \tag{8.181}$$

For the flexible polymer, $[\eta^*(\omega)]$ decreases as the frequency becomes comparable to that of the bending mode.[28] The experimental situation is summarized in ref. 29 and theoretical analysis is in ref. 37.

Appendix 8.I Derivation of the Smoluchowski equation by the Kirkwood theory

In this appendix we derive the kinetic equation based on the Kirkwood theory described in Section 3.8, and using the model shown in Fig. 8.2. The position vector of the n-th segment $(-N/2 < n < N/2)$ is written as

$$\boldsymbol{R}_n = \boldsymbol{R} + nb\boldsymbol{u}. \tag{8.I.1}$$

Let $\boldsymbol{V}_n$ and $\boldsymbol{F}_n$ be the velocity of this segment and the force acting on it. The velocity $\boldsymbol{V}_n$ is expressed by the velocity $\boldsymbol{V}$ of the centre of mass and the angular velocity $\boldsymbol{\omega}$

$$\boldsymbol{V}_n = \boldsymbol{V} + nb\boldsymbol{\omega} \times \boldsymbol{u} \tag{8.I.2}$$

which can be also written in terms of the mobility tensor $\boldsymbol{H}_{nm}$ as

$$\boldsymbol{V}_n - \boldsymbol{\kappa} \cdot \boldsymbol{R}_n = \sum_m \boldsymbol{H}_{nm} \cdot \boldsymbol{F}_m \tag{8.I.3}$$

From the definition of $\boldsymbol{H}_{nm}$ (eqn (3.106)) and eqn (8.I.1), it follows that

$$\boldsymbol{H}_{nm} = (\boldsymbol{uu} + \boldsymbol{I})h_{nm} \quad n \neq m \tag{8.I.4}$$

with

$$h_{nm} = 1/8\pi\eta_s\,|n-m|\,b. \tag{8.I.5}$$

In eqn (8.I.4), we neglected the term h_{nn}. The validity of this approximation is discussed later.

Now the total force acting on the centre of mass is given by $\sum_n \boldsymbol{F}_n$. Equating this with that given by the thermodynamic potential we get

$$\sum_n \boldsymbol{F}_n = -\frac{\partial}{\partial \boldsymbol{R}}(k_B T \ln \Psi + U). \tag{8.I.6}$$

Similarly the balance in torque is written as

$$\sum_n nb\boldsymbol{u} \times \boldsymbol{F}_n = -\mathcal{R}(k_B T \ln \Psi + U). \tag{8.I.7}$$

Our aim is to express $\boldsymbol{V}$ and $\boldsymbol{\omega}$ in terms of the quantity on the right-hand side of eqns (8.I.6) and (8.I.7). For that purpose we solve eqn (8.I.3) for $\boldsymbol{F}_n$,

$$\boldsymbol{F}_n = \sum_m (\boldsymbol{H}^{-1})_{nm} \cdot (\boldsymbol{V}_m - \boldsymbol{\kappa} \cdot \boldsymbol{R}_m) \tag{8.I.8}$$

where $(\boldsymbol{H}^{-1})_{nm}$ is the inverse of $\boldsymbol{H}_{nm}$, i.e.,

$$\sum_m \boldsymbol{H}_{nm} \cdot (\boldsymbol{H}^{-1})_{mk} = \delta_{nk}\boldsymbol{I}. \tag{8.I.9}$$

From eqn (8.I.4), $(\boldsymbol{H}^{-1})_{nm}$ is written as

$$(\boldsymbol{H}^{-1})_{nm} = (h^{-1})_{nm}\left(\boldsymbol{I} - \frac{\boldsymbol{uu}}{2}\right) \tag{8.I.10}$$

where $(h^{-1})_{nm}$ is the inverse of h_{nm},

$$\sum_m h_{nm}(h^{-1})_{mk} = \delta_{nk}. \tag{8.I.11}$$

We substitute eqn (8.I.8) into eqns (8.I.6) and (8.I.7), and use eqn (8.I.2) to obtain

$$\sum_{n,m} (\boldsymbol{H}^{-1})_{nm}\cdot[(\boldsymbol{V} + mb\boldsymbol{\omega} \times \boldsymbol{u}) - \boldsymbol{\kappa} \cdot (\boldsymbol{R} + mb\boldsymbol{u})] = -\frac{\partial}{\partial \boldsymbol{R}}(k_B T \ln \Psi + U) \tag{8.I.12}$$

$$\sum_{n,m} nb\boldsymbol{u} \times (\boldsymbol{H}^{-1})_{nm} \cdot [\boldsymbol{V} + mb\boldsymbol{\omega} \times \boldsymbol{u} - \boldsymbol{\kappa} \cdot (\boldsymbol{R} + mb\boldsymbol{u})] = -\mathcal{R}(k_B T \ln \Psi + U). \tag{8.I.13}$$

Using eqn (8.I.10), the right-hand side is rewritten as

$$\sum_{n,m} (h^{-1})_{nm}\left(\boldsymbol{I} - \frac{\boldsymbol{uu}}{2}\right) \cdot (\boldsymbol{V} - \boldsymbol{\kappa} \cdot \boldsymbol{R}) = -\frac{\partial}{\partial \boldsymbol{R}}(k_B T \ln \Psi + U) \tag{8.I.14}$$

$$\sum_{n,m} (h^{-1})_{nm} nmb^2 \boldsymbol{u} \times \left(\boldsymbol{I} - \frac{\boldsymbol{uu}}{2}\right) \cdot (\boldsymbol{\omega} \times \boldsymbol{u} - \boldsymbol{\kappa} \cdot \boldsymbol{u}) = -\mathcal{R}(k_B T \ln \Psi + U). \tag{8.I.15}$$

(Here we used the property $\sum_{n,m} n(h^{-1})_{nm} = 0$, which follows from $h_{nm} = h_{-n-m}$.) Defining

$$\zeta_t = \sum_{n,m} (h^{-1})_{nm}, \tag{8.I.16}$$

$$\zeta_r = b^2 \sum_{n,m} (h^{-1})_{nm} nm, \tag{8.I.17}$$

we rewrite eqns (8.I.14) and (8.I.15) as

$$\zeta_t\left(\boldsymbol{I}-\frac{\boldsymbol{uu}}{2}\right)\cdot(\boldsymbol{V}-\boldsymbol{\kappa}\cdot\boldsymbol{R})=-\frac{\partial}{\partial\boldsymbol{R}}(k_BT\ln\Psi+U) \tag{8.I.18}$$

$$\zeta_r\boldsymbol{u}\times(\boldsymbol{\omega}\times\boldsymbol{u}-\boldsymbol{\kappa}\cdot\boldsymbol{u})=\zeta_r(\boldsymbol{\omega}-\boldsymbol{u}\times\boldsymbol{\kappa}\cdot\boldsymbol{u})=-\mathcal{R}(k_BT\ln\Psi+U), \tag{8.I.19}$$

which is solved for $\boldsymbol{V}$ and $\boldsymbol{\omega}$ giving

$$\boldsymbol{V}=\boldsymbol{\kappa}\cdot\boldsymbol{R}-\frac{1}{\zeta_t}(\boldsymbol{I}+\boldsymbol{uu})\cdot\frac{\partial}{\partial\boldsymbol{R}}(k_BT\ln\Psi+U), \tag{8.I.20}$$

$$\boldsymbol{\omega}=\boldsymbol{u}\times\boldsymbol{\kappa}\cdot\boldsymbol{u}-\frac{1}{\zeta_r}\mathcal{R}(k_BT\ln\Psi+U). \tag{8.I.21}$$

Comparing eqn (8.I.20) with eqn (8.24), we have

$$\zeta_\perp=\zeta_t,\qquad \zeta_\parallel=\zeta_t/2. \tag{8.I.22}$$

Substituting this into the conservation equation (8.25), we finally get

$$\frac{\partial\Psi}{\partial t}=D_r\mathcal{R}\cdot\left[\mathcal{R}\Psi+\frac{\Psi}{k_BT}\mathcal{R}U\right]-\mathcal{R}\cdot(\boldsymbol{u}\times\boldsymbol{\kappa}\cdot\boldsymbol{u}\Psi)$$
$$+D_t\frac{\partial}{\partial\boldsymbol{R}}\cdot(\boldsymbol{I}+\boldsymbol{uu})\cdot\left[\frac{\partial\Psi}{\partial\boldsymbol{R}}+\frac{\Psi}{k_BT}\frac{\partial U}{\partial\boldsymbol{R}}\right]-\frac{\partial}{\partial\boldsymbol{R}}\cdot\boldsymbol{\kappa}\cdot\boldsymbol{R}\Psi \tag{8.I.23}$$

where

$$D_r=k_BT/\zeta_r \quad\text{and}\quad D_t=k_BT/\zeta_t. \tag{8.I.24}$$

Equation (8.I.23) agrees with eqn (8.26).

Next we obtain the expression for the stress tensor. According to eqn (3.134),

$$\sigma^{(p)}_{\alpha\beta}=-\nu\sum_n\langle F_{n\alpha}R_{n\beta}\rangle\,. \tag{8.I.25}$$

Substituting eqns (8.I.1), (8.I.2), and (8.I.10) into eqn (8.I.8), we obtain the force $\boldsymbol{F}_n$ as

$$\boldsymbol{F}_n=\sum_m(h^{-1})_{nm}\left(\boldsymbol{I}-\frac{\boldsymbol{uu}}{2}\right)\cdot[\boldsymbol{V}+mb\boldsymbol{\omega}\times\boldsymbol{u}-\boldsymbol{\kappa}\cdot(\boldsymbol{R}+mb\boldsymbol{u})]. \tag{8.I.26}$$

Here we consider the case when the total force acting on the polymer is zero,

$$\sum_n\boldsymbol{F}_n=0, \tag{8.I.27}$$

which gives $\boldsymbol{V}=\boldsymbol{\kappa}\cdot\boldsymbol{R}$, and

$$\boldsymbol{F}_n=\sum_m(h^{-1})_{nm}\left(\boldsymbol{I}-\frac{\boldsymbol{uu}}{2}\right)\cdot[mb\boldsymbol{\omega}\times\boldsymbol{u}-\boldsymbol{\kappa}\cdot mb\boldsymbol{u}]. \tag{8.I.28}$$

From eqns (8.I.25) and (8.I.28), it follows that

$$\sigma_{\alpha\beta}^{(p)} = -\nu \sum (h^{-1})_{nm} nmb^2 \langle (\boldsymbol{\omega} \times \boldsymbol{u})_\alpha u_\beta - (\boldsymbol{\kappa} \cdot \boldsymbol{u})_\alpha u_\beta + \tfrac{1}{2} u_\alpha u_\beta \boldsymbol{u} \cdot \boldsymbol{\kappa} \cdot \boldsymbol{u} \rangle. \tag{8.I.29}$$

Finally, substituting eqns (8.I.17) and (8.I.21) we get

$$\sigma_{\alpha\beta}^{(p)} = \nu \langle (e_{\alpha\mu\nu} u_\nu \mathscr{R}_\mu (k_B T \ln \Psi + U)) u_\beta \rangle + \frac{\nu}{2} \zeta_r \langle u_\alpha u_\beta u_\mu u_\nu \kappa_{\mu\nu} \rangle. \tag{8.I.30}$$

The first term on the right-hand side of eqn (8.I.30) is rewritten using integration by parts (eqn (8.10) to give

$$\begin{aligned} \langle u_\beta e_{\alpha\mu\nu} u_\nu \boldsymbol{\mathscr{R}}_\mu \ln \Psi \rangle &= \int \mathrm{d}\boldsymbol{u} e_{\alpha\mu\nu} u_\beta u_\nu \mathscr{R}_\mu \Psi \\ &= e_{\alpha\mu\nu} \int \mathrm{d}\boldsymbol{u} (-\Psi) \mathscr{R}_\mu u_\beta u_\nu \\ &= 3 \langle u_\alpha u_\beta - \tfrac{1}{3} \delta_{\alpha\beta} \rangle. \end{aligned} \tag{8.I.31}$$

Hence

$$\begin{aligned} \sigma_{\alpha_b}^{(p)} = 3\nu k_B T \langle u_\alpha u_\beta - \tfrac{1}{3} \delta_{\alpha\beta} \rangle - \nu \langle (\boldsymbol{u} \times \boldsymbol{\mathscr{R}} U)_\alpha u_\beta \rangle \\ + \frac{\nu}{2} \zeta_r \kappa_{\mu\nu} \langle u_\alpha u_\beta u_\mu u_\nu \rangle. \end{aligned} \tag{8.I.32}$$

The first two terms represent the elastic stress, and the last term is the viscous stress.

Finally we calculate the friction constants ζ_r and ζ_t using eqns (8.I.16) and (8.I.17). Since h_{nm} decreases quickly with $|n-m|$, we may approximate it by

$$h_{nm} \simeq \bar{h} \delta_{nm} \tag{8.I.33}$$

with

$$\bar{h} = 2 \int_1^{N/2} \mathrm{d}m h_{0m} = \frac{\ln(N/2)}{4\pi\eta_s b}, \tag{8.I.34}$$

$$(h^{-1})_{nm} = \delta_{nm} / \bar{h}. \tag{8.I.35}$$

Therefore from eqns (8.I.16), (8.I.17), and (8.I.35)

$$\zeta_t = \frac{N}{\bar{h}} = \frac{4\pi\eta_s Nb}{\ln(N/2)} = \frac{4\pi\eta_s L}{\ln(L/2b)} \tag{8.I.36}$$

and

$$\zeta_r = 2b^2 \int_0^{N/2} \mathrm{d}m \frac{m^2}{\bar{h}} = \frac{\pi\eta_s (Nb)^3}{3 \ln(N/2)} = \frac{\pi\eta_s L^3}{3 \ln(L/2b)}. \tag{8.I.37}$$

References

1. Yamakawa, H., *Modern Theory of Polymer Solutions*. Harper & Row, New York (1971).
2. Tsvetkov, V. N., and Andreeva, L. N., *Adv. Polym. Sci.* **39,** 95 (1981).
3. See for example, Blumstein, A., (ed.) *Liquid Crystalline Order in Polymers*. Academic Press, New York (1978); and Ciferri, A., Krigbaum, W. R., and Meyer, R. B., (eds.) *Polymer Liquid Crystals*. Academic Press, New York (1982).
4. Kirkwood, J. G., and Auer, P. L., *J. Chem. Phys.* **19,** 281 (1951).
5. Jeffery, G. B., *Proc. R. Soc. London* **A102,** 161 (1922).
6. Happel, J., and Brenner, H., *Low Reynolds Number Hydrodynamics,* Chap. 5. Prentice-Hall, Englewood Cliffs, New Jersey (1965).
7. Perrin, F., *J. Phys. Radium.* **5,** 497 (1934); **7,** 1 (1936).
8. Burgers, J. M., *Ver. Kon. Ned. Akad. Wet.* (*1*). **16,** 113 (1938). See also *Second Report on Viscosity and Plasticity,* pp. 113. North-Holland, Amsterdam (1938).
9. Broersma, S., *J. Chem. Phys.* **32,** 1626 (1960).
10. Yoshizaki, T., and Yamakawa, H., *J. Chem. Phys.* **72,** 57 (1980).
11. Bremner, H., *Int. J. Multiphase Flow* **1,** 195 (1974).
12. Rallison, J. M., *J. Fluid Mech.* **84,** 237 (1978).
13. Berne, B. J., and Pecora, R., *Dynamic Light Scattering*. Wiley, New York, (1976).
14. Maeda, H., and Saito, N., *J. Phys. Soc. Jpn* **27,** 984 (1969); *Polymer J.* **4,** 309 (1972).
15. Schaefer, D. W., Benedek, G. B., Schofield, P., and Bradford, E., *J. Chem. Phys.* **55,** 3884 (1971).
16. Wilcoxon, J., and Shurr, J. M., *Biopolymers* **22,** 849 (1983).
17. Maeda, T., and Fujime, S., *Macromolecules* **17,** 1157 (1984); Kubota, K., Urabe, H., Tominaga, Y., and Fujime, S., *Macromolecules* **17,** 2096 (1984).
18. Rallison, J. M., and Leal, L. G., *J. Chem. Phys.* **74,** 4819 (1981).
19. A classical textbook on dielectric relaxation is Debye, P., *Polar Molecules*. Dover, New York (1945). A recent review on dielectric properties including rodlike polymers is given by Mandel, M., and Odijk, T., *Ann. Rev. Phys. Chem.* **35,** 75 (1984).
20. Fredericq, E., and Houssier, C., *Electric Dichroism and Electric Birefringence*. Oxford Univ. Press, London (1973). A recent review is given by Watanabe, H., and Morita, A., *Adv. Chem. Phys.* **56,** 255 (1984).
21. Wada, A., *J. Chem. Phys.* **30,** 329 (1959); **31,** 495 (1959).
22. Sakamoto, M., Kanda, H., Hayakawa, R., and Wada, Y., *Biopolymers* **15,** 879 (1976); **18,** 2769 (1979).
23. Bur, A. J., and Roberts, D. E., *J. Chem. Phys.* **51,** 406 (1969).
24. Peterlin, A., and Stuart, H. A., *Hand und Jahrbuch der Chemischen Physik,* Vol 81. Akademische Verlagses, Leipzig (1943).
25. Benoit, H., *Ann. Phys.* **6,** 561 (1951); Tinoco, I. Jr., and Yamaoka, K., *J. Phys. Chem.* **63,** 423 (1959).
26. Doi, M., *J. Polym. Sci. Phys. ed.* **19,** 229 (1981).
27. Nemoto, N., Schrag, J. L., Ferry, J. D., and Fulton, R. W., *Biopolymers* **14,** 409 (1975).

28. Ookubo, N., Komatsubara, M., Nakajima, H., and Wada, Y., *Biopolymers* **15,** 929 (1976).
29. Ferry, J. D., *Viscoelastic Properties of Polymers* (3rd edn), Chap. 9. Wiley, New York (1980).
30. See for example Bird, R. B., *et al. Dynamics of Polymeric Liquids* Vol 2, Chap. 11. Wiley, New York (1977).
31. Hinch, E. J., and Leal, L. G., *J. Fluid Mech.* **52,** 683 (1972).
32. Hinch, E. J., and Leal, L. G., *J. Fluid Mech.* **76,** 187 (1976).
33. Stewart, W. E., and Sørensen, J. P., *Trans. Soc. Rheol.* **16,** 1 (1972).
34. Kratky, O., and Porod, G., *Rec. Trav. Chim.* **68,** 1106 (1949).
35. Saito, N., Takahashi, K., and Yunoki, Y., *J. Phys. Soc. Jpn* **22,** 219 (1967).
36. Yamakawa, H., *Ann Rev. Phys. Chem.* **25,** 179 (1974); *Pure Appl. Chem.* **46,** 135 (1976).
37. Yamakawa, H., *Ann. Rev. Phys. Chem.* **35,** 23 (1984).

9

SEMIDILUTE SOLUTIONS OF RIGID RODLIKE POLYMERS

9.1 Semidilute and concentrated solutions of rodlike polymers

In Chapter 8, we discussed the dynamics of a single rodlike polymer. Let us now consider the interaction between the polymers at finite concentration.

Solutions of slender rodlike polymers of length L and diameter b may be classified into four concentration regimes (Fig. 9.1). Let ρ be the weight of polymers in unit volume of the solution, then the number of polymers per volume is given by

$$\nu = \frac{\rho}{M} N_A \tag{9.1}$$

where M is the molecular weight.

(i) Dilute solution (Fig. 9.1*a*). A dilute solution is defined as one having a sufficiently low concentration that the average distance between the polymers $\nu^{-1/3}$ is much larger than L, i.e.,

$$\nu \lesssim \nu_1 \simeq 1/L^3 \tag{9.2}$$

In such a solution each polymer can rotate freely without interference by other polymers. The effect of the interaction can be expressed by a power series expansion with respect to ν as in the case of flexible polymers.

(ii) Semidilute solution (Fig. 9.1*b*). If $\nu \gg \nu_1$ the rotation of each polymer is severely restricted by other polymers, so that the dynamics of the polymers will be entirely different from that in dilute solution. However, the static properties will not be affected seriously until the concentration reaches another characteristic concentration ν_2. This is easily seen if one considers that the polymers are mathematical lines with no thickness. The equilibrium distribution of such polymers is entirely independent of each other at all concentrations. Thus the effect of the interaction becomes important in static properties only for polymers with finite diameter. Indeed the excluded volume of rigid rods is shown to be of the order of bL^2 (see Fig. 9.2)), so that the static properties are unaffected if νbL^2 is small. We call the concentration regime

$$\nu_1 \lesssim \nu \ll \nu_2 \simeq 1/bL^2 \tag{9.3}$$

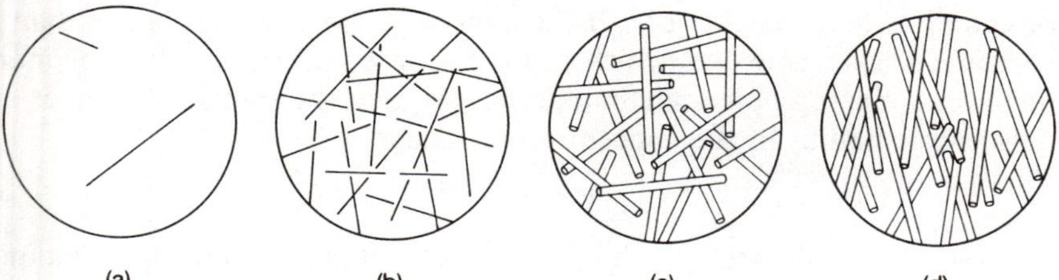

Fig. 9.1. Four concentration regimes of rodlike polymers: (*a*) dilute solution, (*b*) semidilute solution, (*c*) isotropic concentrated solution, and (*d*) liquid crystalline solution.

semidilute. (Note that this definition of semidilute is different from that used for flexible polymers.) In this concentration regime, the effect of intermolecular interaction can be neglected in the static properties, while the dynamical properties are completely changed by the constraint that polymers cannot cross each other. We call such an interaction the entanglement interaction, although rods do not entangle with each other literally.

(iii) Concentrated (isotropic) solution (Fig. 9.1*c*). The static correlation of the polymers becomes important at a higher concentration $\nu \gtrsim \nu_2$. Here, as will be shown later, polymers tend to orient in the same

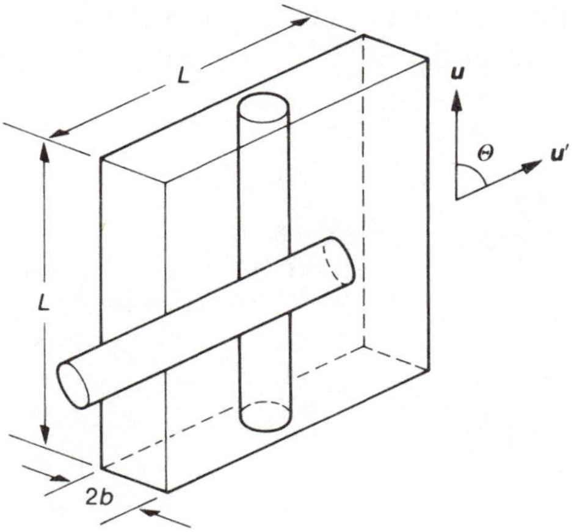

Fig. 9.2. Excluded volume between the rods in the direction $\boldsymbol{u}$ and $\boldsymbol{u}'$. For a given position of the rod in the direction $\boldsymbol{u}$, the centre of mass of the other rod is not allowed in the parallelepiped region shown in the figure. The volume of this region is $2bL^2\,|\sin\Theta|$ where Θ is the angle between $\boldsymbol{u}$ and $\boldsymbol{u}'$.

direction as their neighbours. If the concentration becomes larger than a certain critical value ν^*, which is of the order of $1/bL^2$, the polymers align on a macroscopic scale in equilibrium, and the solution becomes an anisotropic liquid. In the concentration regime

$$\nu_2 \lesssim \nu \lesssim \nu^* \tag{9.4}$$

the solution is still isotropic, but the excluded volume interactions among the polymers are important in both the static and dynamic properties. Such a solution is called a concentrated isotropic solution.

(iv) Liquid crystalline solution (Fig. 9.1*d*). The anisotropic solution above ν^* is called a liquid crystalline solution. This solution shows a range of interesting properties quite distinct from those of isotropic solutions, and will be discussed in the next chapter.

In this chapter we shall discuss the semidilute solution. This is in fact an ideal system for studying the entanglement effect since the dynamics of rodlike polymers is much simpler than that of the flexible polymers and the excluded volume effect is negligibly small.

9.2 Entanglement effect in rodlike polymers

9.2.1 *Tube model*

In the semidilute region, the dominant interaction is caused by the topological constraint that the polymers cannot cross each other. This effect can be treated[1,2] by the same model as that for flexible polymers. In the situation shown in Fig. 9.1*b*, the motion along a polymer is almost

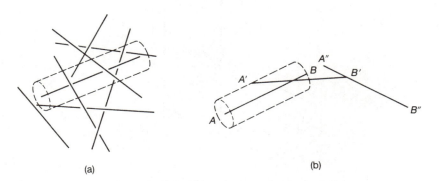

Fig. 9.3. (*a*) Tube model for the rodlike polymer. (*b*) The mechanism of rotation in semidilute solution. The polymer can change its direction when it disengages from a tube, e.g., by moving from AB to $A'B'$ and then to $A''B''$. Reproduced from ref. 10.

free, while the motion perpendicular to the polymer is severely limited by surrounding polymers. Such a characteristic feature of the Brownian motion can be again represented by a tube which surrounds the polymer (Fig. 9.3). The tube radius a corresponds to the average distance that the polymer can move perpendicularly to its own axis without being hindered by other polymers.

Let us now study how the tube constraint affects the translational and the rotational Brownian motion.

9.2.2 *Translational diffusion*

The translational motion is easily analysed. The motion of the polymer parallel to the tube axis, which is nearly parallel to $\boldsymbol{u}$, is not hindered by the tube, so that the parallel component $D_{\parallel}$ will be nearly equal to the diffusion constant in dilute solution $D_{\parallel 0}$.† On the other hand, the perpendicular motion is limited to within the distance a, so that the perpendicular component $D_{\perp}$ can be regarded as zero provided the small-scale motion of order a is neglected. Thus

$$D_{\parallel} \simeq D_{\parallel 0}, \qquad D_{\perp} = 0. \tag{9.5}$$

9.2.3 *Rotational diffusion*

The rotational diffusion can be discussed using the following model (Fig. 9.3*b*). As long as the polymer stays in a certain tube, its direction is essentially fixed in the direction of the tube axis. If the polymer moves the distance $L/2$ along itself, it disengages from the old tube and goes into a new tube which is generally tilted from the old tube with an angle of about $\varepsilon \simeq a/L$. If the polymer leaves this tube, it again changes direction by order ε. Thus overall rotation of the polymer is attained by repetition of this process. If τ_d is the average time necessary for the polymer to leave a certain tube, the polymer repeats the process t/τ_d times in a time interval t. Since the direction $\boldsymbol{u}(t)$ changes about ε per step, the mean square displacement of $\boldsymbol{u}(t)$ is estimated as

$$\langle [\boldsymbol{u}(t) - \boldsymbol{u}(0)]^2 \rangle \simeq \frac{t}{\tau_d}\, \varepsilon^2. \tag{9.6}$$

Comparing this with eqn (8.1), the rotational diffusion constant is estimated as

$$D_r \simeq \frac{\varepsilon^2}{\tau_d} \simeq \left(\frac{a}{L}\right)^2 / \tau_d. \tag{9.7}$$

† Strictly speaking, $D_{\parallel}$ is different from $D_{\parallel 0}$ due to the hydrodynamic interaction among polymers. However, this effect gives only a logarithmic correction.[3,4]

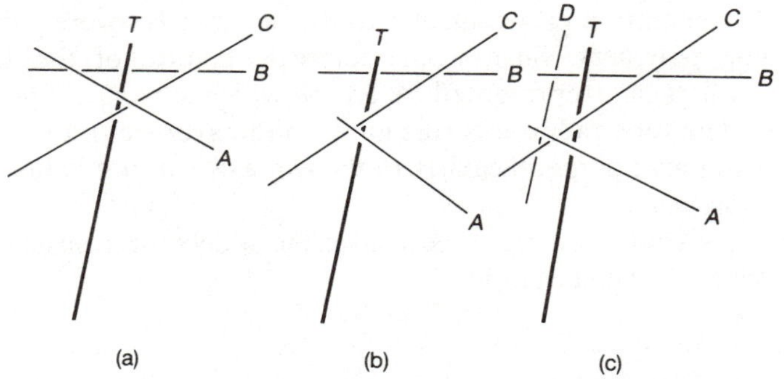

Fig. 9.4. Constraint release process. The constraint on the test polymer T (thick line) imposed by a surrounding polymer A can be released if the polymer A moves away as shown in (b), or a new constraint can be created by an incoming polymer D as shown in (c).

The time τ_d is estimated as the time necessary for the polymer to move the distance L:

$$\tau_d \simeq L^2/D_{\|} \simeq L^2/D_{\|0}. \tag{9.8}$$

Since $L^2/D_{\|0} \simeq D_{r0}^{-1}$ (see eqns (8.16) and (8.27)), eqn (9.7) is written as

$$D_r \simeq D_{r0}\left(\frac{a}{L}\right)^2 \tag{9.9}$$

i.e., the rotational diffusion constant in the semidilute region becomes smaller than that in dilute solution by a factor $(a/L)^2$.

In the above estimation, it is implicitly assumed that the tube is fixed during the time τ_d. Actually this assumption is not correct because during the time τ_d, the surrounding polymers constituting the tube will also move a distance of the order of L, and the constraints imposed by them will be released. At the same time, new constraints are created by incoming rods (see Fig. 9.4). Thus the tube itself changes with correlation time τ_d. However, this effect only changes the numerical coefficients and does not change the functional form in the result of eqn (9.9) because the characteristic time and the step length associated with this process are again τ_d and ε.[1] (Note that this situation is entirely different from that of the flexible polymers, for which the release of tube constraints has negligible effect if the polymer is sufficiently long.)

9.2.4 Estimation of the tube radius and the rotational diffusion constant

To complete the analysis, we express the tube radius a in terms of ν and L. A crude estimation of a is done as follows. We consider a tube of radius r around the test polymer and calculate the average number $N(r)$

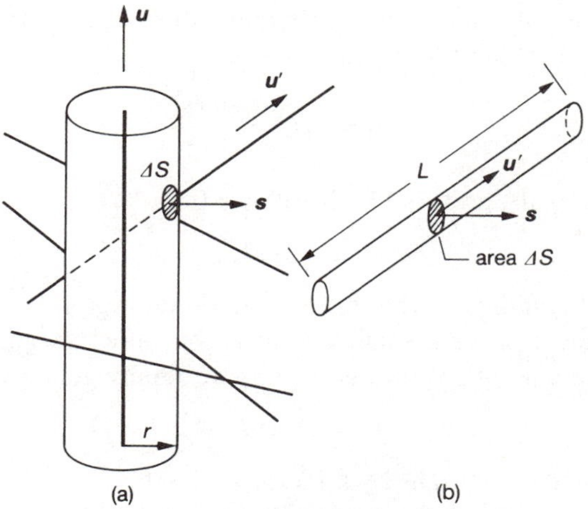

Fig. 9.5. (*a*) Tube enveloping the test polymer (thick line). (*b*) The polymers in the direction $\boldsymbol{u}'$ intersect the area ΔS if their centres of mass are in the region shown here. The volume of this region is $\Delta SL\,|\boldsymbol{u}'\cdot\boldsymbol{s}|$, where $\boldsymbol{s}$ is the unit vector normal to the region ΔS.

of the surrounding polymers which intersect this tube. If $r \simeq a$, the number $N(r)$ will be of order unity. Hence a is estimated by

$$N(a) \simeq 1. \tag{9.10}$$

To calculate $N(r)$, we consider a small region of area ΔS on the surface, and count the number $\Delta N(r)$ of surrounding polymers which penetrate this area (see Fig. 9.5). Let $\boldsymbol{s}$ be a unit vector normal to this region. As shown in Fig. 9.5, the surrounding polymers in the direction $\boldsymbol{u}'$ intersect the region ΔS if their centres of mass are in the region of volume $L\Delta S\,|\boldsymbol{u}'\cdot\boldsymbol{s}|$. Hence

$$\Delta N = \int \Psi_s(\boldsymbol{u}')\nu L\Delta S\,|\boldsymbol{u}'\cdot\boldsymbol{s}|\,\mathrm{d}\boldsymbol{u}' \tag{9.11}$$

where $\Psi_s(\boldsymbol{u}')$ is the orientational distribution function of the surrounding polymers.

First we consider that the distribution of the surrounding polymers is isotropic. In this case ΔN is independent of $\boldsymbol{s}$, and given by

$$\Delta N = \nu L\Delta S\int_0^{\pi} \mathrm{d}\theta\tfrac{1}{2}\,|\cos\theta|\sin\theta = \frac{\nu L}{2}\Delta S. \tag{9.12}$$

Summation of ΔN over the entire surface element gives $\nu LS/2$, ($S \simeq 2\pi rL$ being the total surface area). This is twice the number of polymers penetrating the tube since most polymers intersect the surface of the tube twice. Hence $N(r)$ is obtained as

$$N(r) = \tfrac{1}{4}\nu LS = \frac{\nu\pi r L^2}{2}. \tag{9.13}$$

From eqns (9.10) and (9.13), it follows that

$$a \simeq 1/\nu L^2. \tag{9.14}$$

Equation (9.14) indicates that in the semidilute region $1/L^3 \ll \nu \ll 1/bL^2$, the tube radius a is much smaller than L, but much larger than b.

Substituting eqn (9.14) into eqn (9.9) we finally get

$$D_r \simeq D_{r0}(\nu L^3)^{-2}. \tag{9.15}$$

Thus D_r is smaller than D_{r0} by a factor $(\nu L^3)^{-2}$.

In the above estimation we neglected the numerical factor entirely. To write eqn (9.15) as an equality, we have to put in a numerical factor β

$$D_r = \beta D_{r0}(\nu L^3)^{-2}. \tag{9.16}$$

The precise value of the numerical factor β is not known. Various data suggest that β is rather large (of the order of 10^3) as will be shown later in comparisons with experiments.

If the distribution of the surrounding polymers is not isotropic, the tube radius becomes a function of $\boldsymbol{u}$, and so does the rotational diffusion constant, which will be written as $\hat{D}_r(\boldsymbol{u})$. This is calculated in Appendix 9.I,

$$\hat{D}_r(\boldsymbol{u}) = D_r\left[\frac{4}{\pi}\int \mathrm{d}\boldsymbol{u}'\, |\boldsymbol{u}\times\boldsymbol{u}'|\, \Psi_s(\boldsymbol{u}')\right]^{-2} \tag{9.17}$$

where D_r is the rotational diffusion constant in the isotropic environment, eqn (9.16).

9.3 Brownian motion in equilibrium

9.3.1 Time correlation functions

Having obtained the translational and rotational diffusion constants, we can write down a kinetic equation for the probability $\Psi(\boldsymbol{R}, \boldsymbol{u}; t)$ that a given test polymer is in the configuration $(\boldsymbol{R}, \boldsymbol{u})$ at time t.†

$$\frac{\partial\Psi}{\partial t} = D_{\parallel}\left(\boldsymbol{u}\cdot\frac{\partial}{\partial\boldsymbol{R}}\right)^2\Psi + \mathcal{R}\cdot\hat{D}_r\cdot\mathcal{R}\Psi. \tag{9.18}$$

† $\hat{D}_r$ must be placed between the two rotational operator $\mathcal{R}$'s because (i) the equilibrium distribution of Ψ must be isotropic and (ii) the integral of $\partial\Psi/\partial t$ over $\boldsymbol{u}$ and $\boldsymbol{R}$ must vanish.

Equation (9.18) describes the Brownian motion of a test polymer in a given environment whose distribution is specified by $\Psi_s(\boldsymbol{u})$. In the isotropic solution, $\hat{D}_r$ becomes a constant given by eqn (9.16). The conditional probability $G(\boldsymbol{R}, \boldsymbol{u}, \boldsymbol{R}', \boldsymbol{u}'; t)$ that the test polymer which was in the configuration $(\boldsymbol{R}', \boldsymbol{u}')$ at time $t = 0$ is in the configuration $(\boldsymbol{R}, \boldsymbol{u})$ at time t satisfies

$$\frac{\partial G}{\partial t} = D_{\|}\left(\boldsymbol{u} \cdot \frac{\partial}{\partial \boldsymbol{R}}\right)^2 G + D_r \mathcal{R}^2 G. \tag{9.19}$$

Since eqn (9.19) has the same form as eqn (8.46), the time correlation functions for the test polymer are immediately obtained from the result of Section 8.4.

(i) Translational motion.
The mean square displacement of the centre of mass is given by

$$\langle (\boldsymbol{R}(t) - \boldsymbol{R}(0))^2 \rangle = 2D_{\|}t \simeq 2D_{\|0}t. \tag{9.20}$$

Thus the diffusion constant D_G is

$$D_G = \lim_{t\to\infty} \frac{1}{6t} \langle (\boldsymbol{R}(t) - \boldsymbol{R}(0))^2 \rangle \simeq \tfrac{1}{3} D_{\|0}. \tag{9.21}$$

Hence the ratio between D_G and D_{G0}, the diffusion constant in dilute solution, is given by

$$\frac{D_G}{D_{G0}} \simeq \frac{D_{\|0}}{D_{\|0} + 2D_{\perp 0}} = \frac{1}{2}. \tag{9.22}$$

Note that in the rodlike polymers, the entanglement does not change the molecular weight dependence of D_G. This is in contrast to the result for flexible polymers, in which the molecular weight dependence changes from $D_{G0} \propto M^{-\nu}$ to $D_G \propto M^{-2}$.

(ii) Rotational motion
The correlation function of $\boldsymbol{u}(t)$ is calculated as

$$\langle \boldsymbol{u}(t) \cdot \boldsymbol{u}(0) \rangle = \exp(-2D_r t) \tag{9.23}$$

Thus the rotational relaxation time $\tau_r = 1/2D_r$ is larger than that in dilute solution by

$$\tau_r / \tau_{r0} = (\nu L^3)^2 / \beta. \tag{9.24}$$

This shows a strong entanglement effect.

It must be mentioned that in eqns (9.21) and (9.23) we have neglected the motion inside the tube. As in the case of flexible polymers, the tube constraint is ineffective in a very short time-scale, so that

$$\langle (\boldsymbol{R}(t)-\boldsymbol{R}(0))^2\rangle = 6D_{G0}t \tag{9.25}$$

and

$$\langle \boldsymbol{u}(t)\cdot\boldsymbol{u}(0)\rangle = \exp(-2D_{r0}t). \tag{9.26}$$

These equations hold if the time-scale is less than

$$\tau_e \simeq \frac{a^2}{D_{\perp 0}} \simeq \frac{\varepsilon^2}{D_{r0}}. \tag{9.27}$$

In Section 9.6, we shall show how to describe both short-time ($t < \tau_e$) and long-time ($t > \tau_e$) behaviour.

9.3.2 *Dynamic light scattering*

Now we shall consider the dynamic light scattering in semidilute solution. In general the light scattering from polymer solutions of finite concentrations includes both the intramolecular and the intermolecular interferences. In the semidilute regime, however, even though the average distance between the polymers is smaller than the polymer size, the intermolecular interference is negligible because there is no correlation between the configurations of different polymers. Therefore the dynamic structure factor is again given by eqn (8.59)

$$g(\boldsymbol{k},t) = \int \mathrm{d}\boldsymbol{u}\int \mathrm{d}\boldsymbol{u}'\left(\frac{\sin(\boldsymbol{K}\cdot\boldsymbol{u})}{\boldsymbol{K}\cdot\boldsymbol{u}}\right)\left(\frac{\sin(\boldsymbol{K}\cdot\boldsymbol{u}')}{\boldsymbol{K}\cdot\boldsymbol{u}'}\right)G_{\boldsymbol{k}}(\boldsymbol{u},\boldsymbol{u}';t)\Psi_{\mathrm{eq}}(\boldsymbol{u}') \tag{9.28}$$

where

$$\boldsymbol{K} = \boldsymbol{k}L/2 \tag{9.29}$$

and

$$\frac{\partial}{\partial t}G_{\boldsymbol{k}}(\boldsymbol{u},\boldsymbol{u}';t) = (D_r\mathcal{R}^2 - D_{\parallel}(\boldsymbol{k}\cdot\boldsymbol{u})^2)G_{\boldsymbol{k}}(\boldsymbol{u},\boldsymbol{u}';t). \tag{9.30}$$

The distinction between the dilute and semidilute regimes is in the magnitude of the ratio between the translational term and the rotational term in the equation for the Green function $G_{\boldsymbol{k}}(\boldsymbol{u},\boldsymbol{u}';t)$. Let r be defined by:

$$r = D_G k^2/D_r. \tag{9.31}$$

In dilute solutions, r is given by

$$r_0 = D_{G0}k^2/D_{r0} \simeq (kL)^2 \tag{9.32}$$

which is usually of the order of unity. In the semidilute regime, on the other hand, r can be quite large because

$$r = D_G k^2/D_r \simeq (kL)^2(\nu L^3)^2 \gg (kL)^2 \quad \text{for} \quad \nu L^3 \gg 1. \tag{9.33}$$

To see the characteristic feature of the semidilute regime, let us consider the extreme case of $r = \infty$ (i.e., $D_r = 0$). Equation (9.30) is then solved by

$$G_k(\boldsymbol{u}, \boldsymbol{u}'; t) = \delta(\boldsymbol{u} - \boldsymbol{u}')\exp(-D_{\parallel}(\boldsymbol{k} \cdot \boldsymbol{u})^2 t). \tag{9.34}$$

Hence

$$g(\boldsymbol{k}, t) = \int \frac{d\boldsymbol{u}}{4\pi} \left(\frac{\sin(\boldsymbol{K} \cdot \boldsymbol{u})}{\boldsymbol{K} \cdot \boldsymbol{u}}\right)^2 \exp(-D_{\parallel}(\boldsymbol{k} \cdot \boldsymbol{u})^2 t)$$
$$= \int_0^1 d\xi \left(\frac{\sin(|\boldsymbol{K}|\,\xi)}{|\boldsymbol{K}|\,\xi}\right)^2 \exp(-D_{\parallel} k^2 \xi^2 t). \tag{9.35}$$

Thus the dynamical structure factor has a very broad distribution of decay rates ranging from 0 to $D_{\parallel} k^2$.

Experimental study of dynamic light scattering has been carried out by several groups.[5–9] Though some qualitative features of the above predictions are indeed observed, clear interpretation of experimental results has been hindered by various factors inherent in real polymers, such as polydispersity, partial flexibility, and association. These effects will be discussed later in connection with rotational motion. An important factor which will not affect the rotational motion, but will be important in the dynamic light scattering is the effect of weak, long-range repulsive force.[9] As in the case of flexible polymers, such interaction tends to keep the segment density homogeneous, and increases the decay rate of $g(\boldsymbol{k}, t)$ with the concentration. This is indeed observed in several systems.[5,9]

9.4 Orientation by external fields

9.4.1 *Linear regime*

So far we have been considering the motion of a test polymer in an isotropic environment. We now consider a slightly different problem: how does the orientational distribution function of polymers change under external fields such as a potential field $U_e(\boldsymbol{u})$ or a velocity gradient $\boldsymbol{\kappa}$. Let $\Psi(\boldsymbol{u}; t)$ be the probability that an arbitrarily chosen polymer is in the direction $\boldsymbol{u}$. Since each polymer feels the external field as in eqn (8.15), the time evolution of $\Psi(\boldsymbol{u}; t)$ can be described by[2,10]

$$\frac{\partial \Psi}{\partial t} = \mathcal{R} \cdot \hat{D}_r \left[\mathcal{R}\Psi + \frac{1}{k_B T}(\mathcal{R} U_e)\Psi \right] - \mathcal{R} \cdot [\boldsymbol{u} \times (\boldsymbol{\kappa} \cdot \boldsymbol{u}\Psi)]. \tag{9.36}$$

An important point here is that in this problem, the environmental distribution function Ψ_s is the same as $\Psi(\boldsymbol{u}; t)$ itself, and $\hat{D}_r$ is now given by

$$\hat{D}_r = D_r \left[\frac{4}{\pi} \int d\boldsymbol{u}' \Psi(\boldsymbol{u}'; t) \, |\boldsymbol{u} \times \boldsymbol{u}'| \right]^{-2}. \tag{9.37}$$

Equations (9.36) and (9.37) give a nonlinear equation for Ψ. The nonlinearity indicates a mean field character of the present theory: it comes from identifying the distribution of the surrounding polymers with that of the test polymer. Equation (9.36) is thus different from the usual Smoluchowski equation, which is always linear in Ψ.

In the linear response regime, however, the nonlinearity of the kinetic equation is not important because there $\hat{D}_r$ can be replaced by D_r since the change in D_r appears only in the higher order perturbation. Therefore the linear response function is given by the same form as that in dilute solution except that D_r is much smaller than D_{r0}. For example, consider a rodlike polymer which has permanent dipole moment μ and isotropic polarizability ($\alpha_\parallel = \alpha_\perp \equiv \alpha_\infty$). The complex polarizability and the dynamic Kerr constant (per polymer) are given in the same form as eqns (8.94) and (8.104),

$$\alpha^*(\omega) = \frac{\mu^2}{3k_B T} \frac{1}{1 + i\omega\tau} + \alpha_\infty \tag{9.38}$$

$$K_{dc}(\omega) = \frac{K_0}{2} \frac{1}{1 + (\omega\tau)^2}, \tag{9.39}$$

with

$$\tau = 1/2D_r. \tag{9.40}$$

The rotational diffusion constant can be obtained from these expressions.

9.4.2 *Nonlinear regime—tube dilation*

The nonlinearity in the kinetic equation becomes important if the external perturbation is large. In this case precise mathematical analysis becomes quite difficult. A convenient approximation is to replace $\hat{D}_r$ by the average $\bar{D}_r$

$$\hat{D}_r \simeq \bar{D}_r \equiv D_r \left[\frac{4}{\pi} \int d\boldsymbol{u} \, d\boldsymbol{u}' \Psi(\boldsymbol{u}; t) \Psi(\boldsymbol{u}'; t) \, |\boldsymbol{u} \times \boldsymbol{u}'| \right]^{-2}. \tag{9.41}$$

Since $\bar{D}_r$ is independent of $\boldsymbol{u}$, the kinetic equation can be written as

$$\frac{\partial \Psi}{\partial t} = \bar{D}_r \mathcal{R} \cdot \left[\mathcal{R}\Psi + \frac{1}{k_B T} (\mathcal{R} U_e) \Psi \right] - \mathcal{R} \cdot [\boldsymbol{u} \times (\boldsymbol{\kappa} \cdot \boldsymbol{u} \Psi)]. \tag{9.42}$$

This can be handled much more easily than eqn (9.36) (see refs 2 and 10). Though the approximation (9.41) is crude, it takes into account the

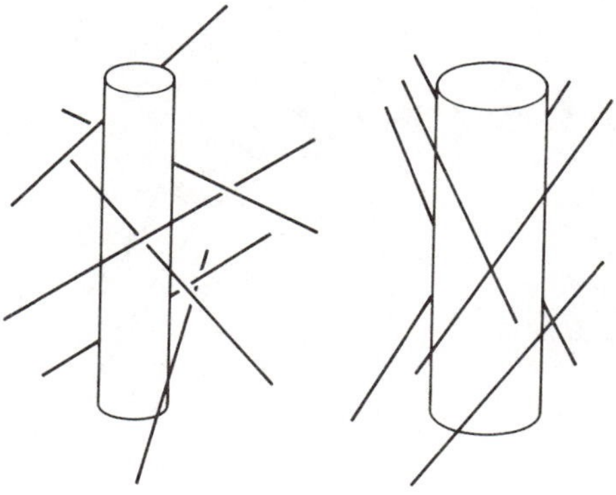

Fig. 9.6. Tube dilation. The tube radius increases when the surrounding polymers are oriented in the same direction as the test polymer.

following effect. As the polymers orient in the same direction, the average diameter of the tube becomes larger, so that the average rotational diffusion constant increases (see Fig. 9.6). This effect is called the tube dilation.

The tube dilation may be seen in, for example, the relaxation of birefringence from the highly oriented state; the initial relaxation rate is larger than the final one. A theoretical analysis of this effect has been done in ref. 2

9.4.3 *Experimental study of the rotational diffusion constant*

The rotational diffusion constant has been measured by relaxation of the Kerr effect,[11,12] and by dynamic light scattering.[5] The experimental results are in accordance with the theoretical predictions

$$D_r/D_{r0} = \beta(\nu L^3)^{-2} \propto \rho^{-2}M^{-4}, \tag{9.43}$$

or, since $D_{r0} \propto \ln(M)/M^3$,

$$D_r \propto \rho^{-2}M^{-7}\ln(M). \tag{9.44}$$

However, the absolute magnitude of D_r has turned out to be quite large: experimental values of β range from 10^3 to 10^4. Such large values of β have also been found by a computer simulation for thin polymers (with no thickness).[14,15] Figure 9.7 shows the concentration-dependence of D_r obtained by dynamic electric birefringence and computer simulation. The solid line indicates eqn (9.43) with $\beta = 1.3 \times 10^3$, which was obtained by Hayakawa *et al.*[13] by detailed study of the tube statistics. Though the preciseness of this value can be questioned, it is clear that the numerical

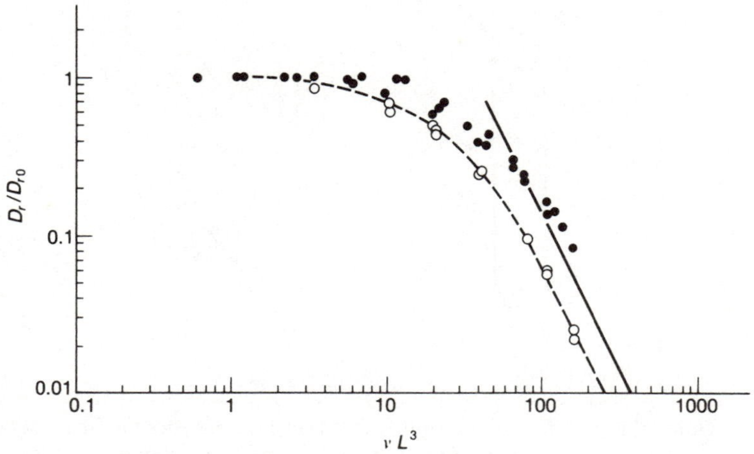

Fig. 9.7. Reduced rotational diffusion constant D_r/D_{r0} is plotted against reduced concentration νL^3. Filled circles: the experimental results[12] of dynamic electric birefringence of poly(γ-benzyl-L-glutamate, molecular weight ranging from 7.3×10^4 to 1.5×10^5). Open circles: the result of computer simulation.[14] The solid line is the theoretical result.[13] (Courtesy of Prof. Hayakawa, Tokyo University.)

factor is large and that the semidilute regime starts at rather large values of νL^3.

As has already been mentioned, precise comparison with experiments is hindered by various effects such as polydispersity and bending of polymers. Quantitative theory for these effects is difficult, but the qualitative aspects have been discussed in refs 16–18

9.5 Viscoelasticity

9.5.1 Expression for the stress tensor

To discuss the viscoelastic properties in semidilute solutions, we first consider the expression for the stress tensor. This is obtained from the principle of virtual work as in Section 8.6.1.

In the semidilute region, the expression for the free energy is the same as that in the dilute solution since the excluded volume effect is negligible:

$$\mathscr{A} = \nu \int \mathrm{d}\boldsymbol{u}\Psi(k_B T \ln \Psi + U_e). \tag{9.45}$$

The change in Ψ by instantaneous deformation $\delta\varepsilon_{\alpha\beta}$ is again given by

$$\delta\Psi = -\mathscr{R} \cdot (\boldsymbol{u} \times \delta\boldsymbol{\varepsilon} \cdot \boldsymbol{u}\Psi). \tag{9.46}$$

Hence the elastic stress is given by precisely the same form as eqn (8.114)

$$\sigma^{(E)}_{\alpha\beta} = 3\nu k_B T \langle u_\alpha u_\beta - \tfrac{1}{3}\delta_{\alpha\beta}\rangle - \nu\langle(\boldsymbol{u}\times\mathcal{R}U_e)_\alpha u_\beta\rangle. \tag{9.47}$$

The viscous stress is obtained from the hydrodynamic energy dissipation under a given velocity gradient. If we assume that the hydrodynamic interaction is completely screened, this is calculated in the same manner as in eqn (8.119):

$$\sigma^{(V)}_{\alpha\beta} = \nu\zeta_{\text{str}}\kappa_{\mu\nu}\langle u_\mu u_\nu u_\alpha u_\beta\rangle \tag{9.48}$$

with

$$\zeta_{\text{str}} = \frac{\pi\eta_s L^3}{4}. \tag{9.49}$$

On the other hand if we neglect the hydrodynamic screening entirely, ζ_{str} is given by the same formula as eqn (8.122)

$$\zeta_{\text{str}} = \frac{\pi\eta_s L^3}{6\ln(L/b)}. \tag{9.50}$$

The actual form will be between the two. Indeed the effective medium theory[4] indicates that

$$\zeta_{\text{str}} \simeq \text{number} * \frac{\eta_s L^3}{\ln(1/\nu b L^2)}. \tag{9.51}$$

It is important to note that ζ_{str} is not affected as seriously as D_r. This is because the viscous stress reflects very fast motions, for which the tube constraint is not effective.†

For simplicity we proceed using eqn (9.50). The stress tensor is given in precisely the same manner as for dilute solutions (see eqn (8.123)).

$$\begin{aligned}\sigma_{\alpha\beta} = {} & 3\nu k_B T\langle u_\alpha u_\beta - \tfrac{1}{3}\delta_{\alpha\beta}\rangle - \nu\langle(\boldsymbol{u}\times\mathcal{R}U_e)_\alpha u_\beta\rangle \\ & + \nu\zeta_{\text{str}}\kappa_{\mu\nu}\langle u_\mu u_\nu u_\alpha u_\beta\rangle + \eta_s(\kappa_{\alpha\beta} + \kappa_{\beta\alpha}).\end{aligned} \tag{9.52}$$

9.5.2 *Linear viscoelasticity*

To calculate the stress, we have to solve

$$\frac{\partial\Psi}{\partial t} = \mathcal{R}\cdot\hat{D}_r\cdot\mathcal{R}\Psi - \mathcal{R}\cdot(\boldsymbol{u}\times\boldsymbol{\kappa}\cdot\boldsymbol{u}\Psi). \tag{9.53}$$

† It has been suggeted[19,20] that better agreement with some experiments can be obtained if ζ_{str} is replaced by $k_B T/D_r \simeq \eta_s L^3(\nu L^3)^2/\ln(L/b)$. However, this agreement is perhaps superficial, caused by non ideal effects of real polymers such as molecular weight distribution, or flexibility. Theoretically there is no reason to believe that ζ_{str} is given by $k_B T/D_r$.

In the linear viscoelastic region, $\hat{D}_r$ can be replaced by D_r, and the solution is obtained in the same way as in Section 8.6.

In steady-state shear flow with shear rate κ, the terms in eqn (9.52) are given by (the case of $U_e = 0$ is being considered)

$$\sigma_{xy}^{(\mathrm{E})} = 3\nu k_B T \langle u_x u_y \rangle = \frac{\nu k_B T}{10 D_r} \kappa = \frac{\pi}{30\beta} \frac{(\nu L^3)^3}{\ln(L/b)} \eta_s \kappa, \tag{9.54}$$

$$\sigma_{xy}^{(\mathrm{V})} = \nu \zeta_{\mathrm{str}} \langle u_x^2 u_y^2 \rangle \kappa = \tfrac{1}{15} \nu \zeta_{\mathrm{str}} \kappa \simeq \frac{\pi}{90} \frac{\nu L^3}{\ln(L/b)} \eta_s \kappa, \tag{9.55}$$

and

$$\sigma_{xy}^{(\mathrm{sol})} = \eta_s \kappa.$$

Hence their ratio is

$$\sigma_{xy}^{(\mathrm{E})} : \sigma_{xy}^{(\mathrm{V})} : \sigma_{xy}^{(\mathrm{sol})} \simeq \beta^{-1} (\nu L^3)^3 : (\nu L^3) : 1. \tag{9.56}$$

In the semidilute regime $\nu L^3 > 1$, the contribution of the elastic stress is much larger than the viscous stress and the solvent stress. Thus in the ideal semidilute region of $1/L^3 \ll \nu \ll 1/bL^2$, the stress can be written as

$$\sigma_{\alpha\beta} = 3\nu k_B T \langle u_\alpha u_\beta - \tfrac{1}{3}\delta_{\alpha\beta} \rangle = 3\nu k_B T S_{\alpha\beta}. \tag{9.57}$$

Since the characteristic time of $S_{\alpha\beta}$ is very large (being of the order of $1/D_r$), the viscoelastic behaviour becomes quite pronounced in the semidilute region.

From eqns (9.54) and (9.57), the steady-state viscosity is given by

$$\eta = \frac{1}{\kappa} \sigma_{xy} = \frac{\nu k_B T}{10 D_r} = \frac{\pi}{30\beta} \eta_s \frac{(\nu L^3)^3}{\ln(L/b)} \tag{9.58}$$

which depends on the molecular weight and concentration as

$$\eta \propto \rho^3 M^6 / \ln(N). \tag{9.59}$$

The strong molecular weight dependence of η is in accordance with experiments,[21–25] though the precise exponent is difficult to extract because of the nonideal effect discussed previously. The numerical factor β can also be obtained from the viscosity and has again turned out to be rather large ranging from 10^3 to 10^4. (Comparison between the theoretical prediction and the experimental results are given in refs 1, 19, 24, and 26) In some cases the concentration dependence of η is stronger than predicted by eqn (9.59) at higher concentration.[22,23,25] One possible explanation for this is the rod jamming effect.[27]

The complex viscosity defined by eqn (8.139) is calculated as

$$\eta^*(\omega) = \frac{3\rho R T}{5M} \frac{\tau}{1 + i\omega\tau} \tag{9.60}$$

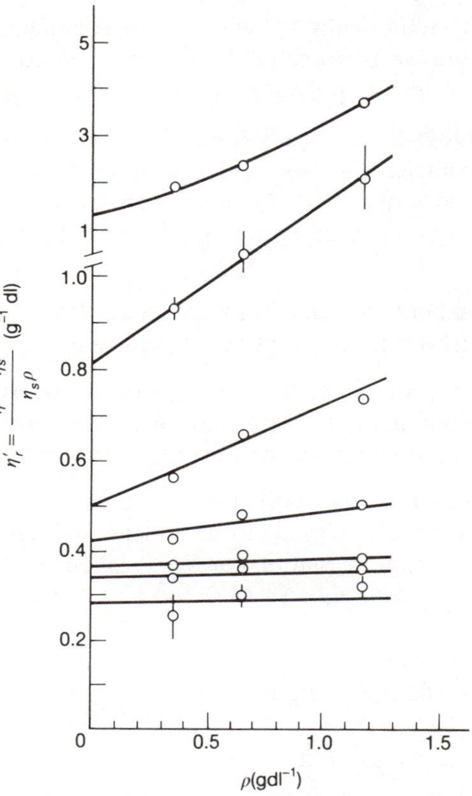

Fig. 9.8. Real part of the complex viscosity $\eta'(\omega)$ of poly(γ-benzyl-L-glutamate) in m-cresole is plotted against concentration ρ at various frequencies (from top to bottom, 0, 2.2, 6.6, 20, 58, 202, and 525 KHz). Here η_s is the viscosity of the pure solvent. Reproduced from ref. 28.

with

$$\tau = 1/6D_r. \tag{9.61}$$

Figure 9.8 shows the real part of the complex viscosity at various frequencies.[28] As the concentration increases, the viscosity at low frequency increases sharply, while the viscosity at high frequency increases only in proportion to the concentration. This is in accordance with eqns (9.54) and (9.55).

9.5.3 Nonlinear viscoelasticity

Since D_r becomes small in the semidilute region, the nonlinear viscoelasticity becomes quite important. To handle the nonlinear viscoelasticity, we need the full solution of eqn (9.53). This has been done by solving eqn (9.53) numerically[10,29] with the approximation (9.41). It turns out that the qualitative features of the nonlinear viscoelasticity are quite similar to those of flexible polymers, for example:

(i) Shear thinning: In the steady shear flow, the viscosity $\eta(\kappa)$, the first normal stress coefficient $\Psi_1(\kappa)$, and the absolute value of the second normal stress coefficients $\Psi_2(\kappa)$ (which is negative) all decrease with the shear rate. The characteristic shear rate for the shear thinning is about D_r.

(ii) Stress overshoot: When a constant shear flow is started, the shear stress shows overshoot if the shear rate is sufficiently high.

(iii) Elongational viscosity: The elongational viscosity first increases slightly with the elongational rate and then decreases. These features agree at least qualitatively with the observed behaviour.[21–24,30–33]

Unlike the case of flexible polymers the constitutive equation cannot be written in a simple closed form unless we use the decoupling approximation. Since the equation for $S_{\alpha\beta}$ is given by the same equation as that in dilute solution, eqns (8.149) and (9.57) give a closed equation for $\sigma_{\alpha\beta}$:

$$\frac{\partial}{\partial t}\sigma_{\alpha\beta} = -6\bar{D}_r\sigma_{\alpha\beta} + \kappa_{\alpha\mu}\sigma_{\beta\mu} + \kappa_{\beta\mu}\sigma_{\alpha\mu} + \frac{1}{3}G_e(\kappa_{\alpha\beta} + \kappa_{\beta\alpha})$$
$$- 2\sigma_{\mu\nu}\kappa_{\mu\nu}\left(\frac{\sigma_{\alpha\beta}}{G_e} + \tfrac{1}{3}\delta_{\alpha\beta}\right) \tag{9.62}$$

where

$$G_e = 3\nu k_B T. \tag{9.63}$$

The qualitative features described above can be checked by this approximate constitutive equation.

9.6 Short time-scale motion

9.6.1 Chopstick model

So far we have neglected the small-scale fluctuation that the polymer makes inside the tube. To include such motion, we consider the following model[34] (see Fig. 9.9). We separate the direction of the polymer and the direction of the tube, and consider two vectors $\boldsymbol{u}$ representing the direction of the polymer and $\boldsymbol{n}$ the direction of the tube. The diffusion constant of $\boldsymbol{n}$ is D_r, while the diffusion constant of $\boldsymbol{u}$ is D_{r0} since the polymer can move freely inside the tube. The condition that $\boldsymbol{u}$ is fluctuating inside the tube is represented by the coupling potential

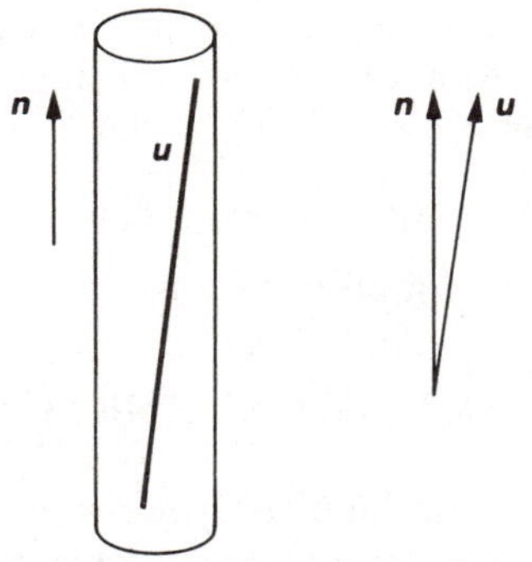

Fig. 9.9. The chopstick model.

between $\boldsymbol{u}$ and $\boldsymbol{n}$.

$$U_c(\boldsymbol{u}-\boldsymbol{n}) = \frac{k_B T}{2\varepsilon^2}(\boldsymbol{u}-\boldsymbol{n})^2. \tag{9.64}$$

Equation (9.64) guarantees that at equilibrium, the deviation between $\boldsymbol{u}$ and $\boldsymbol{n}$ is of order ε:

$$\langle(\boldsymbol{u}-\boldsymbol{n})^2\rangle = \frac{\int \mathrm{d}\boldsymbol{u}(\boldsymbol{u}-\boldsymbol{n})^2 \exp\left(-\frac{(\boldsymbol{u}-\boldsymbol{n})^2}{2\varepsilon^2}\right)}{\int \mathrm{d}\boldsymbol{u} \exp\left(-\frac{(\boldsymbol{u}-\boldsymbol{n})^2}{2\varepsilon^2}\right)} \simeq 2\varepsilon^2. \tag{9.65}$$

The kinetic equation is now given for the two-vector distribution function $\Phi(\boldsymbol{u}, \boldsymbol{n}; t)$ as

$$\frac{\partial \Phi}{\partial t} = D_r \mathcal{R}_n \cdot \left(\mathcal{R}_n \Phi + \frac{\Phi}{k_B T}\mathcal{R}_n U_c(\boldsymbol{u}-\boldsymbol{n})\right)$$
$$+ D_{r0}\mathcal{R}_u \cdot \left(\mathcal{R}_u \Phi + \frac{\Phi}{k_B T}\mathcal{R}_u(U_c(\boldsymbol{u}-\boldsymbol{n}) + U_e(\boldsymbol{u}))\right) \tag{9.66}$$

where $\mathcal{R}_u$ and $\mathcal{R}_n$ are written as

$$\mathcal{R}_u = \boldsymbol{u} \times \frac{\partial}{\partial \boldsymbol{u}}, \qquad \mathcal{R}_n = \boldsymbol{n} \times \frac{\partial}{\partial \boldsymbol{n}} \tag{9.67}$$

and for simplicity the tube dilation is neglected.

The model described by eqn (9.66) is easy to visualize: the two vectors $\boldsymbol{u}$ and $\boldsymbol{n}$ move together like a pair of chopsticks. The external field, represented by the potential $U_e(\boldsymbol{u})$, affects the rapidly moving vector $\boldsymbol{u}$, which drags the slowly moving vector $\boldsymbol{n}$ through the coupling potential $U_c(\boldsymbol{u}-\boldsymbol{n})$.

An important property of the kinetic equation (9.66) is that at equilibrium, the distribution of $\boldsymbol{u}$ is not affected by the vector $\boldsymbol{n}$. In fact, the equilibrium solution of eqn (9.66) is

$$\Phi_{\text{eq}}(\boldsymbol{u}, \boldsymbol{n}) \propto \exp[-(U_e(\boldsymbol{u}) + U_c(\boldsymbol{u} - \boldsymbol{n}))/k_B T]. \tag{9.68}$$

Hence the equilibrium distribution of $\boldsymbol{u}$ is

$$\Psi_{\text{eq}}(\boldsymbol{u}) = \int \mathrm{d}\boldsymbol{n}\, \Phi_{\text{eq}}(\boldsymbol{u}, \boldsymbol{n}) \propto \exp[-U_e(\boldsymbol{u})/k_B T] \tag{9.69}$$

which is the same as the isolated polymer without the tube constraint. This must be so since the entanglement interaction does not affect the static properties.

9.6.2 *Local equilibrium approximation*

Although the kinetic equation (9.66) is conceptually simple, it is not easily handled mathematically. However, if we focus our attention on slow motion, a simple treatment is possible.

If the external field changes slowly compared to $\tau_e \simeq \varepsilon^2/D_{r0}$, the distribution of the vector $\boldsymbol{u}$ can be assumed to be in a local equilibrium for given $\boldsymbol{n}$. Then $\Phi(\boldsymbol{u}, \boldsymbol{n}; t)$ can be written as

$$\Phi(\boldsymbol{u}, \boldsymbol{n}; t) = \bar{\Psi}(\boldsymbol{n}; t)\psi_{\text{eq}}(\boldsymbol{u}, \boldsymbol{n}) \tag{9.70}$$

where ψ_{eq} represents the local equilibrium distribution of $\boldsymbol{u}$ for given $\boldsymbol{n}$, i.e.,

$$\psi_{\text{eq}}(\boldsymbol{u}, \boldsymbol{n}) = \frac{\exp\left[-\dfrac{U_c(\boldsymbol{u}-\boldsymbol{n}) + U_e(\boldsymbol{u})}{k_B T}\right]}{\displaystyle\int \mathrm{d}\boldsymbol{u} \exp\left[-\dfrac{U_c(\boldsymbol{u}-\boldsymbol{n}) + U_e(\boldsymbol{u})}{k_B T}\right]} \tag{9.71}$$

$$= \exp\left[-\frac{U_c(\boldsymbol{u}-\boldsymbol{n}) + U_e(\boldsymbol{u}) - \bar{U}_e(\boldsymbol{n})}{k_B T}\right]. \tag{9.72}$$

Here

$$\bar{U}_e(\boldsymbol{n}) \equiv -k_B T \ln \int \mathrm{d}\boldsymbol{u} \exp\left[-\frac{U_c(\boldsymbol{u}-\boldsymbol{n}) + U_e(\boldsymbol{u})}{k_B T}\right] \tag{9.73}$$

and $\psi_{\text{eq}}(\boldsymbol{u}, \boldsymbol{n})$ is normalized such that

$$\int \mathrm{d}\boldsymbol{u}\, \psi_{\text{eq}}(\boldsymbol{u}, \boldsymbol{n}) = 1. \tag{9.74}$$

To determine $\bar{\Psi}(\boldsymbol{n}; t)$ we substitute eqn (9.70) into eqn (9.66). The result is

$$\frac{\partial \bar{\Psi}}{\partial t}\psi_{\text{eq}} = D_r \mathcal{R}_n \cdot \left[\psi_{\text{eq}} \mathcal{R}_n \bar{\Psi} + \bar{\Psi}\mathcal{R}_n \psi_{\text{eq}} + \frac{\bar{\Psi}\psi_{\text{eq}}}{k_B T}\mathcal{R}_n U_c(\boldsymbol{u}-\boldsymbol{n})\right]$$

$$= D_r \mathcal{R}_n \cdot \left[\psi_{\text{eq}} \mathcal{R}_n \bar{\Psi} + \frac{\bar{\Psi}\psi_{\text{eq}}}{k_B T}\mathcal{R}_n \bar{U}_e(\boldsymbol{n})\right] \tag{9.75}$$

Integrating this over $\boldsymbol{u}$, we have

$$\frac{\partial \bar{\Psi}}{\partial t} = D_r \mathcal{R}_n \cdot \left[\mathcal{R}_n \bar{\Psi} + \frac{1}{k_B T}(\mathcal{R}_n \bar{U}_e(\boldsymbol{n}))\bar{\Psi}\right]. \tag{9.76}$$

This equation corresponds to eqn (9.36). The difference between eqn (9.36) and (9.76) is in the potential U_e: the potential for the tube $\bar{U}_e(\boldsymbol{n})$ is not equal to $U_e(\boldsymbol{n})$. To the lowest order in ε, we can show (see Appendix 9.II)

$$\bar{U}_e(\boldsymbol{n}) = U_e(\boldsymbol{n}) + \tfrac{1}{2}\varepsilon^2\left[\mathcal{R}_n^2 U_e(\boldsymbol{n}) - \frac{1}{k_B T}(\mathcal{R}_n U_e(\boldsymbol{n}))^2\right]. \tag{9.77}$$

Similarly the average of an arbitrary function of $\boldsymbol{u}$ is given by (see Appendix 9.II).

$$\langle A(\boldsymbol{u})\rangle = \int \mathrm{d}\boldsymbol{n}\int \mathrm{d}\boldsymbol{u} A(\boldsymbol{u})\bar{\Psi}(\boldsymbol{n};t)\psi_{\text{eq}}(\boldsymbol{u},\boldsymbol{n})$$

$$= \int \mathrm{d}\boldsymbol{n}\bar{A}(\boldsymbol{n})\bar{\Psi}(\boldsymbol{n};t), \tag{9.78}$$

where

$$\bar{A}(\boldsymbol{n}) \equiv \int \mathrm{d}\boldsymbol{u} A(\boldsymbol{u})\psi_{\text{eq}}(\boldsymbol{u},\boldsymbol{n})$$

$$= A(\boldsymbol{n}) + \tfrac{1}{2}\varepsilon^2\left[\mathcal{R}_n^2 A(\boldsymbol{n}) - \frac{2}{k_B T}(\mathcal{R}_n A(\boldsymbol{n}))\cdot(\mathcal{R}_n U_e(\boldsymbol{n}))\right] + O(\varepsilon^4). \tag{9.79}$$

Equations (9.77) and (9.79) indicate that the formulation given in Sections 9.2 and 9.3 neglects terms of the order of ε^2.

9.6.3 Example

As an example, consider the case when the polymer has a permanent dipole moment μ along its axis, but whose polarizability is isotropic ($\alpha_{\|} = \alpha_{\perp}$). The potential of such a polymer in an electric field E in the z direction is

$$U_e(\boldsymbol{u}) = -\mu E u_z. \tag{9.80}$$

Hence the effective potential for the *tube* calculated with eqn (9.77) is

$$\bar{U}_e(\boldsymbol{n}) = -\mu E n_z + \tfrac{1}{2}\varepsilon^2\left[2\mu E n_z - \frac{\mu^2}{k_BT}E^2(1-n_z^2)\right]$$

$$= -(1-\varepsilon^2)\mu E n_z + \frac{\mu^2\varepsilon^2}{2k_BT}E^2(n_z^2-1). \tag{9.81}$$

Equation (9.81) indicates that the *tube* has not only the effective dipole moment

$$\mu_{\text{tube}} = (1-\varepsilon^2)\mu \tag{9.82}$$

but also the anisotropic polarizability

$$\Delta\alpha_{\text{tube}} = -\frac{\mu^2\varepsilon^2}{k_BT}. \tag{9.83}$$

Since $\Delta\alpha_{\text{tube}} < 0$, this effect tends to orient the tube perpendicular to the electric field.

Equation (9.79) gives

$$\langle u_z\rangle = (1-\varepsilon^2)\langle n_z\rangle - \varepsilon^2\frac{\mu E}{k_BT}\langle n_z^2-1\rangle \tag{9.84}$$

$$\langle u_z^2 - u_x^2\rangle = (1-3\varepsilon^2)\langle n_z^2-n_x^2\rangle + \frac{2\varepsilon^2\mu E}{k_BT}\langle n_z(1-n_z^2+n_x^2)\rangle. \tag{9.85}$$

The average of the quantities on the right-hand sides of eqns (9.84) and (9.85) are calculatd from

$$\frac{\partial\bar{\Psi}}{\partial t} = D_r\mathcal{R}_n\cdot\left[\mathcal{R}_n\bar{\Psi} + \frac{\mathcal{R}_n\bar{U}_e}{k_BT}\bar{\Psi}\right]. \tag{9.86}$$

The calculation is tedious, but straightforward. Here we only quote the results.[34] The complex polarizability is given by

$$\alpha^*(\omega) = \frac{\mu^2}{3k_BT}\left(\frac{1-2\varepsilon^2}{1+i\omega\tau} + 2\varepsilon^2\right) \quad \text{with} \quad \tau = 1/2D_r, \tag{9.87}$$

and the dynamic Kerr constant by

$$K_{\text{dc}}(\omega) = \frac{K_0}{2}\left(\frac{1+\varepsilon^2}{1+(\omega\tau)^2} - \varepsilon^2\right) \quad \text{with} \quad \tau = 1/2D_r. \tag{9.88}$$

Equation (9.88) shows that K_{dc} becomes negative for $\omega > 1/(\varepsilon\tau)$. The negative birefringence comes from the negative anisotropic polarizability of the tube, and has indeed been observed by Mori *et al.*[12] The importance of the short time-scale motion was also noted by Moscicki *et al.*[35] for the dielectric relaxation.

The local equilibrium approximation breaks down if the frequency of the external field becomes comparable with the characteristic time of the fluctuation $\tau_e \simeq \varepsilon^2/D_{r0}$. The motion of the polymer in such a short time-scale can be treated by eqn (9.66)[34] or a rotation model limited in a cone-like region.[36] Both treatments gives qualitatively similar results: e.g., $\alpha^*(\omega)$ and $K_{\mathrm{dc}}(\omega)$ approach zero as $\omega \to \infty$ with the relaxation time τ_e. Recent studies have been reviewed by Moscicki.[37]

Appendix 9.I Tube dilation by orientational ordering

Here we calculate the rotational diffusion constant for the case when the orientational distribution of the surrounding polymer is not isotropic. The number $N(r)$, which is given by the sum of eqn (9.11) over the entire surface, is written as

$$N(r) = \tfrac{1}{2}\nu SL\langle\langle|\boldsymbol{u}'\cdot\boldsymbol{s}|\rangle_{\boldsymbol{u}'}\rangle_{\boldsymbol{s}} \tag{9.I.1}$$

where $\langle\ldots\rangle_{\boldsymbol{u}'}$ stands for the average over the distribution of the surrounding polymer,

$$\langle\ldots\rangle_{\boldsymbol{u}'} = \int \mathrm{d}\boldsymbol{u}'\Psi_s(\boldsymbol{u}')\ldots, \tag{9.I.2}$$

and the average $\langle\ldots\rangle_{\boldsymbol{s}}$ means the average over the surface of the tube

$$\langle\ldots\rangle_{\boldsymbol{s}} = \frac{1}{S}\int \mathrm{d}S\ldots \tag{9.I.3}$$

First we calculate $\langle|\boldsymbol{u}'\cdot\boldsymbol{s}|\rangle_{\boldsymbol{s}}$. For a slender tube, the surface area is mostly on the long side of the tube and $\boldsymbol{s}$ is perpendicular to $\boldsymbol{u}$, then $|\boldsymbol{s}\cdot\boldsymbol{u}'|$ is written as

$$|\boldsymbol{s}\cdot\boldsymbol{u}'| = |\boldsymbol{s}\times(\boldsymbol{u}\times\boldsymbol{u}')| = |\boldsymbol{u}\times\boldsymbol{u}'|\,|\sin\psi| \tag{9.I.4}$$

where ψ is the angle between $\boldsymbol{s}$ and $\boldsymbol{u}\times\boldsymbol{u}'$. Since the distribution of ψ is uniform between 0 to 2π

$$\langle|\boldsymbol{s}\cdot\boldsymbol{u}'|\rangle_{\boldsymbol{s}} = |\boldsymbol{u}\times\boldsymbol{u}'|\int_0^{2\pi}\frac{\mathrm{d}\psi}{2\pi}|\sin\psi| = \frac{2}{\pi}|\boldsymbol{u}\times\boldsymbol{u}'|. \tag{9.I.5}$$

From eqns (9.I.1) and (9.I.5)

$$N(r) = \frac{\nu SL}{\pi}\langle|\boldsymbol{u}\times\boldsymbol{u}'|\rangle_{\boldsymbol{u}'} = \frac{1}{\pi}\nu 2\pi r L^2\cdot\frac{2}{\pi}\int \mathrm{d}\boldsymbol{u}'\,|\boldsymbol{u}\times\boldsymbol{u}'|\,\Psi_s(\boldsymbol{u}') \tag{9.I.6}$$

which gives

$$\hat{D}_r = D_r\left[\frac{4}{\pi}\int \mathrm{d}\boldsymbol{u}'\,|\boldsymbol{u}\times\boldsymbol{u}'|\,\Psi_s(\boldsymbol{u}')\right]^{-2}. \tag{9.I.7}$$

The factor $4/\pi$ is put so that eqn (9.I.7) agrees with eqn (9.16) in the case of an isotropic distribution.

Appendix 9.II Effective potential for the tube

Let $\langle \ldots \rangle_{\boldsymbol{u}}$ be defined by

$$\langle A(\boldsymbol{u})\rangle_{\boldsymbol{u}} = \int \mathrm{d}\boldsymbol{u} \exp\left(-\frac{(\boldsymbol{u}-\boldsymbol{n})^2}{2\varepsilon^2}\right) A(\boldsymbol{u}) \bigg/ \int \mathrm{d}\boldsymbol{u} \exp\left(-\frac{(\boldsymbol{u}-\boldsymbol{n})^2}{2\varepsilon^2}\right). \quad (9.\text{II}.1)$$

We evaluate $\langle A(\boldsymbol{u})\rangle_{\boldsymbol{u}}$ for small ε. To that end we expand $A(\boldsymbol{u})$ with respect to the small vector $\boldsymbol{v} \equiv \boldsymbol{u} - \boldsymbol{n}$

$$A(\boldsymbol{u}) = A(\boldsymbol{n}) + v_\alpha \frac{\partial}{\partial n_\alpha} A(\boldsymbol{n}) + \tfrac{1}{2} v_\alpha v_\beta \frac{\partial}{\partial n_\alpha}\frac{\partial}{\partial n_\beta} A(\boldsymbol{n}) \quad (9.\text{II}.2)$$

and evaluate $\langle A(\boldsymbol{u})\rangle_{\boldsymbol{u}}$ as

$$\langle A(\boldsymbol{u})\rangle_{\boldsymbol{u}} = A(\boldsymbol{n}) + \langle v_\alpha \rangle_{\boldsymbol{u}} \frac{\partial}{\partial n_\alpha} A(\boldsymbol{n}) + \tfrac{1}{2} \langle v_\alpha v_\beta \rangle_{\boldsymbol{u}} \frac{\partial}{\partial n_\alpha}\frac{\partial}{\partial n_\beta} A(\boldsymbol{n}). \quad (9.\text{II}.3)$$

By symmetry, $\langle v_\alpha \rangle_{\boldsymbol{u}}$ is written as

$$\langle v_\alpha \rangle_{\boldsymbol{u}} = C n_\alpha. \quad (9.\text{II}.4)$$

To calculate the constant C, we multiply both sides of eqn (9.II.4) by n_α and sum over α. The result is

$$C = \langle \boldsymbol{v} \cdot \boldsymbol{n} \rangle_{\boldsymbol{u}} = \langle \boldsymbol{u} \cdot \boldsymbol{n} - 1 \rangle_{\boldsymbol{u}}. \quad (9.\text{II}.5)$$

Let θ be the angle between $\boldsymbol{u}$ and $\boldsymbol{n}$, then C is

$$\begin{aligned} C &= \left[\int_0^\pi \mathrm{d}\theta \sin\theta (\cos\theta - 1) \exp\left(-\frac{1-\cos\theta}{\varepsilon^2}\right)\right] \\ &\quad \times \left[\int_0^\pi \mathrm{d}\theta \sin\theta \exp\left(-\frac{1-\cos\theta}{\varepsilon^2}\right)\right]^{-1} \\ &= -\left[\int_0^1 \mathrm{d}x x \exp(-x/\varepsilon^2)\right]\left[\int_0^1 \mathrm{d}x \exp(-x/\varepsilon^2)\right]^{-1} \end{aligned} \quad (9.\text{II}.6)$$

where $x = 1 - \cos\theta$. Since the integrand in eqn (9.II.6) decreases quickly with x, the upper limit of the integral can be replaced by infinity. Hence

$$C = -\left[\int_0^\infty \mathrm{d}x x \exp(-x/\varepsilon^2)\right]\left[\int_0^\infty \mathrm{d}x \exp(-x/\varepsilon^2)\right]^{-1} = -\varepsilon^2, \quad (9.\text{II}.7)$$

i.e.,

$$\langle v_\alpha \rangle_{\boldsymbol{u}} = -\varepsilon^2 n_\alpha. \tag{9.II.8}$$

Likewise, it can be shown

$$\langle v_\alpha v_\beta \rangle_{\boldsymbol{u}} = \varepsilon^2(\delta_{\alpha\beta} - n_\alpha n_\beta). \tag{9.II.9}$$

Thus

$$\langle A(\boldsymbol{u}) \rangle_{\boldsymbol{u}} = A(\boldsymbol{n}) - \varepsilon^2 n_\alpha \frac{\partial}{\partial n_\alpha} A(\boldsymbol{n}) + \frac{\varepsilon^2}{2}(\delta_{\alpha\beta} - n_\alpha n_\beta)\frac{\partial}{\partial n_\alpha}\frac{\partial}{\partial n_\beta} A(\boldsymbol{n}). \tag{9.II.10}$$

By direct calculation, one can prove that

$$\langle A(\boldsymbol{u}) \rangle_{\boldsymbol{u}} = A(\boldsymbol{n}) + \tfrac{1}{2}\varepsilon^2 \mathscr{R}_{\boldsymbol{n}}^2 A(\boldsymbol{n}). \tag{9.II.11}$$

Now let us calculate the effective potential for the tube. Using the definition in eqn (9.73), $\bar{U}_e(\boldsymbol{n})$ is written as

$$\bar{U}_e(\boldsymbol{n}) = -\ln\langle \exp(-U_e(\boldsymbol{u}))\rangle_{\boldsymbol{u}} + \text{terms independent of } \boldsymbol{n}. \tag{9.II.12}$$

(Here $k_B T$ is put equal to unity for simplicity.) Using the formula (9.II.11), we have

$$\begin{aligned}\langle \exp(-U_e(\boldsymbol{u}))\rangle_{\boldsymbol{u}} &= (1 + \tfrac{1}{2}\varepsilon^2 \mathscr{R}_{\boldsymbol{n}}^2)\exp(-U_e(\boldsymbol{n})) \\ &= \exp(-U_e(\boldsymbol{n}))(1 + \tfrac{1}{2}\varepsilon^2[-\mathscr{R}_{\boldsymbol{n}}^2 U_e(\boldsymbol{n}) + (\mathscr{R}_{\boldsymbol{n}} \cdot U_e(\boldsymbol{n}))^2]).\end{aligned} \tag{9.II.13}$$

Hence

$$\begin{aligned}\bar{U}_e(\boldsymbol{n}) &= -\ln\langle \exp(-U_e(\boldsymbol{u}))\rangle_{\boldsymbol{u}} \\ &= U_e(\boldsymbol{n}) + \tfrac{1}{2}\varepsilon^2[\mathscr{R}_{\boldsymbol{n}}^2 U_e(\boldsymbol{n}) - (\mathscr{R}_n U_e(\boldsymbol{n}))^2]\end{aligned} \tag{9.II.14}$$

which is eqn (9.77).

Likewise $\bar{A}(\boldsymbol{n})$ can be calculated as

$$\begin{aligned}\bar{A}(\boldsymbol{n}) &\equiv \frac{\displaystyle\int \mathrm{d}\boldsymbol{u}\, \exp(-U_c(\boldsymbol{u}-\boldsymbol{n}) - U_e(\boldsymbol{u}))A(\boldsymbol{u})}{\displaystyle\int \mathrm{d}\boldsymbol{u}\, \exp(-U_c(\boldsymbol{u}-\boldsymbol{n}) - U_e(\boldsymbol{u}))} \\ &= \frac{\langle \exp[-U_e(\boldsymbol{u})]A(\boldsymbol{u})\rangle_{\boldsymbol{u}}}{\langle \exp[-U_e(\boldsymbol{u})]\rangle_{\boldsymbol{u}}} \\ &= \frac{(1 + \tfrac{1}{2}\varepsilon^2 \mathscr{R}_{\boldsymbol{n}}^2)A(\boldsymbol{n})\exp(-U_e(\boldsymbol{n}))}{(1 + \tfrac{1}{2}\varepsilon^2 \mathscr{R}_{\boldsymbol{n}}^2)\exp(-U_e(\boldsymbol{n}))} \\ &= A(\boldsymbol{n}) + \tfrac{1}{2}\varepsilon^2[\mathscr{R}_{\boldsymbol{n}}^2 A(\boldsymbol{n}) - 2(\mathscr{R}_{\boldsymbol{n}} A(\boldsymbol{n}))\cdot(\mathscr{R}_{\boldsymbol{n}} U_e(\boldsymbol{n}))].\end{aligned} \tag{9.II.15}$$

This is eqn (9.79).

References

1. Doi, M., *J. Physique* **36,** 607 (1975).
2. Doi, M., and Edwards, S. F., *J. Chem. Soc. Faraday II* **74,** 560 (1978).
3. Doi, M., in *Theory of Dispersed Multi Phase Flow* (ed. R. E. Meyer) p. 35. Academic Press, New York (1983).
4. Muthukumar, M., and Edwards, S. F., *Macromolecules* **16,** 1475 (1983).
5. Zero, K. M., and Pecora, R., *Macromolecules* **15,** 87 (1982).
6. Maguire, J. F., *J. Chem. Soc. Faraday Trans. II* **77,** 513 (1981).
7. Jamieson, A. M., Southwick, J. G., and Blackwell, J., *J. Polym. Sci. Phys.* **20,** 1513 (1982); *Faraday Symp.* **18,** 131 (1983).
8. Statman, D. and Chu, B., *Macromolecules* **17,** 1537 (1984); Kubota, K., and Chu, B., *Biopolymers* **22,** 1461 (1983).
9. Russo, P. S., Karasz, F. E., and Langley, K. H., *J. Chem. Phys.* **80,** 5312 (1984).
10. Doi, M., and Edwards, S. F., *J. Chem. Soc. Faraday II* **74,** 918 (1978).
11. Maguire, J. F., McTague, J. P., and Rondelez, F., *Phys. Rev. Lett.* **45,** 1891 (1980); **47,** 148 (1981).
12. Mori, Y., Ookubo, N., Hayakawa, R., and Wada, Y., *J. Polym. Sci.* **20,** 2111 (1982); *Polymer Preprints Japan,* 2369 (1983).
13. Teraoka, I., Mori, Y., Ookubo, N., and Hayakawa, R., *Phys. Rev. Lett.* **55,** 2712 (1985).
14. Doi, M., Yamamoto, I., and Kano, F., *J. Phys. Soc. Jpn* **53,** 3000 (1984).
15. Castelow, D., private communication.
16. Marrucci, G., and Grizzuti, N., *J. Polym. Sci. Lett.* **21,** 83 (1983); *J. Non-Newtonian Fluid Mech.* **14,** 103 (1984).
17. Odijk, T., *Macromolecules* **16,** 1340 (1983); **17,** 502 (1984).
18. Doi, M., *J. Polym. Sci. Polym. Symp.* **73,** 93 (1985).
19. Jain, S., and Cohen, C., *Macromolecules* **14,** 759 (1981).
20. Dahler, J. S., Fesciyan, S., and Xystris, N., *Macromolecules* **16,** 1673 (1983).
21. Hermans, J. Jr., *J. Colloid Sci.* **17,** 638 (1962).
22. Papkov, S. P., Kulichikhin, V. G., Kalmykova, V. D., and Malkin, A. Y., *J. Polym. Sci.* **12,** 1753 (1974).
23. Helminiak, T. E., and Berry, G. C., *J. Polym. Sci. Polym. Symp.* **65,** 107 (1978).
24. Chu, S. G., Venkatraman, S., Berry, G. C., and Einaga, Y., *Macromolecules* **14,** 939 (1981); Venkatraman, S., Berry, G. C., and Einaga, Y., *J. Polym. Sci.* **23,** 1275 (1985).
25. Enomoto, H., Einaga, Y., and Teramoto, A., *Macromolecules* **17,** 1573 (1984).
26. Matheson, R. R., Jr., *Macromolecules* **13,** 643 (1980).
27. Edwards, S. F., and Evans, K. E., *J. C. S. Faraday Trans. II* **78,** 113 (1982).
28. Ookubo, N., Komatsubara, M., Nakajima, H., and Wada, Y., *Biopolymers* **15,** 929 (1976).
29. Kuzuu, N. Y., and Doi, M., *Polymer J.* **12,** 883 (1980).
30. Baird, D. G., Ciferri, A., Krigbaum, W. R., and Salaris, F., *J. Polym. Sci. Phys. ed.* **17,** 1649 (1979); Baird, D. G., and Ballman, R. L., *J. Rheol.* **23,** 505 (1979).
31. Chu, S. G., Venkatraman, S., Berry, G. C., and Einaga, Y., *Macromolecules* **14,** 939 (1981).

32. Kiss, G., and Porter, R. S., *J. Polym. Sci. Phys. ed.* **18,** 361 (1980).
33. Asada, T., and Onogi, S., *Polym. Eng. Rev.,* **3,** 323 (1983).
34. Doi, M., *J. Polym. Sci.* **20,** 1963 (1982).
35. Moscicki, J. K., Williams, G., and Aharoni, S. M., *Macromolecules* **15,** 642 (1982); *Polymer* **22,** 1361 (1981).
36. Warchol, M. P., and Vaughan, W. E., *Adv. Mol. Relaxation and Interaction Processes* **13,** 317 (1978). See also Wang, C. C., and Pecora, R., *J. Chem. Phys.* **72,** 5333 (1980).
37. Moscicki, J. K., *Adv. Chem. Phys.* **58,** 631 (1985).

10

CONCENTRATED SOLUTIONS OF RIGID RODLIKE POLYMERS

10.1 Introduction

As mentioned in Section 9.1, in a solution of rodlike polymers of length L and diameter b, the excluded volume effect becomes important in the concentration regime

$$\nu b L^2 \gtrsim 1. \tag{10.1}$$

The most spectacular aspect in such a solution is that above a certain concentration, the polymers spontaneously align in a certain direction, and the solution becomes anisotropic without any external fields. Such a solution is called a liquid crystal.

The term 'liquid crystal' is generally used to represent the state of matter which has the fluidity of a liquid and the molecular ordering of a crystal. Historically, liquid crystals were first found for low molecular weight materials, and a great deal of work has been done for those materials, much of which is summarized in monographs.[1,2]

Polymeric liquid crystals, on the other hand, have not attracted much attention until recently, when the possibility of spinning high strength fibres from liquid crystalline state stimulated rapid growth of interest in both technological and scientific research. This can be seen in various books published recently.[3–5] At this stage our understanding of polymeric liquid crystals is far less complete than that of low molecular weight materials. Nevertheless it is possible to discuss some characteristic aspects of the material as a natural extension of the theory given in previous chapters.

The type of liquid crystal which will be discussed in this chapter is called 'nematic'. In this phase the directions of the molecules are ordered, being almost parallel to each other, while their positional arrangement is random, as in ordinary liquids (see Fig. 9.1).

On the macroscopic level, nematics have uniaxial symmetry around a certain direction denoted by a unit vector $\boldsymbol{n}$ called the director. There is reflection symmetry with respect to the plane normal to $\boldsymbol{n}$ (thus the state designated by $\boldsymbol{n}$ is the same as that designated by $-\boldsymbol{n}$).

On the microscopic level, nematics are characterized by the fact that the equilibrium distribution function $\Psi(\boldsymbol{u})$ for the molecular direction $\boldsymbol{u}$ is not isotropic. The anisotropy is conveniently represented by the tensor

$$S_{\alpha\beta} = \langle u_\alpha u_\beta - \tfrac{1}{3}\delta_{\alpha\beta} \rangle \tag{10.2}$$

which is zero in the isotropic phase, but nonzero in the nematic phase. (Notice that this is the simplest quantity representing the anisotropy of $\Psi(\boldsymbol{u})$ since the vector average $\langle \boldsymbol{u} \rangle$ vanishes identically due to the reflection symmetry.) The tensor $S_{\alpha\beta}$ is symmetric ($S_{\alpha\beta} = S_{\beta\alpha}$) and traceless ($S_{\alpha\alpha} = 0$). At equilibrium it depends only on the vector $\boldsymbol{n}$. The most general form of such a tensor is

$$S_{\alpha\beta} = S(n_\alpha n_\beta - \tfrac{1}{3}\delta_{\alpha\beta}) \tag{10.3}$$

where S is a certain scalar. From eqns (10.2) and (10.3) S can be written as

$$S = \tfrac{3}{2} n_\alpha n_\beta S_{\alpha\beta} = \tfrac{3}{2}\langle (\boldsymbol{u} \cdot \boldsymbol{n})^2 - \tfrac{1}{3} \rangle. \tag{10.4}$$

The right-hand side takes the value unity when all the polymers are parallel to $\boldsymbol{n}$, and zero when their direction is completely random. Thus S represents how perfectly the polymers are oriented along $\boldsymbol{n}$ and is called the orientational order parameter.

If there is no external field, the director $\boldsymbol{n}$ in nematics is entirely arbitrary. Thus the equilibrium state of nematics is not unique and can be changed by infinitesimal perturbation. This property, generally called 'broken symmetry'[6,7] in statistical mechanics, necessitates a special treatment in the mathematical handling of the kinetic equation, and introduces a new type of constitutive equation, unique to the ordered fluid.

10.2 The phase transition of rigid rods

10.2.1 Free energy for a given orientational distribution function

First we shall study why rodlike polymers form a nematic phase above a certain concentration. The statistical mechanical theory for the rigid rodlike molecules was first given by Onsager[8] and substantial development was made by Flory,[9] who accounted for various factors in real polymers in the framework of the lattice theory. Here we shall describe the theory following the reasoning of Onsager since it is more conveniently extendable to dynamics.

Consider N rodlike polymers in a volume V. Let $\boldsymbol{R}_i$ and $\boldsymbol{u}_i$ ($i = 1, 2, \ldots, N$) be the position and the direction of the i-th polymer. The probability distribution function for the whole system is given by

$$P(\{\boldsymbol{R}_i, \boldsymbol{u}_i\}) \propto \exp\Big(-\sum_{i>j} u(i,j)/k_B T\Big) \tag{10.5}$$

where $u(i, j)$ is the interaction energy between the polymers in the configurations $(\boldsymbol{R}_i, \boldsymbol{u}_i)$ and $(\boldsymbol{R}_j, \boldsymbol{u}_j)$.

To study the orientational distribution function at equilibrium, we calculate the partition function $Z[\Psi]$ for given orientational distribution function Ψ; $Z[\Psi]$ is given by

$$Z[\Psi] = \frac{1}{N!} \int_{\Psi} \prod \mathrm{d}\boldsymbol{u}_i \prod \mathrm{d}\boldsymbol{R}_i \exp\left(-\sum_{i>j} u(i, j)/k_B T\right) \tag{10.6}$$

where the subscript Ψ under the integral symbol indicates that the integration should be done under the constraint that the orientational distribution of polymers is Ψ. To take into account such a constraint, we divide the surface of the sphere $|\boldsymbol{u}| = 1$ into small cells of areas Δ, and evaluate the integral under the condition that the number of polymers which are in the a-th cell is

$$n_a = N\Psi(\boldsymbol{u}_a)\Delta. \tag{10.7}$$

Obviously

$$\sum_a n_a = N = \nu V. \tag{10.8}$$

If we define

$$\langle \ldots \rangle_{\Psi} \equiv \int_{\Psi} \prod \mathrm{d}\boldsymbol{u}_i \prod \mathrm{d}\boldsymbol{R}_i \ldots \bigg/ \int_{\Psi} \prod \mathrm{d}\boldsymbol{u}_i \prod \mathrm{d}\boldsymbol{R}_i, \tag{10.9}$$

eqn (10.6) is written as

$$Z[\Psi] = Z_0[\Psi] Z_1[\Psi] \tag{10.10}$$

with

$$Z_0[\Psi] = \frac{1}{N!} \int_{\Psi} \prod \mathrm{d}\boldsymbol{u}_i \prod \mathrm{d}\boldsymbol{R}_i \tag{10.11}$$

and

$$Z_1[\Psi] = \left\langle \exp\left(-\sum_{i>j} u(i, j)/k_B T\right)\right\rangle_{\Psi}. \tag{10.12}$$

The evaluation of $Z_0[\Psi]$ is easy. Since the number of ways of distributing N polymers into the cells is $N!/\prod n_a!$, and each integral over $\boldsymbol{R}_i$ and $\boldsymbol{u}_i$ gives V and Δ respectively

$$Z_0 = \frac{1}{N!} \frac{N!}{\prod_a n_a!} (V\Delta)^N. \tag{10.13}$$

The free energy (per unit volume) corresponding to this partition function is

$$\mathscr{A}_0[\Psi] = -\frac{k_B T}{V}\ln Z_0 = -\frac{k_B T}{V}\left[N\ln(V\Delta) - \sum_a n_a(\ln(n_a) - 1)\right]$$

$$= -\frac{N}{V}k_B T\left[\ln(V/N) + 1 - \sum_a \Psi(\boldsymbol{u}_a)\Delta\ln(\Psi(\boldsymbol{u}_a))\right]$$

$$= \nu k_B T\left[\ln \nu - 1 + \sum_a \Psi(\boldsymbol{u}_a)\Delta\ln(\Psi(\boldsymbol{u}_a))\right], \qquad (10.14)$$

or replacing the summation by an integral:

$$\mathscr{A}_0[\Psi] = \nu k_B T\left[\ln \nu - 1 + \int d\boldsymbol{u}\Psi(\boldsymbol{u})\ln \Psi(\boldsymbol{u})\right]. \qquad (10.15)$$

To evaluate $Z_1[\Psi]$, we assume that the collisions between the polymers occur independently of each other and therefore approximate eqn (10.12) by

$$\left\langle\prod_{i>j}\exp(-u(i,j)/k_B T)\right\rangle_\Psi \simeq \prod_{i>j}\langle\exp(-u(i,j)/k_B T)\rangle_\Psi. \qquad (10.16)$$

This approximation is justified if the concentration of the polymers is low enough. Systematic improvement of this approximation can be done by the virial expansion of the free energy.[10] However, for polymers of long aspect ratio ($L/b \gg 1$), the higher order terms are not needed because the transition is shown to occur at very low concentration.

Each term in the product of the right-hand side of eqn (10.16) gives

$$\langle\exp(-u(i,j)/k_B T)\rangle_\Psi = \frac{1}{V^2}\int d\boldsymbol{u}_i\, d\boldsymbol{u}_j\, d\boldsymbol{R}_i\, d\boldsymbol{R}_j$$
$$\times\exp(-u(i,j)/k_B T)\Psi(\boldsymbol{u}_i)\Psi(\boldsymbol{u}_j). \qquad (10.17)$$

The integrand depends only on $\boldsymbol{R}_i - \boldsymbol{R}_j$ so that

$$\langle\exp(-u(i,j)/k_B T)\rangle_\Psi = \frac{1}{V}\int d\boldsymbol{u}_i\, d\boldsymbol{u}_j\, d\boldsymbol{R}_i$$
$$\times\exp(-u(\boldsymbol{u}_i, \boldsymbol{R}_i, \boldsymbol{u}_j, 0)/k_B T)\Psi(\boldsymbol{u}_i)\Psi(\boldsymbol{u}_j)$$
$$= 1 - \frac{1}{V}\int d\boldsymbol{u}_i\, d\boldsymbol{u}_j \beta(\boldsymbol{u}_i, \boldsymbol{u}_j)\Psi(\boldsymbol{u}_i)\Psi(\boldsymbol{u}_j) \qquad (10.18)$$

where

$$\beta(\boldsymbol{u}, \boldsymbol{u}') = \int d\boldsymbol{R}[1 - \exp(-u(\boldsymbol{u}, \boldsymbol{R}, \boldsymbol{u}', 0)/k_B T)]. \qquad (10.19)$$

Equation (10.12) thus becomes

$$Z_1[\Psi] = \left[1 - \frac{1}{V}\int d\boldsymbol{u}\, d\boldsymbol{u}'\beta(\boldsymbol{u}, \boldsymbol{u}')\Psi(\boldsymbol{u})\Psi(\boldsymbol{u}')\right]^{N(N-1)/2}. \tag{10.20}$$

In the limit of $V \to \infty$ with $\nu = N/V$ finite, this becomes

$$Z_1[\Psi] = \exp\left[-\frac{V\nu^2}{2}\int d\boldsymbol{u}\, d\boldsymbol{u}'\beta(\boldsymbol{u}, \boldsymbol{u}')\Psi(\boldsymbol{u})\Psi(\boldsymbol{u}')\right]. \tag{10.21}$$

The free energy corresponding to Z_1 is

$$\mathscr{A}_1[\Psi] = -\frac{k_B T}{V}\ln Z_1 = \tfrac{1}{2}\nu^2 k_B T\int d\boldsymbol{u}\, d\boldsymbol{u}'\beta(\boldsymbol{u}, \boldsymbol{u}')\Psi(\boldsymbol{u})\Psi(\boldsymbol{u}'). \tag{10.22}$$

The total free energy of the system is thus given by

$$\begin{aligned}\mathscr{A}[\Psi(\boldsymbol{u})] &= \mathscr{A}_0[\Psi(\boldsymbol{u})] + \mathscr{A}_1[\Psi(\boldsymbol{u})] \\ &= \nu k_B T\left[\ln \nu - 1 + \int d\boldsymbol{u}\Psi(\boldsymbol{u})\ln \Psi(\boldsymbol{u})\right. \\ &\quad \left. + \tfrac{1}{2}\nu\int d\boldsymbol{u}\int d\boldsymbol{u}'\Psi(\boldsymbol{u})\Psi(\boldsymbol{u}')\beta(\boldsymbol{u}, \boldsymbol{u}')\right].\end{aligned} \tag{10.23}$$

For rigid, rodlike polymers, $\beta(\boldsymbol{u}, \boldsymbol{u}')$ is given by (see Fig. 9.2)

$$\beta(\boldsymbol{u}, \boldsymbol{u}') = 2bL^2\,|\boldsymbol{u} \times \boldsymbol{u}'|, \tag{10.24}$$

so that $\beta(\boldsymbol{u}, \boldsymbol{u}')$ has a minimum when $\boldsymbol{u}$ is parallel or antiparallel to $\boldsymbol{u}'$. Thus $\mathscr{A}_1[\Psi]$ decreases as the polymers orient in the same direction. This causes the nematic phase when the effect of the excluded volume becomes strong. Notice that without the excluded volume interaction, the equilibrium state is always isotropic even though the topological and hence dynamical effects still exist.

10.2.2 Equilibrium distribution

The equilibrium distribution is determined by the condition that $\mathscr{A}$ is minimum for all variations of Ψ. Since $\Psi(\boldsymbol{u})$ satisfies the normalization condition,

$$\int d\boldsymbol{u}\Psi(\boldsymbol{u}) = 1 \tag{10.25}$$

the minimum is found by introducing a Lagrangian multiplier λ and solving

$$\frac{\delta}{\delta\Psi}\left[\mathscr{A}[\Psi(\boldsymbol{u})] - \lambda\int d\boldsymbol{u}\Psi(\boldsymbol{u})\right] = 0 \tag{10.26}$$

which gives

$$\ln \Psi(\boldsymbol{u}) + v \int d\boldsymbol{u}' \Psi(\boldsymbol{u}')\beta(\boldsymbol{u}, \boldsymbol{u}') = \text{constant}. \tag{10.27}$$

The constant is determined from the condition (10.25). It is convenient to write eqn (10.27) in the following form

$$\Psi(\boldsymbol{u}) = \frac{1}{z} \exp\left[-\frac{U_{\text{scf}}(\boldsymbol{u}, [\Psi])}{k_B T} \right], \tag{10.28}$$

with U_{scf} defined by

$$U_{\text{scf}}(\boldsymbol{u}, [\Psi]) = v k_B T \int d\boldsymbol{u}' \beta(\boldsymbol{u}, \boldsymbol{u}')\Psi(\boldsymbol{u}') \tag{10.29}$$

and

$$z = \int d\boldsymbol{u} \exp\left[-\frac{U_{\text{scf}}(\boldsymbol{u}, [\Psi])}{k_B T} \right]. \tag{10.30}$$

Equation (10.28) indicates that the orientational distribution of a polymer at equilibrium is a Boltzmann distribution under the potential U_{scf}. Thus U_{scf} is regarded as a mean field potential acting on the polymer.

The nonlinear integral equation (10.28) cannot be solved analytically. Onsager assumed that the equilibrium distribution has the following form:

$$\Psi(\boldsymbol{u}) = \frac{\alpha}{4\pi \sinh \alpha} \cosh(\alpha \boldsymbol{u} \cdot \boldsymbol{n}) \tag{10.31}$$

where $\boldsymbol{n}$ is an arbitrary unit vector, and α is a parameter to be determined from the condition that $\mathcal{A}$ be minimized. Equation (10.31) represents a state in which the polymers are oriented toward $\pm\boldsymbol{n}$. The parameter α represents the degree of ordering: $\alpha = 0$ corresponds to the isotropic state, and $\alpha = \infty$ the completely ordered state.

If eqn (10.31) is substituted into eqn (10.23), $\mathcal{A}$ is expressed as a function of α and v. The result is schematically shown in Fig. 10.1. If v is small, $\mathcal{A}(\alpha, v)$ has only one minimum at $\alpha = 0$, which corresponds to the isotropic state. If v exceeds a certain critical value v_1^*, another minimum appears at positive α, which corresponds to the nematic state. If v exceeds another critical value v_2^*, the minimum at $\alpha = 0$ disappears, and there is only one minimum at a positive value of α.

In Fig. 10.2*a*, the values of the minima (or local minima) of $\mathcal{A}$ are plotted as a function of v. Between v_1^* and v_2^*, $\mathcal{A}(\alpha, v)$ has double values, and the thermodynamically favourable state is the one with a lower value of $\mathcal{A}(\alpha, v)$. It is important to note that the minimum of $\mathcal{A}(\alpha, v)$ does not immediately indicate an equilibrium state because the free energy can be even more lowered by macroscopic phase separation.

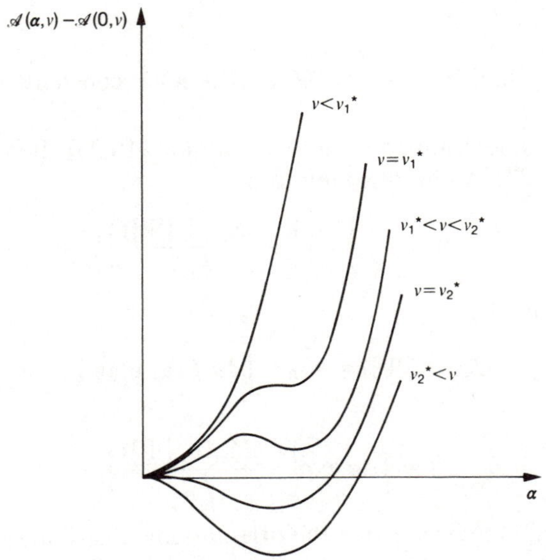

Fig. 10.1. Free energy as a function of the order parameter α at various concentrations.

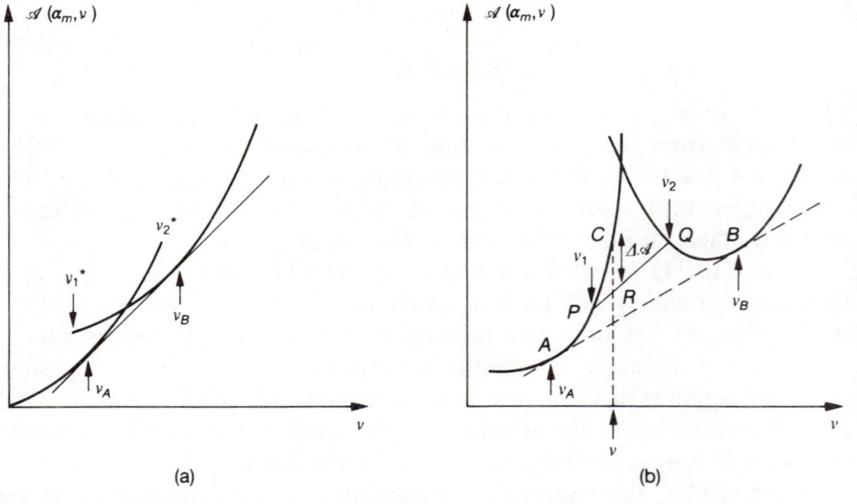

Fig. 10.2. (*a*) Free energy as a function of concentration. For $\nu_1^* < \nu < \nu_2^*$, the values of the two local minima are plotted. (*b*) Graphical meaning of eqn (10.34). If a system of concentration ν is separated into two phases of concentration ν_1 and ν_2, the total free energy changes by $\Delta\mathscr{A}$ shown in the figure. (In this case $\Delta\mathscr{A} < 0$). The total free energy becomes minimum when the line PQ coincides with AB which is tangent to the curve at both points A and B.

Suppose that a system of concentration ν and volume V is separated into two phases of concentration ν_1 and ν_2 with volumes V_1 and V_2 respectively, then the total free energy is given by

$$\mathscr{A}_{\text{tot}} = V_1\mathscr{A}_1 + V_2\mathscr{A}_2. \tag{10.32}$$

The values of V_1 and V_2 are determined from the condition

$$\nu_1 V_1 + \nu_2 V_2 = \nu V \quad \text{and} \quad V_1 + V_2 = V. \tag{10.33}$$

From eqns (10.32) and (10.33), it follows that

$$\mathscr{A}_{\text{tot}} = \frac{\nu_2 - \nu}{\nu_2 - \nu_1} V\mathscr{A}_1 + \frac{\nu - \nu_1}{\nu_2 - \nu_1} V\mathscr{A}_2. \tag{10.34}$$

As explained in Fig. 10.2*b*, in the concentration region between ν_A and ν_B the free energy becomes minimum if the system is separated into two phases; one is the isotropic phase of concentration ν_A and the other is the nematic phase of concentration ν_B. Therefore it is concluded that:

(i) If $\nu < \nu_A$, the solution is isotropic.

(ii) If $\nu_A < \nu < \nu_B$, phase separation takes place, i.e., the isotropic phase of concentration ν_A coexists with the nematic phase of concentration ν_B.

(iii) If $\nu_B < \nu$, the solution becomes nematic.

Notice that ν_A, ν_B, ν_1^*, and ν_2^* are all proportional to $1/bL^2$; the difference among them being only in the numerical factor. In the subsequent discussion we shall use ν_2^* ($\equiv$ the concentration above which the isotropic phase becomes unstable) as the characteristic concentration of the phase transition and write ν_2^* simply as ν^*.

The values of ν_A, ν_B, and ν^* are obtained by numerical calculation:

$$\nu_A = \frac{4.25}{bL^2}, \qquad \nu_B = \frac{5.72}{bL^2}, \quad \text{and} \quad \nu^* = \frac{16}{\pi b L^2} = \frac{5.1}{bL^2}. \tag{10.35}$$

A more elaborate variational method[11] gives

$$\nu_A = \frac{4.19}{bL^2}, \qquad \nu_B = \frac{5.37}{bL^2}, \quad \text{and} \quad \nu^* = \frac{4.44}{bL^2}. \tag{10.36}$$

Note that for a slender rod ($L \gg b$), the volume fraction $\phi = \nu(\pi b^2 L)/4$ at these concentration is quite low. For example

$$\phi^* = \nu^* \frac{\pi b^2 L}{4} \simeq 4\frac{b}{L} \tag{10.37}$$

which decreases as M^{-1}. This agrees, at least qualitatively, with experiment.[1,10]

The advantage of Onsager's theory is that it becomes exact in the limit of $L/b \to \infty$. On the other hand, real polymers suffer from other effects which are not included in Onsager's theory, such as an attractive energy, flexibility of the chain, and molecular weight distribution. These effects are discussed in refs 12 and 13, and a comparison between the improved theory and the experimental results is made in ref. 12.

10.3 The kinetic equation

10.3.1 Dynamical mean field theory

Having discussed the static properties, we shall now consider the dynamics of the concentrated solution. A natural generalization of the static mean field theory to dynamics is to assume[14–16] that each molecule does Brownian motion under the mean field potential U_{scf}. For rigid rodlike polymers U_{scf} is given by

$$U_{\text{scf}}(\boldsymbol{u}) = \nu k_B T \int d\boldsymbol{u}' \beta(\boldsymbol{u}, \boldsymbol{u}')\Psi(\boldsymbol{u}'; t). \tag{10.38}$$

Introducing this into the kinetic equation (9.42), we have[17]

$$\frac{\partial \Psi}{\partial t} = \bar{D}_r \mathcal{R} \cdot \left[\mathcal{R}\Psi + \frac{\Psi}{k_B T} \mathcal{R}(U_{\text{scf}} + U_e) \right] - \mathcal{R} \cdot (\boldsymbol{u} \times \boldsymbol{\kappa} \cdot \boldsymbol{u}\Psi). \tag{10.39}$$

(Here the preaveraging approximation (9.41) is used.)

Equations (10.38) and (10.39) give a nonlinear integro-differential equation for Ψ, and its mathematical handling is not easy. A guidance of how to proceed is obtained from the phenomenological theory in nematics. De Gennes[1,18] showed that the dynamics of nematics is essentially described by the Landau theory of phase transition and proposed a phenomenological nonlinear equation for the order parameter tensor $S_{\alpha\beta}$†

$$\frac{\partial}{\partial t} S_{\alpha\beta} = -L \frac{\partial A}{\partial S_{\alpha\beta}}. \tag{10.40}$$

Here L is a phenomenological kinetic coefficient and A is a free energy. Near the transition point, $S_{\alpha\beta}$ is small so that A can be expanded with respect to $S_{\alpha\beta}$ as in the series

$$A = a_2 \operatorname{Tr}(\boldsymbol{S} \cdot \boldsymbol{S}) + a_3 \operatorname{Tr}(\boldsymbol{S} \cdot \boldsymbol{S} \cdot \boldsymbol{S}) + a_4 \operatorname{Tr}(\boldsymbol{S} \cdot \boldsymbol{S} \cdot \boldsymbol{S} \cdot \boldsymbol{S}) + a_4'(\operatorname{Tr}(\boldsymbol{S} \cdot \boldsymbol{S}))^2 \tag{10.41}$$

where $a_2, \ldots, a_4'$ are constants.

† Since the $S_{\alpha\beta}$ satisfy the constraint $S_{\alpha\alpha} = 0$, $\mathcal{A}$ must be supplemented by a term $-\lambda S_{\alpha\alpha}$ with a Lagrangian multiplier λ, and eqn (10.40) must read

$$\frac{\partial}{\partial t} S_{\alpha\beta} = -L \frac{\partial A}{\partial S_{\alpha\beta}} - \lambda \delta_{\alpha\beta}.$$

We thus aim at deriving a closed equation for $S_{\alpha\beta}$ from eqn (10.39). For simplicity, we first consider the case when there is no external field ($U_e = 0$, $\boldsymbol{\kappa} = 0$).

First we approximate $\beta(\boldsymbol{u}, \boldsymbol{u}')$, which is a decreasing function of $|\boldsymbol{u} \cdot \boldsymbol{u}'|$, as‡

$$\beta(\boldsymbol{u}, \boldsymbol{u}') = \text{const} - \beta_1 b L^2 (\boldsymbol{u} \cdot \boldsymbol{u}')^2 \tag{10.42}$$

where β_1 is a numerical constant.

Substitution of eqn (10.42) into eqn (10.38) gives

$$U_{\text{scf}} = \text{const} - \beta_1 \nu b L^2 k_B T u_\alpha u_\beta \langle u_\alpha u_\beta \rangle. \tag{10.43}$$

We shall write this in the following form:

$$U_{\text{scf}} = \text{const} - \tfrac{3}{2} U k_B T u_\alpha u_\beta S_{\alpha\beta} \tag{10.44}$$

where U is a parameter proportional to $\nu b L^2$. Using the approximation (10.44), we write eqn (10.39) as

$$\frac{\partial \Psi}{\partial t} = \bar{D}_r \mathcal{R} \cdot [\mathcal{R}\Psi - \Psi \mathcal{R}(\tfrac{3}{2} U S_{\alpha\beta} u_\alpha u_\beta)]. \tag{10.45}$$

We now use the same procedure as in Section 8.7, i.e., we multiply both sides of eqn (10.45) by $(u_\alpha u_\beta - \frac{1}{3}\delta_{\alpha\beta})$ and integrate over $\boldsymbol{u}$. The result, after integration by parts, is

$$\frac{\partial S_{\alpha\beta}}{\partial t} = -6\bar{D}_r S_{\alpha\beta} + 6\bar{D}_r U[S_{\alpha\mu}\langle u_\mu u_\beta \rangle - S_{\mu\nu}\langle u_\alpha u_\beta u_\mu u_\nu \rangle]. \tag{10.46}$$

Using the decoupling approximation

$$S_{\mu\nu}\langle u_\mu u_\nu u_\alpha u_\beta \rangle = S_{\mu\nu}\langle u_\mu u_\nu \rangle \langle u_\alpha u_\beta \rangle, \tag{10.47}$$

we finally get

$$\frac{\partial}{\partial t} S_{\alpha\beta} = F_{\alpha\beta}(\boldsymbol{S}) \tag{10.48}$$

where

$$F_{\alpha\beta} = -6\bar{D}_r \left[\left(1 - \frac{U}{3}\right) S_{\alpha\beta} - U\left(S_{\alpha\mu} S_{\beta\mu} - \frac{\delta_{\alpha\beta}}{3} S_{\mu\nu}^2\right) + U S_{\alpha\beta} S_{\mu\nu}^2 \right]. \tag{10.49}$$

Equations (10.48) and (10.49) give a special form of the Landau–de Gennes theory equation (10.40) with the phenomenological parameters

‡ This type of interaction has been used by Maier and Saupe[19] in the theory of the phase transition of low-molecular-weight nematics. An advantage of this interaction is that the nonlinear integral eqn (10.28) can be solved exactly (see ref. 1, p. 43).

determined as

$$L = 6\bar{D}_r, \qquad a_2 = \frac{1}{2}\left(1 - \frac{U}{3}\right), \qquad a_3 = -\frac{U}{3}, \qquad a_4 = 0, \qquad a_4' = \frac{U}{4}. \tag{10.50}$$

Though the validity of the Landau–de Gennes theory is usually limited only to the weakly ordered state, eqn (10.49) can also be used for the highly ordered state. Indeed, as will be shown later, the order parameter S given by eqn (10.49) remains in the physical range ($-1/2 < S < 1$) over the entire range of U. (This is a consequence of the fact that the decoupling approximation (10.47) becomes correct in the highly oriented state.)

It must be noted that $\bar{D}_r$ in eqn (10.49) will depend on $S_{\alpha\beta}$ because of the tube dilation. To include the effect qualitatively, we use an approximation similar to eqn (10.42):

$$\left(\frac{D_r}{\bar{D}_r}\right)^{1/2} = \frac{4}{\pi}\int d\boldsymbol{u}\, d\boldsymbol{u}'\Psi(\boldsymbol{u})\Psi(\boldsymbol{u}')\,|\boldsymbol{u}\times\boldsymbol{u}'| \simeq \beta_2 - \beta_3 S_{\alpha\beta}S_{\alpha\beta} \tag{10.51}$$

where β_2 and β_3 are certain numerical constants, chosen so that both sides of eqn (10.51) agree with each other in the isotropic state ($\Psi(\boldsymbol{u}) = 1/4\pi$, $S_{\alpha\beta} = 0$) and in the completely ordered state ($\Psi(\boldsymbol{u}) = \delta(\boldsymbol{u} - \boldsymbol{n})$, $S_{\alpha\beta} = n_\alpha n_\beta - \delta_{\alpha\beta}/3$). This gives

$$\beta_2 = 1 \quad \text{and} \quad \beta_3 = \tfrac{3}{2}. \tag{10.52}$$

Hence

$$\bar{D}_r = D_r[1 - \tfrac{3}{2}S_{\alpha\beta}S_{\alpha\beta}]^{-2}. \tag{10.53}$$

Equation (10.48) with eqns (10.49) and (10.53) give a closed equation for $S_{\alpha\beta}$.[17] It should be emphasized that this kinetic equation holds in both isotropic and nematic phases. We shall now study the consequence of this equation.

10.3.2 Relaxation of the order parameter

Suppose that for $t < 0$, the polymers are oriented by an external field. If the field is switched off at $t = 0$, the polymers will return to the equilibrium state. We shall first study this relaxation process using the kinetic equation (10.48).

Let $\boldsymbol{n}$ be the direction towards which the polymers have been oriented by the external field. This direction will coincide with the director if the system is in the nematic phase. Since the system will retain uniaxial symmetry around $\boldsymbol{n}$ during the relaxation process, the order parameter tensor is written as

$$S_{\alpha\beta}(t) = S(t)(n_\alpha n_\beta - \tfrac{1}{3}\delta_{\alpha\beta}). \tag{10.54}$$

Substituting eqn (10.54) into eqn (10.48) we get

$$\frac{\partial S}{\partial t} = -6\bar{D}_r\left[\left(1-\frac{U}{3}\right)S - \frac{U}{3}S^2 + \tfrac{2}{3}US^3\right] \qquad (10.55)$$

or

$$\frac{\partial S}{\partial t} = -6\bar{D}_r \frac{\partial}{\partial S} A(S, U) \qquad (10.56)$$

where

$$A(S, U) = \frac{1}{2}\left(1-\frac{U}{3}\right)S^2 - \frac{U}{9}S^3 + \frac{U}{6}S^4. \qquad (10.57)$$

Since $\bar{D}_r > 0$, eqn (10.56) indicates that the change in S occurs in the direction of decreasing $A(S, U)$. In particular, the equilibrium value of S is determined from the minimum of $A(S, U)$. Thus $A(S, U)$ plays the role of the free energy. Indeed, as shown in Fig. 10.3, $A(S, U)$ behaves similarly to $\mathscr{A}(\alpha, v)$ (S and U corresponding to α and v respectively).

Standard analysis of eqn (10.57) shows:

(i) For $U < U_1^* \equiv 8/3$, A has only one minimum at $S = 0$, so that the system finally becomes isotropic, whatever its initial state.

(ii) For $U_1^* < U < U^* \equiv 3$, there are two local minima, one at $S = 0$ and the other at

$$S_{eq} = \frac{1}{4} + \frac{3}{4}\left(1 - \frac{8}{3U}\right)^{1/2}. \qquad (10.58)$$

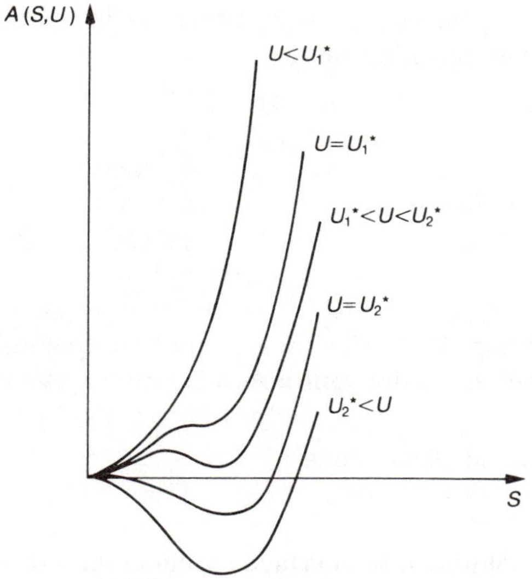

Fig. 10.3. Free energy of $A(S, U)$ plotted against S for various values of U.

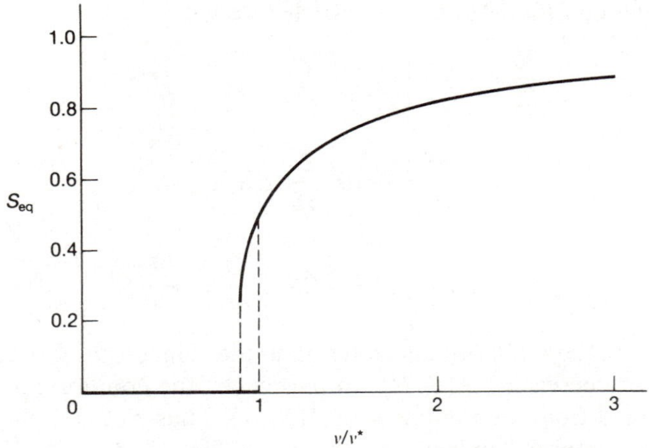

Fig. 10.4. The equilibrium order parameter S_{eq} plotted against ν/ν^*.

In this case the system can become either an isotropic state or a nematic state depending on the initial value of S.

(iii) For $U > U^*$, the point $S = 0$ becomes unstable and the system always approaches the nematic state.

Though this analysis does not tell where phase separation takes place (since the system is assumed to be homogeneous in this treatment), the general behaviour of the phase transition is described by the approximate kinetic equation (10.48).

Equation (10.58) gives the equilibrium order parameter S_{eq} as a function of the concentration. Since

$$\frac{\nu}{\nu^*} = \frac{U}{U^*} = \frac{U}{3}, \tag{10.59}$$

eqn (10.58) is rewritten as

$$S_{eq} = \frac{1}{4} + \frac{3}{4}\left(1 - \frac{8\nu^*}{9\nu}\right)^{1/2}. \tag{10.60}$$

This is plotted in Fig. 10.4. The equilibrium order parameter starts at 1/4 at $\nu = (8/9)\nu^*$ and gradually approaches 1 with increasing concentration.

10.4 Pretransitional phenomena

10.4.1 Introduction

In concentrated solutions, the excluded volume interaction tends to align the polymers in the same direction, and gives rise to the nematic phase if the effect is sufficiently strong. The same effect exists also in the isotropic region, where, though the macroscopic ordering is not achieved, the interaction tends to align the polymers on a microscopic scale. This 'short-range ordering' increases as the concentration increases, and causes various anomalies in the physical properties of isotropic solutions near the transition point. It is called the pretransitional effect.[1,18]

A simple example of the pretransitional effect is seen in the problem discussed in the previous section. Suppose that the external field is weak, then S is small in the isotropic phase, so that eqn (10.55) can be approximated by

$$\frac{\partial S}{\partial t} = -6D_r\left(1 - \frac{U}{3}\right)S = -\frac{S}{\tau} \tag{10.61}$$

where

$$\tau = \frac{1}{6D_r(1 - U/3)} \propto \frac{1}{1 - \nu/\nu^*}. \tag{10.62}$$

Thus the relaxation time diverges as the concentration approaches ν^*. In reality, the true divergence cannot be observed because the phase separation takes place before we reach the concentration ν^*. However, an indication of the divergence is expected to appear near the phase separation concentration ν_A.

Here we shall give two examples of the pretransitional phenomena. Though a rigorous calculation is possible using the kinetic equation (10.39), we shall use the approximate kinetic equation (10.48) for convenience of demonstration.

10.4.2 Magnetic birefringence

A magnetic field can orient the polymer and thus cause birefringence. According to eqn (8.83), the anisotropic part of the refractive index $\hat{n}_{\alpha\beta}$ is written as

$$\Delta n_{\alpha\beta} = AS_{\alpha\beta}. \tag{10.63}$$

Let $\chi_\parallel$ (and $\chi_\perp$) be the magnetic susceptibility of the polymer parallel (and perpendicular) to the magnetic field. In the presence of the magnetic field H, such a polymer feels the potential (see eqn (8.81))

$$U_e = -\tfrac{1}{2}\chi_a(\boldsymbol{u}\cdot\boldsymbol{H})^2 \tag{10.64}$$

where $\chi_a \equiv \chi_\parallel - \chi_\perp$. The kinetic equation is thus

$$\frac{\partial \Psi}{\partial t} = \bar{D}_r\mathcal{R}\cdot\left[\mathcal{R}\Psi - \Psi\mathcal{R}\left(\tfrac{3}{2}US_{\alpha\beta}u_\alpha u_\beta + \frac{\chi_a}{2k_BT}(\boldsymbol{H}\cdot\boldsymbol{u})^2\right)\right] \tag{10.65}$$

which gives

$$\frac{\partial S_{\alpha\beta}}{\partial t} = -6\bar{D}_r S_{\alpha\beta} + 6\bar{D}_r U[S_{\alpha\mu}\langle u_\mu u_\beta\rangle - S_{\mu\nu}\langle u_\alpha u_\beta u_\mu u_\nu\rangle]$$
$$+\frac{\bar{D}_r\chi_a}{k_BT}[H_\alpha H_\mu\langle u_\beta u_\mu\rangle + H_\beta H_\mu\langle u_\alpha u_\mu\rangle - 2H_\mu H_\nu\langle u_\alpha u_\beta u_\mu u_\nu\rangle]. \quad (10.66)$$

The decoupling approximation (10.47) and

$$\langle u_\alpha u_\beta u_\mu u_\nu\rangle H_\mu H_\nu = \langle u_\alpha u_\beta\rangle\langle u_\mu u_\nu\rangle H_\mu H_\nu \quad (10.67)$$

gives the following kinetic equation for $S_{\alpha\beta}$,

$$\frac{\partial}{\partial t} S_{\alpha\beta} = F_{\alpha\beta}(\boldsymbol{S}) + M_{\alpha\beta}(\boldsymbol{S}) \quad (10.68)$$

where $F_{\alpha\beta}(\boldsymbol{S})$ is given by eqn (10.49) and

$$M_{\alpha\beta} = \frac{\bar{D}_r\chi_a}{k_BT}[\tfrac{2}{3}(H_\alpha H_\beta - \tfrac{1}{3}\boldsymbol{H}^2\delta_{\alpha\beta}) - \tfrac{2}{3}\boldsymbol{H}^2 S_{\alpha\beta} + H_\alpha H_\mu S_{\beta\mu} + H_\beta H_\mu S_{\alpha\mu}$$
$$-\tfrac{2}{3}\delta_{\alpha\beta}H_\mu H_\nu S_{\mu\nu} - 2S_{\alpha\beta}S_{\mu\nu}H_\mu H_\nu]. \quad (10.69)$$

For a weak magnetic field, eqn (10.68) gives, to the lowest order of $\boldsymbol{H}$,

$$\frac{\partial}{\partial t} S_{\alpha\beta} = -\frac{S_{\alpha\beta}}{\tau} + \frac{2D_r\chi_a}{3k_BT}(H_\alpha H_\beta - \tfrac{1}{3}\boldsymbol{H}^2\delta_{\alpha\beta}) \quad (10.70)$$

where τ is given by eqn (10.62). (Note $\bar{D}_r$ can be replaced by D_r in this case.) Consider for example that a constant magnetic field parallel to the z axis is switched on at $t = 0$, i.e.,

$$H(t) = \begin{cases} 0 & t<0, \\ H & t>0. \end{cases} \quad (10.71)$$

Equation (10.70) gives

$$\frac{\partial}{\partial t} S_{zz}(t) = -\frac{1}{\tau} S_{zz}(t) + \frac{4D_r\chi_a}{9k_BT} H^2 \quad (10.72)$$

which is solved by

$$S_{zz}(t) = (1 - \exp(-t/\tau))S_{zz}(\infty) \quad (10.73)$$

with

$$S_{zz}(\infty) = \frac{4D_r\chi_a}{9k_BT}\tau H^2 \propto \frac{H^2}{1-(\nu/\nu^*)}. \quad (10.74)$$

Thus not only the relaxation time τ but also the steady-state value $S_{zz}(\infty)$ diverge as ν approaches ν^*. The magnetic birefringence in rodlike polymers has been studied by Nakamura and Okano,[20] and the results of eqns (10.62) and (10.74) have been confirmed.

10.4.3 Viscoelasticity

The pretransitional effect can be also observed by viscoelasticity. Firstly we derive the constitutive equation from the kinetic equation. Again, eqn (10.39) in the presence of the velocity gradient $\boldsymbol{\kappa}$ is rewritten to give a kinetic equation for $S_{\alpha\beta}$ (see eqn (8.149)):

$$\frac{\partial}{\partial t}S_{\alpha\beta} = F_{\alpha\beta}[\boldsymbol{S}] + G_{\alpha\beta}[\boldsymbol{S}] \tag{10.75}$$

where

$$G_{\alpha\beta} = \tfrac{1}{3}(\kappa_{\alpha\beta} + \kappa_{\beta\alpha}) + \kappa_{\alpha\mu}S_{\mu\beta} + \kappa_{\beta\mu}S_{\mu\alpha} - \tfrac{2}{3}\delta_{\alpha\beta}\kappa_{\mu\nu}S_{\mu\nu} - 2\kappa_{\mu\nu}S_{\mu\nu}S_{\alpha\beta}. \tag{10.76}$$

The stress tensor is calculated from the change of the free energy equation (10.23) under a virtual deformation. By the same calculation described in Section 8.6.1, we have (see eqn (8.114))

$$\begin{aligned}\sigma_{\alpha\beta} &= 3\nu k_B T S_{\alpha\beta} - \nu\langle(\boldsymbol{u}\times\mathcal{R}U_{\text{scf}})_\alpha u_\beta\rangle \\ &= 3\nu k_B T[S_{\alpha\beta} - U(S_{\alpha\mu}\langle u_\mu u_\beta\rangle - S_{\mu\nu}\langle u_\mu u_\nu u_\alpha u_\beta\rangle)]\end{aligned} \tag{10.77}$$

which is rewritten by the decoupling approximation (10.47) as

$$\sigma_{\alpha\beta} = 3\nu k_B T\left[S_{\alpha\beta}\left(1-\frac{U}{3}\right) - U(S_{\alpha\mu}S_{\mu\beta} - \tfrac{1}{3}\delta_{\alpha\beta}S^2_{\mu\nu}) + US_{\alpha\beta}S_{\mu\nu}S_{\mu\nu}\right]. \tag{10.78}$$

Equations (10.75) and (10.78) determine the stress for a given velocity gradient and can be regarded as a constitutive equation. It should be emphasized that this constitutive equation holds in both the isotropic and the nematic phases since no presumption has been required about the equilibrium state.

Let us now consider the linear viscoelasticity of the isotropic phase. Since $S_{\alpha\beta}$ is small in this regime, eqn (10.75) can be written as

$$\frac{\partial}{\partial t}S_{\alpha\beta} = -\frac{1}{\tau}S_{\alpha\beta} + \tfrac{1}{3}(\kappa_{\alpha\beta} + \kappa_{\beta\alpha}) \tag{10.79}$$

which gives

$$S_{\alpha\beta}(t) = \frac{1}{3}\int_{-\infty}^{t} dt'\,\exp(-(t-t')/\tau)(\kappa_{\alpha\beta}(t') + \kappa_{\beta\alpha}(t')). \tag{10.80}$$

Hence to the first order in $\kappa_{\alpha\beta}$,

$$\sigma_{\alpha\beta}(t) = 3\nu k_B T\left(1-\frac{U}{3}\right)S_{\alpha\beta}(t)$$
$$= \int_{-\infty}^{t} \mathrm{d}t' G(t-t')(\kappa_{\alpha\beta}(t') + \kappa_{\beta\alpha}(t')) \tag{10.81}$$

where

$$G(t) = G_e \exp(-t/\tau) \tag{10.82}$$

and

$$G_e = \nu k_B T\left(1-\frac{U}{3}\right). \tag{10.83}$$

Note that the instantaneous modulus G_e becomes small near the transition point due to the pretransitional effect. The viscosity η_0 and the steady-state compliance are calculated as

$$\eta_0 = \int_0^\infty \mathrm{d}t\, G(t) = \nu k_B T\left(1-\frac{U}{3}\right)\tau = \frac{\nu}{6}k_B T\frac{1}{D_r}, \tag{10.84}$$

$$J_e^{(0)} = \frac{1}{\eta_0^2}\int_0^\infty \mathrm{d}t\, G(t)t = \frac{1}{\left(1-\frac{U}{3}\right)\nu k_B T} = \frac{M}{(1-(\rho/\rho^*))\rho RT}. \tag{10.85}$$

The steady-state compliance has a minimum at $\rho = \rho^*/2$ and then increases again by the pretransitional effect. This has indeed been observed by Berry and coworkers.[21] On the other hand the viscosity is unaffected by the pretransitional effect. (This is a result of the cancellation of the effect on the relaxation time τ and on the rigidity modulus G_e.) Experimental results on the viscoelasticity of the isotropic solution have been reviewed by Baird.[22]

10.5 Linear viscosity in the nematic phase

10.5.1 Introduction

We shall now study the flow properties of the nematic phase. From the microscopic viewpoint, no special consideration may seem necessary since the constitutive equation (10.75) and eqn (10.78) applies both for the isotropic and the nematic states. This is not the case. The rheological properties of solutions of rodlike polymers are changed entirely when the system becomes nematic.

Figure 10.5 shows the viscosity as a function of concentration. In the isotropic phase, the viscosity increases with increasing concentration (in

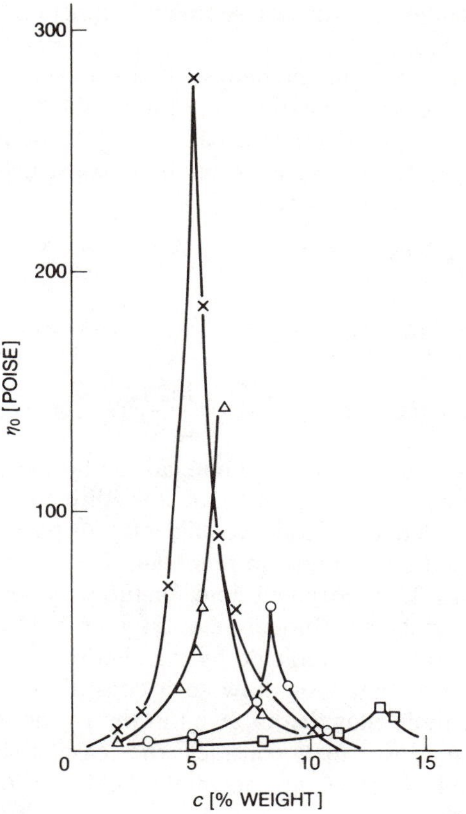

Fig. 10.5. Concentration-dependence of the viscosity at 20°C for solutions of poly(para-benzamide) of various molecular weights in *N,N*-dimethylacetamide/LiCl; (□) 10 900; (○) 16 000; (△) 23 800: (×) 69 000 (adapted from ref. 23). This figure is reproduced from ref. 22.

agreement with the prediction of the previous chapter), whilst it *decreases* in the nematic phase.

A more fundamental difference is observed in the normal stress difference in steady shear flow:

$$\kappa_{\alpha\beta} = \begin{cases} \kappa & \text{if } \alpha = x \text{ and } \beta = y, \\ 0 & \text{otherwise.} \end{cases} \tag{10.86}$$

In the isotropic phase, the first normal stress difference $N_1 \equiv \sigma_{xx} - \sigma_{yy}$ is proportional to κ^2 for small κ, whereas it is proportional to $|\kappa|$ in the nematic phase[22] and written as

$$\sigma_{xx} - \sigma_{yy} = \eta_N |\kappa| \quad \text{for} \quad \kappa \to 0. \tag{10.87}$$

This indicates that the stress is not an analytic function of the shear rate at $\kappa = 0$.

For low-molecular-weight nematics, it has been known that the constitutive equation of nematics is entirely different from that for isotropic liquids.[1,2] A phenomenological theory for the hydrodynamics of nematics (of low molecular weight) has been constructed by Ericksen[24] and Leslie.[25] Their equation reads

$$\boldsymbol{\sigma} = \alpha_1(\boldsymbol{nn}:\boldsymbol{A})\boldsymbol{nn} + \alpha_2\boldsymbol{nN} + \alpha_3\boldsymbol{Nn} + \alpha_4\boldsymbol{A} + \alpha_5\boldsymbol{nn}\cdot\boldsymbol{A} + \alpha_6\boldsymbol{A}\cdot\boldsymbol{nn} \tag{10.88}$$

and

$$\boldsymbol{n}\times(\boldsymbol{h} - (\alpha_3 - \alpha_2)\boldsymbol{N} - (\alpha_6 - \alpha_5)\boldsymbol{A}\cdot\boldsymbol{n}) = 0 \tag{10.89}$$

where

$$\boldsymbol{A} = \tfrac{1}{2}(\boldsymbol{\kappa} + \boldsymbol{\kappa}^+), \qquad \boldsymbol{N} = \frac{d\boldsymbol{n}}{dt} - \tfrac{1}{2}(\boldsymbol{\kappa} - \boldsymbol{\kappa}^+)\cdot\boldsymbol{n}, \tag{10.90}$$

and $\boldsymbol{n}\times\boldsymbol{h}$ is the torque per unit volume caused by the external magnetic field or the spatial inhomogeneity of the director. The coefficients $\alpha_1, \ldots, \alpha_6$ are called the Leslie coefficients. A physical argument for deriving these equations is given in refs 1 and 2.

The Ericksen–Leslie theory will hold for the polymeric nematics if the velocity gradient is small. Indeed the singular behaviour in the first normal stress difference is predicted by this theory.[1,2]

In this section we shall show how such constitutive equations can be derived from the molecular theory given in the previous section. For the sake of simplicity, we first consider the case that the system is homogeneous and there is no magnetic field (so that $\boldsymbol{h} = 0$ in the Ericksen–Leslie theory).

The reason for the peculiar constitutive equation for nematics is easily found. Consider that the velocity gradient is both weak and independent of time. The order parameter of the steady state is given by

$$F_{\alpha\beta}[\boldsymbol{S}] + G_{\alpha\beta}[\boldsymbol{S}] = 0. \tag{10.91}$$

Suppose we look for the solution in the form

$$S_{\alpha\beta} = S_{0\alpha\beta} + S_{1\alpha\beta} \tag{10.92}$$

where $S_{0\alpha\beta}$ is the equilibrium order parameter, and $S_{1\alpha\beta}$ denotes the correction to first order in $\boldsymbol{\kappa}$. As shown in Section 10.3.2, $S_{0\alpha\beta}$ is written as

$$S_{0\alpha\beta} = S(n_\alpha n_\beta - \tfrac{1}{3}\delta_{\alpha\beta}) \tag{10.93}$$

with

$$S = \frac{1}{4} + \frac{3}{4}\left(1 - \frac{8\nu^*}{9\nu}\right)^{1/2}. \tag{10.94}$$

(Here S_{eq} is written as S for notational simplicity.) The peculiar feature of the nematic phase is that the unperturbed state $S_{0\alpha\beta}$ is not known because $\boldsymbol{n}$ is arbitrary in equilibrium. Therefore the standard perturbation method which has been used for the isotropic state fails in the nematic. This gives the constitutive equation characteristic to the nematic phase.[26,27]

10.5.2 Perturbation scheme

Our problem is to find a perturbation method which determines the unperturbed state specified by $\boldsymbol{n}$, uniquely. To see the formal aspect clearly, let us consider the problem in terms of the original kinetic equation. The equation to be solved is

$$F[\Psi] + G[\Psi] = 0 \tag{10.95}$$

where

$$F[\Psi] = \bar{D}_r \mathcal{R} \cdot \left[\mathcal{R}\Psi + \frac{\Psi}{k_B T} \mathcal{R} U_{\text{scf}}[\Psi] \right] \tag{10.96}$$

and

$$G[\Psi] = \bar{D}_r \mathcal{R} \cdot \left(\frac{\Psi}{k_B T} \mathcal{R} U_e \right) - \mathcal{R} \cdot (\boldsymbol{u} \times \boldsymbol{\kappa} \cdot \boldsymbol{u} \Psi). \tag{10.97}$$

We assume that the solution of eqn (10.95) has the following form,

$$\Psi = \Psi_0 + \Psi_1 + \dots \tag{10.98}$$

where Ψ_0 is the solution in the equilibrium state, which includes the director $\boldsymbol{n}$ to be determined, and Ψ_1 is the first-order perturbation in G.

Substituting eqn (10.98) into (10.95) and comparing the first order in G, we get

$$H[\Psi_1] + G[\Psi_0] = 0 \tag{10.99}$$

where $H[\Psi]$ is a linear operator obtained from the expansion of $F[\Psi]$:

$$F[\Psi_0 + \Psi_1] - F[\Psi_0] = H[\Psi_1] + O(\Psi_1^2). \tag{10.100}$$

Note that since Ψ_0 includes the director $\boldsymbol{n}$, the operator H depends on $\boldsymbol{n}$.

Let $\psi^{(i)}$ and $\phi^{(i)}$ be the right- and left-handed eigenfunction of the operator H:

$$H[\psi^{(i)}] = -\lambda^{(i)}\psi^{(i)}, \qquad H^+[\phi^{(i)}] = -\lambda^{(i)}\phi^{(i)}. \tag{10.101}$$

(H^+ being the adjoint operator of H.) They will be orthonormal:

$$\delta_{ij} = \int \mathrm{d}\boldsymbol{u} \phi^{(i)}\psi^{(j)}. \tag{10.102}$$

We look for the solution of eqn (10.99) in the form

$$\Psi_1 = \sum_i a_i \psi^{(i)}. \tag{10.103}$$

Substituting eqn (10.103) into eqn (10.99), we get

$$\sum_i \lambda^{(i)} a_i \psi^{(i)} = G[\Psi_0] \tag{10.104}$$

or using eqn (10.102),

$$\lambda^{(i)} a_i = \int d\boldsymbol{u} \phi^{(i)} G[\Psi_0]. \tag{10.105}$$

Now a crucial point is that another condition is needed for this equation to have a solution, because some of the eigenvalues of the operator H are zero. This is a consequence of the fact that the equilibrium state in nematics is continuously degenerate. If fact, if Ψ_0 and Ψ'_0 denote the equilibrium states having directors $\boldsymbol{n}$ and $\boldsymbol{n}' = \boldsymbol{n} + \delta\boldsymbol{n}$, respectively, then $\delta\Psi_0 \equiv \Psi'_0 - \Psi_0$ corresponds to an eigenfunction of zero eigenvalue, since

$$H[\delta\Psi_0] = F[\Psi'_0] - F[\Psi_0] = 0. \tag{10.106}$$

Thus to obtain the steady-state solution, we must have

$$\int d\boldsymbol{u} \phi^{(i)} G[\Psi_0] = 0 \tag{10.107}$$

for the eigenfunction $\phi^{(i)}$ which corresponds to zero eigenvalue. This equation determines the director $\boldsymbol{n}$ in the unperturbed state.

10.5.3 Approximate calculation

Let us now go back to the approximate kinetic equation and carry out the above procedure. If we substitute eqn (10.92) into eqn (10.91) and compare the first-order in $\boldsymbol{\kappa}$, we get

$$H_{\alpha\beta,\mu\nu} S_{1\mu\nu} + G_{\alpha\beta}[S_0] = 0. \tag{10.108}$$

The matrix $H_{\alpha\beta,\mu\nu}$, which corresponds to the operator $H[\Psi]$, is given by

$$H_{\alpha\beta,\mu\nu} = -6\bar{D}_r \Big\{ \Big[\Big(1 - \frac{U}{3}\Big) + \tfrac{2}{3}US + \tfrac{2}{3}US^2 \Big] \delta_{\alpha\mu}\delta_{\beta\nu} - US(\delta_{\beta\mu} n_\alpha n_\nu + \delta_{\alpha\nu} n_\beta n_\mu) + 2US^2 (n_\alpha n_\beta - \tfrac{1}{3}\delta_{\alpha\beta})(n_\mu n_\nu - \tfrac{1}{3}\delta_{\mu\nu}) \Big\}. \tag{10.109}$$

The eigenvector $\phi^{(i)}_{\alpha\beta}$, which corresponds to the eigenfunction $\phi^{(i)}(\boldsymbol{u})$ of zero eigenvalue now satisfies

$$H_{\mu\nu,\alpha\beta} \phi^{(i)}_{\mu\nu} = 0. \tag{10.110}$$

The solution of eqn (10.110) is given by

$$\phi^{(i)}_{\alpha\beta} = n_\alpha m^{(i)}_\beta + n_\beta m^{(i)}_\alpha \tag{10.111}$$

where $\boldsymbol{m}^{(i)}$ is an arbitrary vector perpendicular to $\boldsymbol{n}$. The condition (10.107) is now written as

$$\phi^{(i)}_{\alpha\beta} G_{\alpha\beta}[\boldsymbol{S}_0] = 0. \tag{10.112}$$

By use of eqns (10.76) and (10.111), this equation is rewritten, after some calculation, as

$$[(1-S)\kappa_{\beta\alpha} n_\beta + (1+2S)\kappa_{\alpha\beta} n_\beta] m^{(i)}_\alpha = 0. \tag{10.113}$$

Since $\boldsymbol{m}^{(i)}$ is perpendicular to $\boldsymbol{n}$, this condition is rewritten as

$$[(1-S)\boldsymbol{\kappa}^+ \cdot \boldsymbol{n} + (1+2S)\boldsymbol{\kappa} \cdot \boldsymbol{n}] \times \boldsymbol{n} = 0 \tag{10.114}$$

which determines $\boldsymbol{n}$.

Given $\boldsymbol{n}$, it is straightforward to calculate the stress. Equation (10.78) can be rewritten using eqn (10.75) as

$$\sigma_{\alpha\beta} = \frac{\nu k_B T}{2\bar{D}_r} \Big[-\frac{\partial}{\partial t} S_{\alpha\beta} + \tfrac{1}{3}(\kappa_{\alpha\beta} + \kappa_{\beta\alpha}) + \kappa_{\alpha\mu} S_{\mu\beta}$$
$$+ \kappa_{\beta\mu} S_{\mu\alpha} - \tfrac{2}{3}\delta_{\alpha\beta}\kappa_{\mu\nu} - 2\kappa_{\mu\nu} S_{\mu\nu} S_{\alpha\beta}. \tag{10.115}$$

In the steady state $\partial S_{\alpha\beta}/\partial t = 0$, so that

$$\sigma_{\alpha\beta} = \frac{\nu k_B T}{2\bar{D}_r} [\tfrac{1}{3}(\kappa_{\alpha\beta} + \kappa_{\beta\alpha}) + \kappa_{\alpha\mu} S_{\mu\beta} + \kappa_{\beta\mu} S_{\mu\alpha}$$
$$- \tfrac{2}{3}\delta_{\alpha\beta}\kappa_{\mu\nu} S_{\mu\nu} - 2\kappa_{\mu\nu} S_{\mu\nu} S_{\alpha\beta}]. \tag{10.116}$$

In the calculation of the first-order perturbation in $\boldsymbol{\kappa}$, $S_{\alpha\beta}$ on the right-hand side can be replaced by the equilibrium value:

$$S_{\alpha\beta} = S(n_\alpha n_\beta - \tfrac{1}{3}\delta_{\alpha\beta}). \tag{10.117}$$

Substituting eqn (10.117) into eqn (10.116) we finally have

$$\sigma_{\alpha\beta} = \frac{\nu k_B T}{2\bar{D}_r} \Big[\frac{1-S}{3}(\kappa_{\alpha\beta} + \kappa_{\beta\alpha}) + S(\kappa_{\alpha\mu} n_\beta n_\mu + \kappa_{\beta\mu} n_\alpha n_\mu)$$
$$-2S^2 \kappa_{\mu\nu} n_\mu n_\nu n_\alpha n_\beta \Big] \tag{10.118}$$

If $\boldsymbol{n}$ is known from eqn (10.114), eqn (10.118) gives the stress in the steady state. It can be easily checked that eqns (10.114) and (10.118) agree with the Ericksen–Leslie equations (10.88) and (10.89) in the case of $\boldsymbol{h} = 0$ and $\mathrm{d}\boldsymbol{n}/\mathrm{d}t = 0$.

10.5.4 Example—shear flow

As an example, consider the shear flow (10.86). From the symmetry of the flow, $\boldsymbol{n}$ is in the x–y plane, and can be written as

$$n_x = \cos\chi, \qquad n_y = \sin\chi, \quad \text{and} \quad n_z = 0. \tag{10.119}$$

Substituting eqn (10.119) into eqn (10.114) we have

$$(1-S)\kappa\cos^2\chi - (1+2S)\kappa\sin^2\chi = 0, \tag{10.120}$$

i.e.,

$$\tan\chi = \left[\frac{1-S}{1+2S}\right]^{1/2}. \tag{10.121}$$

Since $0 < S < 1$, χ is smaller than 45.

The viscosity η is defined by

$$\eta = \sigma_{xy}/\kappa. \tag{10.122}$$

This is calculated from eqn (10.118) as

$$\eta = \frac{\nu k_B T}{2\bar{D}_r}\left(\frac{1-S}{3} + Sn_y^2 - 2S^2 n_x^2 n_y^2\right) \tag{10.123}$$

$$= \frac{\nu k_B T}{6\bar{D}_r}\frac{(1-S)^2(1+2S)(1+3S/2)}{(1+S/2)^2}. \tag{10.124}$$

Using eqns (10.53), (10.93), and $D_r \propto \nu^{-2}$, we have

$$\bar{D}_r = D_r^{(\mathrm{ref})}\left(\frac{\nu^{(\mathrm{ref})}}{\nu}\right)^2(1-S^2)^2 \tag{10.125}$$

where $\nu^{(\mathrm{ref})}$ is a certain reference concentration in the isotropic state. We choose $\nu^{(\mathrm{ref})}$ to be ν^*. The rotational diffusion constant at ν^* is given by (see eqn (9.16)).

$$D_r^* = \beta D_{r0}(\nu^* L^3)^{-2}. \tag{10.126}$$

Then

$$\bar{D}_r = D_r^*\left(\frac{\nu^*}{\nu}\right)^2(1-S^2)^{-2}. \tag{10.127}$$

From eqns (10.124) and (10.127),

$$\frac{\eta}{\eta^*} = \left(\frac{\nu}{\nu^*}\right)^3\frac{(1-S)^4(1+S)^2(1+2S)(1+3S/2)}{(1+S/2)^2} \tag{10.128}$$

where

$$\eta^* = \frac{\nu^* k_B T}{6D_r^*} \simeq \eta_s(\nu^* L^3)^3 \simeq \eta_s\left(\frac{L}{b}\right)^3. \tag{10.129}$$

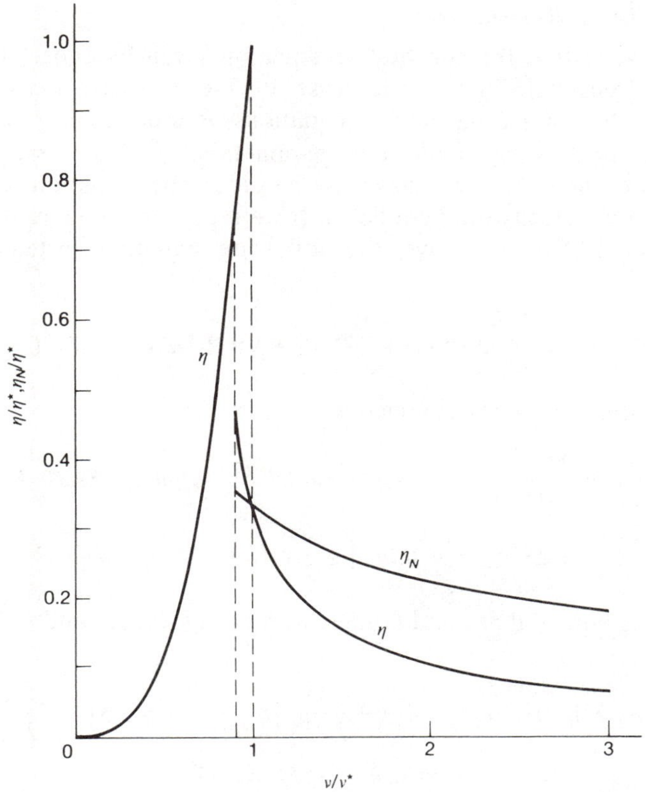

Fig. 10.6. The steady-state viscosity for the shear stress η and that for the normal stress η_N of the solution of rodlike polymers is plotted against concentration. In the isotropic phase, η_N vanishes identically.

Equations (10.94) and (10.128) give the concentration dependence of η. (Note that eqn (10.128) holds also in the isotropic solution if S is put to zero.) The result, shown in Fig. 10.6, indicates that the viscosity takes a maximum near the phase transition point, in agreement with the experimental results.

The viscosity η_N for the normal stress is calculated similarly:

$$\frac{\eta_N}{\eta^*} = 3\left(\frac{\nu}{\nu^*}\right)^3 \frac{S(1-S)^{7/2}(1+S)^2(1+2S)^{1/2}}{(1+S/2)}. \tag{10.130}$$

This vanishes in the isotropic phase ($S=0$), but does not vanish in the nematic phase ($S>0$).

10.5.5 The Leslie coefficients

We have seen that the constitutive equation given by eqns (10.75) and (10.78) agrees with the special case of the Ericksen–Leslie theory. Therefore, by comparing the two equations, it is possible to express the Leslie coefficients by molecular parameters.[26,27] To carry out this programme, however, we have to consider the situation with both magnetic and velocity gradient fields. If we repeat the same calculation as in Section 10.5.3, we have the following equation instead of eqn (10.114):[26]

$$\left((1-S)\boldsymbol{\kappa}^{+}\cdot\boldsymbol{n}+(1+2S)\boldsymbol{\kappa}\cdot\boldsymbol{n}+\frac{\bar{D}_r\chi_a}{k_BT}(2+S)(\boldsymbol{H}\cdot\boldsymbol{n})\boldsymbol{H}\right)\times\boldsymbol{n}=0. \tag{10.131}$$

Also the stress tensor is obtained as

$$\sigma_{\alpha\beta}=\frac{\nu k_BT}{2\bar{D}_r}\left[\frac{1-S}{3}(\kappa_{\alpha\beta}+\kappa_{\beta\alpha})+S(\kappa_{\alpha\mu}n_\beta n_\mu+\kappa_{\beta\mu}n_\alpha n_\mu)\right.$$
$$\left.-2S^2\kappa_{\mu\nu}n_\mu n_\nu n_\alpha n_\beta\right]+\frac{\nu}{2}\chi_a H_\mu n_\mu(H_\alpha n_\beta-H_\beta n_\alpha). \tag{10.132}$$

Comparing eqns (10.88) and (10.89) with eqns (10.131) and (10.132), we have†[27]

$$\alpha_1=-2S^2\bar{\eta},\qquad \alpha_2=-S\left(1+\frac{3S}{2+S}\right)\bar{\eta},\qquad \alpha_3=-S\left(1-\frac{3S}{2+S}\right)\bar{\eta},$$
$$\alpha_4=\tfrac{2}{3}(1-S)\bar{\eta},\qquad \alpha_5=2S\bar{\eta},\quad\text{and}\quad \alpha_6=0, \tag{10.133}$$

where

$$\bar{\eta}=\frac{\nu k_BT}{2\bar{D}_r} \tag{10.134}$$

Experimentally it is not easy to measure all of the Leslie coefficients. What is often measured is the Miesovicz viscosity defined by

$$\eta_a=\tfrac{1}{2}\alpha_4,\qquad \eta_b=\tfrac{1}{2}(\alpha_3+\alpha_4+\alpha_6),\qquad \eta_c=\tfrac{1}{2}(-\alpha_2+\alpha_4+\alpha_5). \tag{10.135}$$

These are calculated from eqn (10.133) as

$$\eta_a=\tfrac{1}{3}(1-S)\bar{\eta}, \tag{10.136}$$

$$\eta_b=\frac{2(1-S)^2}{3(2+S)}\bar{\eta}, \tag{10.137}$$

$$\eta_c=\frac{2(2S+1)^2}{3(2+S)}\bar{\eta}. \tag{10.138}$$

† That α_1 is proportional to S^2 for small S is a general result that can be derived from the Landau–de Gennes theory.[30] On the other hand, that α_6 vanishes is an accidental result of the decoupling approximation, and is not generally true.

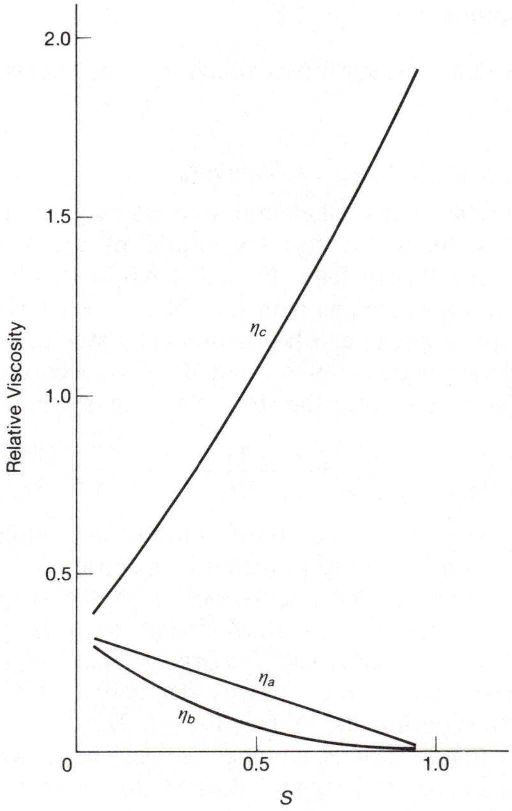

Fig. 10.7. The Miesovicz viscosities η_a, η_b, η_c plotted as functions of S.

These are plotted as a function of S in Fig. 10.7. The relative magnitude of η_a, η_b, and η_c are in qualitative agreement with experimental results[31,32] (see also the reviews[33,34] for low-molecular-weight nematics).

According to the Ericksen–Leslie theory, steady shear flow is possible only when $\lambda = (\alpha_2 + \alpha_3)/(\alpha_2 - \alpha_3)$ is larger than unity.[1,2] The result (10.133) satisfies this condition for all values of S. On the other hand, more accurate analysis of the original kinetic eqn (10.39) without using the decoupling approximation indicates[28,29] that λ becomes less than unity in the highly ordered state. Though this result is disturbing, its relevance to experiment still remains to be seen since the value of λ is close to unity and sensitive to various effects which are not included in the theory.

10.6 Future problems

Here we shall briefly discuss topics which have not been covered in the previous sections.

10.6.1 Nonlinear viscoelasticity in nematics

The Ericksen–Leslie theory takes into account only the first-order effect of the velocity gradient. If the magnitude of the velocity gradient becomes larger, the theory fails. Indeed it has been observed that the steady state viscosity decreases with the shear rate (shear thinning).[23,35] Such nonlinear phenomena can be theoretically studied by the constitutive equation (given by eqns (10.75) and (10.78)), according to which the characteristic shear rate κ_c for the shear thinning is estimated by

$$\kappa_c \simeq 6\bar{D}_r \frac{\partial^2 A}{\partial^2 S}\bigg|_{S=S_{\text{eq}}} = 6D_r\left[\frac{9}{4}\left(\frac{\nu}{\nu^*}-\frac{8}{9}\right)+\frac{3}{4}\left(\frac{\nu}{\nu^*}\left(\frac{\nu}{\nu^*}-\frac{8}{9}\right)\right)^{1/2}\right]. \quad (10.139)$$

Note that this increases with concentration: the nonlinear effect is expected to be less pronounced at higher concentration.

The result of more detailed calculation[36] is shown in Fig. 10.8, where the steady-state viscosity at various shear rates is plotted against concentration. The characteristic feature is that as the shear rate increases the maximum of the viscosity decreases. This agrees at least qualitatively with experiments[37–40] (see Fig. 10.9).

On the other hand, curious nonlinear effects have been observed in some polymeric nematics. Kiss and Porter[38] found that with increasing shear rate, the first normal stress difference becomes negative and then becomes positive again. This effect is not explained by the present constitutive equation and some other physical reason appears to be needed.

10.6.2 Spatial inhomogeneity and domain structure

In the theory described so far, it has been assumed that the system is homogeneous, i.e., the velocity gradient and the director are independent of position. If the director $\boldsymbol{n}$ varies with position, there will be an elastic energy which tends to minimize the spatial gradient of the director. This effect is analysed in detail in the classical theory of low-molecular-weight nematics,[41,42] according to which the elastic energy is written as[1,2]

$$\mathscr{A}_{\text{el}} = \tfrac{1}{2}K_1(\operatorname{div}\boldsymbol{n})^2 + \tfrac{1}{2}K_2(\boldsymbol{n}\cdot\operatorname{curl}\boldsymbol{n})^2 + \tfrac{1}{2}K_3(\boldsymbol{n}\times\operatorname{curl}\boldsymbol{n})^2. \quad (10.140)$$

The constants K_1, K_2, and K_3 have been calculated by a straightforward generalization of Onsager's theory.[43,44] The result indicates that near the

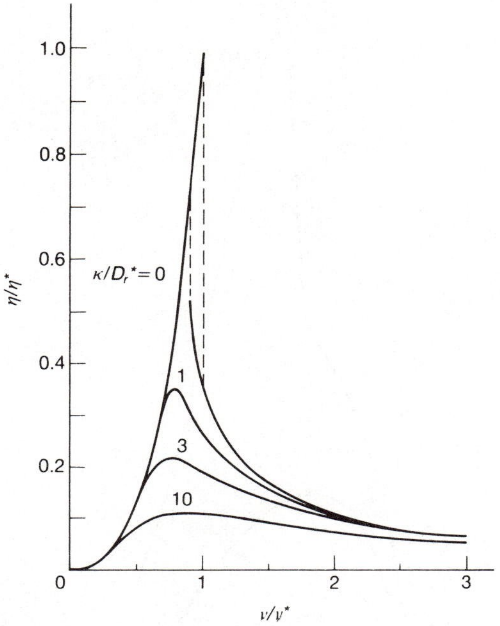

Fig. 10.8. Concentration-dependence of the steady-state viscosity at various shear rates.

transition point

$$K_1 \simeq K_2 \simeq K_3 \simeq \frac{k_B T}{b}. \qquad (10.141)$$

Though the elastic energy equation (10.140) is important in many nonlinear flow properties of low-molecular-weight nematics, its effect is less important in polymeric nematics since the stress is usually dominated by the viscosity in polymeric nematics.

Polymeric nematics often take up a certain domain structure at equilibrium.[45,46] Various patterns have been observed by polarization microscopy. Curiously enough, such structures seem to correspond to the minimum of free energy since even after the system is brought to a homogeneous state by shearing, the domain structure is spontaneously recovered when the flow is stopped.[47] The formation and destruction of the domain structure are quite important in the macroscopic flow properties, and indeed dominate the phenomena at low stress level. At present, however, the physical origin of the domain structure is not known and the flow properties at low stress levels are poorly understood.[48]

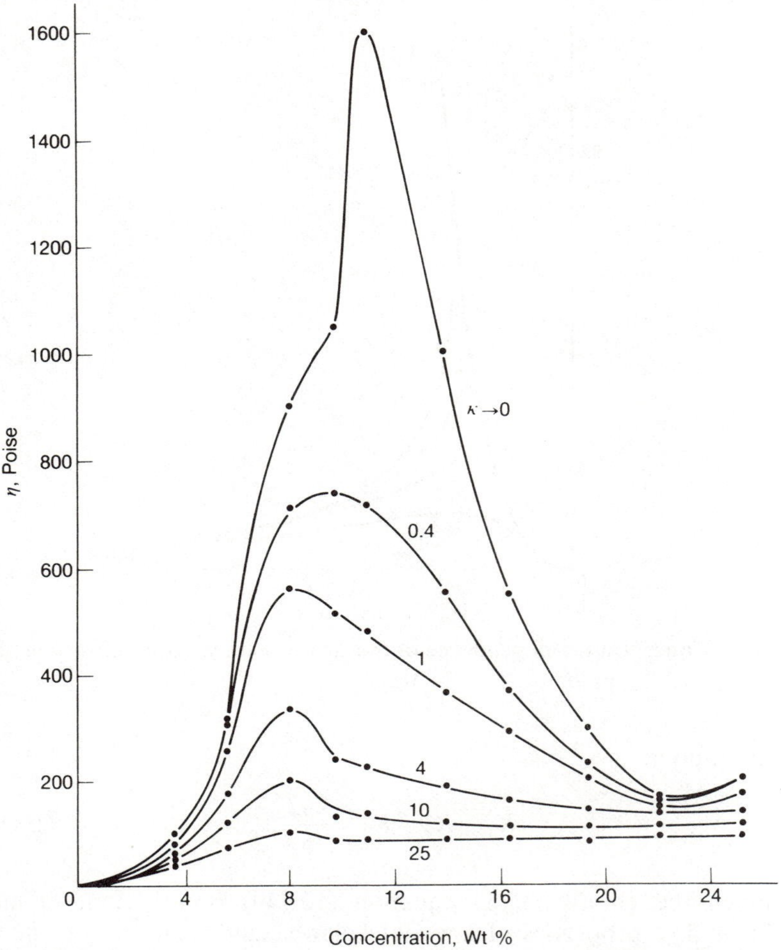

Fig. 10.9. The steady-state viscosity of poly-γ-benzyl-glutamate (molecular weight 350 000) at various shear rates is plotted against concentration. The shear rate (in sec^{-1}) is indicated by the number in the figure. Reproduced from ref. 38.

10.6.3 Thermotropic liquid crystals

Much of the current research on polymeric liquid crystals is directed towards thermotropic liquid crystals[3–5] which are formed when certain polymer melts are cooled. Many of those polymers includes chemical groups similar to low-molecular-weight nematics either along the main chain or in the side chains. These materials are technologically interesting because of their ease of processing. Theoretically, these materials are expected to show curious behaviours which are a mixture of their polymeric and nematic character. Except for a few attempts,[49] the problem is mostly unexplored.

References

1. de Gennes, P. G., *The Physics of Liquid Crystals*. Clarendon Press, Oxford (1974).
2. Chandrasekhar, S., *Liquid Crystals*. Cambridge Univ. Press, Cambridge (1977).
3. Blumstein, A., (ed.) *Liquid Crystalline Order in Polymers*. Academic Press, New York (1978).
4. Ciferri, A., Krigbaum, W. R., and Meyer, R. B., *Polymer Liquid Crystals*. Academic Press, New York (1982).
5. *Adv. Polymer Sci.* Vols 59, 60, 61. Springer (1984).
6. Brout, R., *Phase Transitions*. Benjamin, New York (1965).
7. Foster, D., *Hydrodynamic Fluctuations, Broken Symmetry, and Correlation Functions*. Benjamin, New York (1975).
8. Onsager, L., *Ann. N.Y. Acad. Sci.* **51,** 627 (1949).
9. Flory, P. J., *Proc. R. Soc. London.* **A234,** 73 (1956).
10. Straley, J. P., *Mol. Cryst. Liq. Cryst.* **22,** 333 (1973); **24,** 7 (1973).
11. Lasher, G., *J. Chem. Phys.* **53,** 4141 (1970).
12. Flory, P. J., *Adv. Polym. Sci.* **59,** 1 (1984).
13. Grosberg, A. Y., and Khokhlov, A. R., *Adv. Polym. Sci.* **41,** 53 (1981).
14. Nordio, P. L., Rigatti, G., and Segre, U., *J. Chem. Phys.* **56,** 2117 (1972).
15. Martin, A. J., Meier, G., and Saupe, A., *Faraday, Symp. Chem. Soc.* **5,** 119 (1971).
16. Hess, S., *Z. Naturforsch.* **31a,** 1034, 1507 (1976); **36a,** 554 (1981).
17. Doi, M., *J. Polym. Sci.* **19,** 229 (1981).
18. de Gennes, P. G., *Phys. Lett.* **30A,** 454 (1969); *Mol. Cryst. Liq. Cryst.* **12,** 193 (1971).
19. Maier, W., and Saupe, A., *Z. Naturforsch.* **13A,** 564 (1958); **14A,** 882 (1959); **15A,** 287 (1960).
20. Nakamura, H., and Okano, K., *Phys. Rev. Lett.* **50,** 186 (1983).
21. Wong, C. P., Ohnuma, H., and Berry, G. C., *J. Polym. Sci. Polym. Symp.* **65,** 173 (1978).
22. Baird, D. G., in *Liquid Crystalline Order in Polymers* (ed. A. Blumstein), p. 237. Academic Press, New York (1978).
23. Papkov, S. P., Kulichikhin, V. G., and Kalmykova, V. D., and Malkin, A. Y., *J. Polym. Sci. Phys.* **12,** 1753 (1974).
24. Ericksen, J. L., *Arch. Ration. Mech. Anal.* **4,** 231 (1960); *Trans. Soc. Rheol.* **5,** 23 (1961).
25. Leslie, F. M., *Quart. J. Mech. Appl. Math.* **19,** 357 (1966); *Arch. Ration. Mech. Anal.* **28,** 265 (1968).
26. Doi, M., *Faraday Symp. Chem. Soc.* **18,** 49 (1983).
27. Marrucci, G., *Mol. Cryst. Liq. Cryst.* **72L,** 153 (1982).
28. Semenov, A. N., *Sov. Phys. JETP* **58,** 321 (1983).
29. Kuzuu, N., and Doi, M., *J. Phys. Soc. Jpn* **52,** 3486 (1983); **53,** 1031 (1984).
30. Imura, H., and Okano, K., *Japan J. Appl. Phys.* **11,** 1440 (1970).

31. Kulichikhin, V. G., Vasil'yeva, N. V., Platonov, V. A., Malkin, A. Y., Belousova, T. A., Khanchich, O. A., and Papkov, S. P., *Polymer Sci. USSR* (*Engl. Transl.*) **21,** 1545 (1979).
32. Krigbaum, W. R., in *Polymer Liquid Crystals* (eds. A. Ciferri, W. R. Krigbaum, and R. B. Meyer), p. 275. Academic Press, New York (1982).
33. Porter, R. S., and Johnson, J. F., in *Rheology* (ed. F. R. Eirich) Vol. 4, p. 317. Academic Press, New York (1967).
34. de Jeu, W. H., *Physical Properties of Liquid Crystalline Materials.* Gordon & Breach, London (1980).
35. Hermans, J. Jr., *J. Colloid. Sci.* **17,** 638 (1962).
36. Doi, M., unpublished work.
37. Kiss, G., and Porter, R. S., *J. Polym. Sci. Polym. Symp.* **65,** 193 (1978).
38. Kiss, G., and Porter, R. S., *J. Polym. Sci. Phys. ed.* **18,** 361 (1980); *Mol. Cryst. Liq. Cryst.* **60,** 267 (1980).
39. Asada, T., Muramatsu, H., Watanabe, R., and Onogi, S., *Macromolecules* **13,** 867 (1980).
40. Berry, G. C., *Discuss. Faraday. Soc.,* to be published (1985).
41. Oseen, C. W., *Trans. Faraday Soc.* **29,** 883 (1933); Zoucher, H., *Trans. Faraday Soc.* **29,** 945 (1933).
42. Frank, F. C., *Discuss. Faraday Soc.* **25,** 19 (1958).
43. Priest, R. G., *Phys. Rev.* **7A,** 720 (1973).
44. Straley, J. P., *Phys. Rev.* **8A,** 2181 (1973).
45. Onogi, S., and Asada, T., in *Rheology* (eds. G. Astarita, G. Marrucci, and L. Nicolais) Vol. 1, p. 127. Plenum Press, New York (1980). Asada, T., and Onogi, S., *Polym. Eng. Rev.,* **3,** 323 (1983).
46. Wissbrun, K. F., *J. Rheol.* **25,** 619 (1981).
47. Asada, T., in *Polymer Liquid Crystals* (eds. A. Ciferri, W. R. Krigbaum, and R. B. Meyer) p. 247. Academic Press, New York (1982).
48. Very recently a tentative theory has been presented: Marrucci, G., *Proc. 9th International Congress on Rheology Mexico,* p. 441 (1984); Wissbrun, K. F., *Discuss. Faraday Soc.,* to be published (1985).
49. de Gennes, P. G., in *Polymer Liquid Crystals* (eds. A. Ciferri, W. R. Krigbaum, and R. B. Meyer) p. 115. Academic Press, New York (1982).

SUBJECT INDEX

AUTHOR INDEX